Technology in Agriculture

Edited by Fiaz Ahmad
and Muhammad Sultan

Published in London, United Kingdom

IntechOpen

Supporting open minds since 2005

Technology in Agriculture
http://dx.doi.org/10.5772/intechopen.92469
Edited by Fiaz Ahmad and Muhammad Sultan

Contributors
Fahim Ullah, Lukman Adewale Adewale Ajao, Joseph Ebosetale Okhaifon, Emmanuel Adewale Adedokun, Habib Bello-Salau, Kallol Das, Pijush Kanti Jhan, Srijan Chandra Das, F. M. Aminuzzaman, Benjamin Yaw Ayim, Huma Qureshi, Rafia Younas, Shiza Gul, Rehan Ahmad, Ali Raza Khan, Mumtaz Khan, Tauseef Anwar, Prashant Kaushik, Daniel D. Mann, Mitchell Green, Olayinka Folorunsho, Avery Simundsson, Franklin Ogidi, Uduak Edet, Ratnmala Bhimanpallewar, M. R. Narasinga Rao, Abel Olajide Olorunnisola, Ghulam Hasnain Tariq, Muhammad Ashraf, Umar Sohaib Hasnain, Paul L Woomer, Welissa M. Mulei, Rachel M. Zozo, Muhammad Sultan, Takahiko Miyazaki, Fiaz Ahmad, Muhammad Bilal, Uzair Sajjad, Aftab Khaliq, Qiu Baijing, Jing Ma, Evans Kituyi, Mohd Tariq, Shipra Choudhary, Harjeet Singh, Mohd Asif Siddiqui, Hirdesh Kumar, Asad Amir, Neelesh Kapoor, Muhammad E.H. Chowdhury, Tawsifur Rahman, Amith M. A. Khandakar, Mohammad Tariqul Islam, Mamun Bin Ibne Reaz, Sawal Hamid Md Ali, Nabil Ibtehaz, Muhammad Salman Khan, Nasser Al-Emadi, Aftab Ullah Khan, Robert M. Merton Stwalley III, Hannah M. Boland, Morgan I. Burgett, Aaron J. Etienne, Shadrack Kwadwo Amponsah, Luiz Carlos Guilherme, Songül Gürsoy, Jose L. Zambrano-Mendoza, Carlos A. Sangoquiza-Caiza, Diego F. Campaña-Cruz, Carlos F. Yánez-Guzmán, Kasem Soytong, Somdej Kanokmedhakul, Jiaojiao Song, Rujira Tongon, Abdur Rehim, Muhammad Amjad Bashir, Qurat-Ul-Ain Raza, Hafiz Muhammad Ali Raza, Limei Zhai, Hongbin Liu, Hongyuan Wang, Norman Roy Scott, Vasile Mocanu, Tudor Adrian Ene, Vasile Adrian Blaj, RAKESH S S, Jogarao Poiba, Krishna Chaitanya A, Kiran Pilli, Ravinder Juttu, Kamalakar Jogula, Subhashree Priyadarshini, Sharan Bhoopal Reddy, Prasanta Kumar Patra, Bairi Raju, J.D. Saritha, Princy Thakur, Surabhi Hota

First published in London, United Kingdom, 2021 by IntechOpen
IntechOpen is the global imprint of INTECHOPEN LIMITED, registered in England and Wales, registration number: 11086078, 5 Princes Gate Court, London, SW7 2QJ, United Kingdom
Printed in Croatia

British Library Cataloguing-in-Publication Data
A catalogue record for this book is available from the British Library

Additional hard and PDF copies can be obtained from orders@intechopen.com

Technology in Agriculture
Edited by Fiaz Ahmad and Muhammad Sultan
p. cm.
Print ISBN 978-1-83881-921-7
Online ISBN 978-1-83881-922-4
eBook (PDF) ISBN 978-1-83881-923-1

For EU product safety concerns:
IN TECH d.o.o., Prolaz Marije Krucifikse Kozulić 3, 51000 Rijeka, Croatia, info@intechopen.com or visit our website at intechopen.com.

Meet the editors

Dr. Fiaz Ahmad is an assistant professor and lecturer in the Department of Agricultural Engineering, Bahauddin Zakariya University, Multan, Pakistan. He obtained a Ph.D. in Agricultural Bioenvironmental and Energy Engineering from Nanjing Agriculture University, China, in 2015, and a postdoctorate in Agricultural Engineering from Jiangsu University, Zhenjiang, China, in 2020. He was awarded a fellowship from the Higher Education Commission of Pakistan for Ph.D. studies and from the Chinese Government for post-doctoral studies. He earned a BSc and MSc (Hons) in Agricultural Engineering from the University of Agriculture, Faisalabad, Pakistan, in 2004 and 2007, respectively. He is the author of more than fifty journal and conference articles. He has supervised six master's students to date. In addition, he is supervising five master's and two doctoral students. Dr. Ahmad has completed three research projects. His research interests include the design of agricultural machinery, artificial intelligence (AI), and plant protection environment. He is currently working on the application of unmanned aerial vehicle (UAV) sprayers for agrochemical applications.

Dr. Muhammad Sultan is an assistant professor in the Department of Agricultural Engineering, Bahauddin Zakariya University, Multan, Pakistan. He obtained a BSc and MSc in Agricultural Engineering from the University of Agriculture, Faisalabad, Pakistan, in 2008 and 2010, respectively. He also obtained a Ph.D. in Energy and Environmental Engineering from Kyushu University, Japan, in 2015. He did postdoctoral research in Energy and Environmental Engineering at Kyushu University, Japan, in 2017, and in Mechatronic Systems Engineering at Simon Fraser University, Canada, in 2019. He has published more than 200 articles in international journals, conferences, books, and book chapters. He has been a reviewer for more than seventy renowned journals and is an editor for five journals. His research focuses on developing energy-efficient temperature and humidity control systems for agricultural applications including greenhouse, fruits/vegetable storage, livestock, and poultry applications. His research includes adsorption heat pumps, desiccant air-conditioning, evaporative cooling, Maisotsenko cycle, adsorption desalination, energy recovery ventilator, atmospheric water harvesting, and wastewater treatment.

Contents

Preface

Food Security, sustainable agriculture, and poverty alleviation are the key themes of the United Nations' 2030 Sustainable Development Goals (UN-SDGs). These are directly aligned with agricultural mechanization, automation and robotics, high-efficiency irrigation systems, farm energy systems, post-harvest handling and processing, and associated sustainable environment. Such agricultural, biological, and environmental engineering studies are the need of the 21st century, particularly from the point of view of the agriculture–energy–food security nexus.

Therefore, this book *Technology in Agriculture* focuses on the wide range and interdisciplinary nature of the research for agricultural engineering technologies. It provides the reader with a comprehensive review of recent developments, perspectives, and opportunities in agricultural engineering technologies and applications. It also discusses biosystem technologies and applications related to farm mechanization, farm energy, smart farming, intelligent agriculture, conservation agriculture, and precision agriculture. The book consists of twenty-five chapters organized into four sections: "Agricultural Equipment Engineering" (Chapters 1–9); "Biosystems Engineering and Applications" (Chapters 10–18); "Energy Systems Engineering" (Chapters 19–21); and "Computer Applications in Agriculture" (Chapters 22–25).

Chapter 1 discusses real-time sensory information for remote supervision of autonomous agricultural machines. Chapter 2 reviews advancements in spraying technology. Chapter 3 presents the approach and impacts of the technologies for the African agricultural transformation program. Chapter 4 describes the implementation of innovations and best practices for resilient agricultural systems in Africa. Chapter 5 reviews works related to soil compaction in agricultural areas under the influence of intensive machinery use. Chapter 6 discusses the potential of wood, bamboo, and other natural reinforced materials for developing agricultural machinery. Chapter 7 discusses the domestic production of fish using recirculating water. Chapter 8 provides updates related to precision vegetable farming technologies. Chapter 9 examines technological solutions and specific equipment for improving degraded grasslands.

Chapter 10 discusses the application of biofertilizers in maize. Chapter 11 considers the past, present, and future applications of nanotechnology for crop production. Chapter 12 examines the application of the fungi *chaetomium* in agriculture. Chapter 13 presents synchrotron-based techniques for soil analysis. Chapter 14 describes the bioinoculants in technological alleviation of climatic stress in plants. Chapter 15 examines the strategies and programs for enhancing nutrient use efficiency. Chapter 16 gives updates related to plant growth stimulators in modernized agriculture. Chapter 17 explains the role of nanoparticles in abiotic stress. Chapter 18 discusses the evolution of the soil-based agriculture and food system.

Chapter 19 discusses adsorption-based atmospheric water-harvesting technology. Chapter 20 reviews the applicability of solar energy technology for agricultural applications. Chapter 21 experimentally investigates a parabolic solar water distillation system.

Chapter 22 provides the key concepts related to precision in agriculture decision-making using a machine learning approach. Chapter 23 uses a deep learning technique to detect tomato leaf diseases. Chapter 24 presents a non-linear fuzzy controller design for intelligent greenhouse management. Finally, Chapter 25 discusses CAN-Bus development, utilization, and future potential in networking massaging for off-road mobile equipment.

The editors are pleased to share emerging research in the field of agricultural engineering technologies and applications in agricultural equipment engineering, biosystem engineering, energy systems engineering, and computers in agriculture. This book is useful not only for agricultural engineering professionals but also for associated agricultural scientists. We would like to thank the authors for their valuable contributions to this book.

Fiaz Ahmad and Muhammad Sultan
Assistant Professor,
Department of Agricultural Engineering,
Bahauddin Zakariya University,
Multan, Pakistan

Section 1

Agricultural Equipment Engineering

Chapter 1

Real-Time Sensory Information for Remote Supervision of Autonomous Agricultural Machines

Daniel D. Mann, Uduak Edet, Mitchell Green, Olayinka Folorunsho, Avery Simundsson and Franklin Ogidi

Abstract

The concept of the driverless tractor has been discussed in the scientific literature for decades and several tractor manufacturers now have prototypes being field-tested. Although farmers will not be required to be physically present on these machines, it is envisioned that they will remain a part of the human-automation system. The overall efficiency and safety to be attained by autonomous agricultural machines (AAMs) will be correlated with the effectiveness of information sharing between the AAM and the farmer through what might be aptly called an automation interface. In this supervisory scenario, the farmer would be able to both receive status information and send instructions. In essence, supervisory control of an AAM is similar to the current scenario where farmers physically present on their machines obtain status information from displays integrated into the machine and from general sensory information that is available due to their proximity to the operating machine. Therefore, there is reason to expect that real-time sensory information would be valuable to the farmer when remotely supervising an AAM through an automation interface. This chapter will provide an overview of recent research that has been conducted on the role of real-time sensory information to the task of remotely supervising an AAM.

Keywords: autonomous agricultural machines, remote supervision, automation interface, visual information, ergonomics

1. Introduction

For several decades, university researchers have devoted time and effort to the pursuit of developing a driverless tractor. The scientific literature contains numerous articles describing various technologies that were evaluated, challenges that were encountered, conceptualizations of what future autonomous agricultural machines (AAMs) might look like, and issues (both technical and non-technical) in need of redress. In the late 1990s, the "agricultural ergonomics laboratory" was established at the University of Manitoba based on the hypothesis that engineers would continue to incorporate increasing levels of technology into agricultural machines in pursuit of the ultimate goal of the fully autonomous machine. Based on the lessons learned when automation was introduced to other industry settings, the human operator would experience a changing role. Thus, there was a need to view agricultural guidance technologies from an ergonomic perspective.

Unlike two decades ago, it is now possible to find autonomous tractors that are either available for sale to farmers or are in final stages of field testing. Conceptually, there are still at least four distinct designs being promoted (**Table 1**). There are advantages associated with each of these four distinct types of AAMs. Those which retain the operator station provide flexibility to the farmer for those instances when it is desired that the human operator be physically present on the AAM; this is perhaps most critical in the early days when AAMs are being introduced to the market. AAMs that resemble current tractors (except for the operator cab) and attach to implements in the same manner as existing tractors will reduce the capital cost associated with transitioning to autonomous agricultural production because the farmer will be able to continue to use existing implements. The integrated tractor reflects the situation where the engineer will be able to optimize the design of the tractor-implement system; it potentially enables design opportunities not present with the current paradigm of a tractor pulling an implement (which is a hold-over from the early concept of a horse pulling an implement). The downside, of course, is that a whole new set of implements will be required as the farmer's existing implements will be incompatible with the integrated tractor. The swarm or fleet concept perhaps reflects the most radical concept, reversing the decades-long trend of building bigger agricultural machines. Perhaps inspired by the insect world, the concept is that a fleet of many small AAMs working in an organized manner can outperform a small number of large-sized AAMs. It is too early to predict whether a single concept will emerge as the industry standard, or whether all of these concepts will survive either in niche applications or in direct competition with one another.

Regardless of how the AAM industry evolves, it would be foolish for designers to neglect how these autonomous machines will interact within the larger human-autonomy system. It is inevitable that the AAM will need to interact with a human supervisor to receive instruction and to request assistance when problems cannot be self-corrected by the AAM. Appropriate principles from the discipline of human factors engineering will be essential to the successful integration of AAMs into production agriculture.

Concept	Example	Main Features
Retain operator station	Monarch Tractor	• Tractor unit attaches to implements in the traditional manner. • Small-sized tractor targeted towards orchard/vegetable production or livestock operations.
Eliminate operator station	CNH, John Deere, Kubota, Autonomous Tractor Corporation	• Tractor unit attaches to implements in the traditional manner. • Large-sized tractors to replace the large-sized tractors currently being used for cereal/oilseed production on the prairies.
Integrated tractor	DOT	• Tractor unit is integrated with the implements that it powers.
Swarm/Fleet	Fendt Xaver	• Tasks completed by a swarm or fleet of extremely small machines working in an organized manner.

Table 1.
Autonomous tractor concepts being developed by various companies world-wide.

2. Supervision of autonomous agricultural machines

Supervision is an activity that is undertaken for the purpose of ensuring that a task is done in such a way that it meets our approval (in terms of safety, in accordance with rules, etc.). We would find it absurd to hire a junior employee and not provide some means for supervision of their work. Even senior employees require supervision to ensure that they are held accountable for their performance. The same need for supervision applies to the AAMs currently being developed by engineers. Autonomous machines, though independent, still require human supervision [1, 2] to help minimize any catastrophe that may arise in case of unexpected situations such as system failure or malfunction that exceeds the capability of the machine [3]. Furthermore, since it is currently difficult to automate high-level reasoning and tasks, it is also beneficial that the human remains in the decision-making loop to assist with planning field operations, allocating resources, and coordinating the autonomous machines. Generally, involving the human (as a supervisor) in an autonomous system has been reported to increase the overall reliability and performance of the system [4].

Supervision can be carried out in proximity (where the supervisor and the system being supervised are collocated) or remotely (where the supervisor performs his/her roles from a distant location without being physically present in the work zone). Currently, supervision of agricultural field machines is mainly performed in proximity (i.e., with the operator seated in a cab on the machine), but it is envisioned that future AAMs will be supervised remotely due to farm labour shortages (i.e., enabling one person to supervise multiple AAMs) and to enhance the overall efficiency of the farmer (i.e., enabling the farmer to complete other farm management tasks while supervising AAMs in the field).

Remote supervision is not novel and has been practiced for decades in different sectors. There is evidence of remote supervision being used in non-agricultural sectors such as military, space exploration, marine, industrial applications, and rescue operations [5–8]. As an example, robots that are used to inspect pipelines for cracks are monitored remotely during operation [9] since these areas are not accessible by humans. Search and rescue robots and military drones have been monitored remotely [6, 10]. In both cases, supervisors make use of some type of interface (which may be portable or stationary) to monitor the robot and to receive status updates. In agriculture, remote supervision has been used in livestock husbandry, crop production, and crop storage [8]. Unmanned aerial vehicles (commonly known as drones) have been remotely monitored while using them to determine weed infested regions on the field. In a hog barn environment, the physiology of pigs (body temperature) and environmental conditions of the barn (air temperature, humidity, and concentration of carbon dioxide, hydrogen sulfide, and ammonia) have been remotely monitored to minimize contact with the pigs and to assist the farmer with decision-making from any location [11].

Several remote supervision concepts for autonomous agricultural field machines have been proposed by academic researchers and manufacturers alike. These concepts differ with respect to the type of human involvement, autonomy level, proximity of the remote supervisor to the autonomous machine, and number of autonomous machines being supervised simultaneously [12]. For example, [13] envisioned the human to only monitor autonomous machines, whereas [14] expected the farmer to both manually operate a field machine while supervising another autonomous machine (which may or may not be the same type of machine). A third supervision concept involved manually controlling the actual operation of a field machine remotely (i.e., teleoperation) [15]. Some researchers [2, 16] proposed that the human would monitor just one machine while others [17, 18] envisioned the supervisor to monitor several machines simultaneously.

Edet and Mann [12] described four remote supervision concepts based on the location of the remote supervisor in relation to the AAM: 1) in-field supervision, 2) edge-of-field supervision, 3) supervision from the farm office, and 4) supervision from outside the farm site. A practical example of the 'in-field' supervision concept is the human-machine, master–slave interaction that involves having both an AAM and a human-driven machine working simultaneously on the same field. Supervision of the AAM would be done from an interface located in the cab of the human-driven machine. In the 'edge-of-field' remote supervision concept, the farmer is not operating any of the machines. This gives the farmer the opportunity to also be involved with the logistics of the operation such as bringing supplies to the field, making repairs, and responding to alerts. The 'supervision from the farm office' concept, on the other hand, makes it more challenging for the farmer to respond to in-field demands in a timely manner; the advantage is that the farmer can attend to other non-field related tasks rather than focusing on monitoring the AAM alone. Supervision from outside the farm site would theoretically allow a farmer to remain engaged in field operations while physically away from the farm for personal or vocational reasons, although it would be challenging to address system malfunctions. This role would need to be delegated to someone else, potentially contracted to an agency that would monitor and service AAMs for a fee.

Each remote supervision concept has corresponding benefits and shortcomings. For example, the 'in-field' concept would likely have the shortest response time, however, if the AAM breaks down or requires assistance, the entire field operation may come to a standstill since the manually-operated machine will also be stopped as its' operator handles the problem. In the 'edge-of-field' and 'from the farm office' remote supervision concepts, the farmer would not be controlling any of the AAMs in operation. Hence, the farmer is available to manage both i) malfunctions that are beyond the capability of the AAM and ii) other logistics associated with the operation without assistance from other farm workers. Of these two remote supervision concepts, 'edge-of-field' supervision may be preferred over 'from the farm office' supervision because of the closer physical presence to the AAM which, in theory, should allow for faster response to malfunctions that require human intervention. Remote supervision concepts that rely on servicing of AAMs being done by professional service technicians in a fee-for-service arrangement may not be accepted by many farmers due to a preference for self-sufficiency and the timely manner in which many farm operations need to be completed.

Generally, a suitable remote supervision concept should: i) require minimal labour to function, ii) enable the farmer to monitor and understand the status of the operating machine in the field, iii) not restrict the movement of the farmer, iv) allow the farmer to perform other farm tasks, iv) enable the farmer to attend to in-field problems in a timely manner, and vi) be cost effective. Other factors that may influence the choice of remote supervision for monitoring the operation of AAMs include the size of the farm, ease of use of the automation interface, type of field operation being conducted, business structure of the farm, the farmer's preference, and future legislation that might relate to the supervision of AAMs. Based on an unranked paired comparison analysis of the concepts, the 'edge-of-field' remote supervision concept was determined to be the most viable remote supervision concept for broadacre grain producers [19].

It can also be deduced that remote supervision of AAMs requires an automation interface since it is the communication link that enables the human supervisor to interact with the AAM. Edet [19] generated the following list of functional requirements for an automation interface; the remote supervisor should be able to:

1. Instruct the AAM to commence operation.

2. Monitor telemetrics of the AAM.

3. See key elements of the AAM in real-time.

4. Visualize the position of the AAM within the field.

5. Receive notifications of important events and anomalies from the AAM.

6. Query the AAM about planned actions.

7. Instruct the AAM to stop or shut down, or to alter plans.

In 2017, an important article entitled "From here to autonomy: lessons learned from human-automation research" was published by a leading expert in human-autonomy teams [20]. The key to a successful human-autonomy team is to assume that there will be instances where the autonomous system will require input from the human supervisor who is part of the human-autonomy team, and to ensure that the autonomous system is designed to most effectively share critical information with the human member of the human-autonomy team. In essence, [20] recommended that there should be shared situation awareness within the human-autonomy team where the human supervisor fully understands the actions being taken by the autonomous system so that appropriate actions can be taken by the human supervisor at any instant. Designing to support shared situation awareness is a non-trivial undertaking for the design engineer. Most autonomous systems require substantial complexity to fully automate the various tasks associated with the overall functioning of the machine. Passive monitoring of automation creates a high workload for the farmer [21] – this likely contradicts one of the reasons for using AAMs in production agriculture in the first place (i.e., to reduce the workload for the farmer). In her "human-autonomy oversight model", [20] recommended that a transparent automation interface be designed so that the human responsible for supervision of the automation will be able to successfully navigate from periods of passive supervision to periods requiring intervention. The next section of the chapter will focus specifically on the automation interface.

3. The automation Interface

3.1 The role of the automation interface

For decades, virtually all textbooks that have been written on the topic of ergonomics or human factors engineering have had chapters devoted to the design of displays and the design of controls. Displays must be designed well to clearly convey machine status information to the operator. The design and arrangement of controls is essential to allow efficient communication of instructions from the operator to the machine. When dealing with an autonomous machine, there is perhaps limited reason for the supervisor to need to communicate short-term actions to the autonomous machine. It is more reasonable to expect that communication in this direction will be reserved primarily to high-level management decisions. However, the flow of information from the autonomous machine to the human supervisor is anticipated to remain important.

The key to a successful system comprised of an autonomous machine and a human supervisor is a well-designed interface that allows for the exchange of information between the autonomous machine and the human supervisor. There are several papers published over the past two decades that touted the importance of an automation interface and have postulated on the features essential to an automation interface. In a paper published two decades ago, [22] explained their expectation that the farm manager would be responsible for overseeing the coordinating process from a computer located in the farm office. They proposed the term of 'tractor mimic display' for the automation interface that would be used to display telemetric data from the tractor unit, show the position of the tractor unit on a map, and display real-time video as seen through steerable cameras placed on the tractor unit. In the same year, [16] published a paper that investigated how humans can supervise AAMs. These authors discussed the challenges associated with designing an autonomous system that avoids both false positives and false negatives. Although they stated the desire to design such to err on the side of false positives (i.e., where a machine sees a problem where there are none), they further suggested the use of humans as 'remote troubleshooters' to classify positives as either true or false. Their system was designed to transmit images of the scene whenever the tractor detects an obstacle in its way; the images were presented to the remote troubleshooter using a 'remote operator interface'. In addition to sharing telemetric data and live video, the interface provided a warning when an obstacle was detected and explained what portion of the image was being classified as an obstacle. In a more recent paper, [23] described work completed to develop a team of robotic tractors for autonomous peat moss harvesting. In manual peat moss harvesting, a team leader supervises a team of three of more tractor operators using radio and/or hand signals. The autonomous peat moss harvesting system mimicked the manual harvesting system in that the human team leader communicated with the autonomous harvesters through a 'team leader user interface'. In this instance, the interface displayed telemetric information from each of a team of autonomous harvesters. Furthermore, a map was used to show the position of each harvester and to provide a visual representation of harvesting progress. Moorehead et al. [24] described a system of autonomous tractors for orchard maintenance. The autonomous system was comprised of tractors (equipped with perception systems and capable of driving autonomously) and a remote supervisor who assigns tasks, responds to requests when the perception system is unable to decide how to deal with a detected obstacle, and tracks the fleet of autonomous tractors. Although the tractors were equipped with cameras and the remote supervisor's interface was designed to display video, it was not intended that the supervisor should monitor the real-time video continuously. Rather, a warning message appears when an obstacle is detected and the tractor has stopped forward motion; the supervisor must then review the available video and decide whether a worker needs to be sent to remove the obstacle or if the warning is a false positive meaning that the tractor may proceed safely.

Based on this brief review of automation interfaces that have been reported in the published literature, there are several common elements that are envisioned for an effective 'automation interface'. First, it is anticipated that the automation interface will provide telemetric data related to the autonomous agricultural machine; such information is necessary to assure the human supervisor that the machine is functioning within normal operating parameters. Second, there is a need to show the location of the autonomous machine within the context of its operating environment (i.e., field, orchard, peat bog, etc). Third, it is envisioned that the autonomous machine will, at times, experience situations which will require human intervention. In these instances, a warning message will be displayed for the supervisor. The autonomous machine will stop until the issue has been resolved by the supervisor and the machine is cleared to resume operation. To enable the supervisor to be able to see what is happening, cameras are necessary on the AAM to transmit real-time video which is viewed on the automation interface.

3.2 Identifying information to include on an automation interface

If the automation interface contains irrelevant information, this may result in overcrowding that could reduce the effectiveness of the interface. On the other hand, omitting essential information may impede the supervisor's ability to perform his or her role effectively. Thus, providing the supervisor with the right information is central in designing an effective user interface [25]. Identifying the right information can be achieved through the completion of a requirement analysis which involves identifying and understanding the goals of the task as well as the role of the user [26]. Endsley [27] noted that the supervisor should have a high situation awareness (i.e., "the perception of the elements in the environment within a volume of time and space, the comprehension of their meaning, and the projection of their status in the near future") of the automated machine to be able to take necessary action in a timely manner. This awareness includes: i) machine location awareness, ii) activity awareness, iii) status awareness, iv) surrounding awareness, and v) overall mission awareness [28].

The results of a requirement analysis that was conducted for the task of supervising an agricultural sprayer have been included to demonstrate this process. Both users and designers were consulted to inform the design of an automation interface for an autonomous agricultural sprayer. **Table 2** provides a summary of the information arising from a survey of the farming community [8] which was structured to determine the types of information that should be included on an automation interface in order to remotely supervise an autonomous agricultural sprayer. Parameters listed under the 'very useful' column were recommended by at least 75% of the respondents; [8] concluded that these pieces of information should be included in an automation interface. In further work towards the requirement analysis, [19] consulted with expert designers of AAMs. Parameters such as fuel level, tire pressure battery status, current location, global field (coverage map), tank level, spray pressure, application rate, nozzle status, and boom height were ranked as being essential information for an automation interface by the majority of designers interviewed.

A majority of the users and designers consulted in the completion of the requirement analysis indicated that live video footage of the autonomous sprayer should be made available to the remote supervisor [8, 19]. Panfilov and Mann [29] had previously investigated the importance of live video footage to the remote supervisor of an autonomous sprayer in an experimental study completed using a simulator in a lab environment. They reported that live video provided a sense of security to the supervisor, but was not typically used to detect malfunctions. They also noted that the supervisor spent only 30% of their time viewing the video. In their experimental study, participants spent the majority of their time monitoring telemetric data displayed using traditional display elements (i.e., gauges, dials, etc.). Finally, they suggested that it might be appropriate to provide real-time video on-demand. Edet [8] reported that some respondents felt that providing only one view of the sprayer was not enough to properly understand the entire spraying operation – they suggested having multiple views of the machine during operation available through an automation interface.

Information	Descriptions			
	Very useful	**Useful**	**Least useful**	**Other suggestions/ comments**
Machine status	Engine temperature, engine speed, fuel level, oil pressure, hose leakage, boom folding (open/close), and agitator	Tire pressure		Auto-steer status, GPS status, slippage
Spraying functions	Boom height, nozzle status, area covered, spray pressure, application rate, travel speed, wind speed, and wind direction	Daily temperature, skip/double application, delivery rate, current task	Humidity and altitude	Field condition, tank level, chemical mix of what is in the tank, sectional control, gallon per hour sprayed, acres covered, area sprayed per coverage, number of fills, and droplet size
Navigation features	Route taken, current location, and overhead view	Distance travel	Compass	Planned route, and coverage map
Warning and notifications	Plugged nozzle, machine breakdown, Obstacle detection, loss of GPS signal, and unexpected shutdown	Tank level drop, fuel level drop, and route change	Task completed, and skip/ double application	Emergency shutdown,

Table 2.
Interface requirements as determined by the farming community through survey. >75: Very useful, 51–75%: Useful, 25–50%: Least useful, and < 25: Not at all useful (adapted from [8]).

4. Practical considerations associated with incorporating real-time visual information in an automation Interface

4.1 Determining the appropriate 'look zones'

The previous section established the need to provide the remote supervisor of an AAM with real-time video showing the machine and its environment [8, 29]. Real-time video helps the supervisor to better understand abnormalities within the AAM [29]. Blackmore [22] also noted that the presence of live video will enable the remote supervisor to understand the machine's environment. These statements are evident in studies carried out by [2, 16, 28, 30, 31]. With these potential benefits, the challenge is to determine where the cameras should be placed to maximize the benefit of the real-time visual information.

Operators of conventional tractor-seeding machines visually monitor seven distinct areas of the machine and its environment (termed 'look zones') [32]. These areas included: i) forward, ii) right side, iii) planter, iv) planter edge, v) display X (located at the top right corner from the operator's seating position), vi) display Y (located close to the arm rest of the operator), and vii) other (located close to the front-left tire of the tractor). Other researchers [33] identified four sectors: i) field ahead, ii) left boom, iii) right boom, and iv) the light bar while investigating the "workload associated with operating an agricultural sprayer equipped with a navigation device." Hence, it can be inferred from these studies that the visual information that is useful to operators can be derived primarily from i) displays located inside the machine's cab, ii) external field cues, and iii) the implement.

Although these studies identified the different regions of importance, they did not describe what information was gained by viewing those regions. Hence, a study was conducted to identify what visual information about the machine and its environment would assist the remote supervisor to make decisions [34]; the primary focus of this study was the high-clearance sprayer. GoPro cameras were mounted at difference locations on a sprayer to record the sprayer and its environment while in operation (**Figure 1**). After collection of this video footage, 29 experienced operators (defined as having at least two years experience as a sprayer operator) were recruited and presented with 10 distinct video clips of the high-clearance sprayer in operation (**Figure 2**). For each video clip, the operators were asked i) to describe what they saw in each video clip, ii) to describe the information gained from the viewing the video clip, and iii) to rank the importance of the visual information perceived using a 5-point Likert scale.

Not all video clips were equally effective at providing information that was relevant to the spraying operation. Among the 10 clips, clips 1 and 2 were considered very important and extremely important, respectively, by the operators while clips 3 through 7 did not provide much information that was considered relevant to the operators. The results of data analysis revealed that experienced operators generally preferred the view from the i) boom and nozzles, ii) the view ahead of the sprayer (front view), and iii) an aerial view of the sprayer. These regions enable them to determine the field/crop variability or conditions, upcoming field information (e.g., headland, obstacles) as well as assess if the sprayer is functioning properly and spraying effectively. Specifically, the information that was perceived from these views included the spray pattern, nozzle height and status (plugged or not), spray pattern/drift, obstacles in front and beside sprayer, poor areas in the field (i.e., crop condition), wet spots, approximate travel speed, headlands, type of crop being sprayed, weather (windy and sunny), location of sprayer in the field, overall picture of the field (aerial view), and if the sprayer was moving and following the right path (i.e., moving straight). This result was found to be independent of how

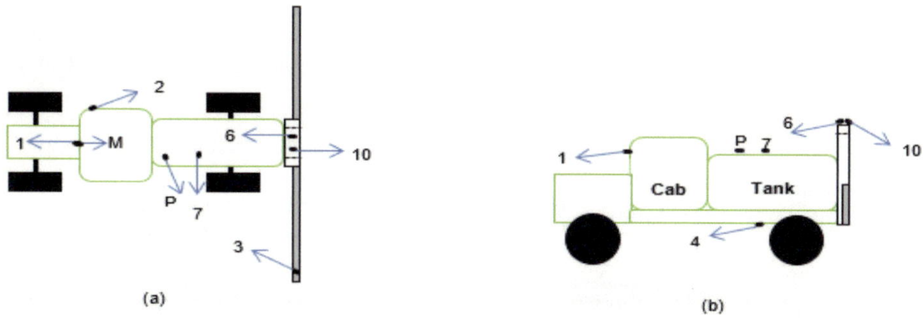

Figure 1.
A schematic showing the various camera positions on the high-clearance sprayer during the spraying operation (a) plan view (b) side view. This figure was originally published in [34].

Figure 2.
Screen shots of clips 1–10. This figure was originally published in [34].

frequently a particular view (or information) was presented. A visual representation of the information gained from the operator's 'look zones' is shown in **Figure 3**.

Camera angle and position influenced what information the operators perceived from the video clips [34]. Participants tended to describe features that were more prominent within the frame of the camera in comparison to less-prominent features. For example, many participants described features that were associated with the spray boom in clip 2 since the clip focused mainly on the right boom. Similarly, clip 9 emphasized the field and correspondingly most participants focused on the relevant information that was gained from the field. This finding suggests that designers can influence which features the user will perceive by positioning the camera such that those specific features are prominent in the camera's field of view.

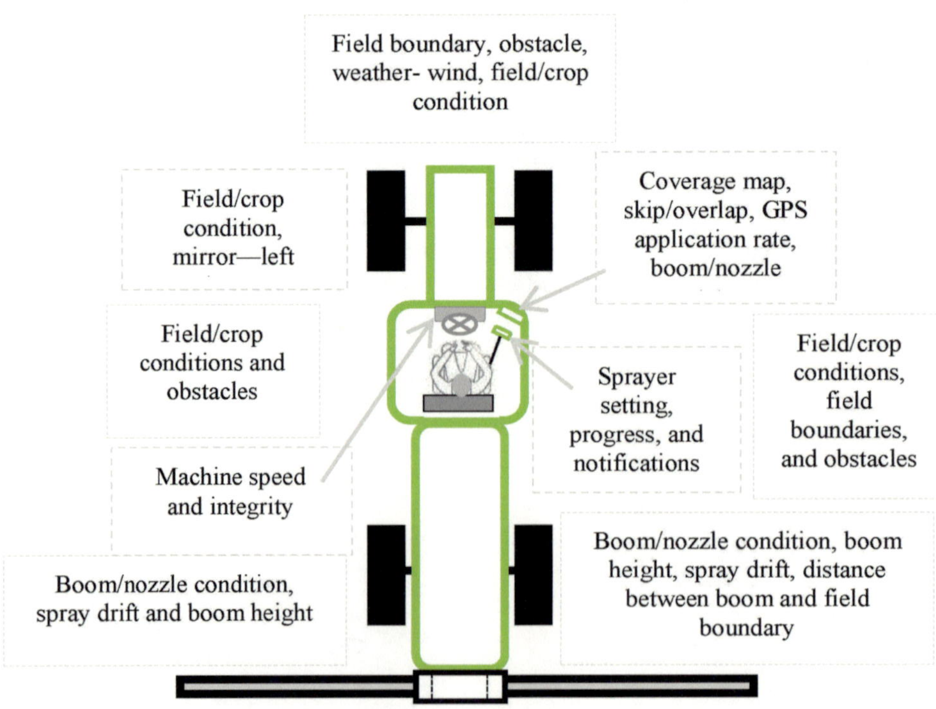

Figure 3.
Visual representation of the information gained from the operator's seated position in each of the look zones. This figure is adapted from a figure originally published in [34].

One of the most suggested views was to have a camera at one end of the boom facing forward. However, when analyzing the videos, this view was found to have minimal difference from the forward view for a sprayer with a long boom. Other views that were suggested included i) a close-up view of the nozzle and its tip, ii) a view from under the sprayer facing backward to see the spray pattern behind the sprayer and wheels, iii) dashboard/displays, and iv) a camera that would focus on the wheel to show how well the sprayer was either following old tire tracks or steering within the rows of a crop.

4.2 Effect of camera placement on the usability of look-ahead visual information

The importance of real-time visual information in the autonomous interface has been demonstrated in the previous sections, however, it is also necessary to consider how camera placement (i.e., camera height and camera tilt angle) influences the usefulness of this visual information. Previous research by [35] investigated the impact of camera placement on guidance performance for a manual guidance task in which the tractor operator relied on visual information provided by an implement-mounted camera that was displayed on a monitor close to the operator's seat. Tang and Mann [35] described a phenomenon that they called 'image velocity' which quantified the rate at which the visual information scrolled across the monitor from top to bottom as the tractor drove forward through the field. Image velocity is based on the camera's optical parameters, placement of the camera on the implement (height and tilt angle), and the tractor velocity. The reader is directed to [35] for a thorough description of how the parameter of image velocity was calculated.

The results published by [35] did not provide definitive evidence of a relationship between image velocity and lateral guidance error, however, trends were observed with lateral error increasing with image velocity. Test participants self-evaluated their performance following each trial; these results showed a decreasing linear trend with increasing image velocity. Participants preferred a tilt angle of 20° below horizontal as this gave them the best look-ahead view (i.e., the greatest look-ahead distance); however, the 30° tilt angle yielded the statistically smallest guidance error. It is unknown how the prior research by [35] will inform the current task of placing cameras on an AAM for the purpose of remotely supervising the machine on an automation interface because their research was focused primarily on trying to minimize lateral error associated with a manual guidance task. Nevertheless, their prior research inspired subsequent studies intended to determine the impact of camera placement on the usability of the visual information for the task of remotely supervising an AAM.

A lab experiment was conducted in which test participants were asked to watch pre-recorded video clips as a means of obtaining real-time visual information from the field (simulating the task of remotely supervising an AAM) [36]. Video footage was pre-recorded for nine unique combinations of camera placement, namely three camera tilt angles (20, 30, and 40°) and three camera heights (0.5, 1.0, and 1.5 m), to yield nine different look-ahead situations. Participants, some of whom were inexperienced agricultural machinery operators recruited from the university student population and some of whom were farmers experienced in operating agricultural machines, were asked to complete two distinct experimental tasks. First, they were asked to choose their preferred look-ahead position after watching two unique video clips playing side-by-side on the screen (**Figure 4**). Second, the participants responded to questions that would help determine the effect of camera placement on the difficulty of detecting and interpreting the randomly-placed frisbees in the video clips watched (**Figure 5**).

Figure 4.
Screenshot of look-ahead videos side by side (part one).

Figure 5.
Screenshot of video footage with frisbees appearing on the soil (part two).

An unranked pairwise comparison was used to analyze the data from part one of this study. This is a decision-making tool in which alternatives are compared to each other, one at a time, to arrive at the best choice. Each alternative is considered relative to other options available, with a value of one assigned to the more desirable option and a value of zero assigned to the less desirable option to arrive at alternative choice coefficients for each option being considered [37, 38]. Using this methodology, participants made pairwise comparisons for all nine look-ahead combinations. For both groups of participants (i.e., university students and experienced sprayer operators), look-ahead videos of 30° were the highest ranked of the nine combinations of height and tilt angle (**Figure 6**).

In the second part of the experimental study, participants were asked to rate each video clip based on i) the level of difficulty associated with detecting randomly placed frisbees and ii) the level of difficulty associated with interpreting randomly placed frisbees (each on a four-point Likert scale with one indicating low difficulty). Look-ahead views associated with a camera tilt angle of 30° were the look-ahead views perceived as creating the least degree of difficulty (**Figures 7** and **8**). Overall, the results of the experimental work completed by [36] suggest that forward-facing cameras on AAMs should be mounted such that they are 30° below horizontal to provide the most useful look-ahead visual information for remote supervision of AAMs.

4.3 Alerting the supervisor of a problem with the autonomous agricultural machine

With reference to an earlier section, the reader is reminded that previous researchers identified the need to warn the supervisor when the AAM experiences an abnormality which it cannot resolve itself. In such situations, there should be a means to communicate the problem to the remote supervisor immediately to increase the operational safety of the system. Different methods have been adopted in non-agricultural devices for similar purposes. They primarily make use of visual, auditory, and tactile (haptic) modalities [39]. Other modalities include olfactory (smell) and gustatory (taste) [39]. Visual, tactile, and auditory modalities have also been adopted in agriculture to inform operators about abnormalities in current agricultural machines. For example, both auditory and visual modalities have been

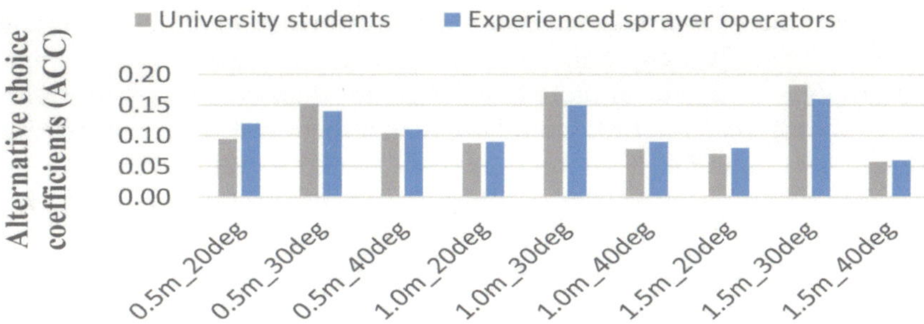

Figure 6.
Alternative choice coefficients calculated based on the unranked paired comparisons of look-ahead videos.

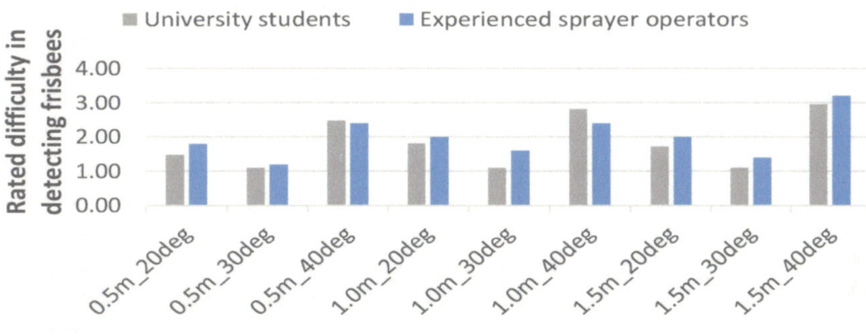

Figure 7.
Rated difficulty in detecting frisbees.

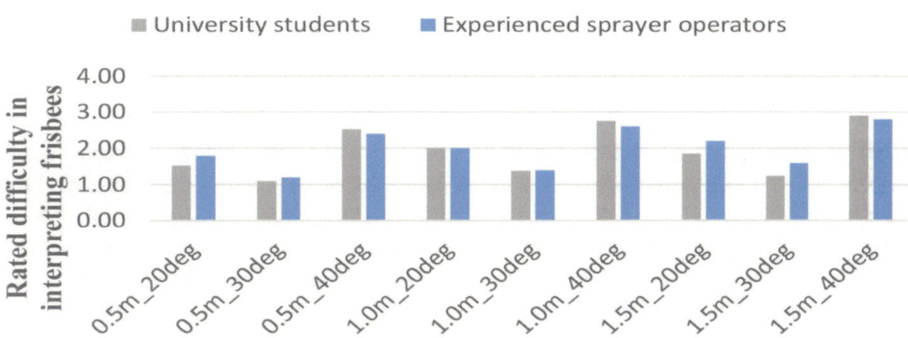

Figure 8.
Rated difficulty in interpreting frisbees.

used to notify operators about plugged nozzles while tactile and visual modality, respectively, have been used to inform operators about lateral deviation of the machine from its desired path.

Visual stimulus can be presented as text, graphics or flashing light [40] while auditory warning can be a continuous or periodic tone or tones (sounds), auditory icon (natural or symbolic), or verbal message [40, 41]. Tactile stimulus, on the other hand, communicates information through the skin (i.e., touch). Each modality has its benefits and shortcomings. For example, auditory modalities are omnidirectional unlike visual modalities that are more effective when the user is stationary. However, it may impede the user's ability to perceive the source of auditory warning. Tactile information is valuable in an environment where noise must be limited [42], but may be less effective if there is minimal contact between the tactile medium and the user's skin. One method that is widely used to assess the effectiveness of these modalities to communicate information to the user is reaction (response) time [43, 44]. This is the time interval between when the warning is communicated (using one or more modalities) to when the user reacts to the warning. A shorter reaction time would imply that the warning is more effective than a longer reaction time. Reaction times have been reported to vary with age, gender, experience, education level, culture, personality types, and intelligence of the user [45].

Researchers [46–48] have also shown that there are benefits to using multiple modalities in comparison to single ones, especially in situations where the primary task or environmental condition overloads one sensory modality. For example, [49] found that drivers responded faster when presented with multimodal warnings in comparison to unimodal warnings when evaluating driver's response time under different situational urgency while [50] noted that unimodal warnings yielded longer reaction time responses in comparison to multiple modalities while investigating the effectiveness of seven warning methods (visual, auditory, tactile, visual and auditory, visual and tactile, auditory and tactile, and no warning) under three different types of interference (in-vehicle device, audio noise, and vibration of the vehicle). On the other hand, no significant differences were experienced between unimodal and bimodal warnings when informative tactile warning and audio-tactile warnings were compared [51] – suggesting that single warning methods can be as effective as multiple warnings, depending on how they are designed or presented to the supervisor. Hence, as agricultural machinery moves towards full automation, it would be useful to distinguish which of these modalities (single or multiple) would be the most effective in alerting the remote supervisor about a problem with the machine – since these modalities vary in their ability to draw the attention of the supervisor.

A study was conducted to assess which of the seven modalities (visual, auditory, tactile, audio-visual, audio-tactile, visual-tactile, visual–auditory-tactile) would be the most effective in providing feedback to the remote supervisor of an autonomous sprayer [52]. They modified an autonomous agricultural machine control interface (AAMCI) simulator that was designed by [29] to include the different warning methods. Their experiment involved participants playing a game on the secondary screen, monitoring the operation of the autonomous sprayer through the AAMCI simulator, and clicking an 'Alert Perceived' button when they are notified of any error. Response time was used to determine the effectiveness of each modality (single or multiple).

One of their sessions was conducted in a quiet environment without having participants play the game. The remaining sessions were conducted i) in a quiet environment, ii) with tractor background noise and iii) with office (call center) background noise, respectively, to replicate the various scenarios of the four remote supervision concepts that were described by [12]. The experimental setup is shown in **Figure 9**. Further details of the experimental procedure can be found in [52]. They noticed that all seven warning modalities were able to accurately warn the participants of the errors, but varied in their effectiveness (i.e., response time). Overall, the visual and tactile (visual-tactile) warning method was found to be the most effective warning among all the seven warning methods since it had the lowest response time regardless of the background noise or environment (**Figure 10**). However, this observation was only statistically significant for the tractor background noise ($p < 0.05$).

The response time obtained when participants were continuously monitoring the autonomous sprayer through the AAMCI was also compared with those obtained when participants had to play the game and monitor the simulation in a quiet environment. Their findings revealed that for all warning methods,

Figure 9.
Experimental setup for the alert study.

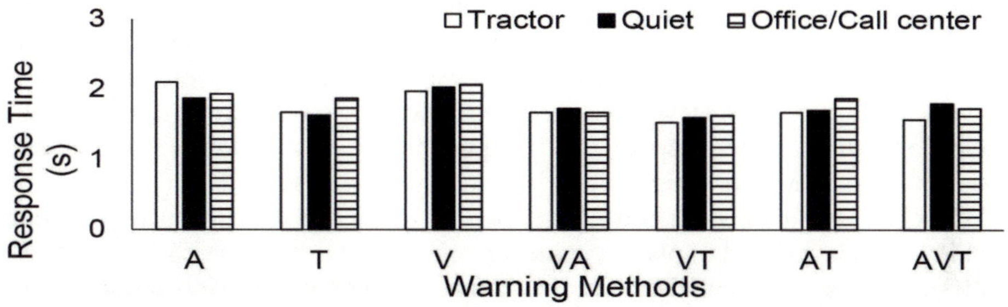

Figure 10.
Effectiveness of the warning methods as a function of response time for participants exposed to tractor noise, quiet environment, and office/call center background noise, respectively (a = audio, T = tactile, and V = visual).

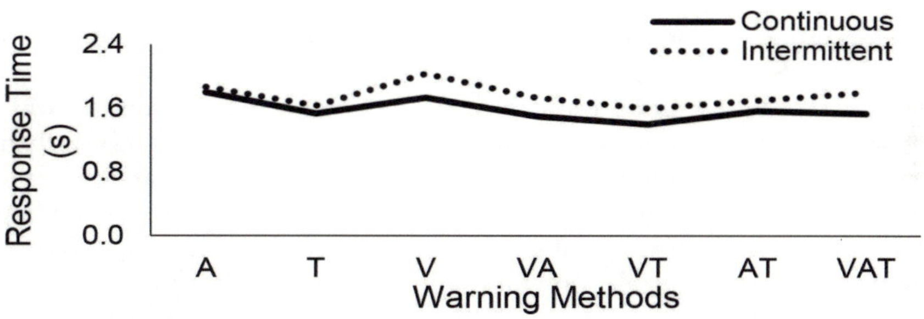

Figure 11.
Comparison of participants' response time in relation to the monitoring levels (a = audio, T = tactile, and V = visual).

participants responded faster when they were monitoring the simulated sprayer (i.e., the interface) continuously in comparison to intermittent monitoring (**Figure 11**). This result was found to be statistically significant (t-test, $\alpha = 0.05$). Despite this result, it was noted that most participants experienced boredom due to low mental workload during the 'No-Game' session (i.e., continuous monitoring) as demonstrated either through yawning, frequent eye blinking, and body posture adjustment.

Overall, the findings from the study may be biased by the fact that the simulation and game may have lacked the type of complexity and workload a remote supervisor may experience while monitoring an actual AAM. Hence, engineers must conduct further analysis during prototype testing to ensure that these results apply in an actual situation of remote supervision of an AAM.

4.4 Latency associated with transmission of real-time visual information

Real-time visual information originates from cameras mounted on AAMs and must be transmitted to the automation interface, perhaps located at the edge of the field, to enable 'edge-of-field' remote supervision. This will require the transference of data through some method of wireless transmission. Conversion of visual data into electronic signals and the time required for data to propagate incurs latency, or delay.

Latency can be described as the difference in time between an action and a response and in the context of autonomous vehicle surveillance can refer to several delay measurements. Glass-to-glass, or capture-to-display, latency is among the most typical to consider for a video being delivered to a user, and refers to the full latency from the occurrence of an event in front of a camera to the time the event can be recognized in the display used to monitor the machine [53]. Delays measured from the beginning of the encoding process to the end of decoding are also critical and are simpler to measure, as they require fewer external tools to evaluate than for glass-to-glass latency. Delays induced by the network, encoding and decoding, camera capture and video display, and the queueing of data packets can all be said to be important elements which comprise transmission latency [54].

The selected method of encoding and decoding is a significant source of latency for video. Networks tend to place restrictions on available bandwidth which must be mitigated to provide consistent video streams for a viewer. Coder-decoder (CODEC) formats, such as the widely used H.264 standard, compress and simplify video streams based on a range of algorithms and protocols. While this compression results in a significant reduction in the size of transmitted information, a trade-off is present where the computations required to reduce the size take varying amounts of time to complete. Compression can tend to be somewhat lossy such as in the case of H.264 [55], sacrificing what is considered an acceptable amount of visual information to produce a reasonably complete image for a viewer. Alternative transmission formats such as MJPEG instead send the video as a steady stream of captured JPEG video frames without such reductions due to compression, which correspondingly reduces image encoding time while resulting in significantly higher bitrates, resulting in demonstrably lower latencies in some comparative experimentation [56]. H.264 has been supplanted somewhat by H.265 video encoding [55], which promises faster rates of encoding and more efficient compression due to larger block sizes for the selection mechanisms used to simplify existing video frames, but still makes up a significant amount of the market today, used by 92% of developers in 2018 [57].

Transmission over the network is a key element of latency. If the required bitrate of video cannot be adequately accommodated by the allotted bandwidth it can result in increasing latency, as successive frames must wait for already queued frames to be received. This latency will theoretically approach infinity or some arbitrary limit, and frames of video will be dropped due to overflow of buffers used in the video stream to hold incoming video frames. The network path taken by video will inevitably introduce further delay, with longer paths with more frequent hops resulting in an increasing latency due to the travel time for data packets. Selection of the transmission level protocol used for the two devices to communicate will also have significant implications for overall latency. While the widely used TCP/IP communication protocol will eventually successfully transmit frames of video, the required two-way acknowledgement of data reception places significant time constraints. As frames are lost due to travel through the network this latency induced due to TCP can steadily increase as successive frames are forced to wait for complete transmission of earlier data [58]. In comparison, a transmission protocol such as UDP makes no guarantees of successful reception due to lacking this same handshake mechanism, but delivers video frames with lower latency due to the reductions in time as a result of not requiring the handshaking process with each transmitted data packet.

Latency has been measured using a range of methods for various applications, which vary depending on the specific delay that is to be measured. Kaknjo [56] measured latency during transmission for a robot utilizing the common method of placing a pulse per second enabled LED in front of a camera to act as an event recognizable by the system. This direct test of the camera system was coupled with utilization of timestamps to measure latency when transmitting video over a larger network. A customized application for WebRTC communication [59] transmits specialized video frames containing a spinning object and a continuously counting timestamp to measure latency between users.

An investigation of transmission latency in an agricultural setting was undertaken at the University of Manitoba [60]. A Raspberry Pi 4 was configured using the GStreamer multimedia application to be able to selectively stream video over cellular internet and a direct radio connection. The Pi 4 was fitted with a cellular header and connected via Ethernet to the radio system and mounted to a riding mower. Open-source GStreamer libraries were then used to overlay timestamps into the video feed from the Raspberry Pi, which could then be decoded and extracted by a laptop acting as a receiver and compared against the laptop time to measure the latency experienced.

In a subsequent experimental study, video transmission latency was measured for three transmission distances (200, 400 & 600 m) and for three resolutions of video (480p20fps@400kbit, 480p25fps@500kbit & 576p20fps@600kbit) using two transmission modes (cellular and radio). Data were collected at four geographical locations within 1 h (driving time) from the university campus. Complete details of the experimental procedure and results can be found in [60].

For the relatively short transmission distances tested (which were selected as representative of the 'edge-of-field' remote supervision concept proposed by [12] there were no obvious differences or trends in transmission latency for either transmission mode (cellular or radio) with a couple of exceptions which can likely be explained by the presence of trees adjacent to one of the test sites which may have interfered with radio transmission. **Figure 12** shows the results of data collection at the Glenlea, Sanford, and university campus testing locations.

As was expected, transmission latency increased with increasing video resolution for both transmission modes (cellular and radio), but with the transmission times below 300 ms in most cases. Despite the more direct transmission path for

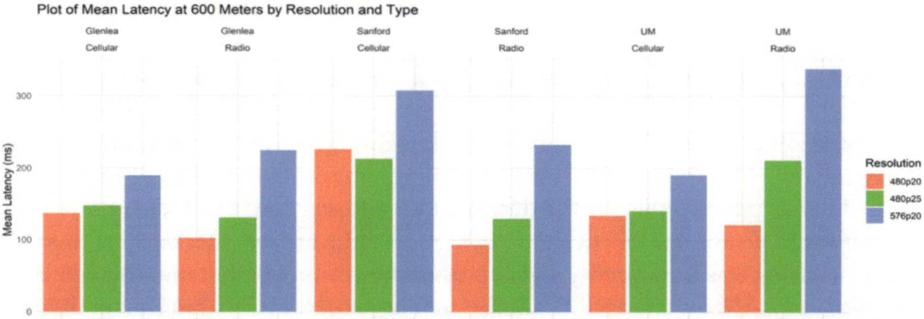

Figure 12.
Graphical depiction of experimental mean latency measurement results.

radio transmission, measured latencies were less for cellular transmission at test sites with strong signal strength. It was observed, however, that a couple of the test sites had poor cellular coverage. At one site, cellular transmission of video was not feasible with transmission latencies up to 86 s observed.

Overall, [60] concluded that it should be feasible to transmit real-time video from an AAM to an automation interface located at the edge of the field using either cellular or radio transmission. Latencies measured fell within acceptable international telecommunications union recommendations for acceptable one-way delay of less than 400 ms. These values were also in line with experiments for a telerobotic surgery simulator [61], where below 300 ms it was observed that performance of surgical tasks did not tend to degrade much with increasing latency. In locations where adequate cellular signal strength exists, cellular transmission is recommended as it causes less transmission latency and would give a greater overall range. Radio transmission of real-time video is recommended only in locations where there is poor cellular coverage.

Despite the promising results reported by [60] related to latency of real-time video transmission, there are several questions that warrant further investigation. First, research is warranted to determine the impact of transmission latency on the usability of the automation interface. Assuming constant latency, is there a magnitude beyond which it becomes impossible to remotely supervise an AAM? A related question is to determine the effect of varying latency on the usability of the automation interface. A second question worthy of further investigation is to determine the quality of video that is required for remote supervision of an AAM. [60] have reported that transmission latency increases with increasing video resolution, suggesting that it is beneficial to use low-resolution video for this application. The effect of video resolution on the usability of the automation interface must be determined. It is anticipated that the optimum video resolution for real-time supervision of AAMs will be a compromise between transmission latency and usability. A third issue is that the techniques and equipment used for transmitting video data have not yet been optimized. With the implementation of elements such as dedicated specific hardware, adaptive bitrate encoding and H.265 compression, it would likely be possible to further reduce latency by reducing the time required for encoding and decoding. Similar studies have been able to obtain latencies under 200 ms with Raspberry Pis [62] in different environments. It is important for developers and product manufacturers to consider these various aspects of video transmission to be able to provide low latency video feeds for end users. CODEC mechanisms should be selected to balance the requirements of bandwidth and latency and appropriate transmission protocols utilized to keep video streams loss tolerant while keeping latency low. Implementing the appropriate mechanisms in video streams will minimize delay for the AAM supervisor.

5. Case study: automation interface for an autonomous plot sprayer

This section presents a case study where knowledge gained from prior research activity related to the role of real-time visual information to the task of remote supervision has been applied to the design of an automation interface for an autonomous plot sprayer. The desire to design an automation interface for this specific machine was initiated by a group of undergraduate students interested in developing an AAM for the agBOT Challenge sponsored by Purdue University. To meet the objectives of the agBOT Challenge, the students would need to design and build an autonomous machine that was designed to autonomously navigate through a cornfield, detect and distinguish between weeds and corn plants, and

Figure 13.
CanAM ATV and setter plot sprayer used by a group of undergraduate University of Manitoba students to produce an autonomous sprayer for the agBOT challenge in 2019.

automatically spray the corn plants with fertilizer while spraying the weeds with herbicide (both in real-time). The students modified a CanAm ATV to navigate autonomously (**Figure 13**). Weed detection was achieved using a ground-facing camera mounted on the front of the ATV feeding data to image processing applications. Modifications were made to a Setter plot sprayer that would allow individual nozzles to be activated to apply herbicide to be applied when weeds were detected. The students desired an automation interface that could be used to remotely supervise the autonomous sprayer during the agBOT Challenge.

The autonomous agricultural sprayer is outfitted with four cameras (one of which is used for the weed and corn detection task) and a variety of sensors for the navigation and spraying tasks. The sensor input is processed by multiple onboard computers. One of these devices is dedicated to processing the visual input for the plant detection task, while the others process the remaining sensors and control actuators that allow the machine to move and spray the plants. The on-board computers communicate with each other through a middleware known as the Robot Operating System (ROS). The computers are also connected to a web server, through the internet, where they dump sensor data in real-time while the machine is running.

The automation interface for the autonomous agricultural sprayer was designed to display both sensor data and live video for a supervisor at a remote location. To enable remote supervision of the machine from anywhere in the world with internet connectivity, the interface is connected to the machine through a web server. The automation interface is shown in **Figure 14**.

From top to bottom, the interface is divided into three prominent sections: the toolbar, the video feeds, and the indicators (icons and graphical elements). The toolbar includes the start button and the emergency stop button. Since all other elements were designed for monitoring purposes, the start and stop buttons serve as the primary controls that the remote supervisor has over the machine. In the current iteration of the interface, the start button initiates the machine's autonomous operations, while the stop button terminates autonomous operations. The notification bar keeps the user informed about the status of the machine, its sensors, and its environment. The text-based notifications are enhanced by a color-coded status indicator – green, yellow, and red – to indicate the corresponding severity.

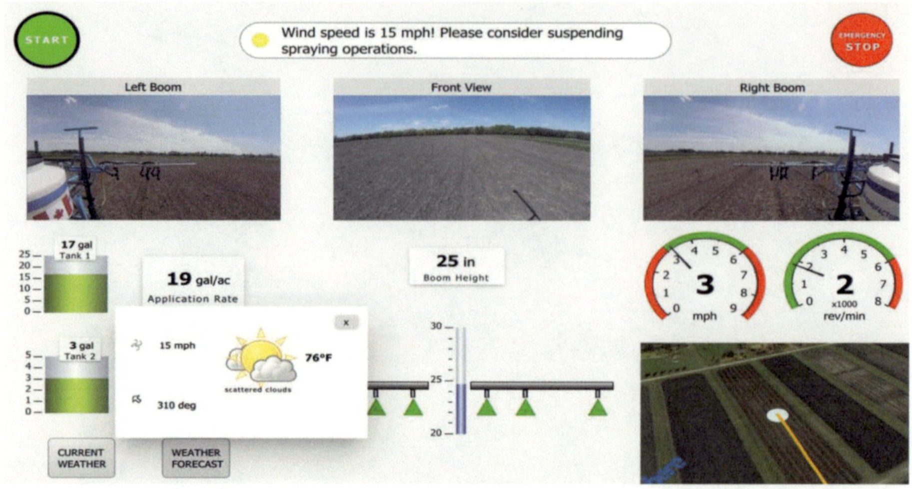

Figure 14.
The automation interface designed for an autonomous plot-sized agricultural sprayer.

Three video feeds were included in the automation interface following the recommendations made by [63]. The videos provide visual and auditory information, although at a lower fidelity than experienced when inside a tractor. The middle video provides a view ahead of the autonomous sprayer to show what is coming. The videos on the left and right sides are from rearward-facing cameras that show the left and right booms of the plot sprayer. These videos allow the remote supervisor to monitor both the machine and the spraying operation and take quick action in the case of an emergency.

Below the video feeds are the icons and graphical elements that display information regarding the state of the vehicle, sprayer, and the environment. The indicators were organized according to two main goals: i) monitoring the machine and ii) monitoring the sprayer, with the most important information placed towards the center of the display. Towards the far right of the interface is a group of indicators for monitoring the vehicle, including the vehicle speed, the engine speed, and a coverage map. In addition to providing up to level 3 situation awareness, the design of the speed indicators follows common design patterns for such indicators in most vehicles and is expected to fit the mental model of most users. The coverage map provides global situation awareness of the spraying operation.

The rest of the indicators, including the tank level, application rate and boom height, are related to the sprayer. The tank level indicates the amount of liquid that is currently in each tank, while the application rate provides information about how much liquid is being sprayed per area of the field from all the nozzles connected to the tank. While the application rate indicator supports only level 1 situation awareness, the tank level indicator was designed to support up to level 3 situation awareness by utilizing a digital display and color-coded value bar. Finally, towards the center of the display is an indicator, which was designed to provide an intuitive understanding of the state of the sprayer boom and the 6 nozzles attached to it. This indicator moves up and down in a similar fashion to the movement of the boom to indicate the height of the boom above the ground. The state of the nozzle is indicated by a green triangle (for an active nozzle) and a red square (for a blocked nozzle). Weather information is also included, which in addition to the information provided by other indicators, allows the remote supervisor to make judgments about the quality of the spraying operation, and project this judgment into the future (e.g., through available weather forecast) to take timely action.

6. Next steps in the design of an automation interface

6.1 Utilization of real-time auditory information

Auditory information can be extremely useful to a human operator, even one with minimal experience, and can provide information about changes in parameters being independently monitored via sensors (rpm, load, etc.) [64]. There is opportunity to consider what role auditory information might play in the task of remotely supervising an AAM through an automation interface.

Based on anecdotal information, it was recognized that machinery operators are often able to detect existing or impending problems from the changes in sound produced by the mechanical components of the machine. Karimi [65] reported that the addition of auditory cues did not improve steering performance (in a simulated agricultural vehicle) perhaps because steering is a purely visual task, however, auditory cues did improve the monitoring task. Donmez [64] investigated the use of sonifications (continuous auditory alerts) during the control of unmanned aerial vehicles and found that visual information supported by sonifications yielded faster reaction times than visual information supported by discrete auditory signals. Though an autonomous controller may not use or interpret sound in the same way as a human does, it is important to evaluate its potential use in control applications given its value in monitoring non-autonomous machine operation.

Though auditory information alone may not provide sufficient information for automated control, it can provide qualitative information on changing parameters, or be an indicator of a change in state. This information can be used directly to trigger certain responses, or can be used in a training set to become a single indicator of a specific state, replacing several other parameters that may have to be combined to glean the same information. Capturing high quality auditory information is generally simple and inexpensive with modern technology and can be captured from multiple locations within a machine or system, making it a good option for a variety of applications.

Classification of sounds with machine learning is already prevalent in music. There are a number of applications available to consumers to classify songs to both organize music and provide recommendations based on previous listening history. These applications use various classification algorithms (Fourier Transform, Mel Frequency Cepstrum Coefficients, etc.) to provide this service. The existence of these classification services implies that machine sounds could be classified in a similar manner to determine the state of operation, unexpected variations in parameters (i.e., malfunctions), and more. In the case where humans are controlling a machine with assistance from automation, sonifications have been shown to be very effective at helping the human operator predict the future state of the machine, and therefore react accordingly [64]. Thus, automation via audio feedback has the potential to not only improve human-machine interaction in semi-autonomous applications, but also to provide input to prompt automated responses in fully autonomous applications.

Classification is a pattern recognition problem. If a classifier can be built to recognize specific characteristics of an input signal that identify, within a certain level of confidence, what grouping or 'state' that input belongs to [66], it can then be classified, and this information can be used to produce an appropriate response. These specific characteristics are referred to as 'features' and can be comprised of any distinctive measurement or structural component of the signal that can be extracted. Multiple features may be needed for classification, but analysis can be performed to determine which features in which combinations produce the quickest classification algorithm with the highest level of confidence.

Two features that have been explored in experimentation are the spectral centroid and formant (dominant) frequency. The spectral centroid is used to detect the 'center of mass' of the spectrum (distribution of values) representing the frequency [67]. Sub band spectral centroids have been used successfully in speech recognition applications [68] and so are a good starting point for machinery audio classification. The formant frequency of a signal represents the concentration of acoustic energy (peak), and has also been used successfully in speech recognition, as well as biomedical signal analysis and musical instrumentation analysis [69].

A classification experiment was performed using video collected from the rear of an S680 John Deere combine harvester near the straw chopper [70]. The video was recorded using a GoPro Hero Session during harvest of canola in a Manitoba field during the 2017 harvest season. From this video, audio clips were extracted corresponding to the operational sounds of the machine. The audio was sampled at a rate of 48 kHz with AAC compression and automatic gain control and converted to.wav file format for analysis. Sound samples underwent a Fourier transform, and then eight features were extracted from each segment for analysis, all based around frequency characteristics. Features in each segment were analyzed to both build and then test a feedforward, pattern recognition neural network. Samples were divided into those used for training (70%), validation (15%), and testing (15%), and three operating modes were selected for classification: 1) Engine running with no threshing, 2) Engine running and threshing engaged, and 3) Engine running, and threshing engaged at 80% capacity.

By varying parameters such as segment size, accuracies of 88–100% were obtained with larger segment sizes (over 2048 segments) producing a consistent classification accuracy of 99%. This sample size allowed for a total of 1970 samples which is sufficient to declare a high degree of confidence in the result. The results of this experiment show that a relatively basic model with audio as a sole input can successfully be used to classify machinery operating modes in real-time. These results are promising enough to justify further study to better understand how to optimally apply this technique in a practical application.

The current study focuses on identifying three broad classes of operation based on a single audio input. However, it is possible that there are a number of operational modes, or even specific events, scenarios, or changes in conditions that can be classified through auditory input. Further research is required to understand what other audio inputs (location and type of sound recording), or combination of audio inputs can be used for classification of a broader range of machinery parameters. It is also critical to understand how various conditions (wind, crop type, machine parameters, etc.) impact classification and what types of calibrations or modifications may be necessary to account for variation in operating conditions. The previous study focused on a single crop type with all recordings taken under identical operating conditions. A robust field prototype would need to account for changing environmental conditions in order to be reliable. It would also be beneficial to investigate other methods of recording and processing sound along with how this input is used to build the classifier.

The current study used manual inspection for feature extraction from raw audio, but it is likely that efficiencies could be gained through automated feature extraction or some level of sound processing to enhance various aspects of the audio that may be more useful for classification. There are many ways to build and train a classification system, and it is likely that a practical system could be optimized with further investigation. One follow-up study that was conducted with the same data explored the use of a 7-layer convolutional neural network (CNN) as a classifier [71]. Through this method, greater accuracies could be achieved when lower numbers of audio samples were used compared with a conventional neural network analysis. Using 5000 samples of audio segments resulted in an accuracy of 95% compared with 78% accuracy achieved with a neural network with the same samples.

It is likely that further investigation would provide insight into optimal audio sampling techniques and feature extraction and analysis to classify a greater variety of operating modes under more variable conditions.

7. Conclusions

The purpose of this chapter is to provide an overview of recent research that has been conducted to understand how to design an effective automation interface for the task of remotely supervising an autonomous agricultural machine (AAM). First of all, it has been assumed that the existence of an automation interface is essential because the owner of the AAM will always want to have some means of monitoring the status of the machine in the field and, in some instances, human input may be required to diagnose problems and/or to make management decisions. Secondly, it has been assumed that the automation interface needs to include real-time visual information showing the AAM within the field environment to complement telemetric data that is displayed using conventional means. Experimental data has supported the important role that is played by real-time visual information, and has provided insight on related issues such as i) where the cameras should be pointed to provide information that supports the supervisory task, ii) how the cameras should be positioned to yield useful look-ahead information, and iii) how to alert the supervisor of system problems, and iv) the latency associated with wireless transmission of live video. Early research results suggest that it may also be possible to use auditory information to provide additional information to the supervisor through the automation interface.

Acknowledgements

The authors would like to acknowledge the financial support from i) the Natural Sciences and Engineering Research Council of Canada (NSERC), ii) the Bell MTS Innovations in Agriculture Graduate Student Fund administered through the Faculty of Agricultural & Food Sciences, University of Manitoba, and iii) the Canadian Agricultural Partnership (CAP) administered by Ag Action Manitoba.

Conflict of interest

The authors declare no conflict of interest.

Author details

Daniel D. Mann*, Uduak Edet, Mitchell Green, Olayinka Folorunsho, Avery Simundsson and Franklin Ogidi
Department of Biosystems Engineering, University of Manitoba, Winnipeg, MB R3T 5V6, Canada

*Address all correspondence to: Danny.Mann@umanitoba.ca

References

[1] Scholtz J. Theory and evaluation of human robot interactions. In: Proceedings of the 36th Annual Hawaii International Conference IEEE; 2003.

[2] Berenstein R, Halevi IB, Edan Y. A remote interface for a human-robot cooperative vineyard sprayer. In: 11th International Conference on Precision Agriculture, Indianapolis, USA; 2012. p. 15-18

[3] Bechar A, Vigneault C. Agricultural robots for field operations: concepts and components. Biosystems Engineering. 2016;149:94-111.

[4] Adamides G, Katsanos C, Parmet Y, Christou G, Xenos M, Hadzilacos T, Edan Y. HRI usability evaluation of interaction modes for a teleoperated agricultural robotic sprayer. Applied Ergonomics. 2017;62:237-246.

[5] Taylor RM. Human automation integration for supervisory control of UAVs. In: Virtual Media for Military Applications Meeting Proceedings RTO-MP-HFM, Paper 12; Neuilly-sur-Seine, France. 2006. p. 12.1-12.10

[6] Cubber GD, Doroftei D, Rudin K, Berns K, Matos A, Serrano D, Sanchez J, Govindaraj S, Bedkowski J, Roda R, Silva E, Ourevitch S. Introduction to the use of robotic tools for search and rescue. In: Search and Rescue Robotics; IntechOpen: London, UK, 2017; p. 1—17. DOI: 10.5772/intechopen.69489

[7] Kulakov FM. Methods of supervisory remote control over space robots. Journal of Computer and Systems Sciences International. 2018;57(5):822-839.

[8] Edet U, Hawley E, Mann DD. Remote supervision of autonomous agricultural sprayers: the farmer's perspective. Canadian Biosystems Engineering. 2018;60:2.19-2.31.

[9] Wong C, Yang E, Yan XT, Gu D. An overview of robotics and autonomous systems for harsh environments. In: 23rd International Conference on Automation and Computing (ICAC); September 2017; IEEE; 2017. p.1-6

[10] Cummings ML. Human supervisory control of swarming networks. In: 2nd annual swarming: autonomous intelligent networked systems conference; June 2004; p.1-9

[11] Duan Y, Ma L, Zhang C, Liu G, Ma X, Li L. A remote monitoring system for physiology and environmental information in pig farming. In: 2014 American Society of Agricultural and Biological Engineers Annual International Conference; 13-16 July 2014; Montreal, Quebec; 2014.

[12] Edet U, Mann DD. Remote supervision of autonomous agricultural machines: concepts and feasibility. Submitted (November 2020) to Canadian Biosystems Engineering.

[13] Gonzalez-de-Santos P, Ribeiro A, Fernandez-Quintanilla C, Lopez-Granados F, Branstoetter M, Tomic S, Pedrazzi S, Peruzzi A, Pajares Kaplanis G, Perez-Ruiz M, Valero C, del Cerro J, Vieri M, Rabatel G, Debilde B. Fleets of robots for environmentally safe pest control in agriculture. Precision Agriculture. 2017;18:574-614.

[14] Zhang X, Geimer M, Noack PO, Grandl L. Development of an intelligent master-slave system between agricultural vehicles. In: 2010 IEEE Intelligent Vehicles Symposium; June 2010; IEEE; 2010. p.250-255

[15] Murakami N, Ito A, Will JD, Steffen M, Inoue K, Kita K, Miyaura S. Development of a teleoperation system for agricultural vehicles. Computers and Electronics in Agriculture. 2008;63(1):81-88.

[16] Stentz A, Dima C, Wellington C, Herman H, Stager D. A system for semi-autonomous tractor operations. Autonomous Robots. 2002;13(1): 87-104.

[17] Noguchi N, Will J, Reid J, Zhang Q. Development of a master–slave robot system for farm operations. Computers and Electronics in Agriculture. 2004;44(1):1-19.

[18] Emmi L, Gonzalez-de-Soto M, Pajares G, Gonzalez-de-Santos P. New trends in robotics for agriculture: integration and assessment of a real fleet of robots. The Scientific World Journal. 2014;Volume 2014, Article ID 404059:1-21. DOI: 10.1155/2014/404059

[19] Edet U. Design of an interface for remotely supervised autonomous agricultural sprayers [thesis]. Winnipeg: University of Manitoba; 2021.

[20] Endsley MR. From here to autonomy: lessons learned from human-automation research. Human Factors. 2017;59(1):5-27.

[21] Grubb PL, Miller LC, Nelson WT, Warm JS, Dember WN. Cognitive failure and perceived workload in vigilance performance. In: Mouloua M, Parasuraman R, editors. Human performance in automated systems: Current research and trends. Hillsdale, NJ: Lawrence Erlbaum; 1994. p.115-121

[22] Blackmore BS, Have H, Fountas S. A proposed system architecture to enable behavioural control of an autonomous tractor. In: Zhang Q, editor. Automation Technology for Off-Road Equipment. St. Joseph: ASABE;2002. p.13-23

[23] Johnson DA, Naffin DJ, Puhalla JS, Sanchez J, Wellington CK. Development and implementation of a team of robotic tractors for autonomous peat moss harvesting. Journal of Field Robotics. 2009;26(6-7):549-571.

[24] Moorehead SJ, Wellington CK, Gilmore BJ, Vallespi C. Automating orchards: a system of autonomous tractors for orchard maintenance. In: IEEE/RSJ Int. Conf. on Intelligent Robots and Systems (IROS) Workshop on Agricultural Robotics; 2012.

[25] Dorais G, Bonasso RP, Kortenkamp D, Pell B, Schreckenghost D, Adjustable autonomy for human-centered autonomous systems. In: Working notes of the Sixteenth International Joint Conference on Artificial Intelligence Workshop on Adjustable Autonomy Systems. 1999. p.16-35

[26] Endsley MR, Bolté B, Jones DG. Designing for situation awareness. An approach to user-centered design. New York: Taylor & Francis; 2003.

[27] Endsley MR. 1988. Design and evaluation for situation awareness enhancement. In: Proceedings of the Human Factors Society 32nd Annual Meeting. Santa Monica, CA: Human Factors Society; 1988. p.97-101

[28] Adamides G, Berenstein R, Ben-Halevi I, Hadzilacos T, Edan Y. User interface design principles for robotics in agriculture: the case of telerobotic navigation and target selection for spraying. In: Proceedings of the 8th Asian Conference for Information Technology in Agriculture (volume 36); September 2012; Taipei, Taiwan; 2012.

[29] Panfilov I, Mann DD. The importance of real-time visual information for the remote supervision of an autonomous agricultural machine. Canadian Biosystems Engineering. 2018;60:2.11-2.18. DOI: 10.7451/CBE.2018.60.2.11

[30] Lewis M. Human Interaction with multiple remote robots. Reviews of Human Factors and Ergonomics. 2013;9(1):131-174.

[31] Ishibashi M, Iida M, Suguri M, Masuda R. Remote monitoring of agricultural robot using web application. IFAC Proceedings. 2013;46(18):138-142.

[32] Sanchez J, Duncan JR. Operator-automation interaction in agricultural vehicles. Ergonomics in Design. 2009;17(1):14-19.

[33] Dey AK, Mann DD. A complete task analysis to measure the workload associated with operating an agricultural sprayer equipped with a navigation device. Applied Ergonomics. 2010;41(1):146-149.

[34] Edet U, Mann DD. Visual information requirements for remotely supervised autonomous agricultural machines. Applied Sciences. 2020;10,2794. DOI: 10.3390/app10082794

[35] Tang P, Mann DD. Factors contributing to guidance performance when using a camera-based guidance aid. Journal of Agricultural Safety and Health. 2003;9(1):47-60.

[36] Folorunsho O. The influence of image velocity on the situation awareness of the remote supervisor of an autonomous agricultural machine [thesis]. Winnipeg, University of Manitoba; 2021.

[37] Niec JH, Margheim W, Umble A. Applying an innovative equipment selection process using the unranked-paired comparison technique. Proceedings of the Water Environment Federation. 2010;15:2021-2032. DOI: 10.2175/193864711802713108

[38] Mulder P. Paired comparison method [Internet]. 2018. Available from: toolshero:https://www.toolshero.com/decision-making/paired-comparison-method/ [Accessed: 2021-02-04]

[39] Laughery KR, Wogalter MS. 2006. Designing effective warnings. Reviews of Human Factors and Ergonomics. 2006;2(1):241-271.

[40] Elbert KK, Kroemer HB, Hoffman ADK. Selection, design, and arrangement of controls and displays. In: Ergonomics: how to design for ease and efficiency. Academic Press; 2018. p.500- 557

[41] Stevens CJ, Brennan D, Petocz A, Howell C. Designing informative warning signals: Effects of indicator type, modality, and task demand on recognition speed and accuracy. Advances in Cognitive Psychology. 2009;5:84-90.

[42] Smith CAP, Clegg BA, Heggestad ED, Hopp-Levine PJ. Interruption management: A comparison of auditory and tactile cues for both alerting and orienting. International Journal of Human-Computer Studies. 2009;67(9):777-786. DOI: 10.1016/j.ijhcs.2009.05.002

[43] Wogalter MS, Conzola VC, Smith-Jackson TL. Research based guidelines for warning design and evaluation. Applied Ergonomics. 2002;33(3):219-230.

[44] Whelan R. Effective analysis of reaction time data. The Psychological Record. 2008;58(3):475-482.

[45] Jain A, Bansal R, Kumar A, Singh KD. A comparative study of visual and auditory reaction times on the basis of gender and physical activity levels of medical first year students. International Journal of Applied & Basic Medical Research. 2015;5(2):124-127.

[46] Burke JL, Prewett MS, Gray AA, Yang L, Stilson FR, Coovert MD, Elliot LR, Redden E. Comparing the effects of visual-auditory and visual-tactile feedback on user performance: a meta-analysis. In: Proceedings of the 8th

International Conference on Multimodal interfaces; November 2006; 2006; p.108-117

[47] Ho C, Reed N, Spence C. Multisensory in-car warning signals for collision avoidance. Human Factors. 2007;49(6):1107-1114.

[48] Baldwin CL, Spence C, Bliss JP, Brill JC, Wogalter MS, Mayhorn CB, Ferris TK. Multimodal cueing: the relative benefits of the auditory, visual, and tactile channels in complex environments. In: Proceedings of the Human Factors and Ergonomics Society Annual Meeting; September 2012; Los Angeles, CA: SAGE Publications; 2012. p. 1431-1435

[49] Politis I, Brewster SA, Pollick F. Evaluating multimodal driver displays under varying situational urgency. In: Proceedings of the SIGCHI Conference on Human Factors in Computing Systems; April 2014. p. 4067-4076

[50] Murata A, Kanbayashi M, Hayami T. Effectiveness of automotive warning system presented with multiple sensory modalities. In: International Conference on Digital Human Modeling and Applications in Health, Safety, Ergonomics and Risk Management; July 2013; Berlin: Springer; 2013. p. 88-97

[51] Ahtamad M, Robert G, Cristy H, Nick R, Charles S. Informative collision warnings: effect of modality and driver age. In: Proceedings of the Eighth International Driving Symposium on Human Factors in Driver Assessment, Training and Vehicle Design 22-25 June 2015; Salt Lake City, Utah. Iowa City, IA: Public Policy Center, University of Iowa; 2015. p. 324-330.

[52] Edet U, Mann DD. Evaluating warning modalities for remotely supervised automated agricultural machines. Submitted (November 2020) to Journal of Agricultural Safety & Health.

[53] Bachhuber C, Steinbach E, Freundl M, Reisslein M. On the minimization of glass-to-glass and glass-to-algorithm delay in video communication. IEEE Transactions on Multimedia. 2018;20(1),238-252. DOI: 10.1109/TMM.2017.2726189

[54] Li M, Junior FEF, Sheng W, Bai H, Fan Z, Liu M. Measurement of latency on visual feedback in an immersive telepresence robotic system. In: Proceedings of the World Congress on Intelligent Control and Automation (WCICA); July 2018; Changsha, China. IEEE; 2018. p. 1757-1762. DOI: 10.1109/WCICA.2018.8630495

[55] Anonymous. H.264: Advanced video coding for generic audiovisual services [Report]. International Telecommunication Union; 2003. Available from https://www.itu.int/rec/T-REC-H.264-200305-S/en

[56] Kaknjo A, Rao M, Omerdic E, Robinson L, Toal D, Newe T. Real-time video latency measurement between a robot and its remote control station: causes and mitigation. Wireless Communications and Mobile Computing. 2018;Article ID 8638019: 1-19. DOI: 10.1155/2018/8638019

[57] Anonymous. Welcome to the 2018 video developer report [Internet]. 2018 Available from https://go.bitmovin.com/hubfs/Bitmovin-Video-Developer-Report-2018.pdf [Accessed: 2021-02-18]

[58] Kumar S, Rai S. Survey on transport layer protocols: TCP & UDP. International Journal of Computer Applications. 2012;46(7): 975-8887.

[59] García B, Gortázar F, López-Fernández L, Gallego M. Analysis of video quality and end-to-end latency in WebRTC. 2016 IEEE Globecom Workshops, GC Wkshps 2016 - Proceedings, 2-7. 2016. DOI: 10.1109/GLOCOMW.2016.7848838

[60] Green MK, Mann DD, Hossain E. Measurement of latency during real-time video transmission for remote supervision of agricultural machines. Submitted (February 2021) to Computers and Electronics in Agriculture.

[61] Perez M, Xu S, Chauhan S, Tanaka A, Simpson K, Abdul-Muhsin H, Smith R. Impact of delay on telesurgical performance: study on the robotic simulator dV-Trainer. International Journal of Computer Assisted Radiology and Surgery. 2016;11(4):581-587. DOI: 10.1007/s11548-015-1306-

[62] Jennehag U, Forsstrom S, Fiordigigli FV. Low delay video streaming on the internet of things using raspberry Pi. Electronics. 2016;5(3). DOI: 10.3390/electronics5030060

[63] Edet U, Mann DD. Visual information requirements for remotely supervised autonomous agricultural machines [Unpublished project report]. Winnipeg, University of Manitoba; 2018.

[64] Donmez B, Cummings ML, Graham HD. Auditory decision aiding in supervisory control of multiple unmanned aerial vehicles. Human Factors. 2009;51(5):718-729.

[65] Karimi D, Mondor TA, Mann DD. Application of auditory signals to the operation of an agricultural vehicle: results of pilot testing. Journal of Agricultural Safety and Health. 2008;14(1):71-78.

[66] Mahana P, Singh G. Comparative analysis of machine learning algorithms for audio signals classification. International Journal of Computer Science and Network Security. 2015;15(6):49-55.

[67] Peeters G. A large set of audio features for sound description (similarity and classification) in the CUIDADO project. Paris: Ircam, Analysis/Synthesis Team. 2004.

[68] Gajic B, Paliwal KK. Robust speech recognition in noisy environments based on subband spectral centroid histograms. IEEE Transactions on Audio, Speech and Language Processing. 2006;14(2):600-608. DOI: 10.1109/TSA.2005.855834

[69] Nisar S, Khan OU, Tariq M. An efficient adaptive window size selection method for improving spectrogram visualization. Computational Intelligence & Neuroscience. 2016; Article ID 6172453:1-13. DOI: 10.1155/2016/6172453

[70] Simundsson A, Mann DD, Thomas G. A neural network to classify auditory signals for use in autonomous harvester control systems. CSBE/SCGAB 2019 Annual Conference; 14-17 July 2019; Vancouver. Winnipeg: CSBE; 2019. Paper No. CSBE19-163.

[71] Thomas G, Simundsson A, Mann DD, Balocco S. Classification of auditory signals from a combine harvester based on mel-frequency cepstral coefficients and machine learning. Submitted (October 2020) to Canadian Biosystems Engineering.

Chapter 2

Advancements of Spraying Technology in Agriculture

Fiaz Ahmad, Aftab Khaliq, Baijing Qiu, Muhammad Sultan and Jing Ma

Abstract

Plant protection activities are most important practices during crop production. Application of maximum pesticide products with the sprayer. The application of fungicides, herbicides, and insecticides is one of the most recurrent and significant tasks in agriculture. Conventional agricultural spraying techniques have made the inconsistency between economic growth and environmental protection in agricultural production. Spraying techniques continuously developed in recent decades. For pesticide application, it is not the only sprayer that is essential, but all the parameters like the type and area of the plant canopy, area of a plant leaf, height of the crop, and volume of plants related to plant protection product applications are very important for obtaining better results. From this point of view, the advancement in agriculture sprayer has been started in last few decades. Robotics and automatic spraying technologies like variable rate sprayers, UAV sprayers, and electrostatic sprayers are growing to Increase the utilization rate of pesticides, reduce pesticide residues, real-time, cost-saving, high compatibility of plant protection products application. These technologies are under the "umbrella" of precision agriculture. The mechanized spraying system, usually implemented by highly precise equipment or mobile robots, which, makes possible the selective targeting of pesticide application on desire time and place. These advanced spraying technologies not only reduces the labour cost but also effective in environmental protection. Researchers are conducting experimental studies on the design, development and testing of precision spraying technologies for crops and orchards.

Keywords: Protection, Ground Sprayer, Aerial Sprayer, Variate Rate Sprayer

1. Introduction

Pesticide applications are considered significant during the plant protection practices in current agriculture. Efficient use of pesticides can helpful to control plant pests and diseases to increase the crop yields [1]. The use of agrochemicals can effectively enhance the quantity and quality of crops, however, it increases the environmental risks in recent years [2]. Pesticide application utilizes the to a significant percentage of the production cost [3]. During chemical plant protection practices, over application of pesticide or inefficient spraying equipment may cause serious issues on human health and the environment [4]. Effective pesticide application is a critical activity during crop production season that requires efficient spraying machinery with proper calibration as well as relevant regulations to reduce off-target spray deposition. Thus, the significant among all the factors which are influencing the degree of off-target spray deposition is the design and application spraying technology [5]. Spray drift is not only the loss of spray but also risk for the environment and residents [6]. Spray drift may influenced of many factors such as equipment and technology, spray characteristics and operator skill and performance [7]. Some other microclimatic factors which also influence the degree of drift such as wind speed, direction, relative humidity and temperature [8]. Spraying technology aims to effectively and economically application of the precise quantity of the chemical to the set target with minimum threat for the environmental pollution [9]. Conventional agricultural pesticide application practices have developed a contradiction among the yield enhancement, cost effectiveness and environmental protection [10]. Therefore, pesticides have to be applied using suitable spraying systems to avoid adverse effects on environment as well as human health [11]. Thus, in recent decades, spraying methods and technologies have been undergoing continuous evolution [1]. Conventional sprayers were very laborious and require very heavy machinery to operating them in the field. They increase the application amount because of no proper application method for spraying nozzles, crop foliage detection, and weather parameters. The canopy and foliage vary from plant to plant and crop to crop. There were very serious issues in spray uniformity and spray loss in the form of spray drift which was not measured with conventional sprayers like backpack sprayers, and PTO driven boom sprayers. Geometric and structural parameters of plants and crops are conventionally attained using time-consuming and costly manual measurements approaches. But with the advancement in spraying technologies measurement of these parameters becomes very easy with using sensors technology. Advancement in spraying shifts the pesticide application technology on Variate rate sprayers, electrostatic sprayers, and UAV sprayers. These technologies change the pesticide application scenario with the use of IoT sensors for pest and weeds detection, plant canopy and foliage measurement, leaf structure calculation, and weather parameters sensing and apply measure amount of pesticide on required part of plant and crops. With the use of advanced sprayers, the effect of pesticide exposure on the environment, water, and soil contamination reduces with the reduction of spray drift and overdose of pesticide application by the control pesticide applied spray nozzles and quick detection of structure and geometry of crops and plants canopy. In conventional sprayers these flow control nozzles, sensor technology would not be used because of this these sprayers are not working efficiently in the field and produce more amount of drift and increase pesticide amount and less effective for pests, and insects. By changing this technology with sensor base variate rate and aerial spraying technology the life of agriculture framer improves with good production of crops, less and effective pesticide use, and real-time application of pesticides.

In recent years variable spray technology has attract researchers as well as farmers and showed great advancement in the development and utilization of technology variable rate sprayers in terms of target detection technology, based on real-time sensors. Sensors are (Ultrasonic sensor, Infrared sensor, LiDAR sensor) used in variable sprayers for the feature detection of the target area. The process of spray device control unit adjusts the real-time application of spray rate and allows accurate sprays which are essential on canopy [10]. To enhance the spraying efficiency and droplet deposition for tree crops, the measurement of the plant canopy is a critical consideration because the geometry of the plant canopies variable in height and width [12]. Variable spraying technologies are most useful for orchard and vineyards as this technology increases the spray deposition rate in the target zone and reduces the off-target application. Variable-rate spray technology is growing rapidly because it reduces pollution during spray application. Electrostatic spraying system was firstly designed and developed in the early 1930s. The aim of Electrostatic sprayers to increase spray deposition and penetration in the canopy. Electrostatic sprayers technology consists of static electric charge in every drop which emits from the tip of the nozzle, which develop the force of attraction between the droplet and plant that has a neutral charge. In this spraying system air injects into the spray nozzle to the high energy charge of the drops causes them to reach the plant very fastly, before the volatilization of the drop. In addition, all the drops which emit from the spray nozzles have the same charge which causes repulsion between the drops and safely uniformly reached the plant leaf even in more hidden regions. This due to electrostatic sprayers has high spraying efficiency and save the amount of pesticide requirement. Electrostatic method of pesticide spraying decreases drift, environmental pollution, and human health risks. Electrostatic sprayers are also called ultra-low volume sprayers because it uses 5 L/ha pesticide solutions and produces fine drops with a diameter between 50 and 120 microns but on the other side in high volume spraying, the application rate is >400 L/ha, droplet sizes vary from 300 to 500 μm and gun sprayers are used and low volume spraying, the application rate is 50 to 400 L/ha and droplet sizes vary 125 to 250 μm. Electrostatic spraying in agriculture is not a new technique but is promoted to work on electrostatic spraying technique development in the technology of production and environmental concerns [13]. Thus, electrostatic spraying system is considered as most advanced spraying machinery for efficient application of pesticides to crops and orchards.

To avoid the health problems of humans with the manual spraying mechanism Unmanned aerial vehicles (UAV) aircraft are used. The equipment and labors difficult to operate but UAVs can be used easily operated in the field. UAVs now widely applied in field for purpose of precision agriculture in the developed countries. Before that, a variety of UAV models are utilizing for civilian and military purposes [14]. Addition of vision and sensor systems are also increase the potential of the UAVs [15] and reduces the spray loss in form of spray drift by accurate spray application on the required field with perfect target detection sensors and a good handling system. Advancement in spray application technologies increase crop production and increase pesticide efficiency on weed and insects in the field. Spray loss in the form of spray drift, weather effects, target detection, and control flow nozzles problem solved by advanced pesticide application technologies. The time-consuming problem for tank refilling is reduced with ultra-low volume sprayers and fine droplet spray nozzles. With time the crop maintaining technology advance from weeds and pest, detection to spray application nozzles. The objective of this chapter to highlight the advanced spraying technologies use for the agriculture pesticide application to improve the spray deposition penetration in the plant canopy, reduces the spray drift, and provide a comprehensive understanding of the spraying process from machine to target for improving fruits, vegetables, and cereals crop production.

2. Ground sprayers

2.1 Crop sprayers

The use of herbicides and pesticides has made sure to improve the yields from crops until now have been an overdose of pesticides and herbicides causing herbicide-resistant weeds and a very big decrease in biodiversity. At the early stage of the crop, when herbicide application takes place the detection of weeds is very difficult to identify because they are so small especially. Using conventional spraying (Knapsack, Boom sprayers) technology for the crops (**Figure 1(a)** and **(b)**), farmers lose money on herbicides and pesticides that are sprayed ineffective. Over spraying, of pesticides and herbicides put a squeezing impact on the agriculture farmers. Farmer spends a lot of money on pesticides, but a huge volume of these chemicals never reaches weeds and pests with the use of conventional sprayers. The maximum amount of spray not reached at the target and spray loss in the form of spray drift and field turns because of no path planning and GPS survey of the field. The reason for this inefficiency lies in the poor precision of broadcast sprayers. In old times farmers were in difficulty because the weeds a pest was damage the crop production. The technologies were not so fast to detect the accurate weed and pest target and apply the plant protection product inefficient way which causes the spray loss in the form of spray drift and environmental contamination. In recent past, a smart spraying technology (Variable sprayer, Drones Sprayers) has been introduces that uses sensors and artificial intelligence technology. This technology has a number of advantages such as when it detects plants, weeds, and pests, then selectively apply chemicals where needed. Firstly, the image is recorded using camera or detected plant by using the sensor, after that deep learning algorithms are used for identification of different plants and diseases which helps to decision support system to fix the target. The algorithms automatically choose the plant/herbicide to spray. **Figure 1(c)** shows a schematic diagram of a smart sprayer. The advance sprayer technologies are designed keeping in future crop production requirement under the sustainable agricultural goals along with protecting the environment and reducing costs for farmers and food production. The current technology (4G LTE, and 5G cellular) in agriculture protection change the conventional methods for crop monitoring and applications of pesticides and weedicides on the target area with high accuracy. In the spread field crops like (vegetables, wheat, rice, and cotton), weeds and pests have very good conditions to grow.

2.2 Orchard sprayers

For the orchards and vineyards, a powerful and effective plant-protection method is extensively adopted to attain higher quantity and quality of the production [16]. For the orchard and vineyard, it was difficult to apply spray on the whole

Figure 1.
Cotton filed using Knapsack sprayer (a) wheat field using boom sprayer (b) schematic diagram of smart sprayer (c).

area of the plant because of the shape and size change from plant to plant. It was very difficult in old times when no proper sprayers design for orchard spraying. By the advances in mechanical equipment during the period from the 1890s to 1940, some improvement in sprayers such as steam power, gasoline engines, pressure regulators, and the adjustable spray gun use to apply pesticide sprays to trees [17] as shows in **Figure 2**. Tree structure such as size, shapes and density of canopy significantly vary during different growth periods and different locations [18] thus, it need special operating parameters of sprayer (flow rate and air flow) along the adjustment facilities to match the geometry of the plant [19]. These parameters cannot be calculated with conventional spraying equipment because Conventional orchard sprayers applied pesticides continuously and do not have variable rate capability which causes environmental pollution [20] and human health hazards with producing a huge amount of spray drift.

To enhance orchard sprayer performance a number of new mechanisms have been introduced such as, an automatic variable-rate (VAR), Electrostatic, air-jet, air assisted and air blast systems. Which facility the effective spray penetration in the plant canopy and reduced the spray loss [19]. Real-time sensor are used for the detection of canopy features (density, size, shape and height) for proper spraying fluid control [21]. Thus, the characterization of the plant and crop is the basic concern for pesticide applications. As the true information of the geometrical features of the crop allows enhancing spraying performance and to reduce environmental and economic impact [22]. For the detection of plant geometry, many sensors are used like ultrasonic sensors, Infrared sensors, LiDAR sensors, and computer vision-based technology used. Ultrasonic sensor detects the target distance from the sprayer however it is sensitive to environmental conditions such as humidity and temperature, and [19]. The infrared sensor is an electronic sensor that detects the target area with measures infrared light radiating from objects in its field of view [23]. LIDAR sensor technology is an accurate remote sensing technique for distance measurements It has good accuracy for the detection and quantification of biological and nonmetallic objects [24]. LIDAR sensor measures the distance of the elapsed time between the transmission of a pulsed laser beam and the reception of its echo from a reflecting object [23]. In computer vision-based technology, the cameras are inserted on sprayers which segregate the physical parameters of the plant-like area, height, density, and color of the plants as shown in **Figure 3.**

Figure 2.
Old sprayers for orchard spraying (a) hand operated sprayer (b) steam power sprayer (c) pressure regulator engine sprayer (d) conventional VRT sprayer [16].

Several types of variable rate sprayers use for orchard sprayings like air-assisted variate rate sprayer, Tunnel sprayer, tower sprayers, axialf fan air blast sprayers, canon air blast sprayers, and Electrostatic sprayers are shown in **Figure 4**. These sprayers apply a measurable amount of pesticide with proper target detection and save the number of pesticides by reducing spray drift. Air assisted sprayers are mostly used for the fruit trees. Variable air assistance (VAA) system consists of incessant real-time air volume control attached on both sides of the sprayer. A double axial fan arrangement is used which permits remote adjustment of air volume. These sprayers apply pesticide parallel to the plant and cover the maximum area of the target but throw the spray with fan circulation away from the target which produces some airborne drift. Variable sprayers are not used in high humidity and temperature conditions. Tunnel sprayers are very famous in small fruit trees (apple,

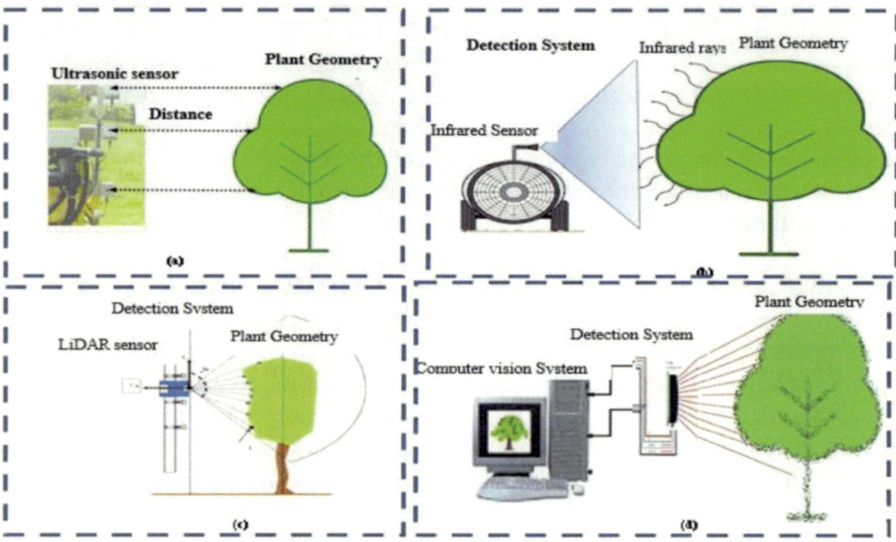

Figure 3.
Sensor based VR sprayers. Ultrasonic sprayer (a) infrared sensor sprayer (b) LiDAR sensor sprayer (c) computer vision based spraying technology (d).

Figure 4.
VAA orchard sprayer (a) tunnel sprayer (b) tower sprayer (c) multi channel air blast sprayer (d) [25].

vineyard) growth for the last few decades. Tunnel sprayer technology has long been recognized as an important tool to reduce both airborne drift and soil contamination [26]. Tunnel sprayer is enclosed target spray application technology. Some tunnels sprayer work on the recirculation principle to recycle the extra spray from the target area. Tunnel sprayers are feasible for working in every weather condition. Tower sprayers are air assisted type sprayers that discharge the spray horizontally with the direction of airflow from the fan into horizontal conductus on the vertical level. Tower sprayers are used for very high plants. Canon air blast sprayers consist of cylindrical outlets that create high air velocity jets that break spray mixture into fine droplets and penetrate the spray into the canopy. The canon sprayer can cover the maximum spray area and often use in orchards where the conventional air blast sprayers are not feasible for the spray to the crops. Due to high air velocity in the air blast sprayer, throw spray can enter into the canopies and improve spray deposition on the plant leaf and reduce the spray drift. Variable rate sprayers produce very fine (150 to 250 μL/m) mist of spray from the nozzles which reduce the pesticide amount and increase spray coverage area, but this size of the droplet is very sensitive to weather parameters and air velocity. In high humidity and low temperature condition very, fine droplets not reached to the target and hanging in the air which cause airborne spray drift, and in low humidity and high temperature conditions these droplets are evaporate into the air before reaching to the target area and cause spray loss and increase pesticide amount which is hazardous for environmental and human health.

2.3 Ultra-low volume sprayers

Ultra-low volume (ULV) spraying is a common and advanced spraying method [27] and considered a most effective and standard technique control of locust using chemicals and is also extensively used by farmers of cotton crops to control pest and insects. ULV sprayer is designed to create very small droplets (50 to 150 $\mu L/m^2$), which help for uniform coverage with low spray volumes. Ultra-low volume (ULV) fungicide application sprayer was first developed as thermal fogging [28]. The objective of ULV sprayer to reduce the fluid application rate, drift, and wastage of chemicals while increasing insect and diseases control. Conventional tractor-mounted boom sprayers apply spray on the upper side of the leaves, However, mostly the sucking insects (aphid, whitefly, jassid, thrips, etc.) have their shelter houses on the bottom side of leaves of the upper half section of the cotton plant which not only get shield from sprays but also attain the shadow of leaves shadow as of umbrella coverage. Therefore, the chemical spray using conventional sprayer do not reach the definite target and cause wastage of the spray material to the ground and air. Various pests and insects need a different droplet numbers per cm2 [29] which can only apply using a ULV sprayer. Vehicle mounted ULV sprayer shows in **Figure 5**. Pesticide droplets deposit on the upper side of the leaf using conventional sprayers may be washed off by rain or in some cases by overhead irrigation. Some researchers have concluded that up to 80% of the total pesticide applied to the plant may finally reach the soil [30].

Thus, conventional spray approaches are considered as mostly inefficient due to higher spectrum of the droplet size (150 to 250 $\mu L/cm^2$) which do not go to the target surface and ultimately become the part of waste substance. Nevertheless, the use of ULV sprayers has drastically transformed spray technology as they develop relatively small droplets [29]. Due to this Volume Application Rate (VAR) of ULV sprayers utilize fluid less than 5 L/ha for field crops or less than 50 L/ha for tree/bush [29]. Electrostatic sprayers are the most emerging technology for ultra-low-volume pesticide application.

Figure 5.
Vehicle mounted ultra-low volume sprayer.

Air-assisted electrostatic sprayers are new development in plant protection machinery which enhanced the pesticides application efficiency to crops, vineyards, orchards, plants, and trees. Electrostatic spraying method spray decreases off-target drift, environmental issues, and human health risks [31]. It is assumed that electrostatic spraying technique has revolutionized the spraying machinery through higher droplet deposition and retention on the plant leaf [32]. It is assumed as one of the appropriate methods to overcome complications associated with agrochemical spraying in the conventional system such as volatility and drift of spray droplets from temperature and wind effects [33]. Electrostatic space charge and induced image charge forces increase the spray uniformity on the target surface, enhance the transfer efficiency, bio-efficacy and adhesion. These electrostatic forces minimize the effect of gravitational force which is the main cause of spray drift [34] as shown in **Figure 6**. The electrostatic spray application increases the spray retention time on leave. There is an interaction between formulation effects on the tenacity of a deposit and the surface of the leaf to which it adheres. Droplets often bounce on waxy leaves (a property that is often influenced by age) and poor retention may occur with water-based formulations, especially those with high dynamic surface tensions. But with the ULV electrostatic sprayers the droplet gets negative charge from the nozzles with air injection and repel each other and reached at the target separately without bonding each other and create a charge on plant leaf which produce adhesion force to the drops for retain for long time on the leaf and reduce spray drift.

2.4 Aerial spraying

Although the aerial spraying has been used since middle of 20th century however, the innovation on unmanned aerial vehicle is assumed as most of are one important development in the field of agricultural spraying and plant protection engineering because of its tremendous merits over the conventional ground sprayers. The crop monitoring and the assessment of pesticide and fertilizer requirement at accurate time and location of crop area is an important parameter to effective utilization of the inputs for the purpose of yield enhancement [35]. Aerial spraying using UAV has gained great interest worldwide [36]. Thus, currently Unmanned aerial vehicles (UAVs) are known as most advanced spraying technology that is helpful for

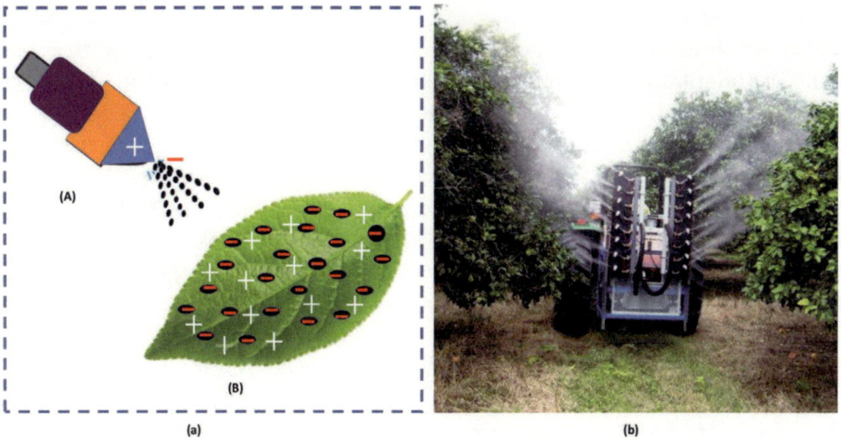

Figure 6.
Electrostatic spraying mechanism (a) variate rate multichannel electrostatic sprayer (b).

effective and precision spraying. Unmanned aerial sprayers are potentially supportive in decreasing negative effect pesticides to the environment and human during the application process of agrochemical at farm level [37]. The use of unmanned aerial vehicles (UAVs) is facilitating the to ease plant production practices and also provide the conform to spray on the tall stalk crop such as maize, cotton and water ponding crop such as rice. In fact, UAV aerial spraying capability is not only limited to crop protection but it has also utilize for the fertilization practices [38]. The idea of aerial spraying by means of UAV is firstly developed founded on the technology of unmanned helicopters that had been developed by Yamaha Corporation (Japan) for rice cultivation [39]. Mostly, chemicals application such as pesticide and fertilizer is applied by using ground sprayers, chemigation, aircraft aerial spraying, and broadcasting method without the real time assessment of the specific conditions, [40]. Unmanned aerial vehicle (UAV) sprayer develops the downwash airflow by the UAV rotor which interacts with the crop canopy and creates a conical vortex shape in the crop plant [41]. The droplet deposition efficiency is one of the major concerns in UAV spraying operation. During the spraying with UAV sprayer, the droplets pervade to the crop canopy However, some droplets drift often occurs, which wastes pesticides, decreases the control effect, and even causes environmental pollution and poisoning [42]. Spray system arrangements equipped on UAV have not yet been optimized to accompaniment spray pattern based on the proper nozzle selection, [43]. Droplet size, weather conditions, and operational parameters of sprayers influence the spray coverage, absorption, and attachment to the target [44]. The effect of climatic condition (temperature, wind direction, wind speed, humidity, etc.) on UAV spraying efficiency should clearly be understood to the applicators [45]. Unmanned aerial vehicles (UAV's) are operated remotely either by using telemetry, where the operator holding visual contact with the aircraft, or autonomously along planned paths using GPS and inertial guidance [39]. UAV sprayers working on the proper path planning with GPS, 4G, or 5G network technology to the target area which provide the proper path to the drone to apply a precise amount of pesticides on plants which reduces the spray drift, save pesticide application amount, and high accuracy results. Old UAV sprayers were very large which was difficult in handling, transport issues, landing problems, and produces more downwash pressure with high air guests by the large size of wings which produce more off-target spray deposition as shows in **Figure 7**. Due to their large size, these types of sprayers do not use in trees and orchards, it was only used in broad field crops.

In comparison, small-size unmanned aerial vehicle (UAV) crop protection operations provides the advantages of a low flight altitude, a flight velocity control, and well field adaptability, mainly for small fields and diversified crop planting zones [48]. Many small sizes of drones like Four-rotor, six-rotor, and eight rotors are used nowadays as shown in **Figure 8** which is easily handled, transport, use senor technology more easily, use in orchards, trees, and crop very sufficiently. Based on the current UAV models, crop protection operation requirements, UAV loading capability, and flight duration, under an identical load, a six-rotor UAV is thought more stable than a four-rotor UAV and consumes less energy than an eight-rotor UAV. Small size drone sprayers provide us the efficient spray deposition on the target, reduce the airborne drift, and easily operate between the orchards and crops. UAVs have high control precision and fast response speed when using a variable spray system. Spray drift is a practical reality during the pesticide spraying operations. **Figure 9** shows the flow field during the spraying operation of unmanned aerial vehicle. Recent studies on UAV sprayer application and spray deposition performance shows in **Table 1**.

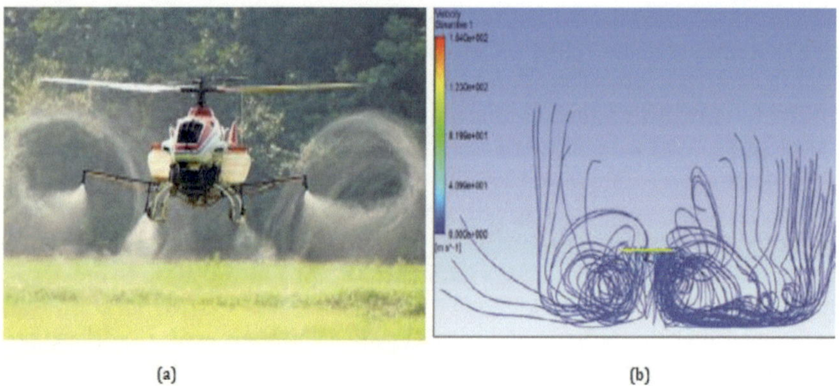

Figure 7.
Rotor wings effect on spray drift (a) [46] streamlines of the flow field under the rotor (b) [47].

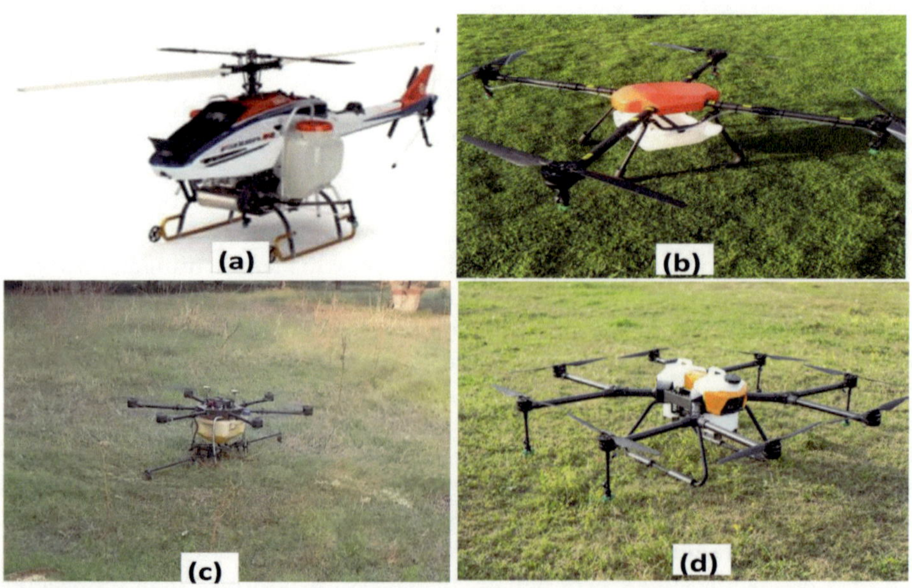

Figure 8.
Single rotor UAV sprayer (a) four rotor UAV sprayer (b) six rotor UAV sprayer (c) eight rotor UAV sprayer (d).

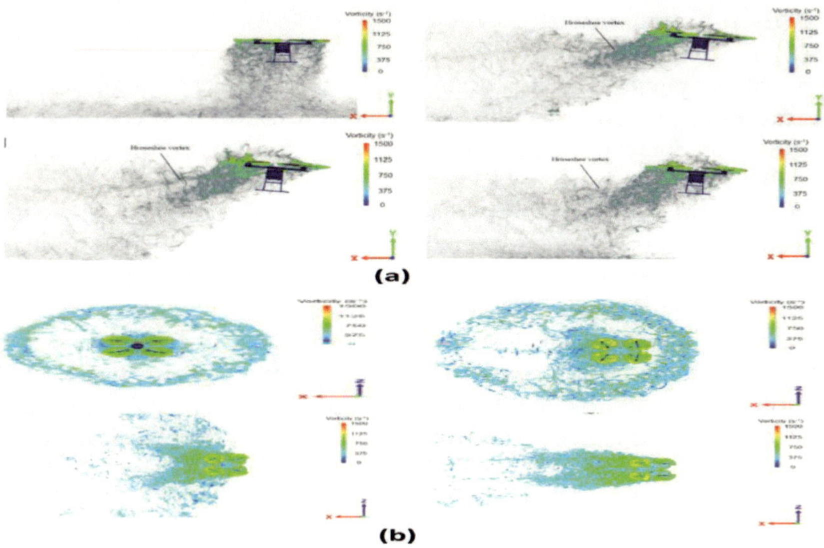

Figure 9.
Quad-rotor drone flow fields with different forward velocities and pitch angles (a) wake structure variations with flight speed (top view) (b) [49].

Reference	Study type	UAV type	Crop	Study Parameters
Shi et al. [47]	Numerical Simulation	Single rotor	—	Flight Hight, Spray deposition
Teske et al. [50]	Numerical Simulation	Single rotor and octocopter		Flow field, Effective drift
Wang et al. [51]	Field study	Single rotor	Wheat	Spray volume, nozzle size, droplet deposition
Wang et al. [52]	Filed and	Multirotor	Rice	Vision based crop detection, fuzzy logic application
Sadam et al. [53]	Field Study	Multirotor		Nozzle opening, Operating height
Liao et al. [54]	Field Study	Multi rotor	Cotton	Volume rate spray adjuvants, flight altitude, flight speed
Lv et al. [55]	Field study	Multi rotor		Flight speed, thermal imaging,
Martin et al. [56]	Field	Multi rotor	Eeed	Spray rate, spray efficiency
Wang et al. [57]	Field	Multirotor	Wheat	Overage area, total deposition
Zhang et al. [58]	Field study	Multirotor	Citrus	Tree shape and flight height
Yanliang et al. [48]	Filed study	Multirotor		Electrostatic spraying, Spray altitude, Spray pressure
Yang et al. [59]	Numerical simulation	Multirotor		CFD simulation, Flow tubulance model

Table 1.
Recent studies on UAV sprayer application and spray deposition performance.

3. Spray drift management

Spray drift is a physical moment of the droplet during the application of spraying liquid to the off target area under influence of climatic factors such as wind speed and temperature [60]. Spray drift may occur to numerous forms as a droplet, dry particles, or vapor. Particle drift enhances when water and other pesticide carriers evaporate rapidly from the droplet lifting tiny particles of concentrated pesticide. Vapor may arise directly from the spray or by evaporation of pesticide from sprayed surfaces [61]. Spray drift is a complex phenomenon due to the accumulative effect of spraying equipment design, crop architecture, atmospheric conditions and the physicochemical properties of the spray mix [62]. Spray drift is occurred due to droplet characteristics infuse by the weather parameters, nozzles types, operating pressure, speed, and height. Drift corresponds to a modification of droplet trajectory induced by the drag force due to external air velocity. The expression of the drag force Fd is given by Eq. (1):

$$Fd \equiv \frac{1}{2}\rho_a C_d A (V_d - V_a)^2 [N] \tag{1}$$

The safe, efficient and efficacious use of pesticides requires the management of pesticide drift and deposition. The sensitivity of drift to numerous factors, including atmospheric conditions and application equipment, makes it difficult to field test the full range of possible meteorological application scenarios. There are two approaches to aerial spray models: empirical and mechanistic [46].

The empirical models do not account for any physical basis and are generally applicable only to situations that are very similar to those for which they are developed. Mechanistic models based on Gaussian dispersion equations and particle tracking models. Gaussian modeling is a classical approach that is used in atmospheric dispersion modeling and lagrangian models track a cohort of droplets in a given drop size category and overlay a random component on the movement of the droplets to account for atmospheric turbulence [63]. Empirical models primarily include field testing and wind tunnel research. Compared with field testing, the wind tunnel environment can accurately control the test conditions, such as the wind speed, airflow direction, temperature. Mechanistic model include CFD, Gaussian modeling and Lagrangian model (AGDISP) [46]. The CFD model provides whole-flow field data, fast and reproducible results, repeatable and controllable conditions, reliable data, and rapid, economical and accurate means, but at the expense of incompatibility [64].

For the drift estimation three models use the German drift curve, Dutch IMAGE drift Calculator, and the Model Proposed by Meli. The equations of these models are discussed below.

German Drift

$$\%\text{Drift} = 60.36 \text{ x z}(-1.2243)(\text{z} < 15\,\text{m}) \tag{2}$$

$$\%\text{Drift} = 298.83 \text{ x z}(-1.8672)(\text{z} > 15\,\text{m}) \tag{3}$$

IMAGE Model

$$\%\text{Drift} = 48 \text{ x e}(-\text{z}/2.7) + 0.45 \text{ x e}(-\text{z}/0.091) \tag{4}$$

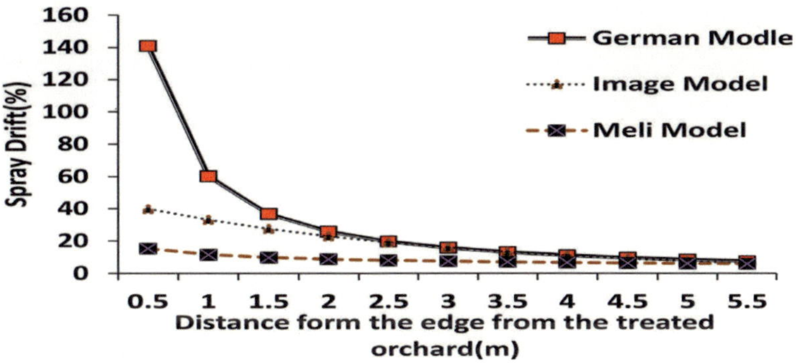

Figure 10.
Graphical representation of drift model values with distance.

Meli Model

$$\%\text{Drift} = 11.45 \text{ x z} (-0.4026) \quad (5)$$

The graphical representation of drift model values shows in **Figure 10**. Its shows that which parameters should adopted for pesticide application for the spray drift man agent.

From the result it shows that spray drift increases with increasing the application distance, height and angle. By reducing the application distance drift reduce immediately and drift curve of three model's constant after 5 m.

4. Conclusion

Plant protection practices are most important activities during crop production. Progress in spraying technology has been increase in recent past. Robotics and automatic spraying technologies like variable rate sprayers, UAV sprayers, and electrostatic sprayers has gained more attention to enhance. These advanced spraying technologies not only reduces the labor cost but also effective in environmental protection. Researchers are conducting experimental studies on the design, development and testing of precision spraying technologies for crops and orchards. Simulation modeling studies are also conducting by the researcher to increase the sprayer's efficiency and to improve the design for better utilization. However, there is still needed to conduct further studies to reduce the spray loss and health risks during the pesticide application to the orchards.

Author details

Fiaz Ahmad[1,2]*, Aftab Khaliq[1], Baijing Qiu[2]*, Muhammad Sultan[1] and Jing Ma[2]

1 Department of Agricultural Engineering, Bahauddin Zakariya University, Multan, Pakistan

2 School of Agricultural Engineering, Jiangsu University, Zhenjiang, China

*Address all correspondence to: fiazahmad@bzu.edu.pk and qbj@ujs.edu.cn

References

[1] Gil, E.; Arnó, J.; Llorens, J.; Sanz, R.; Llop, J.; Rosell-Polo, J.R.; Gallart, M.; Escolà, A. Advanced technologies for the improvement of spray application techniques in Spanish viticulture: An overview. Sensors (Switzerland) 2014, 14, 691-708, doi:10.3390/s140100691.

[2] Yarpuz Bozdogan, N. The importance of personal protective equipment in pesticide applications in agriculture. Curr. Opin. Environ. Sci. Heal. 2018, 4, 1-4, doi:10.1016/j.coesh.2018.02.001.

[3] Kim, K.-H.; Kabir, E.; Jahan, S.A. Exposure to pesticides and the associated human health effects. Sci. Total Environ. 2017, 575, 525-535.

[4] Damalas, C.A.; Eleftherohorinos, I.G. Pesticide exposure, safety issues, and risk assessment indicators. Int. J. Environ. Res. Public Health 2011, 8, 1402-1419.

[5] Bahlol, H.Y.; Chandel, A.K.; Hoheisel, G.A.; Khot, L.R. The smart spray analytical system: Developing understanding of output air-assist and spray patterns from orchard sprayers. Crop Prot. 2020, 127, 104977, doi:10.1016/j.cropro.2019.104977.

[6] Zhang, B.; Tang, Q.; Chen, L.; Zhang, R.; Xu, M. Numerical simulation of spray drift and deposition from a crop spraying aircraft using a CFD approach. Biosyst. Eng. 2018, 166, 184-199.

[7] Antuniassi, U.R. Evolution of agricultural aviation in brazil. Outlooks Pest Manag. 2015, 26, 12-15.

[8] Rojo Baio, F.H.; Antuniassi, U.R.; Castilho, B.R.; Teodoro, P.E.; da Silva, E.E. Factors affecting aerial spray drift in the Brazilian Cerrado. PLoS One 2019, 14, doi:10.1371/journal.pone.0212289.

[9] Baio, F.H.R.; Silva, E.E.; Vrech, M.A.; Souza, F.H.Q.; Zanin, A.R.; Teodoro, P.E. Vegetation indices to estimate spray application rates of crop protection products in corn. Agron. J. 2018, 110, 1254-1259.

[10] Dou, H.; Zhang, C.; Li, L.; Hao, G.; Ding, B.; Gong, W.; Huang, P. Application of variable spray technology in agriculture. IOP Conf. Ser. Earth Environ. Sci. 2018, 186, doi:10.1088/1755-1315/186/5/012007.

[11] Mohanty, M.K.; Behera, B.K.; Jena, S.K.; Srikanth, S.; Mogane, C.; Samal, S.; Behera, A.A. Knowledge attitude and practice of pesticide use among agricultural workers in Puducherry, South India. J. Forensic Leg. Med. 2013, 20, 1028-1031.

[12] Ladd Jr, T.L.; Reichard, D.L.; Collins, D.L.; Buriff, C.R. An automatic intermittent sprayer: A new approach to the insecticidal control of horticultural insect pests. J. Econ. Entomol. 1978, 71, 789-792.

[13] Giles, D.; Blewett, T. Effects of Conventional and Reduced-Volume, Charged-Spray Application Techniques on Dislodgeable Foliar Residue of Captan on Strawberries. J. Agric. Food Chem. 1991, 39, 1646-1651.

[14] Mogili, U.R.; Deepak, B.B.V.L. Review on Application of Drone Systems in Precision Agriculture. Procedia Comput. Sci. 2018, 133, 502-509, doi:10.1016/j.procs.2018.07.063.

[15] Gupte, S.; Mohandas, P.I.T.; Conrad, J.M. A survey of quadrotor unmanned aerial vehicles. In Proceedings of the 2012 Proceedings of IEEE Southeastcon; IEEE, 2012; pp. 1-6.

[16] R. D. Fox; R. C. Derksen; H. Zhu; R. D. Brazee; S. A. Svensson A History of Air-Blast Sprayer Development and

Future Prospects. Trans. ASABE 2008, 51, 405-410, doi:10.13031/2013.24375.

[17] Brann Jr, J.L. Apparatus for application of insecticides. Annu. Rev. Entomol. 1956, 1, 241-260.

[18] Chen, Y.; Zhu, H.; Ozkan, H.E.; Derksen, R.C.; Krause, C.R. An experimental variable-rate sprayer for nursery and orchard applications. In Proceedings of the 2011 Louisville, Kentucky, August 7-10, 2011; American Society of Agricultural and Biological Engineers, 2011; p. 1.

[19] Li, L.; He, X.; Song, J.; Liu, Y.; Zeng, A.; Yang, L.; Liu, C.; Liu, Z. Design and experiment of variable rate orchard sprayer based on laser scanning sensor. Int. J. Agric. Biol. Eng. 2018, 11, 101-108, doi:10.25165/j.ijabe.20181101.3183.

[20] Yang, Z.; Niu, M.; Li, J.; Xu, X.; Xu, J.; Chen, Z. Design and experiment of an electrostatic sprayer with online mixing system for orchard. Trans. Chinese Soc. Agric. Eng. 2015, 31, 60-67.

[21] Wang, W.Z.; Hong, T.S.; Li, J.; Zhang, F.G.; Lu, Y.C. Review of the pesticide precision orchard spraying technologies. Trans. CSAE 2004, 20, 78-80.

[22] Rosell, J.R.; Sanz, R. A review of methods and applications of the geometric characterization of tree crops in agricultural activities. Comput. Electron. Agric. 2012, 81, 124-141.

[23] Zhang, Z.; Wang, X.; Lai, Q.; Zhang, Z. Review of Variable-Rate Sprayer Applications Based on Real-Time Sensor Technologies. Autom. Agric. - Secur. Food Supplies Futur. Gener. 2018, doi:10.5772/intechopen.73622.

[24] Bietresato, M.; Boscariol, P.; Gasparetto, A.; Mazzetto, F.; Vidoni, R. On the design of a mechatronic mobile system for laser scanner based crop monitoring. In Proceedings of the Proceedings of the 14th Mechatronics forum international conference, Karlstad, Sweden, June 16-18 2014; 2014.

[25] Jadav, C. V.; Jain, K.K.; Khodifad, B.C. Spray of Chemicals as Affected by Different Parameters of Air Assisted Sprayer: A Review. Curr. Agric. Res. J. 2019, 7, 289-295, doi:10.12944/carj.7.3.03.

[26] Pergher, G.; Petris, R. a Novel, Air-Assisted Tunnel Sprayer for Vineyards: Optimization of Operational Parameters and First Assessment in the Field. J. Agric. Eng. 2009, 40, 31, doi:10.4081/ija.2009.4.31.

[27] Maas, W. ULV application & formulation techniques. NV. ULV Appl. Formul. Tech. NV 1971.

[28] Niekerk, J.M. Van; Mavuso, Z.S. Evaluation of ultra-low volume (ULV) fungicide applications for the control of diseases on avocado fruit – Results from the 2009 / 10 season. 2011, c, 71-76.

[29] Ali, M.A.; Nasir, A.; Khan, F.H.; Khan, M.A. Fabrication of ultra low volume (ULV) pesticide sprayer test bench. Pakistan J. Agric. Sci. 2011, 48, 135-140.

[30] Courshee, R.J. Some aspects of the application of insecticides. Annu. Rev. Entomol. 1960, 5, 327-352.

[31] Patel, M.K. Technological improvements in electrostatic spraying and its impact to agriculture during the last decade and future research perspectives - A review. Eng. Agric. Environ. Food 2016, 9, 92-100, doi:10.1016/j.eaef.2015.09.006.

[32] Patel, M.K.; Kundu, M.; Sahoo, H.K.; Nayak, M.K. Engineering in Agriculture, Environment and Food. 2015.

[33] Ficker, T. Electrification of human body by walking. J. Electrostat. 2006, 64, 10-16.

[34] Shrimpton, J.S. Electrohydrodynamics of charge injection atomization: Regimes

and fundamental limits. At. Sprays 2003, 13.

[35] Gayathri Devi, K.; Sowmiya, N.; Yasoda, K.; Muthulakshmi, K.; Kishore, B. Review on application of drones for crop health monitoring and spraying pesticides and fertilizer. J. Crit. Rev. 2020, 7, 667-672, doi:10.31838/jcr.07.06.117.

[36] Zhang, Y.; Li, Y.; He, Y.; Liu, F.; Cen, H.; Fang, H. Near ground platform development to simulate UAV aerial spraying and its spraying test under different conditions. Comput. Electron. Agric. 2018, 148, 8-18, doi:10.1016/j.compag.2017.08.004.

[37] Ahmad, F.; Qiu, B.; Dong, X.; Ma, J.; Huang, X.; Ahmed, S.; Ali Chandio, F. Effect of operational parameters of UAV sprayer on spray deposition pattern in target and off-target zones during outer field weed control application. Comput. Electron. Agric. 2020, 172, 105350, doi:10.1016/j.compag.2020.105350.

[38] Muhammad, M.N.; Wayayok, A.; Mohamed Shariff, A.R.; Abdullah, A.F.; Husin, E.M. Droplet deposition density of organic liquid fertilizer at low altitude UAV aerial spraying in rice cultivation. Comput. Electron. Agric. 2019, 167, doi:10.1016/j.compag.2019.105045.

[39] Giles, D.; Billing, R. Deployment and performance of a UAV for crop spraying. Chem. Eng. Trans. 2015, 44, 307-312.

[40] Lan, Y.; Shengde, C.; Fritz, B.K. Current status and future trends of precision agricultural aviation technologies. 1-6.

[41] Guo, S.; Li, J.; Yao, W.; Zhan, Y.; Li, Y.; Shi, Y. Distribution characteristics on droplet deposition of wind field vortex formed by multi-rotor UAV. PLoS One 2019, 14, 1-16, doi:10.1371/journal.pone.0220024.

[42] Zhang, H.; Zheng, J.; Zhou, H.; Dorr, G.J. Droplet deposition distribution and off-target drift during pesticide spraying operation. Nongye Jixie Xuebao 2017, 48, 114-122.

[43] Moltó, E.; Chueca, P.; Garcerá, C.; Balsari, P.; Gil, E.; van de Zande, J.C. Engineering approaches for reducing spray drift. Biosyst. Eng. 2017, 154, 1-2.

[44] Qin, W.-C.; Qiu, B.-J.; Xue, X.-Y.; Chen, C.; Xu, Z.-F.; Zhou, Q.-Q. Droplet deposition and control effect of insecticides sprayed with an unmanned aerial vehicle against plant hoppers. Crop Prot. 2016, 85, 79-88.

[45] Songchao, Z.; Xinyu, X.; Zhu, S.; Lixin, Z.; Yongkui, J. Downwash distribution of single-rotor unmanned agricultural helicopter on hovering state. Int. J. Agric. Biol. Eng. 2017, 10, 14-24.

[46] Chen, H.; Lan, Y.; Fritz, B.K.; Hoffmann, W.C.; Liu, S. Review of agricultural spraying technologies for plant protection using unmanned aerial vehicle (UAV). 2021, 14, 38-49, doi:10.25165/j.ijabe.20211401.5714.

[47] Shi, Q.; Mao, H.; Guan, X. Numerical Simulation and Experimental Verification of the Deposition Concentration of an Unmanned Aerial Vehicle. Appl. Eng. Agric. 2019, 35, 367-376.

[48] Zhang, Y.L.; Lian, Q.; Zhang, W. Design and test of a six-rotor unmanned aerial vehicle (UAV) electrostatic spraying system for crop protection. Int. J. Agric. Biol. Eng. 2017, 10, 68-76, doi:10.25165/j.ijabe.20171006.3460.

[49] Wen, S.; Han, J.; Ning, Z.; Lan, Y.; Yin, X.; Zhang, J.; Ge, Y. Numerical analysis and validation of spray distributions disturbed by quad-rotor drone wake at different flight speeds. Comput. Electron. Agric. 2019, 166, 105036, doi:10.1016/j.compag.2019.105036.

[50] Teske, M.E.; Wachspress, D.A.; Thistle, H.W. Prediction of aerial spray

release from UAVs. Trans. ASABE 2018, 61, 909-918.

[51] Wang, G.; Lan, Y.; Qi, H.; Chen, P.; Hewitt, A.; Han, Y. Field evaluation of an unmanned aerial vehicle (UAV) sprayer: effect of spray volume on deposition and the control of pests and disease in wheat. Pest Manag. Sci. 2019, 75, 1546-1555, doi:10.1002/ps.5321.

[52] Wang, L.; Lan, Y.; Yue, X.; Ling, K.; Cen, Z.; Cheng, Z.; Liu, Y.; Wang, J. Vision-based adaptive variable rate spraying approach for unmanned aerial vehicles. Int. J. Agric. Biol. Eng. 2019, 12, 18-26, doi:10.25165/j.ijabe.20191203.4358.

[53] Hussain, S.; Cheema, M.J.M.; Arshad, M.; Ahmad, A.; Latif, M.A.; Ashraf, S.; Ahmad, S. Spray uniformity testing of unmanned aerial spraying system for precise agro-chemical applications. Pakistan J. Agric. Sci. 2019, 56, 897-903, doi:10.21162/PAKJAS/19.8594.

[54] Liao, J.; Zang, Y.; Luo, X.; Zhou, Z.; Lan, Y.; Zang, Y.; Gu, X.; Xu, W.; Hewitt, A.J. Optimization of variables for maximizing efficacy and efficiency in aerial spray application to cotton using unmanned aerial systems. Int. J. Agric. Biol. Eng. 2019, 12, 10-17.

[55] Lv, M.; Xiao, S.; Tang, Y.; He, Y. Influence of UAV flight speed on droplet deposition characteristics with the application of infrared thermal imaging. Int. J. Agric. Biol. Eng. 2019, 12, 10-17, doi:10.25165/j.ijabe.20191203.4868.

[56] Martin, D.; Singh, V.; Latheef, M.A.; Bagavathiannan, M. Spray Deposition on Weeds (Palmer Amaranth and Morningglory) from a Remotely Piloted Aerial Application System and Backpack Sprayer. Drones 2020, 4, 59.

[57] Wang, G.; Lan, Y.; Yuan, H.; Qi, H.; Chen, P.; Ouyang, F.; Han, Y. Comparison of spray deposition, control efficacy on wheat aphids and working efficiency in the wheat field of the unmanned aerial vehicle with boom sprayer and two conventional knapsack sprayers. Appl. Sci. 2019, 9, 1-16, doi:10.3390/app9020218.

[58] Zhang, P.; Deng, L.; Lyu, Q.; He, S.L.; Yi, S.L.; De Liu, Y.; Yu, Y.X.; Pan, H.Y. Effects of citrus tree-shape and spraying height of small unmanned aerial vehicle on droplet distribution. Int. J. Agric. Biol. Eng. 2016, 9, 45-52, doi:10.3965/j.ijabe.20160904.2178.

[59] Yang, F.; Xue, X.; Cai, C.; Sun, Z.; Zhou, Q. Numerical simulation and analysis on spray drift movement of multirotor plant protection unmanned aerial vehicle. Energies 2018, 11, doi:10.3390/en11092399.

[60] Al Heidary, M.; Douzals, J.P.; Sinfort, C.; Vallet, A. Influence of spray characteristics on potential spray drift of field crop sprayers: A literature review. Crop Prot. 2014, 63, 120-130, doi:10.1016/j.cropro.2014.05.006.

[61] William, E. Smith, 2004. Reducing Herbic. Spray Drift. New Agric. Tamworth Agric. Inst.

[62] Foqué, D.; Braekman, P.; Pieters, J.G.; Nuyttens, D. A vertical spray boom application technique for conical bay laurel (*Laurus nobilis*) plants. Crop Prot. 2012, 41, 113-121.

[63] Teske, M.E.; Thistle, H.W.; Schou, W.C.; Miller, P.C.H.; Strager, J.M.; Richardson, B.; Ellis, M.C.B.; Barry, J.W.; Twardus, D.B.; Thompson, D.G. A review of computer models for pesticide deposition prediction. Trans. ASABE 2011, 54, 789-801.

[64] Fesal, S.N.M.; Fawzi, M.; Omar, Z. A numerical analysis of flat fan aerial crop spray. In Proceedings of the IOP Conference Series: Materials Science and Engineering; IOP Publishing, 2017; Vol. 243, p. 12044.

Chapter 3

A New Paradigm in the Delivery of Modernizing Agricultural Technologies across Africa

Paul L. Woomer, Welissa M. Mulei and Rachel M. Zozo

Abstract

This Chapter describes the approach and impacts of the Technologies for African Agricultural Transformation (TAAT) Program. TAAT is an operational framework based upon collaboration between the African Development Bank, the International Institute of Tropical Agriculture, and many other partners. This Program is designed to deliver modernizing agricultural technologies as a means of achieving food and nutritional security, and to boost employment and agricultural exports across Africa. TAAT consists of nine Commodity Compacts that have assembled technology toolkits for use in development programs and six specialized Enablers that help them to do so. These commodities are rice, maize, wheat, sorghum, millet, cassava, sweet potato, common beans, fish, and small livestock. The Enablers provide policy support, youth empowerment, capacity development, irrigation and soil fertility expertise, and control of invasive pests. Together these Compacts and Enablers design and conduct collaborative agricultural development projects in partnership with national counterparts. To date, TAAT has staged 88 interventions in 31 African countries, including the incorporation of customized technology toolkits within country loan projects of major development banks. Over three years, these efforts have reached about 10.6 million adopter households and increased food supply by 12 million tons worth over US $763 million, resulting in substantial improvements in smallholder farmer's food supply (0.75 MT $yr.^{-1}$) or income (\$128 $yr.^{-1}$). Environmental gains in terms of carbon offset average 0.74 MT CO_2e $yr.^{-1}$ per adopter household, an outcome indicative of positive combined rural development and climate actions. This Chapter describes how these technology toolkits are designed, deployed and evaluated, and how TAAT is becoming a leading mechanism for agricultural innovation delivery across Africa. This evaluation is limited to eight critical field crops and does not consider animal enterprises or the strategic roles of TAAT Enablers, two other important activities within the larger Program.

Keywords: African Development Bank, agricultural transformation, COVID-19, food security, IITA, strategic investment, TAAT Program, technology delivery

1. Introduction

Agriculture is a major economic activity and source of income across Africa, but its untapped potential contributes to persistent poverty, rural stagnation and deteriorating food security. Also, increased food demand and changing consumption habits have led to rising food imports, including staple foods, and the value of these imports now exceeds US $35 billion per year [1]. At the same time, failure to commercialize agriculture is associated with massive rural unemployment; where over 120 million Africans are out of work, more than 672 million live on less than US $2 per person per day, and 60% of those unemployed are young people. Nonetheless, agricultural modernization serves as the critical engine for African economic growth and this transformation requires that a host of production and processing technologies become widely introduced and adopted by the region's smallholder farmers and entrepreneurs.

Technologies for African Agricultural Transformation (TAAT; https://taat-africa.org) is a recent program led by the International Institute of Tropical Agriculture (IITA) that has pioneered new approaches to deploying proven technologies to African farmers. TAAT arose as a joint effort of IITA and the African Development Bank (AfDB) and is a crucial component of the latter's Feed Africa Strategy [1]. TAAT is organized around 15 "Compacts" that represent priorities and partnerships to achieve Africa's potential in achieving food security and advancing its role in global agricultural trade. Nine of these Compacts relate to specific priority value chains of rice (*Oryza sativa*), wheat (*Triticum* spp.), maize (*Zea mays*), sorghum (*Sorghum bicolor*) and pearl millet (*Pennisetum glaucum*), cassava (*Manihot esculenta*), sweet potato (*Ipomoea batatas*), common bean (*Phaseolus vulgaris*), fish, and small livestock [2]. Weaknesses in producing these commodities are viewed as responsible for Africa's food and nutritional insecurity, the need for excessive importation of food, and the unrealized expansion of food exports. Six of these Compacts serve as Enablers, providing specialist services to the Commodity Compacts in policy support, capacity development, youth empowerment, soil management, water management, and response to invasive pests. Together these Compacts design interventions in collaboration with national programs to introduce technologies and management innovations that respond to country priorities for agricultural development. In many cases, these targets are addressed by implementing projects resulting from sovereign country loans awarded by development banks; and TAAT's role in the planning and execution of these loan projects is becoming a vital element of their success. This Chapter describes the range of technologies advanced through TAAT and the impact of their deployment through various mechanisms over the past three years (mid-2018 through early-2021).

TAAT's portfolio of proven technologies deemed ready for deployment is open-ended, meaning that all holders are welcome to work through the Program; however their inclusion is subject to scrutiny through different mechanisms [3]. The most abundant technology holders are Consultative Group on International Agricultural Research (CGIAR) Centers through their commitment to producing new knowledge and technologies necessary to meet several Sustainable Development Goals (CGIAR, 2018). Many of these innovations result from the 16 programs and platforms of the CGIAR Collaborative Research Program (CRP), mainly those eight addressing Agri-Food Systems [4]. The CGIAR has paid special attention to compiling and categorizing these innovations and projecting their impact [4, 5]. A review by TAAT in early-2019 indicated a 63% overlap between CGIAR technologies deemed ready for next users and the elements of the TAAT toolkits [3, 6], with the discrepancy mainly attributable to technologies related to CGIAR mandate commodities outside of TAAT's nine priority value chains.

TAAT is well-positioned to advance agricultural modernization across Africa, both as a developmental model and a mechanism forward. As a model, TAAT allows an open-ended flow of proven agricultural technologies to stream toward rural development planning. It redirects the technical expertise within its Compacts to the needs of others' programs in ways that correct for flawed designs of the past and that provides the technical assistance required to assure that more aggressive actions toward agricultural modernization result in the future. At the same time, it reorients research institutions to the expectation that simply novel ideas and innovative piloting are insufficient. Ultimately success is judged and rewarded by the widespread adoption and commercial appeal of the proven technologies that result. TAAT consists of four components: 1) Enabling Environment, 2) Rural Technology Delivery Infrastructure, 3) Technology Deployment, and 4) Program Management [2]. The first component promotes a common transformation agenda, proven technologies and supporting policies; the second organizes technology access and partnership around that agenda; the third conducts collaborative technology delivery at national and site levels; and the last creates an administrative and reporting structure for the first three. This simple model is proving to be quite effective. Furthermore, recent developments in "Scaling Readiness" permit innovations to be classified based upon their stage of development and range of applicability, serving as both a means to prioritize commitments and predict impact trajectories [7].

As a mechanism, TAAT represents a dynamic platform where those committed to advancing transformative agricultural technologies connect with those who need them most, particularly within government programs expected to improve their agricultural production, strengthen rural economies and advance the lives and livelihoods of farming households. It is a way for the development community to buy into technical advances under the assurance that offered technologies achieve expected levels of adoption and promised impacts. TAAT was quick to pivot in response to the COVID-19 pandemic [8] and its disruption of planting, harvesting, and marketing; and took quick action to mobilize improved seed and planting materials to national partners. In the larger view, TAAT recognizes that the world's most vulnerable, including women, youth, and the most impoverished farmers, are the hardest hit by the pandemic, that the African economy is particularly vulnerable to it [9], and that it has a vital role to play in support of Africa's recovering food systems. This Chapter describes the technologies that TAAT promotes, how they are advanced, and the impacts achieved by this Program over its first three years, with attention to eight priority crop commodities.

2. An enabling environment

An enabling environment is necessary for modernizing technologies to become better recognized and mobilized. The Enabling Environment offered as a Program component through TAAT is partly the result of its carefully considered design and implementation process. The design of TAAT started through dialog between IITA and AfDB and culminated in endorsement and buy-in from Regional Member Countries of the Bank. TAAT arose as a service to the AfDB Feed Africa Strategy by international centers holding proven but under-deployed technologies of importance to Africa's agricultural future [1]. Three alternative models were evaluated based upon alignment among different agro-ecologies, within recognized priority developmental interventions, and along key agricultural value chains, with the last option deemed the most promising in terms of ready linkage to national programs and the private sector. In the process, a comprehensive list of candidate technologies were examined, particularly those from the CGIAR, as being ready for the next users [4, 5]. A key step in this process was submitting letters of intent by national authorities from 25 African countries to participate in the Program [10]. Negotiations led to the release of US $40 million to the TAAT Program by the African Development Fund, primarily for three-year operations (2018 to 2021) of the 15 TAAT Compacts and the Program Management Unit; and an additional $7 million raised from the Bill and Melinda Gates Foundation in support of technology brokerage by the TAAT Clearinghouse. TAAT was officially launched through two events, one held at IITA, Ibadan, Nigeria to formalize its technical approaches, and another at AfDB HQ in Abidjan, Cote d'Ivoire to finalize administrative requirements and to announce the Program to the larger international community; both conducted during January 2018 [2]. The Program began its field operations shortly afterwards.

As a result of these preparations, the Enabling Environment fostered through TAAT then focused upon three areas; the promotion of agricultural modernization as a regional priority, the removal of bottlenecks preventing regional integration of its technologies, and an understanding of which trade and regulatory policies are required to advance its goals. Agricultural modernization is crucial to the achievement of numerous Sustainable Development Goals (SDGs) [11], starting with Goal 1-No Poverty, Goal 2-Zero Hunger, and Goal 13-Climate Action but extending into several other Goals as well (Section 3.2). TAAT also serves as a replicable mechanism within Goal 17-Partnerships to Achieve the Goals. Technologies associated with the various TAAT commodity Compacts are extremely relevant to this Chapter and appear in **Table 1**. This list covers only crop technologies, but information on TAAT's approaches to aquaculture, poultry, and small livestock are available elsewhere [12].

Note that all African countries are eligible to participate in TAAT based upon their willingness to enact policies that facilitate the uptake of food production technologies, to align national research and extension systems to it, and to participate in agreed-upon activities. This participation is often performed in conjunction with sub-regional programs and sovereign country developmental loans. This approach ensures that TAAT interfaces with a wide variety of agricultural value chain development programs to secure the deployment and adoption of recommended food production technologies by farmers, particularly dynamic seed systems, commercializable accompanying technologies, and backstopping information and extension campaigns.

The need to accelerate and harmonize technology release and registration across regions shapes TAAT's policy approaches, particularly when current situations restrict technology scaling across national boundaries within regional economic blocks. Policies and protocols for the release and registration of crop varieties are streamlined so that improved varieties available in one country are utilized in others within the same agroecological zone. In many cases, the policies needed to reinforce agricultural transformation are in place, such as the duty-free importation of agricultural equipment and inputs and reducing regulatory obstacles to the seed sector. However, in some cases regional cooperation is unenforced or lacking. It is not the purpose of this Chapter to review national and regional policies related to the fast-track release of technologies within and between countries, but the provision for policy support can indeed lead to faster uptake of modernizing technologies, and this became a primary concern of the Program's Enabling Environment.

3. A regional technology delivery infrastructure

TAAT operates a Regional Technology Delivery Infrastructure (RTDI) that offers a menu of proven food production technologies in nine priority commodities to all Program partners and stakeholders. These technologies are bundled into "technology toolkits" [3] that may be configured into elements of country projects and offered through knowledge sharing and extension campaigns. These technologies include improved varieties, seed systems innovations, accompanying soil fertility and pest management strategies, harvest and post-harvest management, digital

Commodity value chain	TAAT Technologies ready for scaling
Rice	New Rice for Africa (NERICA), Hybrid Rice Varieties for Africa (ARICA), Aromatic rice varieties (ORYLUX), Engineered irrigation, Deep urea placement, Foliar micronutrient addition, Motorized weeders, RiceAdvice digital support, Axial flow threshers, GEM parboiling and flour production
Wheat	Heat and drought tolerant varieties, Yellow stem rust resistant varieties, Hessian fly resistance, Sahel "winter" production, Raised bed furrow irrigation, Hermetic grain storage, Conservation agriculture and zero tillage, Integrated pest, disease and weed management, Expansion of harvesting combine fleets, Flour milling and blending systems
Maize	Drought-tolerant varieties (DTMA, WEMA), Imazapyr resistance for striga management (IR-maize), Golden maize (Vitamin A fortification), TEGO seed licensing mechanism, Contract mechanization (Hello Tractor), Specialized pre-plant fertilizer blending and N topdressing, Maize-legume rotation and intercropping, Pre-emergent herbicides, Fall Armyworm control (e.g. Fortenza Duo®), Aflatoxin management (AflaSafe®)
Millet & Sorghum	Rust-resistant varieties, striga-tolerant varieties, Community-based seed production, Zai pits and contour bunds (water harvesting), Conservation agriculture, Fertilizer microdosing, Parasitoid wasps for Head Miner, Mobile stover choppers, Warrantage cereal banking, Flour production and wheat and barley substitution
Cassava	Resistance to Cassava Mosaic Virus, Golden Cassava (Vitamin A fortified), High dry matter and starch content varieties, Semi Autotrophic Hydroponics (rapid propagation system), Stem cutting enterprise, Integrated weed, pest and soil management, Cassava Business Connector (App), Chipping and livestock feed production, Mobile processing plant, High Quality Cassava Flour and industrial starches
Sweet Potato	Orange Fleshed Sweet Potato (OFSP, high Vitamin A), Drought- and virus-tolerant varieties, Purple sweet potato (high in antioxidants), Community-based cutting production, Greenhouse production of vines and cuttings, Raised bed production, Specially blended fertilizers, Relay intercropping with legumes, Silage production from vines, Puree production and products
Beans	High Iron Bean (HIB, also high Zn), Climbing bean with higher nitrogen fixation potential, Seed inoculation with rhizobia, Chemical seed dressing, Specialized fertilizer blends, Low-cost staking, Mechanical weeder & herbicide (integrated management), Hermetic grain storage, flour and flour products, Pre-cooked beans

Table 1.
Technologies ready for bundling and scaling arranged by TAAT commodity value chain.

applications and value addition processes. These technologies represent Regional Public Goods that offer broad public interest and recognizable benefits, strong alignment with development bank strategic orientation and regional objectives, and higher impact through cooperation [10]. For the most part, these technologies are available to all stakeholders for free. TAAT is not a research program intended to develop new technologies, instead it serves as a mechanism to advance proven approaches that warrant massive up-scaling. In some cases, applied research is needed to customize toolkits for specific conditions and clients.

3.1 The TAAT top 100 technologies

Compilation and interpretation of the technologies described in **Table 1** resulted from the assembly and analysis of a TAAT Top 100 Technologies database. This database consists of 10 leading technologies for each of TAATs priority commodities and contains three main categories and 13 descriptive variables (data not presented). To a large extent, these technologies are based upon achievements of the CGIAR [4, 5] where many of the technologies from its CRPs are described as either ready for next user or in an advanced stage of innovation. In terms of technical approach

(**Figure 1a**), the technologies are divided among those involving improved genetics and plant and animal breeding (23%), those based upon the distribution of digital information (3%) [13], production input products of proven efficacy (21%), crop and animal management technologies of utility within agricultural extension messaging and campaigns (27%) and the availability of appropriately designed labor-saving

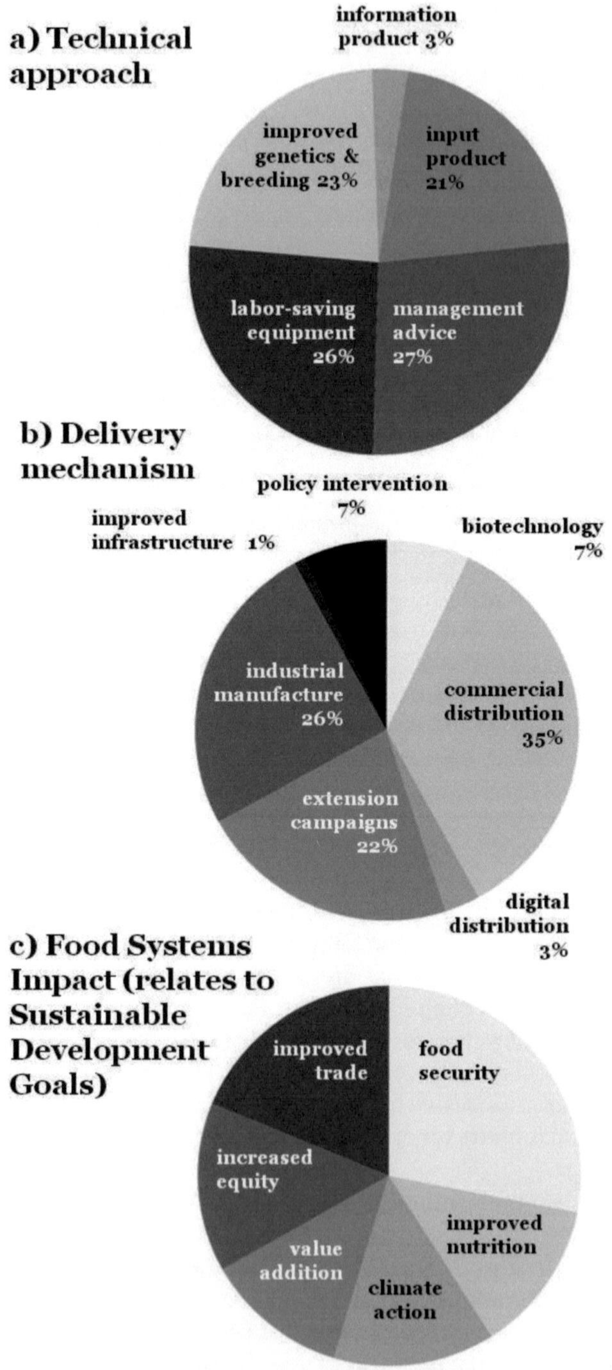

Figure 1.
The technical approach (a), delivery mechanism (b) and food systems impacts (c) of the TAAT top 100 technologies.

equipment (26%). These approaches are balanced, suggesting enormous scope for the deployment of diverse and accompanying technologies. This symmetry may be attributable to how the various Compacts were encouraged by TAAT to design their technology toolkits along entire value chains in the first place [3, 6].

Within the TAAT Theory of Change, a modern African farmer is encouraged to acquire the best varieties for their growing and market conditions, raise them with recommended accompanying production inputs, conduct their production through sound management advice, and do so in a labor- and cost-effective manner [2, 3, 6, 12]. It also encourages clients to add value to agricultural products in a way that addresses upper-end markets. TAAT is also examining the policy implications of its Top 100 transformative technologies. While this area remains under development, preliminary results suggest that very few technologies (12%) have no immediate policy implications or may otherwise be advanced through the national legislative process involving enactment or repeal of laws (23%) (data not presented). In contrast, technology deployment is best facilitated through the agendas of African regional organizations (46%) or requires progressive regulatory adjustments (rather than lawmaking) by national authorities (58%).

3.2 Technology delivery strategy

Of equal importance to TAAT and its mission are the mechanisms through which its targeted technologies are delivered (**Figure 1b**). TAAT operates under the assumption that there are seven separate delivery mechanisms in support of its technologies but also recognizes that they are complementary and overlapping. These delivery mechanism categories and the proportion of technologies belonging to them are applications reliant upon biotechnologies (7%); commercial production and distribution along supply chains (35%), digital applications (3%), government and private extension services (22%), industrial manufacture and distribution (26%), inclusion within infrastructure development projects (1%) and through policy reform (7%). The main sources of overlap in delivery strategy exist between biotechnologies, industrial advancement, and commercial distribution.

The technologies and their relationship to food systems (**Figure 1c**) advanced by TAAT clearly relate to the Strategic Development Goals [11]. Of the Top 100 Technologies: 76% intend to increase farm productivity, improve food security, and reduce hunger (relates to SDG 2-No Hunger); 35% improve the quality of food, household nutrition, and diets (relates to SDGs 2 and 3-Good Health); 37% represent climate-smart innovation and reduced risk of environmental extremes (impact upon SDG 13-Climate Action); 33% add value and provide opportunity for cottage-scale and commercial processing (relates to SDG 1); 39% improve equity by providing special opportunity to women, youth or vulnerable stakeholders (relates to SDG 5-Gender Equity); and 51% impact upon trade imbalances and provide opportunity for agribusiness and exports (relates to SDG 8-Economic Growth). The total percentages are more than 100% because a given technology can contribute to more than one SDG. In this regard, the TAAT technology portfolio represents a balanced approach to food systems improvement.

Of the Top 100 transformative technologies, 76% were extensively deployed by TAAT Compacts to African farmers. Of these same technologies, 64% are sufficiently established for widespread adoption, whereas the remaining 36% require further fine-tuning or site adjustment (**Table 2**). From these characteristics, it is evident that almost 2/3 of the transformative technologies advanced by TAAT are mature in that they are ready for immediate adoption by clients. This applicability is in keeping with its mandate as a leading element of the AfDB Feed Africa Strategy. At the same time, the technology identification process by TAAT must also be

Technology category	No.	Comment
Included within current Compact workplans	76	Over 3/4 of these Top 100 transformative technologies are currently advanced through TAAT's country-level deployment
Based upon Advanced R&D technologies	27	Many of the technologies advanced by TAAT require further refinement and site-specific adjustment
Based upon Established technologies	49	About 2/3 of the technologies currently advanced by TAAT are fully established and ready for widespread adoption
Plans for future inclusion	24	Several additional Top 100 technologies warrant attention during future stages of TAAT
Based upon Advanced R&D technologies	9	There are some key areas of technical Research & Development that require future attention among TAAT partners
Based upon Established technologies	15	Several additional established technologies could be rapidly included within existing TAAT technology toolkits

Table 2.
Categories of current TAAT technologies, their proportions and current status.

forward-looking, and 24 of them show potential to make future contributions to the TAAT agenda. Nine of these candidates are at an advanced stage of Research and Development but requires additional attention, and 15 of them exist as innovations that are ready for the next users but remain weakly distributed. In this way, the nine candidates represent a research agenda for the institutions leading these Compacts in the following areas and include 1) Imazapyr resistance in maize for striga (*Striga spp.*) management; 2) Golden maize for vitamin A fortification; 3) Improved, hardy, duo-purpose breeds of poultry for smallholders; 4) release of purple-fleshed sweet potato that is high in antioxidants; 5) Hessian fly resistance in wheat; 6) Aflatoxin management of maize and other crops using Aflasafe® atoxigenic fungi; 7) refinement of integrated vegetable-aquaculture systems; 8) water quality protection and aeration of fish ponds, and 9) expansion of harvesting combine fleets of wheat. Over 50% of these topics are related to improved crop varieties and animal breeds, making the CGIAR particularly well-positioned to contribute to their achievement through genetic innovation [4, 5].

Of the 15 underutilized off-the-shelf technologies, one is a biotechnology (preventative antibiotics for poultry bio-security), another is a commercial product (pond lining in fishpond construction), and four are well-established management interventions related to different forms of crop rotation and value addition to manure and composts. Six technologies involve underutilized industrial devices and applications including semi-automated poultry hatcheries, mechanized de-feathering and egg sorting, motorized weeders and choppers for field operations, hide curing and secondary leatherworks, and sorghum flour milling and blending for wheat and barley substitution. Three underutilized technologies have strong policy dimensions related to the clustering fish farmers within aquacultural parks, and the humane slaughtering and meat inspection of both poultry and small ruminants. RTDI requires that TAAT technologies be recognized and appreciated through a variety of information resources offered by the Program [2, 3] including numerous adoptive case studies [12].

3.3 Relationship to resilience and climate action

Agricultural resilience describes the ability to absorb and overcome constraints to agricultural production and livelihoods, whether short-term or long-term, sudden or chronic. This ability responds to complex and accumulating economic, social,

	frequency
a. Resilience factors (technologies offer …)	
Overall climate-smart effects	53%
Increased carbon sequestration	34%
Increased drought tolerance	30%
Management of biotic interactions	25%
Greenhouse gas emissions reduction	14%
Resistance to extreme weather events	14%
b. Resilience pathway (delivered through …)	
System-level transformation	13%
Increased crop biomass	9%
Improved water use	8%
Improved biotic relations	8%
Reduced gaseous loss	4%
Drought tolerant germplasm	3%
Enterprise diversification	3%
Increased animal hardiness	2%
Improved animal sheltering	2%
Reduced fuel use	1%

Table 3.
Relationship of the technologies advanced by TAAT to agricultural and environmental resilience.

environmental, and institutional shocks and stresses. Resilience is often considered within the context of climate change and its relationship to food security [14]. In this sense, the TAAT technologies are classified in terms of their potential contributions to resilience as related to 1) increased carbon sequestration, 2) reduction of GHG emissions, 3) performance during drought, 4) protection against extreme weather events, and 5) fostering beneficial biodiversity. These factors were evaluated to assess their roles in climate-smart agricultural practice. The relationship of these technologies was examined in two ways; the types of resilience they infer, and the pathway through which their benefits are achieved (**Table 3**) with one of five different sorts of climate-smart features expressed among 53% of the different technologies (**Table 3a**). Within the context of agricultural transformation, the pathways to adoption are extremely relevant. This analysis revealed ten different pathways ranging from 1–13% of the individual technologies. These pathways included (in descending order): system-level transformation, increased crop biomass, improved water use, improved biotic relations, reduced gaseous loss, drought-tolerant germplasm, enterprise diversification, increased animal hardiness, improved animal sheltering, and reduced fuel use (**Table 3b**). This diversity of pathways reinforces the advantages and complexity of bundling technologies into toolkits for application within climate-responsive rural development projects [3, 15].

4. Technology deployment

The TAAT Program mobilizes a wide range of agricultural technology providers to assemble flexible menus of proven food production technologies related to

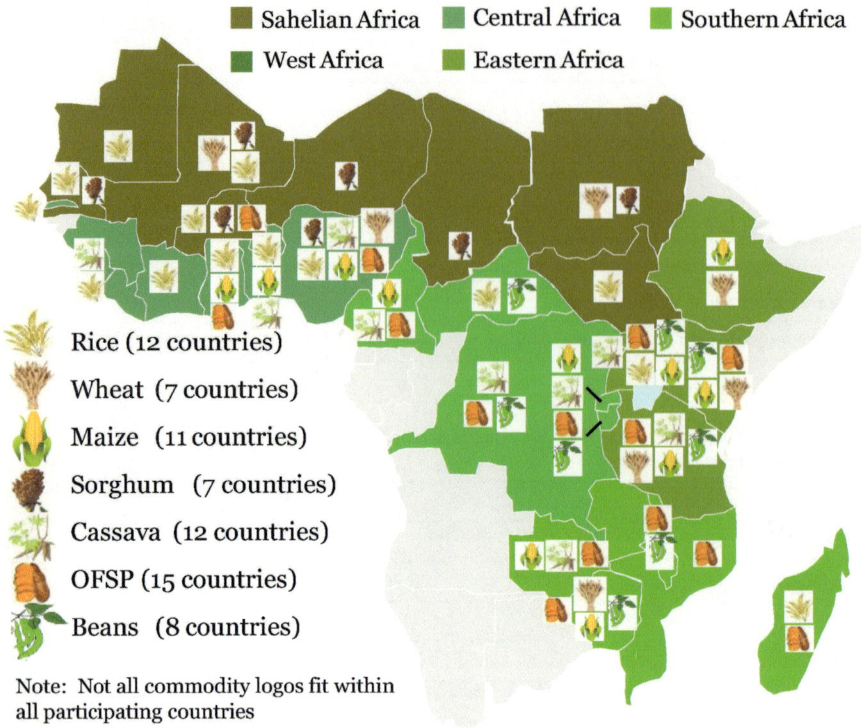

Figure 2.
TAAT activities include commodity compact interventions across 31 African countries.

improved seed and breeding systems and their accompanying management and input products. These bundles are referred to as toolkits [3] intended for large-scale deployment through promotional and outreach campaigns, usually in conjunction with the country projects of development banks and humanitarian organizations. For the most part, these efforts are linked to country-level extension campaigns and with existing community-based innovation platforms. This approach allows for efficient outreach actions. As a result of these efforts, TAAT Commodity Compacts have performed 88 technology deployment interventions in 31 countries (**Figure 2**) over the past three years (mid-2018 through early-2021). Some highlights of these interventions grouped by their respective commodity Compacts follow, and further information about them is available through https://taat-africa.org/compact-briefs/.

4.1 The rice compact

This Compact is coordinated by AfricaRice and operates in 12 countries, with concentrated activities in West Africa and the Sahel (10 countries). Its focus is to reduce massive rice imports into Africa and produce a quality product that rivals those imports. In all, this Compact worked through 41 partnerships and delivered 162 MT of certified seed through its cooperators. It introduced climate-smart rice varieties with potential yields reported to be 4–7 MT ha^{-1}, but its reported yield results were far less. This discrepancy suggests that these improved rice varieties were inconsistently accompanied by recommended Good Agricultural Practices (**Table 1**). In some cases, however, outstanding yields were achieved with harvests up to 4.2 MT ha^{-1}. A total of 69.1 MT of breeder seeds was produced for multiplication into certified seeds by seed companies. In some cases, deep placement of large urea pellets was introduced to reduce the loss of ammonia to the atmosphere. In terms of rice utilization, a parboiling system and substitution of firewood with rice husk briquettes as cooking fuel resulted in substantial energy savings.

4.2 The wheat compact

This Compact is coordinated by The International Center for Agricultural Research in the Dry Areas and operates in seven countries. It focuses upon reducing massive wheat imports into Africa and advancing cool-season, irrigated production in the Sahel. The improved varieties delivered through TAAT have two key characteristics; some varieties are heat tolerant, allowing wheat production across a broader range of growing conditions, including lower elevations. Others are tolerant of stem rust, a constraining fungal disorder. This Compact operates through 129 partnerships and delivered 131,211 MT of certified seed through its cooperators. As a result of this strategy, wheat self-sufficiency is becoming a realizable national agenda item. In terms of outreach, grain and seed production, and value addition have impacted the lives of about 1.8 million beneficiaries, with 44% of them being women and youth.

4.3 The maize compact

This Compact is co-managed by The African Agricultural Technology Foundation and IITA and operates in 11 countries. It focuses on greater reliance upon 35 high-yielding and drought-tolerant maize hybrids and their adoption by commercial seed companies. It builds upon TEGO®, an innovative mechanism to accelerate the commercial acceptance of new maize varieties. The Maize Compact operates through 60 partnerships reaching 5,337,003 beneficiaries. It delivered 20,111 MT of certified seed through its seed system network. It also deployed 6,598 tons of certified seed treated with Fortenza Duo™ to counter the invasion of Fall Armyworm, an outbreak that directly preceded the initiation of TAAT's activities in 2018.

4.4 Millet and Sorghum compact

This Compact is coordinated by The International Crops Research Institute for the Semi-Arid Tropics and operates in seven countries of the Sahel. Promotion of pearl millet and sorghum is combined within a single Program unit because both of these crops are critical staple cereals of Africa's cultivated drylands, with millet assuming importance in the driest cultivated areas of the Sahel. This effort primarily focuses upon the achievement of food security among some of the most vulnerable farmers in the world. In all, this Compact worked through 16 partnerships reaching 83,629 beneficiaries and delivered 1,391 MT of certified seed through its cooperators. The Compact also addresses the substitution of sorghum and millet grains for imported ones, particularly in blended porridges and brewing.

4.5 Cassava compact

This Compact is directed by IITA and operates in 12 countries spread over Central, East, Southern and West Africa. Its focus is upon the distribution of improved virus tolerant varieties, accompanying sound management practices, and the elevation of cassava as an agro-processed export commodity. The Compact delivered 20,457,130 cuttings of this vegetatively cultivated crop through its cooperators. Its integrated approaches have resulted in substantial improvement of cassava yield. For example, in two countries, Tanzania and Nigeria, farmers relying upon TAAT technologies increased yield from an average of 8 MT to 29 MT per ha, similar to levels reported in commercial plantations in S.E. Asia. Critical to that increase is adopting improved vegetative propagation systems that ensure virus-free planting material (see Section 7.1). One of these technologies, Semi-Autotrophic Hydroponics, has produced and distributed 47,065 plantlets used to establish cutting operations. One problem confronted by the Compact is the low cost and weak marketability of cassava. Post-harvest (e.g. tuber waxing to extend shelf life) and agro-industrial processing (e.g. manufactured starch) technologies address this problem. These efforts intend that African countries become suppliers of cassava products on international markets.

4.6 Orange fleshed sweet potato compact

This Compact is coordinated by The International Potato Center and operates in 15 countries across Central, East, Southern and West Africa. Its focus is upon the introduction of Orange Fleshed Sweet Potato (OFSP), rich in pro-vitamin A, its accompanying propagation and management technologies, and the use of OFSP in human nutrition and localized value-added processing. This Compact worked through 18 partnerships and delivered 32,995,950 cuttings and vines of this vegetatively cultivated crop through its cooperators. This Compact promoted 93 improved varieties with high-yielding, virus-tolerant, and climate-resilient characteristics based on their adaptation within different agro-ecological zones [16] and through national approval mechanisms. Localized vegetative propagation systems that provide affordable, virus-free cuttings and vines are central to this approach and the Compact established more than 500 propagation enterprises. Value addition to OFSP is also pivotal, with about 18,659 individuals trained in enterprise development [17], 555 small-scale enterprises established, and over 40 industrial food processors now offering OFSP products. It also promotes sweet potato vines as a source of nutritious livestock feed and silage.

4.7 High Iron bean compact

This Compact is directed by the Alliance of Bioversity International and the International Center for Tropical Agriculture (ABC) and operates in eight countries of Central, East and Southern Africa. It focuses on the distribution and use of improved High Iron Bean varieties that may also be high in zinc. It operates through 10 partnerships that reached 1,150,929 beneficiaries. To date, 5,006 MT of seed, both bush and climbing types, were disseminated through public-private partnerships and community-based actions. A suite of improved agricultural practices accompany these varieties that increase yields from 0.8 MT to 1.25 MT per ha for bush types, and even more for climbers. Many of its beneficiaries also engage in value-adding pursuits, including the production of pre-cooked beans and the milling of bean flour.

Not included in this section is a description of the TAAT Aquaculture and Small Livestock Compacts led by WorldFish and the International Livestock Research Institute, respectively (see Section 7.4).

5. Larger outcomes and impacts

The deployment of technologies by TAAT has resulted in substantial impacts among eight of the Program's commodities; rice, wheat, maize, sorghum, millet, cassava, sweet potato and bean (**Table 4**). Productivity has increased an average

TAAT commodity compact	Increased productivity[1]	Crop coverage[2]	Increased production	Increased value[3]
	MT ha^{-1} (%)	km^2	MT y^{-1}	\$ y^{-1} x 10^-6
Rice	0.21 (10%)	7,058	148,223	55.9
Wheat	0.80 (36%)	18,000	1,440,000	290.9
Maize	0.75 (50%)	8,412	630,930	107.3
Millet	1.0 (133%)	237	23,765	4.5
Sorghum	1.75 (140%)	1001	175,172	37.7
Cassava	27.5 (262%)	3,441	9,462,750	217.6
Sweet Potato[4]	5.3 (106%)	198	104,927	34.6
High Iron Bean	0.45 (56%)	588	26,471	14.7
Total (*mean*[5])	*1.33* (58%)	38,936	12,012,238	763.2

[1]*Based upon improved vs. baseline yields (% increase in parentheses).*
[2]*1 km^2 = 100 ha.*
[3]*Based upon January 2021 international prices.*
[4]*Promotion of the Orange Fleshed Sweet Potato.*
[5]*Mean (0.133) is normalized by dry matter content and coverage-weighting, Standard Error of the Mean ± 0.76.*

Table 4.
Increased crop productivity, production and value resulting from technology deployment by the TAAT program.

of 58% ± 25% in response to these technologies, ranging between only 10% for rice and 262% for cassava. When adjusted for moisture contents and coverage, these gains amount to 1.33 ± 0.76 MT ha^{-1}. These technologies were extended across 3.9 million ha and resulted in an additional 12.0 million tons of food worth over US $763 million. At the beneficiary household level (**Table 5**), these technologies were applied to an average 0.37 ± 0.16 ha with the least areas devoted to bean (0.13 ha per household) and the largest to millet (1.92 ha). In large part, these scaling dimensions reflect the design and complexity of the technology packages themselves, with open-pollinated seed-based cereals appearing the least complicated and vegetatively propagated crops (cassava and sweet potato) more so. Wetland rice represents an intermediate case because of its complex field preparation and irrigation requirements.

The return in terms of increased food and income reflects the areas devoted to those crops, the magnitude of yield increase, and the commodity value. On average, adoption of TAAT technologies resulted in an average of 0.75 ± 0.41 MT of additional food per household per year (**Table 5**), with bean offering the least (0.05 MT) and cassava the most (3.6 MT) when expressed on a dry weight basis. Among the five kinds of cereal, food supply increased by an average of 0.94 MT per household, a substantial amount in terms of household food security. Alternatively, these gains could be marketed for an average of \$128 ± \$72 per household with the greatest returns realized in the order sweet potato >sorghum or millet> cassava >wheat>bean >rice > maize (**Table 5**).

In all, the various technology toolkits reached 10,616,372 households. If we assume equal division of project resources and services between these commodities, the expenditures by the three-year Program reached these beneficiaries at an average cost of \$3.28 per household. This estimate does not take into account co-financing and in-kind contributions of TAAT's many partners. The Benefit to Cost Ratio of TAAT's program investment is about 22:1, an impressive return on developmental investment to that part of the Program directly supporting these eight crop commodities and accompanying Program services. The COVID-19 pandemic coincided

TAAT Commodity compact	Beneficiary households (hh)	Adoption area	Increased food supply[1]	Increased Income[2]
	no	ha hh^{-1}	MT y^{-1}	\$ hh^{-1}
Rice	2,257,987	0.31	0.06	25
Wheat	1,777,845	1.01	0.70	164
Maize	5,174,965	0.16	0.11	21
Millet	12,403	1.92	1.69	364
Sorghum	71,217	1.41	2.16	529
Cassava	817,314	0.42	3.62	266
Sweet Potato[3]	54,641	0.36	0.67	634
High Iron Bean	450,000	0.13	0.05	33
Total *(mean)*	10,616,372	*0.37 ± 0.16*	*0.75 ± 0.41*	*128 ± 72*

[1]Dry matter yield increase takes different Compact adoption areas and moisture contents into account, overall mean is weighted by coverage, ± Standard Error of the Mean.
[2]Based upon January 2021 international prices, overall mean is weighted by the number of adopters, ± Standard Error of the Mean across eight commodities.
[3]Promotion of the Orange Fleshed Sweet Potato.

Table 5.
Adoption and household impacts resulting from technology deployment by the TAAT program.

with TAAT's third year of operations, and it is difficult to assess to what extent the momentum of the Program was impeded by the contagion despite its timely adjustments to it (see Section 6).

Projections of carbon sequestration resulting from TAAT interventions may also be made (**Table 6**). These projections are based upon Compact reports of increased yield, coverage (**Table 4**), numbers of adopters (**Table 5**), and assumptions concerning biomass yield, moisture content, Harvest Index, crop

TAAT Commodity Compact	Increased system CO_2e	Annual increased CO_2e[1]	Value of annual increase in CO_2e[2]	Annual reduction per adopter[3]
	MT ha^{-1}	MT y^{-1}	\$ y^{-1} x 10^6	MT
Rice	1.06	186,882	4.11	0.08
Wheat	2.92	1,753,606	38.6	1.9
Maize	3.10	869.284	19.1	1.5
Millet	5.71	22,611	0.50	1.8
Sorghum	9.27	154,668	3.40	2.2
Cassava	26.1	1,498,159	33.0	1.8
Sweet Potato	5.24	25,944	0.57	0.1
High Iron Bean	1.63	48,079	1.06	0.1
Total *(mean)*	*4.8 ± 2.3*	4,559,213	100.3	*0.74 ± 0.44*

*[1]Mean weighted by coverage from **Table 5**, ± Standard Error of the Mean.*
[2]Based upon US \$22 per MT CO_2e.
*[3]Based upon the annual increase of CO_2e and overall mean weighted by beneficiary households from **Table 6**, ± Standard Error of the Mean.*

Table 6.
Carbon sequestration resulting from TAAT intervention.

carbon content, CO_2e:crop C ratio, planning horizons, and the price of CO_2e. This approximation allows for the expression of realizable gains of CO_2e associated with increased biomass and residual benefits in terms of CO_2e gain per ha and as total average gain per project year. Realizable gains were achieved based upon increased focus upon climate-smart field practices and products within the technology toolkits employed by country agricultural development projects (see Section 3.3). This approach results in estimated CO_2e gains averaging 4.8 ± 2.3 MT ha^{-1} across these eight commodities and a total of 4.6 million MT of CO_2e per year worth about US $100 million. When the number of adopters is considered, this amounts to per capita emissions reductions of 0.74 ± 0.44 MT CO_2e per household per year, very similar to the targets established by Branca et al. [18] and Lipper et al. [19] when like sorts of climate-smart technologies are described. The possibility of organizing small-scale African farmers into a legion devoted to carbon sequestration is an exciting option for the future, but one that does not greatly benefit individual climate-smart practitioners from the standpoint of direct financial benefit. The gains they achieve are worth only $16 per household per year at current prices of CO_2e (calculated from **Table 6**). For this reason, the benefits of climate-smart technologies are perhaps better promoted in terms of improved yield and land care and then factored in terms of realizing national commitments rather than being presented to farmers as a separate and tangible income opportunity.

6. COVID-19 pandemic disruption and adjustments

The COVID-19 pandemic has strongly affected TAAT's plans for technology deployment, and attempts at adjustments to the contagion were only partly successful. TAAT's deployment strategy was initially based upon travel to partnering countries by technology holders and physical gathering among stakeholders adopting their technologies. Most of this travel and participation within the missions of development banks was curtailed during 2020, prompting rounds of discussion on alternative ways forward. The TAAT Policy Support Compact offered a quick response through a position paper describing the threat to food security and economic growth, and raising some useful policy support options [8]. Another response considered the implications of COVID-19 safety precautions at the farm level and among potential adopters of TAAT technologies. This latter approach formed the basis for COVID-safe agricultural practices distributed to many project partners and stakeholders.

COVID-safe agricultural practices mitigate the risk of contracting the disease among farming household members while providing essential services as food producers during the pandemic. COVID-safe farming considers that contagion within a farm household is a function of the frequency of encounters with others and the risks they pose. There are two basic risks of exposure, external risk based upon exposure to others while away from the farm and internal risk from infected persons coming onto the farm. There are four main mitigation measures related to COVID-safe agricultural practice: 1) adherence to recommended social precautions, 2) adjustments in the management of farm labor, 3) increased use of small-scale machinery allowing for greater social distancing, and 4) greater reliance upon digital applications. Social precautions include personal distancing, wearing protective gear, and avoiding unnecessary contacts. The farm household should practice these precautions as the pandemic spreads deeper into rural communities. Greater reliance upon digital applications for farm planning, routine diagnostics and market intelligence may allow fewer outside contacts.

The gloomiest forecasts of food system disruption due to the pandemic [20] have fortunately not come to pass. Causes for these reduced impacts across Africa are still under study but they may be related to quick and decisive lockdown actions by governments, inherent resistance to viral infection within communities, and reduced frequency and range of travel among the poor. In many ways, small-scale farmers already operate in a socially distanced fashion. Nonetheless, the effect of the pandemic on TAAT's plans to widely disseminate its proven technologies through ever-growing partnership mechanisms during its third year was profound, and it is intriguing to consider how the Program can "build back better" in the aftermath of the pandemic [21]. TAAT's widespread multiplication of improved crop varieties preceded the pandemic by two years and effectively carried over into it to great advantage among earlier beneficiaries.

7. Technologies and African agricultural transformation

The promotion and adoption of improved technologies have an essential role in the modernization of African agriculture. In this sense, technologies include production input products and equipment distributed through commercial channels, improved management strategies advanced through extension and information campaigns, and to a lesser extent, agro-industrial processes that add value to commodities. This very applied concept of technologies overlaps with offered innovations and proposed solutions but are not necessarily synonymous with them. Innovations often lead to technologies of these sorts but may also include partnership arrangements, policy interventions, and novel operating assumptions beyond the reach or influence of farmers [7]. Solutions may also be expressed as broader knowledge management systems, disruptive paradigm shifts, and alternative guiding ideologies [15, 22–24]. From the practical context of how African smallholders and agricultural entrepreneurs can change their operations, the narrower definition of technology seems more appropriate. At the same time, development strategies are necessarily influenced by larger issues and regard new technologies as only one element of a more complex equation. Nonetheless, the need to establish proven technologies from research products is an important function and one that increasingly drives the agendas of research institutions and their donors toward more impact-oriented designs [4, 5].

7.1 Technology bundles

Throughout its short project lifetime, TAAT has recognized several trends in terms of technology selection and deployment. Among crop commodities, improved varieties usually serve as a lead technology that is backstopped by accompanying products and managements [3, 12]. The need for genetic innovation as a replacement for traditional varieties is well established, particularly lines that are better suited to biotic and edaphic stress, including climate-induced extremes. This trend explains why TAAT's priority technologies listed by commodity in **Table 1** all start with improved germplasm and then include additional products and managements that ensure their increased genetic potential. Genetic improvement was hard-earned across Africa and the result of concentrated breeding efforts over the past decades, resulting in disease-tolerant rice, wheat, maize, cassava, and other crops. At the same time, breeders also focused upon biofortification to develop crops with higher nutritive value, such as sweet potatoes, maize, and cassava rich in pro-vitamin A or beans containing more iron and zinc. Technology promotion by

TAAT has resulted in much wider recognition of these new varieties and its technology toolkits link these genetic advantages to regional and country rural development programs, as well as nest them into climate adaptation efforts. There is also the advantage that commercial and community-based mechanisms often co-exist and aid in the accelerated deployment of these new varieties, and this entered into TAAT's scaling strategy.

At the same time, excessive focus upon improved varieties and their breeding objectives may also pose disadvantages. For the most part, TAAT's advancement of hybrid rice has yielded little gains because of its complexity. So too, over-reliance upon drought tolerance in maize can lead to lower yields during years of better rainfall because of its quicker maturity. Across much of Africa, parasitic striga limits maize production more than drought, and TAAT's efforts have not adequately addressed this massive constraint. Similarly, the trait of higher iron and zinc contents in common beans appears inadequately combined with disease resistance and greater symbiotic capacities in biological nitrogen fixation. Vegetatively propagated crops such as cassava and sweet potato are more challenging to scale than crops propagated by seed, limiting early deployment by greater complexity and expense (see Sections 4.5 and 4.6). Nonetheless, the interactions within TAAT toolkits allow for customization to specific conditions in ways that can guide the design and implementation of development projects, including optimal scaling of new crop varieties, and this is a particular strength of the TAAT Program.

This strength requires that TAAT also be flexible in terms of its partnership arrangements [25] if its toolkits are to become widely recognized and adopted. An important partnership mechanism exists through the assisted design of sovereign country loans during their planning stage and buy-in from existing loan projects. TAAT had limited success in this area over the past three years because the formulation and approval process of these loan projects is longer than the project has been operating. Besides, existing loan projects are understandably unable or reluctant to redirect funding toward technical directions and partners not considered in their original, approved designs. Nonetheless, TAAT has experienced substantial buy-in to its technologies as their potency becomes more widely recognized with the strongest interest expressed by loan projects for rice, wheat, maize, and cassava in numerous countries, and through the emergent agendas within key regional projects addressing climate change adaptation.

7.2 Private sector engagement

The Program experienced difficulties to date in the adoption of technologies by the private sector through expansion of their product lines, in large part because companies move cautiously into the manufacture and distribution of new products. At the same time, Regional Public Goods offered for free do not guarantee that competitors cannot quickly move into a new commercial opportunity paved through others' initial investment. This situation is particularly true for improved crops bred by research organizations and later released by national authorities but must then be offered without licensing cost to seed companies as an original donor condition. Alternatively, many seed companies prefer to breed, produce, and market their own proprietary lines. Mechanisms to overcome this reluctance are being explored, including the right to give different branded product names to the same varieties. The manufacture of accompanying input product technologies such as fertilizer blends, biotechnologies, or small-scale equipment is less subject to this constraint, in large part because investors have the latitude of developing trade secrets around their adapted manufacturing process. TAAT is in the process of organizing a technology investor's forum to promote commercial opportunities among its proven technologies in order to overcome these constraints.

7.3 Youth-led technology dissemination

TAAT established an Enabler activity devoted to the role of youth as technology adopters, the so-called ENABLE-TAAT Compact. This inclusion responds to concerns of the aging farmer population across Africa and expectations that youth-led entrepreneurship drives future economic growth [26]. This Compact builds upon IITA's Youth Agripreneur Movement and the operations of its experiential learning and enterprise incubation approaches [27]. The Compact serves to reinforce the priority commodities of TAAT (**Table 1**) and test the youth appeal of its technology toolkits. ENABLE -TAAT operates under the understanding that entrepreneurial youth are especially attracted to higher-value, readily marketed and value-added commodities. Youth seek to avoid the drudgery associated with small-scale production through greater reliance upon equipment and automation. They also hold greater access to and expertise in handheld communication devices and are better positioned to rely upon them. In this regard, youth are particularly attracted to the cereal seed industry, biofortified crops especially orange-fleshed sweet potato, cassava processing, flour blending, and raising poultry and fish [28]. In some cases, they served not only as technology adopters but also contributed to the refinement of those technologies, as illustrated by the development of more efficient vegetative propagation, intercropping and animal feed systems. In all, ENABLE-TAAT guided 71,562 youth toward TAAT commodities and technologies and assisted in establishing 166 new youth-led agribusinesses. During its activities, this Compact also registered 5,582 youth as local champions in six countries, trained 2,829 youth in TAAT technologies and reached 197,360 persons through social media outreach. Youth were also quick to recognize and apply the COVID-safe agricultural practices described in Section 6. Another important indicator of success by the ENABLE-TAAT Compact is the magnitude of interest it attracted from local authorities and the buy-in to its expanded activities by additional donors such as the International Fund for Agricultural Development and the Mastercard Foundation.

7.4 Livestock and animal enterprise as technology targets

Both animal enterprise and inland aquaculture are priority commodities within TAAT, and they have their own technology toolkits. These commodities and toolkits are not described in detail in this Chapter but warrant a brief description because of their importance as farm enterprises. The technology bundles of the Small Animal Compact consist of strategies that combine improved animal breeds, nutritious feeding regimes, and better veterinary care [3]. TAAT's original design grouped poultry and small livestock (goats and sheep) into a single Compact, but different technology toolkits were subsequently developed for each. Program design also excluded dairy and beef, in large part because of their lengthy production cycles. Commercial poultry is separated between broiler and layer production with different breeds, feeds, and housing requirements associated with each. A contrasting technology is the rearing of dual-purpose, free-ranging breeds that produce both meat and eggs for small-scale farmers. Entrepreneurial youth are strongly attracted to commercial poultry production, particularly eggs, because of its longevity and even cash flow. However, producers experience difficulties regarding the disproportionately high costs of feed and the risks of disease. Aquaculture focuses upon distributing improved varieties of tilapia and catfish through localized hatchery operations and then raising them in various containment systems under improved feed regimes. In short, improved animal and fish production technologies are key to Africa's agricultural transformation, and this opportunity is incompletely addressed within this Chapter as it is focused primarily upon specific field crops.

8. Conclusions and next steps

The TAAT Program is a novel and successful means to popularize and deliver agricultural technologies needed by small-scale African farmers. The deployment strategy in Section 4 and the immediate impacts from that deployment presented in Section 5 support this claim. TAAT achieves increased agricultural productivity and diversification, leading to improved food and nutrition security; job creation through expanded commercialization and industrialization; and improved soil, land, and water management practices due to Good Agricultural Practices; among other benefits (Section 4). The validity of its technology bundling and deployment approaches is reinforced by an across-the-board 58% increase in crop productivity resulting in 0.75 MT (dry weight) increased household food supply among over 10 million adopting farmers. The Program and the technologies it advances were expected to strongly interface with national partners, sovereign country loans, donor organizations and development banks, and this too has occurred. TAAT continued to operate throughout 2020 despite the disruption of the COVID-19 pandemic, in large part because the germplasm-led technologies it advocates were already distributed to numerous countries before the onset of the pandemic. Their increased productive capacity strategically contributed to food systems into the contagion [3, 12, 21]. Gains in agricultural production across crop commodities were linked to corresponding increases in system carbon stocks, suggesting a future connection between technology scaling and climate action. These factors signal considerable success of a project initially designed around the need to deploy a suite of proven, modernizing agricultural technologies to small-scale African farmers.

The growth of TAAT will also influence its role as a longer-term deployment mechanism of technologies and an improved approach to delivery partnership. In its first three-year Phase initiated through IITA and the African Development Bank, TAAT did not receive the buy-ins as anticipated from other development banks, and this reluctance limited the scope of its direct activities. One result was the failure to establish additional commodity Compacts as initially intended, particularly one's focusing upon grain legumes, vegetables, and the restoration of tree plantations. TAAT recognizes the need for these additional Compact alliances but could not support them given its available resources. One very positive signal of success, however, is the considered adoption and expansion of its approach within the Consultative Group on International Agricultural Research (CGIAR), a global partnership that unites international organizations engaged in agricultural research. Nine of its 15 Centers currently collaborate within TAAT, and the ongoing "One CGIAR" reform process [29] examines TAAT as a lead regional mechanism for technology delivery in Africa and as a model for delivery alliances elsewhere. This success was further reinforced by the recent "High-level Dialogue on Feeding Africa: Leadership to Scale-up Successful Innovations" (29 & 30 April 2021) conducted as a global virtual event organized by AfDB, The International Fund for Agricultural Development and others. During this event, 17 African heads of state and numerous other dignitaries recognized the need for Africa's agricultural transformation and they pledged support to it (see https://www.afdb.org/en/events/

high-level-virtual-dialogue-feeding-africa-leadership-scale-successful-innovations). In this respect, technology refinement, characterization, bundling and delivery to small-scale farmers of the tropics will remain an essential ongoing effort involving many researchers, development specialists, extensionists, and stakeholders into the future.

Acknowledgements

The authors appreciate information and assistance provided by Kennedy M. Kago of the IITA TAAT Clearinghouse and Ms. Eniola Olanrewaju of the IITA Youth in Agribusiness Department. Also appreciated are the seven TAAT crop Commodity Compact leaders, their teams, and national partners too numerous to identify that provided information on productivity increases and technology adoption. This Chapter was prepared by the Technologies for African Agricultural Transformation (TAAT) Clearinghouse, a facility operated by IITA in Cotonou, Benin and Nairobi, Kenya. The Clearinghouse receives financial support from the Bill and Melinda Gates Foundation. The African Development Fund of the African Development Bank supports the larger TAAT Program as a component of its Feed Africa Strategy. The contributions of all these individuals and organizations are gratefully acknowledged.

Author details

Paul L. Woomer*, Welissa M. Mulei and Rachel M. Zozo
International Institute of Tropical Agriculture, Nairobi, Kenya

*Address all correspondence to: plwoomer@gmail.com

References

[1] African Development Bank (AfDB). Feed Africa: Strategy for agricultural transformation in Africa 2016-2025. Abidjan, Cote d' Ivoire: AfDB; 2016. 79 p.

[2] Technologies for African Agricultural Transformation (TAAT). TAAT in 2018: Preparing for African agricultural transformation. Ibadan, Nigeria: TAAT, International Institute of Tropical Agriculture (IITA); 2019. 65 p.

[3] TAAT Clearinghouse. TAAT technology toolkits and their strategic deployment. Cotonou, Benin: TAAT Clearinghouse; 2018. 18 p. Clearinghouse Technical Report Series 001. Available from https://taat-africa.org/taat-technology-toolkits-and-their-strategic-deployment.

[4] Consultative Group on International Agricultural Research (CGIAR). CGIAR performance report 2017: Transforming the global food system. Montpellier, France: CGIAR; 2018. Available from https://www.cgiar.org/cgiar-system-annual-performance-report-2017.

[5] CGIAR. CGIAR performance report 2018: Food systems innovations for impact. Montpellier, France: CGIAR; 2019. Available from https://www.cgiar.org/annual-report/performance-report-2018.

[6] TAAT Clearinghouse. TAAT Clearinghouse establishment and first year operations. Cotonou, Benin: TAAT Clearinghouse; 2019. 16 p. Clearinghouse Technical Report Series 003. Available from https://taat-africa.org/taat-clearinghouse-establishment-and-first-year-operations.

[7] Sartas M, Schut M, Proietti C, Thiele G, Leeuwis C. Scaling readiness: Science and practice of an approach to enhance impact of research for development. Agric Syst. 2020; 184: 12 p. Available from https://doi.org/10.1016/j.agsy.2020.102874

[8] Willy DK, Diallo Y, Affognon H, Nang'ayo F, Waithaka M, Wossen T. COVID-19 pandemic in Africa: Impacts on agriculture and emerging policy responses for adaptation and resilience building. Nairobi: African Agricultural Technology Foundation (AATF); 2020. 12 p. TAAT Policy Compact Working Paper No. 01.

[9] Laborde D, Martin V, Vos R. Poverty and food insecurity could grow dramatically as COVID-19 spreads. In: Swinnen J, McDermott J, editors. COVID-19 and global food security [Internet]. Washington, DC: International Food Policy Research Institute (IFPRI); 2020 [cited 2020 Dec 06]. p. 16-19. Part One: Food security, poverty, and inequality, Chapter 2]. Available from https://doi.org/10.2499/p15738coll2.133762_02

[10] African Development Bank (AfDB). Technologies for African Agricultural Transformation: Framework program in support of Feed Africa. Abidjan, Cote d' Ivoire: AfDB; 2017. 26 p.

[11] United Nations. Transforming our world: The 2030 Agenda for Sustainable Development [Internet]. 2015 [(A/RES/70/1; cited 2020 Dec 06). United Nations, New York. 41 p. Available from https://sustainabledevelopment.un.org/content/documents/21252030%20Agenda%20for%20Sustainable%20Development%20web.pdf

[12] TAAT. Pathways to Transformation: A compendium of 2019 TAAT success stories[Internet]. 2020 [cited 2020 Dec 07]. IITA, Ibadan, Nigeria. 94 p. Available from https://taat-africa.org/wp-content/uploads/2021/02/2019-TAAT-Annual-Report-Dec-21.pdf

[13] Tsan M, Totapally S, Hailu M, Addom BK. The digitalization of African agriculture report 2018-2019. 1st edition. Wageningen. Technical Centre for Agricultural and Rural Cooperation (CTA). 2019 [cited 2020 Dec 07].Available from https://www.cta.int/en/digitalisation-agriculture-africa.

[14] Meuwissen MP, Feindt PH, Spiegel A, Termeer CJ, Mathijs E, Mey Y, et al. A framework to assess the resilience of farming systems. Agric Syst. 2019; 176: 10 p. Available from https://doi.org/10.1016/j.agsy.2019.102656

[15] Barrett CB, Benton TG, Fanzo J, Herrero M, Nelson RJ, Bageant E, et al. Socio-technical innovation bundles for agri-food systems transformation, Report of the international expert panel on innovations to build sustainable, equitable, inclusive food value chains [Internet]. Ithaca, NY, and London. Cornell Atkinson Center for Sustainability and Springer Nature. 2020 [cited 2020 Dec 09]. 172 p. Available from https://hdl.handle.net/10568/110864

[16] Kapinga R, Tumwegamire S, Ndunguru J, Andrade MI, Agili S, Mwanga RO, et al. Catalogue of orange-fleshed sweet potato varieties for Sub-Saharan Africa [Internet]. Lima, Peru. International Potato Center (CIP). 2010 [cited 2020 Dec 09]. 40p. Available from https://cipotato.org/wp-content/uploads/2014/08/005374.pdf

[17] Stathers T, Mkumbira J, Low J, TagwireyiJ, Munyua H, Mbabu A, et al. Orange-fleshed Sweetpotato (OFSP): Investment guide [Internet]. Nairobi, Kenya. CIP. 2015 [cited 2020 Dec 08]. Available fromhttp://www.sweetpotatoknowledge.org/wp-content/uploads/2016/02/OFSP-Investment-Guide-F.pdf

[18] Branca G, Lipper L, McCarthy N, Jolejole MC. Food security, climate change, and sustainable land management. A review. Agron Sustain Dev. 2013; 33: 635-650. doi 10.1007/s13593-013-0133-1

[19] Lipper L, Thornton P, Campbell BM, Baedeker, T, Braimoh A, Bwalya M, et al. Climate-smart agriculture for food security. Nature Clim Change. 2014; 4, 13:1068-1072. http://dx.doi.org/10.1038/nclimate2437

[20] Ayanlade A, Raden M. COVID-19 and food security in Sub-Saharan Africa: implications of lockdown during agricultural planting seasons. Sci Food. 2020;4:13. 6 p. doi.org/10.1038/s41538-020-00073-0

[21] CGIAR. Responding to COVID-19:CGIAR's contribution to global response, recovery and resilience[Internet]. No date [cited 2020 Dec 08].48 p. Available from https://cgspace.cgiar.org/bitstream/handle/10568/108548/CGIAR-Responding-to-COVID-19.pdf.

[22] Rockström J, Karlberg L, Wani SP, Barron J, Hatibu N, Oweis T, et al. Managing water in rainfed agriculture: The need for a paradigm shift. Agr Water Manage. 2009; 96 (12):1683-1860. doi:10.1016/j.agwat.2009.09.009

[23] Munang R, Mgendi R, Alverson K, O'Brien-Onyeka M, Ochieng C, Molua E, et al. Ecosystem Based Adaptation (EbA) for food security in Africa –Towards a comprehensive strategic framework to upscale and out-scale EbA-driven agriculture in Africa [Internet]. United Nations Environment Programme (UNEP), Nairobi. 2015 [cited 2020 Dec 09]. 50 p. Available from https://www.researchgate.net/publication/275816304_Ecosystem_Based_Adaptation_EBA_for_food_security_in_Africa_- Towards_a_comprehensive_Strategic_Framework_to_Upscale_and_Out-scale_EbA-driven_agriculture_in_Africa_United_Nations_Environment_Programme_U.

[24] Caron P, Ferrero y de Loma-Osorio G, Nabarro, D, Hainzelin E, Guillou, M, Andersen, I, et al. Food systems for sustainable development: proposals for a profound four-part transformation. Agron Sustain Dev. 2018; 38,41: 12 p. https://doi.org/10.1007/s13593-018-0519-1

[25] Igbinnosa MA. Effective and functional partnership for successful upscaling of technologies across agricultural value chains to feed Africa. AJSAD [Internet]. 2020 [cited 2020 Dec 09]; 1(3):45-59. Available fromhttps://www.ijaar.org/articles/ajsad/v1n3/ajsad-v1n3-jul-sep20-p25.pdf

[26] Alliance for a Green Revolution in Africa (AGRA). Africa agriculture status report 2015: youth in agriculture in Sub-Saharan Africa [Internet]. 2015 [Issue No. 3; cited 2020 Dec 09]. AGRA, Nairobi. 229 p.https://www.slideshare.net/bunmyajilore/africaagriculturestatusreport2015.

[27] Ohanwusi E, Woomer PL, Adenmosun A. IITA youth agripreneurs (IYA) in 2017: Improved approaches and broader activities [Internet]. Ibadan, Niger*ia. IITA*. 2016. [cited 2020 Dec 09]. 20 p. Available from https://youthagripreneurs.org/wp-content/uploads/2018/09/IYA-report-final-pdf.pdf

[28] Owoeye M, Ohanwusi E, Adenmosun A, Nathaniel M, and Woomer PL. Agribusiness planning by IITA Youth Agripreneurs (IYA). Ibadan, Niger*ia. IITA*. 2016. 37 p

[29] CGIAR. Towards One CGIAR. [Internet]. 2020. [cited 2020 Dec 10]. Available from https://storage.googleapis.com/cgiarorg/2020/05/04215189-one-cgiar-newsletter_22may.pdf

Chapter 4

Modalities for Scaling up Implementation of Innovations and Best Practices for Resilient Agricultural Systems in Africa

Evans Kituyi

Abstract

Climate change is already impacting negatively on Africa's agriculture and threatens to significantly reverse the gains realized in food security as the 1.5 degC warming threshold set by the Paris Agreement fast approaches. This is happening at a time when a wide range of tested and viable technologies, innovations and best practices exist with the potential to scale up climate resilient food production across the region's diverse agricultural systems. A framework and modalities are proposed to support stakeholders in identifying and scaling up appropriate technologies, innovations and best practices for climate-resilient food production in different farming systems. These provide a much needed solution for Africa's policymakers who are currently grappling with options to meet their citizens' food security today even as they ponder over how they will feed their rapidly growing populations, expected to reach 2 billion by 2030 under worsened climate conditions.

Keywords: food security, climate change, technologies, innovations, best practices, scaling up

1. Introduction

Agriculture is the largest productive sector in Africa employing between 60 and 70% of the total labour force as well as supplying up to 50–70% of household incomes [1]. Over 80% of the food produced is from smallholders in a mainly rain-fed sector where women constitute 50–60% of the total workforce and which contributes to about 20–25% of the overall GDP of the continent [2]. Livestock production accounts for about 30% of the gross value of agricultural production, with 92% of that coming from the production of beef cattle, dairy cattle, goats, sheep and chickens [3]. Pastoralism is practiced in more than 75% of African countries by between 200 and 500 million people, including nomadic communities, transhumant herders, and agro-pastoralists [4]. The sector's output has increased since 2000, mainly due to expansion of agricultural land [1]. However, food production in sub-Saharan Africa needs to increase by at least 60% by 2030 to feed the growing population then [5]. To achieve this, the huge unexploited potential in the continent's agriculture sector needs to be exploited.

A number factors have for long stood in the way of Africa realizing the enormous potential inherent in the agricultural sector for poverty reduction and wealth creation. Two among these stand out. First, is the failure to scale up and out the many proven technologies, innovations and best practices for sustainable food production. These include failure to embed agricultural innovation within local socio-ecological structures and practices, political instability in some regions hindering engagement with farmers, unclear profitability of innovation to end-users, lack of participation by users in technology development and decision-making, and unfavorable policies and legislation such as that on land tenure and intellectual property ownership among others [6].

Secondly, government-led agricultural extension services on the continent have been ineffective owing largely to inadequate support from government. Yet these services should play a critical role of bringing the farming community information on new technologies, innovations and practices, which they can adopt to increase productivity, incomes and standards of living. Whereas technology and skills transfer has been more successful in programs driven by Africa's private sector--in export-oriented sub-sectors such as sugar, coffee and tea, this has remained lackluster or non-existent [7] when it comes to non-cash crop systems, where the smallholders belong [8].

As a consequence of not optimizing the continent's agricultural potential, there has been persistent poverty among its people, projected to prevail beyond 2030 while it is eliminated in other parts of the world. Similarly, the food security situation has been deteriorating in all sub-regions of Africa. Of the total undernourished population in 2018, 17 million were in Northern Africa and 239 million in sub-Saharan Africa [1] and is projected to reach 320 million by 2025 with no policy interventions [9]. This could however be an underestimate, considering additional stress on food systems recently brought about by the unprecedented locust outbreak that severely ravaged the Horn of Africa 2019 to 2020 with significant crop and livestock losses, compounded by the COVID-19 pandemic since early 2020.

The urgency of this chapter is driven by (a.) the Intergovernmental Panel on Climate Change (IPCC) Special Report on Impacts of 1.5degC Warming, predicted to commence towards the end of this decade, and (b.) the challenge to African policymakers produce enough to feed the 2 billion people in 2030 under the envisaged harsher climate then. The objective of this chapter, therefore, is to develop and promote a versatile framework for identifying and scaling up appropriate technologies, innovations and best practices for climate-resilient food production accessible to smallholders across diverse farming systems on the continent in the short to medium term.

2. Impacts of climate change on African agricultural systems

The Intergovernmental Panel on Climate Change (IPCC) Special Report on the Impacts of 1.5 degrees global warming above pre-industrial levels (SR1.5) released in October 2018 was categorical that warming in sub-Saharan Africa will be greater than the global average and that rainfall will decline in certain areas. There's general concurrence among most circulation models (GCMs) that temperatures are increasing across the region. However, many models vary widely regarding predicted precipitation changes. Climatic changes are not uniform across the region but temperatures are expected to increase in all locations.

The forecasts for Sub-Saharan Africa suggest that higher temperatures, increase in the number of heat waves and increasing aridity will affect the rain-fed agricultural systems [10]. Schlenker and Lobell [11] estimated that in sub-Saharan Africa, crop production may be reduced by 17–22% due to climate change by 2050.

Farmers in sub-Saharan Africa practice a wide range of crop and livestock production activities varying across and within the major agro-ecological zones by mainly smallholders who occupy 80% of all arable land in Africa with each owning less than 2 hectares [2]. **Table 1** below presents the major farming systems in sub-Saharan Africa. Smallholder farmers remain vulnerable to climate change. Not only do changes in temperature, rainfall and the frequency or intensity of extreme weather events affect their crop and animal productivity but also significantly degrades their household's food security, incomes and overall well-being [13].

FAO [14] reports that food and agriculture needs to produce 49% more food by 2050 yet climate impacts such as extreme weather events, spreading pests and diseases, loss of biodiversity, degrading ecosystems, and water scarcity will worsen as the planet warms. Climate change is very likely to have an overall negative effect on yields of major cereal crops across Africa, with strong regional variability in the degree of yield reduction [15]. Cereal production growth for a range of crops in SSA is projected to decline by a net of 3.2% in 2050 as a result of climate change, with the largest negative yield impacts projected for wheat, followed by sweet potatoes and yams [16]. Using an ensemble of GCMs and a crop model,

Farming systems	Land area (% of region)	Agric. Popn. (% of region)	Principal source of livelihood
Tree Crop	3	6	Cocoa, coffee, oil palm, rubber, yams, maize, off-farm work
Forest Based	11	7	Cassava, maize, beans, cocoyams
Rice-Tree Crop	1	2	Rice, banana, coffee, maize, cassava, legumes, livestock, off-farm work
Highland Perennial	1	8	Banana, plantain, enset, coffee, cassava, sweet potato, beans, cereals, livestock, poultry, off-farm work
Highland Temperate Mixed	2	7	Wheat barley, tef, peas, lentils, broadbeans, rape, potatoes, sheep, goats, livestock, poultry, off-farm work
Root Crop	11	11	Yams, cassava, legumes, off-farm work
Cereal-Root Crop Mixed	13	16	Maize, sorghum, millet, cassava, yams, legumes, cattle
Maize Mixed	10	15	Maize, tobacco, cotton, cattle, goats, poultry, off-farm work
Large Commercial and Smallholder	5	4	Maize, pulses, sunflower, cattle, sheep, goats, remittances
Agro-Pastoral Millet/ Sorghum	8	8	Sorghum, pearl millet, pulses. Sesame, cattle, sheep, goats, poultry, off-farm work
Pastoral	15	8	Cattle, camels, sheep, goats, remittances
Sparse (Arid)	17	1	Irrigated maize, vegetables, date palms, cattle, off-farm work
Coastal Artisanal Fishing	2	3	Marine fish, coconuts, cashew, banana, yams, fruit, goats, poultry, off-farm work
Urban Based	<1	3	Fruit, vegetables, dairy, cattle, goats, poultry, off-farm work

Source: IPC [12].

Table 1.
Major farming systems of sub-Saharan Africa.

Thornton *et al.* [17] estimated the mean yield losses of 24% for maize and 71% for beans when global warming exceeded 4°C. If Nitrogen stress is considered, some studies [18] found yield decreases higher than 50% for maize in the Sahelian region and around 10–20% in other Sub-Saharan regions, and an overall negative trend of 5–50% not considering Nitrogen stress.

High-temperature sensitivity thresholds for important crops such as maize, wheat and sorghum have been observed, with large yield reductions once the threshold is exceeded [13]. Cassava, millet and sorghum yields are projected to be slightly higher under climate change, probably owing to their higher tolerance to high temperatures and drought stress. Cassava appears to be more resistant to high temperatures and unstable precipitation than cereal crops [19], while multiple-cropping systems appear to reduce the risk of crop failure compared to single-cropping systems [20]. Climate change extremes can alter the ecology of plant pathogens, and higher soil temperatures can promote fungal growth that kills seedlings [21].

World prices are a key indicator of the effects of climate change on agriculture and specifically on food affordability and security. Food prices will likely increase for all major staple crops, when maize, rice and wheat prices in 2050 are projected to be 4%, 7%, and 15% higher than under the historic climate scenario, respectively [16]. Consequently, the per capita calorie availability across sub-Saharan Africa is projected to decline by 1.3% or 37 kilocalories per capita per day.

Climate change also increases the number of malnourished children in 2030 and 2050 being higher by 1 million children in 2030 and still higher by 0.6 million children by 2050, respectively. With respect to trade, little change in net cereal imports is expected as a result of climate change but at sub-regional level, eastern Africa is projected to experience the largest increase (15%) in net cereal imports due to climate change as a result of declining maize yields [16].

The impacts of climate change on grazing systems include changes in herbage growth (due to carbon dioxide concentration, rainfall and temperature) and changes in composition of pasture and in herbage quality [22]. Increases in carbon dioxide concentrations and precipitation will tend to increase rangeland net primary production, though this increase in production will be modified positively or negatively by increased temperature leading to differences in production among species. The proportion of browse in rangelands may increase in combination with more competition if dry spells are more frequent [23]. For example, in a future East Africa with a warmer and wetter climate, tropical broadleaf growth may increase more than other grazing grass species and changes in net primary productivity in African rangelands will likely be mainly negative [22].

Over shorter time horizons, climate risk in pastoral landscapes will be affected by increased variability of rainfall—spatially and temporally [23]. Climate risk will also increase through increased frequency of extreme events such as drought, flooding and extreme highs and lows in temperature. According to [22] this may have negative effects on herd dynamics, stocking density and the productivity of pastoral production systems. Herrero et al. [24] reports that in Arid and Semi-Arid Lands (ASAL) of Kenya, the loss of animals and subsequent loss of milk and meat production to 2030 due to increased drought frequency would lead to over USD 630 million and outbreaks of rift valley fever associated with increased rainfall and flooding.

Incomes and food security for rangeland communities will also be affected. In much of Africa, livestock are an important risk management asset for millions of people. Ericksen et al. [25] also mapped out where the vulnerability of livestock keepers in the rangelands may change in the coming decades. This includes areas that are food insecure and vulnerable to the impacts of future climate change across the tropics and sub-tropics. Using length of growing period (LGP) as a proxy for the number of grazing days, Herrero et al. [22] project substantial reductions in

LGP across African rangeland systems, who also project an average maximum temperature flip above 35°C across Africa. This is a critical threshold for rangeland vegetation and heat tolerance in some livestock species.

Increasing food availability and rural incomes through increased agricultural productivity is recommended by Ringler et al. [16] as the most potent solution for reducing malnutrition particularly in sub-Saharan Africa. This could include expansion of the production area and intensification on existing croplands [26, 27]. But rapid urbanization is emerging as a major challenge to intensification, despite the high potential in available technologies to increase farm output [28]. Sustainable intensification of small-scale farming is therefore a vital option towards climate resilient food production [27].

But given the highly diverse Sub-Saharan Africa's agro-ecology, there is no single universal agricultural technology or practice that can achieve climate resilience and agricultural sustainability in the region [29] and adaptation strategies will vary for different locations [30]. The solution lies in the adoption of climate smart agro-ecology principles and landscape approaches that enhance resilience of agro-ecosystems and diversify production through cultivating crops and raising livestock that are stress-tolerant [29, 31]. Such agricultural techniques include agroforestry, mulching, intercropping, crop rotations, integrated crop–livestock management, conservation agriculture, improved water management, adopting and using innovative practices such as weather forecasts and climate risk insurance, among others [32, 33].

The last 30 years have witnessed evolution in the development and promotion of low-cost agricultural technologies in sub-Saharan Africa that are suitable for small-scale agriculture [34]. Many of these technologies are meant to address challenges of decreasing productivity such as poor soil fertility, degrading farmlands and climate variability. The question then remains; what are the best modalities and strategies of scaling up these best innovations and technologies? This is the main purpose of the following section of this paper.

Future changes in temperature and precipitation in the different agricultural systems may render some of the innovations and practices more or less effective that the present day performance. Considering the considerable costs involved in adopting technologies, innovations and practices for scale up and out, it is important that location-specific screening of the options is carried out against certified future scenarios of temperature and rainfall and other considerations. This is elaborated in Section 5 on modalities for scaling up and out.

The April 2020 flooding in the Lake Victoria basin and the backflow of the lake that has led to thousands of people displaced and tens of villages submerged has wiped out many gains made through farming practices recently promoted by NGOs projects. Thiery *et al.* [35] predicted this phenomenon but which unfortunately was not considered by local development planners. Similarly, the IPCC SR1.5 predicts severe impacts on water availability and the beef industry in southern African countries of Namibia and Botswana when the 1.5-degree C threshold is reached within the decade (from 2028). Africa's low-lying deltas will not be affected until after 2040, and planning for scaling out and up for the transfer of any technologies, innovations and best practices in these areas should check details in the report.

3. Sustainable agricultural technologies, innovations and best practices

There is still little consensus around the term "climate smart agriculture" as controversy still exists among key sustainability actors including agricultural scientists and development practitioners. Hence for this chapter, the term "sustainable agricultural production" is preferred, and it embraces all practices—both adaptation and mitigation solutions. Practices that contribute to sustainable agricultural production and resilience of farming systems in a wide range of contexts include, but are not restricted to: agroforestry, improved soil management such as through conservation agriculture, improved water management such as water harvesting and drip irrigation, integrated livestock and grassland management, improved nutrient management such as micro-fertilization and improved crop varieties. These are elaborated in detail in the Climate-Smart Agriculture Source Book [31].

Adoption of improved integrated production management practices also depends strongly on other factors such as investment costs and returns on investment, availability of labor, access to machinery and priority value chains. As a basic indicator of success, food security has to increase during the shift to more integrated practices [36].

Changes in technologies and practices that favor resilience and sustainable agricultural production must also be supported by a supportive enabling environment. The next section on modalities for scaling up technologies, innovations and best practices elaborates the enabling environments supportive of scaling up and out. Barriers to scaling up and out climate resilient agricultural technologies and innovation differ between contexts and over time [37, 38].

Identifying the main barriers or drivers in any particular context from an array of contributing factors is a key first step and the scaling process should adapt to these. Scaling may also be challenging in particularly diverse or unique agro-ecosystems and socio-cultural settings where climate smart technologies and innovations have to be significantly adapted to work in each setting. Key barriers to scaling up and out include:

a. Lack of technical options for the specific need and context considered and/or awareness of these options by the beneficiaries;

b. Lack of adequate institutional, human and financial resources for capacity building and extension services;

c. Lack of finance at macro-and micro-level within public/government budgets, local organizations and individuals, and aversion of private sector investments for small holders;

d. Lack of political will to address problems mainly among the most vulnerable communities;

e. Lack of awareness of innovative climate-smart approaches such as payments for ecosystem services and insurances;

f. Insecure tenure and access to resources including land and water;

g. High investment risk associated with adoption of certain technologies in given locations; and

h. Loss or high turnover of individual “champions” that drive the interest and processes in specific situations.

4. Scaling up technologies, innovations and best practices

4.1 Conceptual framework for scaling up

Numerous modalities have been developed for effectively scaling up and out agricultural technologies and innovations by organizations working directly with farmers and other stakeholders in the innovation system. Designed to facilitate scaling up and out on the ground, these frameworks are informed more by experiential knowledge than academic theory, but resonate well with the theoretical literature. An evaluation of some of these by Thomas *et al* [38] and Neufelt *et al*. [39] reveals that, despite broad structural differences as well as the number of steps in each and the order in which they are presented, the different modalities or operational frameworks had much in common with each other and the theoretical literature. Using these lessons as building blocks for scaling up strategies for climate smart innovations in the widest possible range of African contexts, the conceptual framework in **Figure 1** is proposed as the consolidation of modalities for successfully scaling up technologies, innovations and best practices for sustainable agricultural production in the context of Africa.

4.2 Modalities for scaling up

The following seven steps lay out the modalities for effective scaling up and out highlighted in **Figure 1**. They are a culmination of the analysis of the wide range of published operational frameworks and modalities for scaling up climate smart innovations in diverse contexts discussed in previous sections.

4.2.1 Step 1: planning adaptively for the scaling process

The need to design scaling into mini-projects through proper planning from the outset is very important and approaches such as Logical Framework Analysis, Theory of Change and Impact Pathway planning may be used [39, 40]. Upscaling is typically considered when climate-smart technologies and innovations have been demonstrated to work at local levels ranging from villages to water basins and landscapes. By identifying barriers to scaling up, these approaches seek to mitigate these risks and overcome barriers as part of the design process from the outset. Furthermore, setting clear milestones that relate to scaling via a well-defined theory of change and impact pathway helps to bring divergent views and options together, cementing a joint understanding and vision of the rationale for scaling up and out.

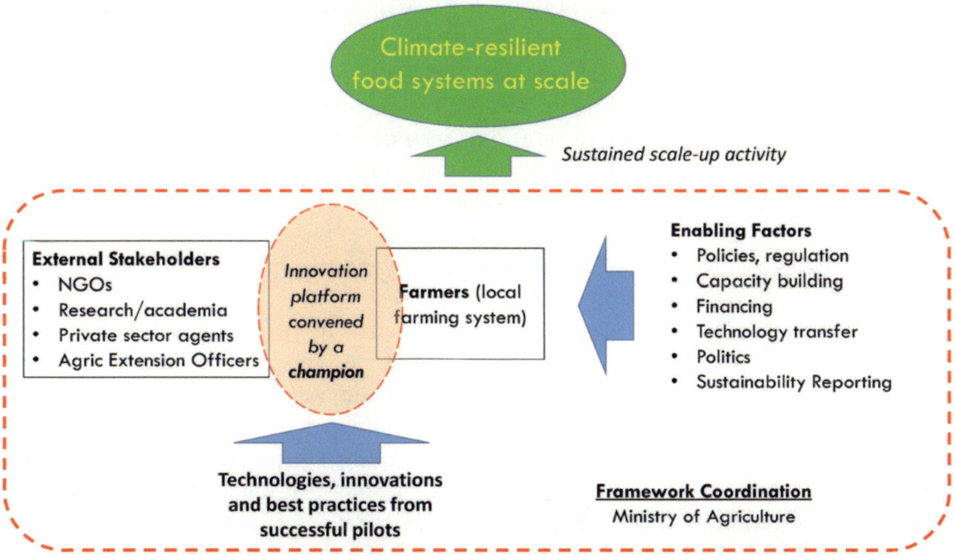

Figure 1.
Conceptual framework for scaling up agricultural practices, technologies and innovations.

4.2.2 Step 2: selection of technologies and innovations based on best-available evidence

Establishing the economic value of land for example can convince policymakers and land managers to invest and re-direct policy and practice towards financially viable climate smart options. However, while economics can drive key decisions, the social and cultural dimensions around changes in farming systems should not be overlooked when introducing new climate innovations [38] and participatory techniques for consider multiple perspectives and dimensions of value are available. Below are three stage process proposed for the identification of technologies, innovations and best practices for scaling up and out.

4.2.2.1 Identifying a portfolio of best practices, technologies and innovations

By the time one arrives to the scaling planning table, one already has a technology, innovation or best practice that has worked well at local scale and would like to disseminate it wider. At this point, it is important to understand that there is no "one-size-fits-all" solution for increasing resilience of a farming system. Instead, portfolios of credible, proven climate-smart best practices and technologies checked for scalability are more likely to realize goals of food security, resilience and increased productivity within a given agricultural system. It is critical that scalability assessment tools are employed in selecting the appropriate practice or a combination of them.

4.2.2.2 Framing the context

The next step involves framing the context for scaling up and out in order to understand the circumstances under which the proposed innovation, technology or best practice worked at one level and the needs and limits of the level to which we are extending to through the process of scaling up and out. In particular, this step clarifies the geographical, historical and cultural contexts within which people's priorities and the possibilities for innovation are set and bounded [41]. There is also evidence that innovation may be suppressed if the dominant culture disapproves of departure from the "normal way of doing things" [42]. Scaling up and out of a technical innovation must comprehend the needs and limits of the level we are extending to and ensure that what is being scaled up is both relevant and possible at that level. For practical reasons, the boundaries for scaling up and out are determined by biophysical, institutional and economic considerations and may be modified to accommodate emerging practical issues such as costs of involving stakeholders across greater distances. Temporal limits also must be set to enable development of realistic work plans.

4.2.2.3 Matching best technologies and practices with farming systems

Once the geographical and social context as well as institutional, administrative boundaries have been determined, the portfolio of climate-smart technologies, innovations and best practices identified in earlier can now be assessed against these boundaries to determine their suitability of adoption. Agro-ecological zones and farming systems are extremely diverse, hence interventions need to be targeted to specific contexts and accommodate both indigenous and scientific knowledge. Heterogeneous qualities of the small-scale agricultural system as regards resource control and access and an array of socio-economic attributes should be considered while designing, delivering, and diffusing the technologies and practices in question. A decision support mechanism must be prioritized to match the practices and technologies with agro-ecological zones in which the farming systems lie. Whereas trade-offs and synergies among these goals are likely, the focus should ultimately be on maximizing synergies [43].

4.2.3 Step 3 mobilize innovation platform and built capacities

The agricultural innovation system is a network of individuals, organizations and enterprises (innovation platform), together with supporting institutions and policies in the agricultural and related sectors. The interactions in the system delivers new products, processes and forms of organization. The policies and institutions (formal and informal) shape the way that these actors interact, generate, share and use knowledge as well as jointly learn. The main actors in scaling up are local and non-local stakeholders.

4.2.3.1 Establishing innovation platform/s

Different stakeholder groups have unique incentives and abilities and will face specific challenges in scaling up climate smart innovations.

Local Stakeholders are the key beneficiaries in a scale-up undertaking. In the context of this paper, these are smallholder farmers (about 80% of sub-Saharan African population) and small and medium enterprises (SMEs) active within a given farming system [44]. These are the most vulnerable to the impacts of climate change and must be the focus of any resilience building programmes. For these farmers and SMEs, a successful agricultural and livelihood transformation depends on their effective integration in value chains [44]. They possess indigenous knowledge characteristic of their respective farming systems and Agro-Ecological Zones (AEZ), knowledge which forms a necessary base for successful climate change adaptation in crop farming, livestock and fisheries.

Non-local stakeholders include donors, government, researchers, civil society and private sector actors. Each of these possesses uniquely diverse roles within an innovation system. Researchers may for instance be interested in situational analyses to frame the context for the new site where new innovation is desired. Research must also be continuous in order to counter evolving climate change realities and contexts. Together with researchers, the civil society and donors may also be active in disseminating the experiences (local innovation and behavioral change) to facilitate further scale up [41].

Perspectives of donor agencies, governments and other non-local actors do not necessarily coincide with local people's perspectives of their own needs. Nyasimi *et al*. [37] emphasized the importance of advocacy in scaling agricultural technologies. By bringing diverse stakeholders together through the innovation platform, an opportunity is availed for exchanging ideas and building mutual understanding and trust. This is a critical ingredient for the innovation process. Similarly, inter-level collaboration is particularly important where local actors may interact with those at regional, national and international levels. It is important to eliminate barriers which may hinder local stakeholders from freely participating and sharing information that is considered culturally sensitive or private to them. Fostering meaningful participation calls for capacity building of the disadvantaged groups, fostering trust among all stakeholders and convincing powerful stakeholders to open the space for the disadvantaged groups to influence the agenda and nominate their own representatives.

4.2.3.2 Capacity building

Capacity-building in the UNFCCC process encompasses activities at the level of individuals, institutions and systems. Scaling up climate smart practices therefore requires capacity building across all levels from farmers, the private sector to national and international policy makers. The intervention selected for scaling up will help define the capacity building needs as well as the boundaries—such as a watershed, national or international scale—within which climate resilient development outcomes are expected. The Subsidiary Body for Implementation (SBI) regularly monitors and reviews progress on the implementation of the frameworks for capacity-building in developing countries.

Quality training coupled with appropriate incentives is an essential component of scaling up and out. The pilot phase develops an effective and efficient programme design. Lessons learnt from this need to be consistently applied during scaling up and out. Training therefore helps to transmit procedural and technical expertise as well as organizational values to new staff joining the ever-expanding innovation system. Similarly, staff of non-local stakeholder agencies needs to learn how to support local stakeholders [43].

Similarly, as scaling up can take significant time (often greater than 10 years) it is important that institutional capacity and incentives are built to maintain scaling beyond the tenure of any individual within an organizations. Training and development of staff responsible for implementing scaling up and out initiatives are critical, as is institutional capacity building, leadership, political support and incentives. As interventions are highly context-dependent, disseminating the principles of scaling may be more important than a specific option thought to fit a particular context. For example, capacity building at grass roots level via farmer-to-farmer visits, peer-to-peer training, training of trainers, development of community-based institutions and best practice competitions [37]. Multi-institutional projects and programs are also a means to ensure capacity is built across the range of actors involved.

Unlike 'traditional' extension services that have transferred *outside* solutions to farmers, new approaches need to be farmer-based, driven by local needs, participatory and considerate of groups such as women, young people and the very poor. Several other extension approaches such as farmer-to-farmer extension, community nurseries and farmer field schools, and rural resource centres can help spread innovation when technical solutions are found to be insufficient [39].

4.2.4 Step 4 create the enabling environment for successful scaling up and out

An enabling environment is necessary for successful innovation and scaling up and out. In additional to adequate finance, appealing to the motives of policy makers at all relevant scales as well as youth and women at the grassroots is particularly important in the context of Africa.

4.2.4.1 Secure a consistent funding mechanism

How well Africa's agriculture deals with climate impacts now and in the future will largely depend on the funding it receives or sets aside for adaptation measures. Meeting this financial challenge in Africa would require innovations, cooperative actions and political will to urgently and adequately address current and projected shortfalls for climate action targeting multiple funding sources. Financing to support up scaling adaptation to climate change could be obtained from public, private and other innovative sources such as crowd-funding and bank guarantees. The approach to scaling up and out will typically need to be adapted to the funding model. For example, international donors each have different priorities, which will influence the selection of climate smart technologies and approaches likely to be promoted in upscaling. Similarly, de-risking investments remains a key concern for the private sector no matter what their objectives are [45].

Financing options could include international sources such as UNFCCC-related funding sources including Adaptation Fund (AF) that finances climate adaptation and resilience activities, the Global Environment Facility (GEF) which finances environment-related projects or programs while supervising the Special Climate Change Fund (SCCF) and the Least Developed Countries Fund (LCDF). The SCCF gives financial support to vulnerable countries for adaptation and technology transfer while the LDCF supports LDCs in financing their efforts to adapt to climate change. The Green Climate Fund (GCF) was also established to finance projects or programs for adaptation to and/or mitigation of climate change. African countries could enlist technical support from Climate Technology Centre and Network (CTCN) to prepare bankable proposals to tap into the GEF/GCF controlled funds.

Regional funds in Africa such as from the Africa Development Bank (AfDB) also provide an additional source of climate finance. However, experience from recent implementation of agriculture-related National Adaptation Plans of Action (NAPAs) of the Least Developed Countries (LDCs) that have largely remained underfunded should be a pointer to the need to intensify the search for innovative local financing mechanisms affordable by most African countries. Accessing climate finance from global sources such as Global Environment Facility (GEF), Special Climate Change Fund and Least Developing Country Fund has largely remained elusive. This is mainly due to the stringent and bureaucratic nature of these funds that local actors need capacity enhancement on how to apply for these funds.

This chapter encourages African countries to source much of their funding from domestic sources. These include from: national budgets through parliamentary appropriation; remittances from the diaspora, which have fast gained a leading position as key foreign exchange earner for most sub-Saharan African countries; NGO funds—this sector controls a significant chunk of donor funds towards climate action; and the private sector, most of whose members in the region are still struggling to understand their clear role in climate resilient development. Given the key role played by SMEs in the critical agricultural value chains, there's need to innovatively target them through awareness-raising on how they may invest in sustainable agriculture practices.

Microfinance institutions, merry-go-round and credit groups may also provide the much-needed alternative particular to smallholders in sub-Saharan Africa. Public and private investment programmes in African countries must be structured in ways that support climate technologies and actions. Such actions can be financed through climate programmes or alternatively climate change considerations can be incorporated into sectoral funding sources or through a special window for inter-sectoral funding of activities that demonstrate climate co-benefits. Decentralized Climate Finance (DCF) is becoming increasingly important, examples being the County Climate Change Fund (CCCF) in Kenya that is currently being scaled up to more regions in Kenya by the National Drought Management Authority (NDMA) after institutionalizing the pilot phase.

4.2.4.2 Gender and youth considerations in planning

Climate change impacts are not gender-neutral. Women in poor countries are vulnerable to the effects of climate change as a result of their critical and differentiated roles and responsibilities in food crop production and dependence on natural resources in their livelihood strategies. In Africa, the main economic sector, agriculture, is highly vulnerable. Women provide more that 60% of the labor force engaged in agriculture, and with increasing impacts of climate change this makes women at greater risk of livelihood insecurity. Women, children, the youth and the poor are particularly more vulnerable as they often have limited resources, receive less education and are not involved in political and household decision-making processes at different levels. Gender transformative adaptation measures will make sure that women, men and youth benefit from actions by increasing participation and identification of potential implementation bottlenecks. Various opportunities exist for removing such bottlenecks. For example, active engagement of women in designing adaptation measures and decision-making processes at different scales (from farm to national level) will improve adoption, thus increasing their adaptive capacity and resilience. Digitizing agriculture has also been found to attract the youth back to agriculture.

4.2.4.3 Public policies and political support

The policy framework, laws, regulations and norms in a given country and region have to be supportive if scaling up of sustainable land and water management options is to succeed [46]. Similarly, scaling up productive technologies can be severely constrained or rapidly advanced by the policy environment [47]. These could be treaties, laws, regulations, statements, administrative actions and funding priorities. Examples of policies that could constrain or advance scaling are those related to farm inputs, regulations regarding food safety and product quality, output markets and trade [6]. For example, a systematic review of published research suggests tenure reform in Africa produces relatively modest agricultural productivity gains.

Climate change adaptation requires decentralized governance where local stakeholders determine key planning decisions and can incorporate local needs and priorities into the local level planning processes. Control of resources by local institutions and organizations as well as secure tenure allows land managers to look towards a future where they can build profitable and climate resilient systems. On the contrary, insecure property rights could hinder the adoption of climate-smart innovations, as there is little incentive to invest time and money into transition management practices. Protection of intellectual property by way of legislation is crucial to private sector engagement, which is fundamental to scalability. Such protection has been found to be responsible for the success of the seed industry across Africa [6].

One key way to ensure that leaders and institutions continue to pay attention to scaling up is to create an effective demand for it through the political system. Social change needs to be embedded in a society and supported by political constituencies which must be created and nurtured. Furthermore, these constituencies must be constantly reminded of the significance of the scaling up process to their political agenda. African leaders in particular will need to deliver on their commitments such as those under the Comprehensive Africa Agriculture Development Programme (CAADP) and the Malabo Declaration on Accelerated Agricultural Growth of the African Union (AU), and the Nationally Determined Contributions under the Paris Agreements, just to name a few.

4.2.5 Step 5 provide tangible early benefits and incentives to stakeholders

Scaling up and out processes can require sustained inputs from both internal and external stakeholders to an innovation platform who can facilitate or hinder attempts to scale up. It is necessary to provide tangible, early benefits to the stakeholders to trigger early support [40]. For example, ensuring contracted horticultural farmers are paid promptly upon delivery of their produce at the end of each season will retain them and win new additional farmers in the area into the programme. Other than incentivizing through sharing early benefits, it is also important to identify disincentives that may slow the pace at which innovations may be scaled and lead to disengagement from stakeholders. To retain stakeholder engagement, it is important to constantly manage expectations during the process of scaling up and out.

4.2.6 Step 6 monitor, evaluate, learn and communicate

Successful scaling up requires regular feedback from monitoring and evaluation systems. It is essential to learn from success and failure alike in order to develop best-practice in scaling up and out [38]. It requires a "learning by doing" culture, one that values adaptation, flexibility and openness to change. While a solid process needs to be laid out, scaling processes need to be adjusted regularly. Therefore, regular monitoring and evaluation as well as feedback from beneficiaries, communities, and field-based staff are important for learning and adjustment to take place. Two types of evaluation are relevant to scaling up. First is the evaluation of the pilot programme to establish whether or not the innovation tested has been successful and what lessons can be gleaned from it. Second, is M&E of the scaling up process [48]. It is necessary to monitor progress towards agreed sustainability targets and evaluate the impacts of climate smart technologies and innovations against measures of sustainability, including sustainable livelihoods [38]. Facilitating learning between different stakeholder groups across scales is critical—designed and implemented in collaboration with stakeholders to enable continuous learning—will improve climate- smart practices and ensure effective scaling up and out.

Local stakeholders require building of their confidence and skills to enable them to share, learn, develop, adapt and apply appropriate knowledge, ideas, methods and tools within a given context. On the other hand, non-local stakeholders need to learn how to support the local stakeholders by pursuing the necessary interdisciplinary research, training and opportunities for knowledge sharing. Scaling up also implies **learning** practically how to facilitate multi-stakeholder processes on a larger scale. Similarly, it implies learning by local stakeholders about the complexities and interconnectivities among the systems on which they depend for their livelihoods. It is this knowledge gained that will enable them modify issues of their concern or meet their priorities effectively. Where good practices and successes are identified, these need to be demonstrated and communicated widely to not only convince non-local stakeholders to learn from these experiences but also to build expertise in scaling up across different contexts.

4.2.7 Step 7 foster institutional leadership to support the scaling process

Scaling up cannot happen in a vacuum and requires leadership and process facilitation. Such leadership can be achieved by engaging a champion from one or more of the stakeholder groups (internal or external) who can lead and link different interests. This champion could be an enthusiastic NGO leader, a member of a farmer group, politician, financier, or a research team leader [38]. Key roles of the champions include inter-level coordination, [6] mobilizing key stakeholders involved in the scale-up and out program, coordination of the decentralized scale up governance framework and can also serve as an intermediary organization between the innovation "originating" and the innovation "adopting" organization to drive the scaling up process [49].

Within a scaling up arrangement for SLWM-related solutions, organizational leadership is provided by the process facilitator by mobilizing participation by different stakeholders through the agricultural innovation system/platform. This strengthens the institutional arrangements linking non-local to local stakeholders [41], which promotes not only local-level innovation involving both scientific and indigenous knowledge but also inter-level coordination [6]. This happens with new groups in new locations as scaling up advances. This task is best performed by the process facilitator (coordinating organization) responsible for the planning and facilitation of the scale up undertaking [49], who strongly recommend an intermediary organization between the originating organization and the adopting organization to effectively drive such a process.

The role of innovation intermediaries is also recognized in the literature. They facilitate interaction among isolated innovation networks, and between farmers and researchers, policy makers and other industry actors. They are also referred to as innovation "brokers" in other literature. Intermediation is presented as a formal professional role in development where intermediaries are hired consultants or web-based platforms for brokering exchange among actors in agriculture and food systems. Examples include AFIDEP which specializes in knowledge translation in the health sector—linking health researchers and ministries of health across Africa. Key partners in scaling up must always be mobilized and brought on board [50]. In most scaling up initiatives, partners were instrumental in helping to keep the momentum and focus [51].

4.3 Incentives and opportunities for scaling up

Incentives aimed at scaling up and out climate smart technologies and innovations need to be designed based on a thorough assessment of stakeholder needs, their local or traditional knowledge and a critical appraisal of existing incentives and their impacts.

4.3.1 Private sector incentives

A number of opportunities exist which the private sector could exploit towards climate resilient smallholder production with the right incentives in place. To realize these opportunities, the private sector needs incentives and co-financing for large-scale public-private partnerships [38]. These could include de-risking investments in land-based projects via state guarantees if projects fail and tax allowances for investing in restoration projects [45]. Secondly, these futuristic opportunities will require innovative partnerships, greater collaboration and connectivity among stakeholders riding on technological innovations along agricultural value chains. Considering the generally low profit margins in agriculture, the increasing interest from the private sector to scale up and out innovations can significantly stimulate such partnerships.

4.3.2 Incentives for farmers and communities

Incentives to encourage farmers should be designed in ways that encourage innovation and testing of interventions. Examples include Farmer Field Schools [52] and farmer competitions which bring prestige to farmers and can strengthen cultural identities enabling knowledge greater exchange and learning. Secondly, resource-poor farmers are unlikely to switch land management practices if there are no rapid returns to their investments—usually within one growing season. It therefore important that any climate smart interventions introduced add tangible value in order to be attractive and adopted. Finally, governments need to provide or improve provision of basic services such as infrastructure, health and education to boost the enabling environment for climate smart technology adoption by farming communities.

4.3.3 Incentives for policymakers

Policymakers will likely respond more readily to evidence that the implementation and scaling up and out of a climate smart option will contribute to current burning agendas such as unemployment, migration from drylands of Africa and Asia to Europe, food security in fragile states among others. Equally important to policymakers is evidence that lack of intervention would lead to increased scarcity of food, water and employment among the youth. Evidence expressed in terms of the indicators required for SDGs, national development and action plans and associated reporting for UNFCCC is likely to receive greater attention than data on areas of land degraded. Sound business cases or proposals for implementing climate smart solutions must demonstrate the multiple benefits obtainable in form of jobs, income creation, improved productivity and provision of other ecosystem services.

4.3.4 Opportunity to implement the Paris agreement

Of all major interventions proposed for promoting the contribution of agriculture to the Sustainable Development Goals (SDGs), scaling up and out of proven technologies, innovations and best practices promises the highest impact in Africa's context. The crucial entry point for agriculture-orientated interventions to improve food and nutrition security on a large scale should be smallholder farmers (over 80% own less than 2 ha) since majority of the food in sub-Saharan Africa (SSA) is produced by smallholder farmers while they are the most vulnerable to food insecurity and poverty [53]. Furthermore, most of the Nationally Determined Contributions (NDCs) submitted at the end of 2020 by African countries to the UNFCCC have identified agriculture as one of the priority areas for intervention towards climate change mitigation and adaptation, with some countries having already developed National Agricultural Action Plans. With the continent having the highest population growth rate, rapid urbanization trends and rising GDP in many countries, which are known to drive changing food consumption patterns, African agricultural systems will need to become resilient and adapt in order to meet growing demand, contribute to the achievement of the SDG-2 in particular, whose goals include ending hunger, achieving food security and improved nutrition by 2030, and promoting sustainable agriculture. It is imperative that those responsible for planning climate actions in different countries consider these agriculture priorities embedded in their NDCs to fast-track in scaling up and out.

5. Conclusion and recommendations

The threat by climate change on agricultural systems is real and calls for urgent interventions to rapidly build resilience among vulnerable communities in Africa. Considering that the smallholder farming community is responsible for almost 80% of the agricultural production, where only 10% is optimally used, greater impact at scale will be realized by targeting smallholder farmers with scale-up interventions. Similarly, considering that over 60% of the actual farmers are women and only 30% of them may access credit, it is imperative that gender-driven approaches be adopted when rolling out the scale-up modalities in order to catalyze transformation to inclusive and climate resilient agricultural systems on a large scale. The Chapter recommends consideration of the following steps by decision makers in any such scale up process:

a. Planning adaptively before rolling out any scaling up activity;

b. Selecting technologies and innovations based on best-available evidence recognizing there is no "one-size-fits-all";

c. Mobilizing innovation platforms by bringing together all stakeholders (local and non-local) and building their capacities to better engage in the scaling process;

d. Creating the enabling environment for successful scale-up/out by *inter alia*, securing consistent funding, integrating gender and youth considerations in planning, pursuing supportive public policies and securing local political support;

e. Providing tangible early benefits and incentives to stakeholders to ensure their sustained engagement and support through seasons of executing the strategies at their respective locations;

f. Consistently monitoring, evaluating, promoting learning and communicating widely the impact of adopting the strategies; and

g. Fostering institutional leadership to support the scaling up and out process.

Acknowledgements

The United Nations Framework Convention on Climate Change (UNFCCC) at its 23rd Conference of Parties (COP23) acknowledged (decision 4/CP.23) the central significance of the impacts of climate change on agriculture by launching the Koronivia Joint Work on Agriculture (KJWA) (UNFCCC, 2018). The KJWA process included the roll-out of five workshops on the status of scientific knowledge concerning agriculture and climate change. At their 50th session, both the Subsidiary Bodies for Implementation (SBI) and for Scientific and Technological Advice (SBSTA) requested the UNFCCC secretariat to organize a workshop on strategies and modalities to scale up implementation of best practices, innovations and technologies that increase resilience and sustainable production in agricultural systems according to national circumstances. The Chapter was commissioned by the African Group of Negotiators Expert Support (www.agnes-africa.org) which is hereby duly acknowledged.

Author details

Evans Kituyi
Dale Agro Ltd., Nairobi, Kenya

*Address all correspondence to: evans.kituyi@gmail.com

References

[1] FAO, ECA and AUC, "Africa Regional Overview of Food Security and Nutrition 2019.," 2020. Accessed: Apr. 21, 2021. [Online]. Available: http://www.fao.org/3/ca7343en/ca7343en.pdf

[2] AfDB, "Feed_Africa: Strategy for Agricultural Transformation in Africa 2016-2025," 2016. Accessed: Apr. 22, 2021. [Online]. Available: https://www.afdb.org/fileadmin/uploads/afdb/Documents/Generic-Documents/Feed_Africa-_Strategy_for_Agricultural_Transformation_in_Africa_2016-2025.pdf

[3] IFAD, "Livestock and Climate Change," 2009. Accessed: Apr. 22, 2021. [Online]. Available: https://www.uncclearn.org/wp-content/uploads/library/ifad81.pdf

[4] Cheikh Mbow, Cynthia Rosensweig, and L.G. Barioni, "Food Security," in *Climate Change and Land: an IPCC special report on climate change, desertification, land degradation, sustainable land management, food security, and greenhouse gas fluxes in terrestrial ecosystems*, 2019. Accessed: Apr. 22, 2021. [Online]. Available: https://www.ipcc.ch/site/assets/uploads/2019/11/08_Chapter-5.pdf

[5] Makhtar Diop, "Foresight Africa 2016: Banking on agriculture for Africa's future," 2016. https://www.brookings.edu/blog/africa-in-focus/2016/01/22/foresight-africa-2016-banking-on-agriculture-for-africas-future/ (accessed Apr. 22, 2021).

[6] T. Ajayi, O. Fatunbi, and 'Yemi Akinbamijo, *Strategies for scaling agricultural technologies in Africa*. Accra, 2018.

[7] Gro Intelligence, "Agricultural Extension Services in Africa," *Gro Intelligence*, 2015. https://gro-intelligence.com/insights/articles/agricultural-extension-services-in-africa (accessed Apr. 22, 2021).

[8] C. P. Msuya *et al.*, "THE ROLE OF AGRICULTURAL EXTENSION IN AFRICA'S DEVELOPMENT, THE IMPORTANCE OF EXTENSION WORKERS AND THE NEED FOR CHANGE," *Int. J. Agric. Ext.*, vol. 5, no. 1, Art. no. 1, May 2017, Accessed: Apr. 22, 2021. [Online]. Available: https://esciencepress.net/journals/index.php/IJAE/article/view/2101

[9] ACB, "Who will feed Africans? Small-scale farmers and agro ecology not corporations," African Centre for Biodiversity, 2017. Accessed: Apr. 22, 2021. [Online]. Available: https://www.acbio.org.za/wp-content/uploads/2017/01/FoE-Africa-report.pdf

[10] Olivia Serdeczny, Sophie Adams, Florent Baarsch, and Dim Coumou, "Climate change impacts in Sub-Saharan Africa: from physical changes to their social repercussions," *Reg. Environ. Change*, vol. 17, no. 6, pp. 1-16, 2017, Accessed: Apr. 22, 2021. [Online]. Available: https://link.springer.com/article/10.1007/s10113-015-0910-2

[11] W. Schlenker and D. B. Lobell, "Robust negative impacts of climate change on African agriculture," *Environ. Res. Lett.*, vol. 5, no. 1, p. 014010, Feb. 2010, doi: 10.1088/1748-9326/5/1/014010.

[12] IPC, "Major Farming Systems of sub-Saharan Africa," 2021. http://www.ipcinfo.org/fileadmin/user_upload/faowater/docs/ruralmaps/farming systems_colour.pdf (accessed Apr. 22, 2021).

[13] Qunying Luo, "Temperature thresholds and crop production: A review," Clim. Change, vol. 109, no. 3, pp. 583-598, 2011, doi: 10.1007/s10584-011-0028-6.

[14] FAO, Ed., *The future of food and agriculture: trends and challenges*. Rome: Food and Agriculture Organization of the United Nations, 2017.

[15] D. Lobell, M. Bänziger, C. Magorokosho, and B. Vivek, "Nonlinear heat effects on African maize as evidenced by historical yield trials," *Nat. Clim. Change*, vol. 1, Apr. 2011, doi: 10.1038/nclimate1043.

[16] C. Ringler, Tingju Zhu, Ximing Cai, Jawoo Koo, and Dingbao Wang, "Climate Change Impacts on Food Security in Sub-Saharan Africa," *IFPRI Int. Food Policy Res. Inst.*, p. 28, 2010.

[17] P. K. Thornton, P. G. Jones, P. J. Ericksen, and A. J. Challinor, "Agriculture and food systems in sub-Saharan Africa in a 4C world," *Philos. Trans. R. Soc. Math. Phys. Eng. Sci.*, Jan. 2011, doi: 10.1098/rsta.2010.0246.

[18] C. Rosenzweig *et al.*, "Assessing agricultural risks of climate change in the 21st century in a global gridded crop model intercomparison," Proc. Natl. Acad. Sci., vol. 111, no. 9, pp. 3268-3273, Mar. 2014, doi: 10.1073/pnas.1222463110.

[19] I. Niang *et al.*, "Chapter 22 Africa. In: Climate Change 2014: Impacts, Adaptation, and Vulnerability. Part B: Regional Aspects. Contribution of Working Group II to the Fifth Assessment Report of the Inter-governmental Panel on Climate Change," 2014, pp. 1199-1265.

[20] K. Waha, L. G. J. van Bussel, C. Müller, and A. Bondeau, "Climate-driven simulation of global crop sowing dates," *Glob. Ecol. Biogeogr.*, vol. 21, no. 2, pp. 247-259, 2012, Accessed: Apr. 22, 2021. [Online]. Available: https://onlinelibrary.wiley.com/doi/abs/10.1111/j.1466-8238.2011.00678.x

[21] J. A. Patz, S. H. Olson, C. K. Uejio, and H. K. Gibbs, "Disease Emergence from Global Climate and Land Use Change," Med. Clin. North Am., vol. 92, no. 6, pp. 1473-1491, Nov. 2008, doi: 10.1016/j.mcna.2008.07.007.

[22] M. Herrero *et al.*, "Greenhouse gas mitigation potentials in the livestock sector," *Nat. Clim. Change*, vol. 6, no. 5, Art. no. 5, May 2016, doi: 10.1038/nclimate2925.

[23] P. Thornton, J. van de Steeg, A. Notenbaert, and M. Herrero, "The Impacts of Climate Change on Livestock and Livestock Systems in Developing Countries: A Review of What We Know and What We Need to Know," Agric. Syst., vol. 101, pp. 113-127, Jul. 2009, doi: 10.1016/j.agsy.2009.05.002.

[24] M. Herrero *et al.*, "Climate variability and climate change and their impacts on Kenya's agricultural sector," ILRI, Nairobi, 2010.

[25] P. Ericksen, P. Thornton, A. Notenbaert, L. Cramer, P. Jones, and M. Herrero, *Mapping hotspots of climate change and food insecurity in the global tropics. 2011. CCAFS Report no. 5. CGIAR Research Program on Climate Change, Agriculture and Food Security (CCAFS). Copenhagen, Denmark. Available online at: www.ccafs.cgiar.org.* 2011.

[26] B. Cohen, "Urbanization in Developing Countries: Current Trends, Future Projections, and Key Challenges for Sustainability," Technol. Soc., vol. 28, pp. 63-80, Apr. 2006, doi: 10.1016/j.techsoc.2005.10.005.

[27] E. Totin *et al.*, "Institutional Perspectives of Climate-Smart Agriculture: A Systematic Literature Review," *Sustainability*, vol. 10, no. 6, Art. no. 6, Jun. 2018, doi: 10.3390/su10061990.

[28] H. C. J. Godfray *et al.*, "Food security: the challenge of feeding 9 billion people," Science, vol. 327, no. 5967, pp. 812-818, Feb. 2010, doi: 10.1126/science.1185383.

[29] J. Hansen *et al.*, "Climate risk management and rural poverty reduction," Agric. Syst., vol. 172, pp. 28-46, Jun. 2019, doi: 10.1016/j.agsy.2018.01.019.

[30] IPCC, "AR5 Climate Change 2014: Impacts, Adaptation, and Vulnerability," 2014. Accessed: Apr. 23, 2021. [Online]. Available: https://www.ipcc.ch/report/ar5/wg2/

[31] FAO, *Climate-smart agriculture sourcebook*. 2014.

[32] V. O. Abegunde, M. Sibanda, and A. Obi, "The Dynamics of Climate Change Adaptation in Sub-Saharan Africa: A Review of Climate-Smart Agriculture among Small-Scale Farmers," *Climate*, vol. 7, no. 11, Art. no. 11, Nov. 2019, doi: 10.3390/cli7110132.

[33] U. Murray, Z. Gebremedhin, G. Brychkova, and C. Spillane, "Smallholder Farmers and Climate Smart Agriculture: Technology and Labor-productivity Constraints amongst Women Smallholders in Malawi," Gend. Technol. Dev., vol. 20, no. 2, pp. 117-148, Jan. 2016, doi: 10.1177/0971852416640639.

[34] M. Schaafsma, H. Utila, and M. A. Hirons, "Understanding trade-offs in upscaling and integrating climate-smart agriculture and sustainable river basin management in Malawi," Environ. Sci. Policy, vol. 80, pp. 117-124, 2018, doi: 10.1016/j.envsci.2017.11.007.

[35] W. Thiery, E. L. Davin, D. M. Lawrence, A. L. Hirsch, M. Hauser, and S. I. Seneviratne, "Present-day irrigation mitigates heat extremes," *J. Geophys. Res. Atmospheres*, vol. 122, no. 3, pp. 1403-1422, 2017, doi: https://doi.org/10.1002/2016JD025740.

[36] GACSA, "Global Alliance for Climate Smart Agriculture: Work Program: 2016 -2017." 2014. Accessed: Apr. 23, 2021. [Online]. Available: http://www.fao.org/3/bp493e/bp493e.pdf

[37] M. Nyasimi, P. Kimeli, G. Sayula, M. Radeny, J. Kinyangi, and C. Mungai, "Adoption and Dissemination Pathways for Climate-Smart Agriculture Technologies and Practices for Climate-Resilient Livelihoods in Lushoto, Northeast Tanzania," *Climate*, vol. 5, no. 3, Art. no. 3, Sep. 2017, doi: 10.3390/cli5030063.

[38] R. J. Thomas *et al.*, "Scaling up sustainable land management and restoration of degraded land," United Nations Convention to Combat Desertification, Working Paper, 2017. Accessed: Apr. 23, 2021. [Online]. Available: https://cgspace.cgiar.org/handle/10568/89515

[39] H. Neufeldt, N. C, H. J, F. K, N. D, and S. P, "Scaling up climate-smart agriculture: lessons learned from South Asia and pathways for success," World Agroforestry Centre (ICRAF), ICRAF Working Paper 209, 2015. doi: 10.5716/WP15720.PDF.

[40] M. S. Reed, *The Research Impact Handbook*. Fast Track Impact, 2016. Accessed: Apr. 23, 2021. [Online]. Available: https://www.goodreads.com/work/best_book/50176571-the-research-impact-handbook

[41] S. Carter and B. Currie-Alder, "Scaling-up natural resource management: insights from research in Latin America," 2006, doi: 10.1080/09614520600562306.

[42] L. Jones, E. Ludi, and S. Levine, "Towards a characterisation of adaptive capacity: a framework for analysing adaptive capacity at the local level," ODI, 2010.

[43] D. Dinesh and S. Vermuelen, "Climate change adaptation in agriculture: practices and technologies Opportunities for climate action in agricultural systems." CCAFS CGIAR, 2016. Accessed: Apr. 23, 2021. [Online]. Available: https://cgspace.cgiar.org/bitstream/handle/10568/71051/SBSTA44-Agricultural-practices-technologies.pdf

[44] AGRA, "Africa Agriculture Status Report: The Hidden Middle: A Quiet

Revolution in the Private Sector Driving Agricultural Transformation," Alliance for a Green Revolution in Africa, Nairobi, 7, 2019. Accessed: Apr. 23, 2021. [Online]. Available: https://agra.org/wp-content/uploads/2019/09/AASR2019-The-Hidden-Middleweb.pdf

[45] A. Cornell *et al.*, "Economics of Land Degradation Initiative: Report for the private sector. Sustainable land management – A business opportunity." 2016. Accessed: Apr. 23, 2021. [Online]. Available: https://www.eld-initiative.org/fileadmin/pdf/ELD-SRPS_08_screen_150dpi.pdf

[46] A. Hartmann and J. F. Linn, "A FRAMEWORK AND LESSONS FOR DEVELOPMENT EFFECTIVENESS FROM LITERATURE AND PRACTICE." WOLFENSOHN CENTER FOR DEVELOPMENT, 2008.

[47] USAID, "Scaling Up the Adoption and Use of Agricultural Technologies: Global Learning and Evidence Exchange (GLEE)," Bangkok, Thailand, 2014.

[48] IFAD, "IFAD's Support to Scaling Up of Results Evaluation Synthesis," 4407, 2017. Accessed: Apr. 23, 2021. [Online]. Available: https://www.ifad.org/documents/38714182/39721352/Scaling+Up+ESR+-+Final+report+for+web.pdf/8b5e9b1e-245c-4d83-a093-7f5fa5f879ea

[49] MSI, "Scaling Up—From Vision to Large-Scale Change A Management Framework for Practitioners." Management Systems International, 2016. Accessed: Apr. 23, 2021. [Online]. Available: https://www.msiworldwide.com/sites/default/files/additional-resources/2018-11/ScalingUp_3rdEdition.pdf

[50] M. Jonasova and S. Cooke, *Thinking Systematically About Scaling Up*. World Bank, 2012. doi: 10.1596/26876.

[51] G. Mansuri and V. Rao, "Community-Based and Driven Development: A Critical Review," Feb. 2004, doi: 10.1596/1813-9450-3209.

[52] FAO, "Farmer Field School Guidance Document: Planning for quality programmes." 2016.

[53] S. Fraval *et al.*, "Food Access Deficiencies in Sub-saharan Africa: Prevalence and Implications for Agricultural Interventions," *Front. Sustain. Food Syst.*, vol. 3, 2019, doi: 10.3389/fsufs.2019.00104.

Chapter 5

Soil Compaction Due to Increased Machinery Intensity in Agricultural Production: Its Main Causes, Effects and Management

Songül Gürsoy

Abstract

In modern agriculture, most of the field operations from sowing to harvesting are done mechanically by using heavy agriculture machines. However, the loads from these heavy machines may induce stresses exceeding soil strength causing soil compaction. Nowadays, soil compaction is considered as a serious form of soil degradation, which may have serious economics and environmental consequences in world agriculture because of its effects on soil structure, plant growth and environmental events. Vehicle load, inflation pressure, number of passes, stress on the soil, and soil properties (e.g. soil water content, soil texture, soil strength, soil bulk density) play an important role on soil compaction. This chapter reviews the works related to soil compaction in agricultural areas. Also, it discusses the nature and causes of soil compaction, the effects of the compaction on soil properties, environment and plant growth, and the possible solutions suggested in the literature.

Keywords: soil compaction, soil degradation, soil properties, plant growth, controlled traffic

1. Introduction

In recent years, soil compaction has been considered as one of the most destructive environmental issues because it affects soil water dynamics, erosion, soil nitrogen and carbon cycling, cultivation energy requirement and effectiveness, pesticide leaching, and crop growth [1]. For example, the European Union has identified soil compaction as one of the main threats to soil that may cause the degradation of soils [2].

In simple terms, soil compaction can be described as the increase of bulk density or decrease in porosity of soil due to externally or internally applied loads. In agricultural production, soil compaction is considered as a complex problem in which soil, crops, weather and machinery interactions have an important role and which may result in dramatic economic and environmental problems. Wheel traffic from the use of heavy machinery and inappropriate soil management can cause the compaction of soil, creating impermeable layers within the soil that restrict water and nutrient cycles. This situation can result in reduced crop growth, yield and quality as well as a decline in the physical, chemical and biological indicators of soil quality such as destroyed soil structure, increased surface water run-off, soil erosion, greenhouse gas emissions, decreased hydraulic conductivity, reduced groundwater recharge and a loss of biodiversity [3, 4].

Mainly, there are two types of soil compaction defined as surface soil compaction and subsurface compaction. Surface soil compaction is also called as soil crusting. This type of soil compaction happens when the surface soil aggregates are broken down through the impact of falling raindrops, runoff, standing water during irrigation, or tillage. Therefore, it restricts water and air entry into soil by causing runoff and erosion of soil and impedes seedling emergence. The subsurface compaction can defined as tillage-induced compaction and wheel traffic-induced compaction depending on where it occurs. Tillage implements such as discs, moldboard plows and sweep type tools cause the tillage-induced compaction. This soil compaction type is mostly called as "hardpan," or "plow pan" and occurs in the layer of soil just below the depth of tillage when soils are cultivated repeatedly at the same depth. Wheel traffic-induced compaction lies beneath the tillage zone and is caused by axle weight load to the soil. This type of soil compaction is the most difficult to eliminate, so prevention is important [4–7].

Until today, many scientists [3, 4, 8–13] have discussed the potential causes of compaction, its main consequences, and strategies to prevent and reduce soil compaction. Also, many field and laboratory studies have been conducted to better understand the mechanics of soil compaction, the factors affecting it and its effects on soil properties and crop growth in many parts of the world. Nawaz et al. [13] reviewed the main causes of the soil compaction, and its effects on soil physical, chemical, biogeochemical processes and biodiversity properties at both macro- and microscales. The authors also discussed the existing models for the soil compaction and proposed new directions for modeling the effects of the compaction on the soil properties. They emphasized that soil productivity was very important for human survival but any form of soil degradation can reduce the soil fertility and ultimately, it lowers the soil productivity. Also, it was underlined in the paper that in spite of hundreds of articles appearing during the last 10 years on the soil compaction, there was an urgent need to apply multidisciplinary approach in the soil compaction studies, addressing diverse effects in different soil compartments. Soane and van Ouwerkerk [14] summarized the studies carried out the impacts of soil compaction on the environment, including its effects on soil water dynamics, erosion, soil nitrogen and carbon cycling, cultivation energy requirement and effectiveness, pesticide leaching, and crop growth. They reported that compaction-induced changes could lead to soil degradation, pollution of the atmosphere and of ground and surface waters, and might increase the mechanical resistance of the soil, resulting in an increase in soil cutting force, fuel consumption, working hours, and abrasion of agricultural instruments. Also, several researchers [15–18] reported that soil compaction negatively influenced the water dynamics, pesticide diffusion, soil erosion, carbon and nitrogen cycle, plant growth, and mechanical operations cost. Although overall effect of the soil compaction on environment, soil properties and the plant yield is mostly reported as negative, moderate compaction can sometimes result in yield increase by increasing the contact between the roots and soil particles which may lead to the rapid exchange of ions between the soil matrix and roots [13]. Batey and McKenzie [19] stated that the effects of soil compaction on crops and soil properties are very complex and should be well documented. Therefore, it is very important to understand what is the causes and harmful level of soil compaction, how soil compaction affects soil properties and plant growth, and to know how to avoid soil compaction as much as possible in agricultural production.

This chapter includes identification of compaction, the main causes of soil compaction in agricultural fields, its effects on soil properties, environment and plant growth, and strategies for preventing compaction in agricultural production.

2. Soil compaction and the main causes of soil compaction in agricultural fields

Soil compaction is defined as the increase of bulk density or decrease in porosity of soil due to externally or internally applied loads. Soil compaction can be identified either in the field, the laboratory or via remote sensing. In field, soil compaction can be detected by observing or measuring the changes in soil structure, soil moisture and soil color, penetrometer resistance, permeability to air or water, waterlogging on the surface or in subsurface layers, and crop or root growth properties [20]. In Laboratory, soil bulk density, pore-size distribution, water permeability and the relative apparent gas diffusion coefficient is measured to determine the permeability of soils to air and water and therefore on the degree of compaction. Also, remote sensing techniques helps to recognize alterations of soil structure, root growth, water storage capacities and biological activity [21].

Soil compaction in agricultural fields is commonly known to be caused by several forces (natural and man-induced) such as raindrop or tillage equipment during soil cultivation, or trampling of livestock animal or from the heavy weight of field equipment such as mostly tractors, heavy cultivation machinery, harvesting equipment [3, 18]. In this case, we can say that high mechanical load, less crop diversification, intensive grazing, low organic matter and tillage at high moisture contents can lead to soil compaction. Shah et al. [12] summarized the main causes of soil compaction as shown in **Figure 1**. Similarly, Ziyaee and Roshan [22] conducted a survey to study the causes of soil compaction problems. They defined common causes of soil compaction as natural processes, earlier planting schedules, overgrazing and animal trampling which directly affect the penetration resistance, besides, increased machinery weight and intensity, and especially excessive tillage at high soil moisture content.

Raindrop is one of the natural causes of surface compaction, and it causes a soil crust (usually less than 1/2 inch thick at the soil surface) that may prevent seedling emergence. This soil crusts reduces water infiltration and increases surface runoff, and thereby sediment and nutrient losses [6, 23].

Using tillage implements at the same depth can cause serious tillage pans or hardpans just below the depth of tillage. This tillage pan, which is generally relatively thin (2,5 to 5 cm), can reduce yield potential by restricting root growth and nutrient uptake of plant in soil [24].

Trafficking by the tyres of heavy farm machines is known to be the major cause of soil compaction in agricultural fields. Soil compaction by wheels is characterized by a decrease in soil porosity localized in the zone beneath the wheel and rut formation at the soil surface [18]. The degree of soil compaction by farm machines depends on not only the characteristics of the agricultural machinery but also the soil properties such as soil moisture, soil type, texture, structure, and moisture [3, 8, 25]. Several researchers [26–28] showed that the compaction rate depended on the soil characteristics, and the weight, the pressure and the vibrations of the agricultural machinery. Different machines, or even the same machines with different tyres, differ in their loading and pressure on the soil [25, 26]. Nawaz et al. [13] reported that the soil compaction by a machine depended on the soil strength which is influenced by the organic matter, water content, soil structure, and texture, and machine properties expressed by axle load, number of tyres, tyre dimensions, tyre velocity, and soil-tyre interaction. Botta et al. [29] reported that tyre sizes and rut depth/tyre width ratio had significant effect on soil compaction. The researchers emphasized that the farmer should pay attention to the axle load, the tyre size and the soil water content at the traffic moment. Also, the size, the inflation and the shape of tire in addition of the tyre's load have significant effects on the ground pressure, indicating traffic-induced soil compaction [30, 31]. Håkansson and Reeder [32] found that the

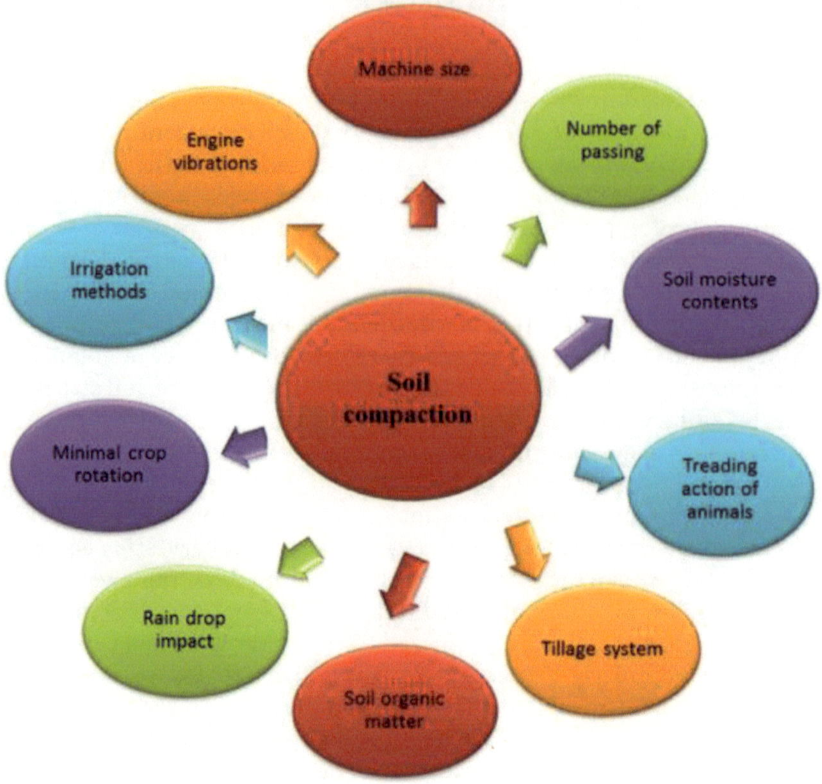

Figure 1.
A summary of the main causes of soil compaction in agriculture fields [12].

vehicles with high axle loads generally caused deep subsoil compaction when trafficked on soils with high moisture contents, and this deep subsoil compaction caused persistent and possibly permanent reductions in crop yields. Soil moisture and axle load causes soil compaction at various depth as shown in **Figure 2**. It is seen in **Figure 2** that high axle load could result in soil compaction to deeper depth. Also, at a given load and tire size, increasing soil moisture content causes much deeper compaction than dry [33]. Da Silva et al. [34] stated that high contact pressures applied to soil, which is described as the ratio of mass of each machine's axle and the contact area of the run, resulted in a greater degree of compaction, in addition to promoting other negative effects. The researchers underlined that it was possible to minimize the effects of soil compaction through the appropriate contact pressure of agricultural machines. They suggested that the use of new technologies and suitable management practices should develop and adopt to characterize machine size and solve the machine-soil problems, especially about the distribution of pressures caused by the wheels in the soil, thus avoiding the negative effects of compression. Similarly, Porterfield and Carpenter [35] stated that high-pressure tire-ground contact caused an increase in the density of the soil. so, the researchers recommended to keep contact pressure low to avoid compaction. The intensity of trafficking, also referred to as the number of passes of agricultural machinery on the soil throughout the life of the crop in addition to the wheel ground contact pressure and absolute wheel load, has a significate effect on the degree of compaction and the depth to which wheel pressure affects the soil [36]. Zhang et al. [37] stated that the increased frequency of passes associated with the small four-wheel tractors, which is smaller mass and lower ground pressure than the medium power tractor, potentially was more detrimental

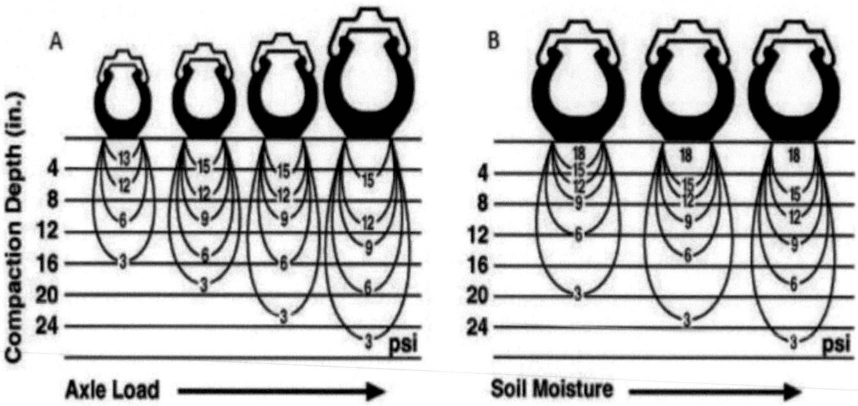

Figure 2.
(A) The effect of axle load on soil compaction depth, (B) The effect of soil moisture content on compaction depth [33].

than those associated with the medium power tractor. Rusanov [38], who reported official standard values of maximum permissible normal stress at a depth of 0.5 m, observed traffic intensity mostly resulted in higher soil stress than the allowable stress in subsoil. Similarly, Botta et al. [39] observed that high traffic frequency (10 and 12 tractor passes in the same tracks) of a light tractor on typical Argiudol soil produced a significant increase in cone index and dry bulk density in the topsoil and subsoil levels. However, Hamza and Anderson [18] reported that the first pass of a wheel was known to cause a major portion of the total soil compaction and subsoil compaction may be induced by repeated traffic with low axle load and the effects can persist for a very long time. In summary, we can say that repeated passes of agricultural machinery at the same locations will increase soil compaction. Also, Shah et al. [12] emphasized that in intensive agriculture, high axle load of heavy tractors and field machines resulted in compacted soil layer, damaging the structure of tilled soil and subsoil and reducing crop and soil productivity. Keller et al. [40], who analyzed the effect of the increase in weight of agricultural vehicles on soil stress and soil bulk density, showed that the increase in machinery weight has resulted in an increase in subsoil compaction levels, and highlighted that future agricultural operations should consider the inherent mechanical limit of soil.

Soil properties (soil texture, soil aggregate properties, moisture content, organic matter content) and frequent use of chemical fertilizers has significant effect on soil compaction [8, 18]. It is mostly stated that the depth to which compressive forces are transmitted depends on the moisture content in soil profile. Batey [3] described the relationship between soil moisture content and compressibility as following: When the soil was dry and firm throughout the profile, there might be no significant compaction effect. However, when the surface layers were moist and soft lying over dry soil, the upper layers might be strongly compressed, and when the surface layers were dry and firm with moist soil below, the compression might be transmitted some way downwards to compress the moister vulnerable soil. Gysi et al. [41] determined that soil moisture content and wheel load significantly influenced the bulk density at a depth of 0.12–0.17 m. The researchers stated that a soil with very low moisture content was less vulnerable to compaction than a soil with high moisture content. The texture, organic matter content, aggregation stability and mineralogy of the soil also have a significant effect on compressibility of a soil by agricultural vehicles. Several researchers [13, 18] stated that soil texture was one of the most important factors in determining the susceptibility of a soil to compaction. Horn and Lebert [42] reported that coarse-textured soils were less susceptible to compaction than those with a fine texture. Moreover, Horn et al. [15] found that the silt loam soils with low colloid contents were more susceptible than medium or fine textured loamy and clayey soils at low water contents while the sandy soils were slightly susceptible to the soil compaction. Soil aggregation and organic matter content are also the most influencing factors that makes soil resistance to compaction [13, 43]. Ellies Sch et al. [44] reported that soils with poor structure and aggregation were extremely vulnerable to the impacts of wheel traffic, while susceptibility to compaction could be reduced with an increase in soil aggregation. Also, soil organic matter is a very important soil property, which can determine the magnitude of soil compaction. The amount of organic matter in soil significantly influence the compressibility degree under axle load of vehicles [13]. Previous works [45–47] has shown that increasing organic matter in soil may reduce compatibility by increasing resistance to deformation and/or by increasing elasticity.

3. The effects of soil compaction on soil properties, environment and plant growth

Soil compaction alters the soil structure and hydrology by changing many aspects of the soil such as strength, gas, water and heat, which affect chemical and biological balances. In turns, all these alterations in the soil influence root and shoot growth and consequently crop production and environmental quality **Table 1** [58]. presents a summary of studies related to the effects of soil compaction on soil, environment and plant growth. In most of these studies, the subsoil compaction negatively affected soil physical conditions, which substantially decreased crop yield. However, some studies [50, 56] showed that moderate compaction had no effect on crop yield or can increase yield. Also, Gürsoy and Türk [59] stated that moderate soil compaction in agriculture production was needed to get good seed/root-soil contact, suitable soil density, timely emergence of seed, root growth and the ability of the plant to absorb the moisture and nutrients from soil.

In Global Land Outlook Working Paper expressed by [60], the consequences regarding the effects of soil compaction on soil properties, environment and crop plant's morphological and physiological growth have been presented as seen in **Figure 3**. Also, Horn et al. [15], who summarize the works carried out about effects of soil compaction on the structure of arable soils, stated that soil compaction caused by traffic of heavy vehicles and machinery resulted in soil structure deterioration, both in the topsoil and in the subsoil. They reported that owing to dynamic loading, soil physical properties such as pore size distribution and pore continuity were negatively affected, which entails decrease in air and water permeability and resulted in increased soil strength. The researchers emphasized that these changes in soil structure may have a negative effect on the soil biota, on physical–chemical equilibria and redox potential, on the soil's filtering and buffering capacity, on ground water recharge and, finally, on crop yield.

The major effect of soil compaction on soil properties is known as increase in soil bulk density and decrease in total porosity as soil aggregates are pressed closer together, resulting in a greater mass per unit volume and less space for air and water in the soil [15]. Changes in the soil pore system due to compaction can adversely affect key soil hydraulic properties and aeration such as saturated hydraulic conductivity and air movement in soil [58]. Ziyaee and Roshan [22] stated that pore space provided a room for air and water to circulate around the mineral particles, providing a healthy environment for plant roots and beneficial microorganisms, however, in compacted soils, the particles were pressed together so tightly that the space for air and water was greatly reduced. The researchers emphasized that lack of pore space resulted in

Land vehicles traffic	Place	Soil texture	Major findings	Reference
Different passes of a tractor pulling a trailer, during harvest operations	Argentina	A fine clay	The first pass of the tractor (4200 kg) resulted in a reduction of grass yield (40.3%) in comparison with the control (no traffic) treatment	[48]
Weighted tractor compaction and a control of no compaction.	Scotland England	An imperfectly drained silty clay loam and a well drained sandy loam	Damage to soil structure through compaction reduced the yields of grassland swards by mechanical (tractor) compaction (14.5% reduction) after three years of these treatments. The soil type had significant effect on yield losses	[49]
Most trafficked rows and least trafficked rows	USA	Loamy soils, fine sandy loam soils and fine sands	Different crops grown in a no till field were not very much influenced by wheel traffic	[50]
Additional load (4.40, 6.40 and 8.40 kg) on the tractor rear wheel and 1, 6, 11 and 16 passes	India	Sandy loam soil	Yields of wheat decreased with an increase in the number of passes/trips at all load and pass combinations.	[51]
Wheel loads (3, 8 and 12 Mg) and single and multiple (4–5) wheel passes	Denmark	Sandy loam soil	High wheel loads caused significant subsoil compaction at>50 cm depth and heavy traffic in dense soils created a potentially restrictive subsoil structure for plant growth	[52]
Axle loads of sugarbeet harvesters and the number of passes	Sweden	Loam soil Sandy loam soil Sand soil	Heavy traffic during harvest of sugarbeet implied a major risk for compaction of the subsoil, which can be seen as a long-term threat to soil productivity.	[53]
The number of passes of 7.0 t roller having length of 1.5 m and diameter of 1.22 m	Pakistan	Well-drained alluvial silt loam	The subsoil compaction adversely affected soil physical conditions, which substantially decreased the yield of wheat.	[54]
Axle load (0, 7 and 17 tons)	Jordan	Clay loam soil	The yield was significantly reduced as a result of compacting loads.	[55]
No-tillage system, the tractor traffic, harvester traffic and soil chiseling management	Brazil	Oxisol soil	Soybean grain yield was reduced due to both soil chiseling and heavy traffic by harvester	[56]
Harvester traffic intensities under direct sowing	Argentina	Fine clay soil	A higher axle load than 79.70 kN caused the cone index peaked in the subsoil to depths below 35 cm and significantly reduced soybean yields.	[57]

Table 1.
Effects of soil compaction on soil, environment and plant growth.

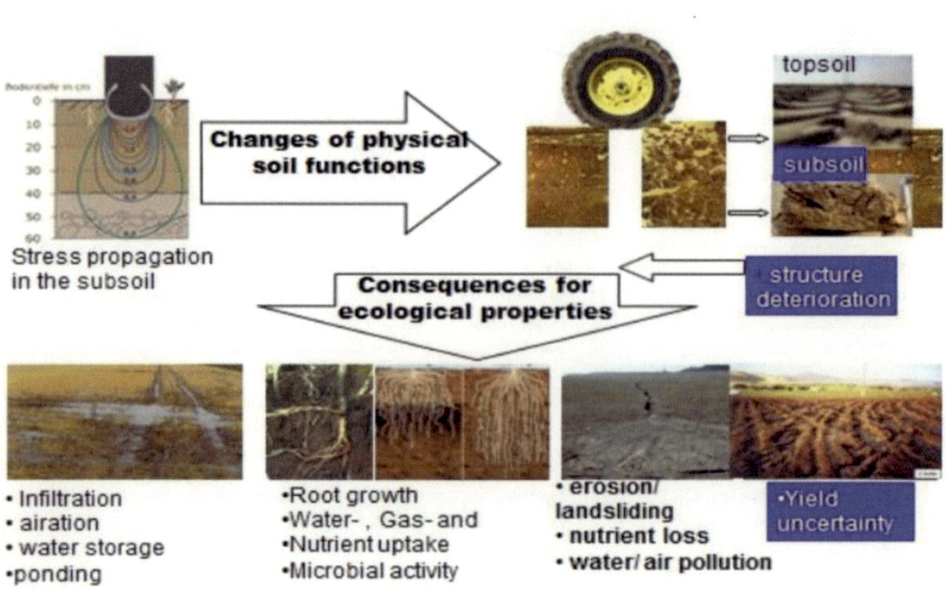

Figure 3.
The consequences of soil compaction regarded with soil, environment and plant growth [60].

the lack of oxygen, which is very important for plant growth, decomposing organic matter, recycling nutrients and aerating the soil. In addition to soil bulk density and porosity, penetration resistance significantly affects plant root growth and crop yield. The previous works [61–64] showed that plant root growth could be slowed down or completely impeded at penetration resistance values of 2 and 3 MPa, respectively. Also, the soil physical properties changed due to soil compaction can alter elements mobility, change nitrogen and carbon cycles and soil biodiversity [13].

Effects of soil compaction on plant growth are complex and depended on many factors. Mainly, it is known that high soil penetration resistance and low oxygen concentration in a compacted soil can reduce crop yield due to decreased root elongation rates and thus limited accessibility to water and nutrients [65]. Also, plants in compacted soils can also suffer from water stress due to reducing water infiltration and increasing runoff [15]. Keller et al. [40] discussed in detail how soil compaction changed soil properties, and how the soil properties changed by soil compaction affected plant growth and environment. They heighted that soil compaction was one of the main causes of the yield decrease observed for major crops in many European countries. In an experiment with barley, Willatt [61] observed that root length density in the upper 0.30 m of soil and rooting depth decreased as the number of tractor passes increased from zero to six. Ishaq et al. [66], who study subsoil compaction effects on root growth, nutrient uptake and chemical composition of wheat and sorghum, determined that root length density of wheat below 0.15 m depth was significantly reduced with increased soil bulk density. They recommended that appropriate measures such as periodic chiseling, controlled traffic, conservation tillage, and incorporating of crops with deep tap root system in rotation cycle should be applied to minimize the risks of subsoil compaction. de Moraes [56], who study the impact of soil compaction on soybean root growth, investigated three soil compaction levels (no-tillage system, areas trafficked by a tractor, and trafficked by a harvester) and soil chiseling management (performed in an area previously cultivated under no-tillage) on soil properties and plant root growth. The researchers observed that root growth was influenced by soil physical conditions during the cropping season and soybean grain yield was reduced due to both compaction (caused by harvester traffic) and excessive loosening (promoted by chiseling) compared the no-tillage system. Obour and Ugarte [67] used a meta-analytical approach to summarize the results from 51 published articles on the impacts of soil compaction attributed to machinery axle load, wheel passes, compaction events and tire inflation pressure on soil bulk density, degree of compactness, penetration resistance, volume of water filled pores at field capacity, air permeability at field capacity, saturated hydraulic conductivity, and grain yield of corn, wheat, barley and soybean. Results from this meta-analysis showed that compaction increased the soil mechanical strength affected by increased soil bulk density, degree of compactness, penetration resistance. Also, compaction decreased hydraulic conductivity characterized by air permeability and saturated hydraulic conductivity from the topsoil down to the subsoil (>40 cm depth), and grain yield of corn, wheat, barley, and soybean. The researchers suggested that soil hydraulic properties might be more sensitive indicators to reflect the impact of soil compaction on soil structure and pore system functions in the soil profile. Also, negative effects of soil compaction on plant root growth and crop yield have been recognized by several researchers [26, 38, 64].

Soil compaction can affect crop yield depending on soil texture and growing season precipitation. Voorhees [5], who presents relative barley grain yield as a function of the degree of compactness of a clay soil in Sweden, reported the effect of soil compaction on grain yield depending on climate change as fallowing: 1) During relatively dry climatic conditions, grain yields were increased by almost 40% with the initial increases in degree of compactness. However, when the degree of compactness approached 90, yields decreased significantly. 2) Similar results were obtained during a normal year but less pronounced. 3) During a wet year, yields decreased with any increase in degree of compactness above 75. The researcher stated that plants could tolerate (might even need) a more highly compacted soil during dry conditions than during wet conditions.

In summary, soil compaction can significantly change the physical, chemical and biological properties of soil depending on climate and initial soil properties such as soil texture, structure, moisture content, organic matter content. These changes in soil properties can have significant effect on the penetration of plant roots, their growth, soil–plant-water relations, the ability of plants to take up nutrients from soil and consciously crop yield. In another sense, the effect of the same compaction degree on plant root growth and yield depends on the crop grown, soil structure, and weather conditions. All literature data shows that the effects of soil compaction on root and plant growth have a complex interaction including many soil, climate and plant properties. This complex interaction requires complex hypotheses to explain the effects of soil, climate and plant properties on root and plant growth.

4. Strategies for avoiding and reducing soil compaction

Many strategies have been used to avoid soil compaction in agricultural fields and to ameliorate compacted soil or alleviate its associated stresses. Strategies can be roughly divided into three groups: measurements to avoid further compaction, remedial treatments, and methods to alleviate soil compaction [33, 68].

The best way to manage soil compaction is to prevent it from happening. This includes reducing axle load, proper inflation and size of tires, minimizing soil tillage, increasing stability of soil structure and conducting field operations at appropriate soil moisture content [6]. The capacity of the soil to resist stress (i.e. the strength against compression) and loading of machine is considered to be the major factors affecting the soil compaction by farm machineries [11]. The soil strength against compression is influenced by the organic matter, water content, soil structure, and texture. Therefore, improving soil structural stability and aggregate can reduce the risk of soil compaction [13]. Soil structural stability and aggregate can improved by increasing soil organic matter content and reducing stress on the soil due to machinery traffic. This can be achieved by using conservation tillage systems [11]. Similarly, several researchers reported that long-term use of conservation tillage systems resulted in lower soil compaction threat because it increased surface organic matter contents, more stable soil structure, and increased hydraulic conductivity due to worm holes and stable biochannels [63, 69]. Loading of machine on soil is expressed by axle load, number of tyres, tyre dimensions, tyre velocity, and soil tyre interaction [13]. The axles load and the contact pressure of tires is known to be the most important parameters affecting soil compaction. A high wheel load may lead to compaction of soil in both the top and deep sublayers, whereas low axle loads will cause compaction in the topsoil and the upper part of the subsoil only. An axle loads or wheel loads describe the weight distribution of machines, depending on the degree of the loading of tank or weight transfer during plowing. Therefore, the weight distribution may vary markedly between wheels on the same axle. In their literature review, Alakkuku et al. [11] recommended single axle load of 4–6 Mg for moist mineral soils to avoid soil compaction below normal primary tillage depth (0.2–0.3 m) and a limit of 8–10 Mg for tandem axle loads on moist soils. The researchers stated that to reduce axle loads, machine weight may be reduced by using new, lighter materials or multiple axles can be used to spread the load. Also, they reported that wheel load should be linked with soil contact pressure recommendations because the wheel load alone does not give any information about the stress level transferred to the soil and the corresponding stress distribution in the soil. Contact area between the wheel and soil and the basic dimensions of wheel such as width and length has significant effect on soil contact pressure of wheel [11]. Also, ten Damme et al. [70] stated that the stress distribution in the contact area between the tyre and the soil is of primary importance for the propagation of stress in the soil. The researchers indicated that tyre design might further help reduce the risk of soil compaction at a specific load if it allows for further reductions of the tyre inflation pressure. This literature findings shows that soil compaction can be avoided by adjusting tire size and tire inflation pressure, or using rubber-belt tracks [11, 71]. Alakkuku et al. [11] stated that low tyre inflation pressure usually provided low ground contact pressure and allowed even pressure distribution, which are advantageous to both soil compaction caused by wheel traffic and to wheel tractive efficiency.

According to our review conclusion, we can say that the wheel load and the soil contact pressure is the major engineering tools for the control of subsoil compaction and to avoid permanent subsoil compaction, the machines and equipment used on the critical field conditions should not be cause higher stress than the bearing capacity (strength) of soils. Also, the compaction risk in given vehicle–soil interactions might be quantitatively estimated by using pre-consolidation stress as an indicator of the bearing capacity of a soil. Alakkuku et al. [11], who reviewed technical choices to minimize the risk of subsoil compaction, presented a framework of machinery–soil system in connection with subsoil compaction as shown in **Figure 4**. The researchers stated that to prevent subsoil compaction, recommendations for wheel load–ground contact pressure combinations should be made available for different soil conditions. Chamen et al. [69], who discussed the machinery usage during field practices to avoid subsoil compaction, stated that the bearing capacity of soils would be improved by increasing their structural stability, such as may be achieved with reduced or no tillage systems. The researchers summarized the preventative strategies suggested for the avoidance of subsoil compaction as follows: (1) no repeated soil loosening as a routine cultivation technique, (2) increased

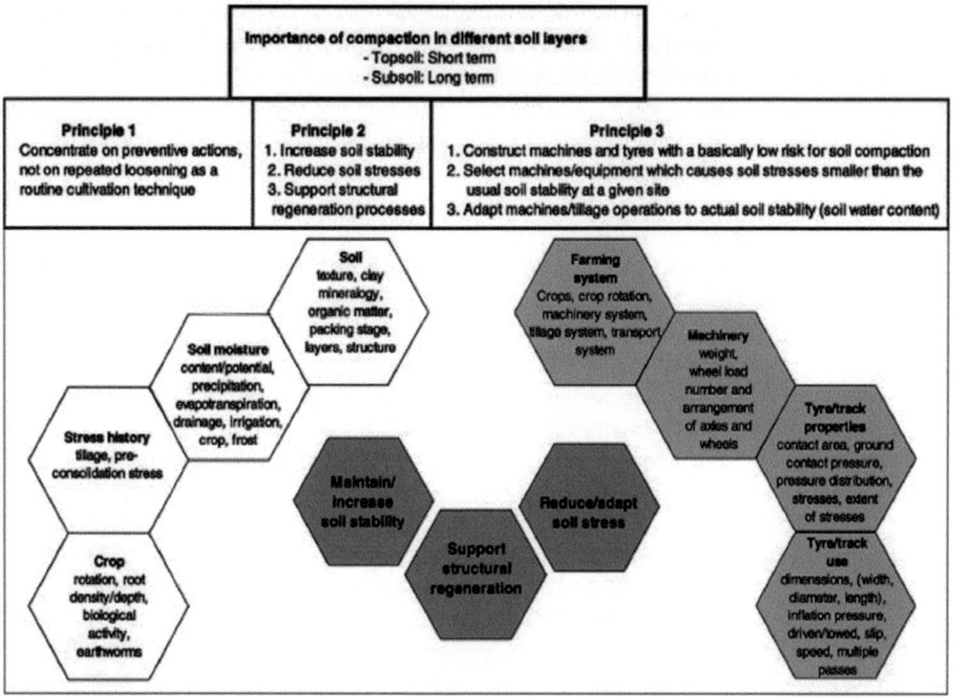

Figure 4.
A framework of machinery–soil system in connection with subsoil compaction [11].

soil stability and reduced soil stress, (3) the selection of machines and field practices with a low risk potential, (4) the assimilation of new, low risk technologies. Also, Kumar et al. [72] stated that the management of soil compaction might be achieved through suitable application of some or all of the following techniques: (1) decreasing the pressure on soil either by reducing axle load or increasing the contact area of wheels, (2) conducting field operations at optimal soil moisture content, (3) decreasing the number of passes of farm machinery and the intensity, (4) restraining traffic to certain areas of the field or controlled traffic, (5) improving soil organic matter through retaining of crop and pasture residues, (6) eliminating soil compaction by deep ripping in the presence of an aggregating agent, (7) including the plants with deep, strong taproots in crop rotations.

In this case, we can suggest the following solutions to prevent and alleviate soil compaction and its detrimental effects:

1. Maintaining an adequate amount of organic matter such as stubble retention, green and brown manure or addition of plant or animal organic matter from external sources in the soil stabilizes soil structure and makes it more resistant to compaction because it acts a buffer preventing or lessening transmission of compaction to subsoil from external loads acting on the earth [18].

2. Controlled traffic farming, which confines all machinery loads to the least possible area of permanent traffic lanes, was recommended by many researchers [73] to reduce the damage to soils caused by heavy or repeated agricultural machinery passes on the land. By controlling traffic, the tracked area will have a slightly deeper compaction but the soil between the tracks will not be compacted. Controlled traffic systems has been stated to have fundamental advantages in maintaining 'good' soil structural conditions of non-trafficked crop beds.

3. Another strategy for managing compaction is properly to decrease the contact pressure of tires and axle loads. Therefore, the machines and equipment used on the field in critical conditions should be adjusted to actual strength of the subsoil by controlling wheel/track loads and using low tyre inflation pressures [11]. Other methods to reduce compaction include the use of dual wheels, rubber tracks and flotation tyres.

4. The conservation agriculture techniques can be practiced to reduce traffic on the soil. A conservation tillage system can reduce the need for vehicle traffic in the field because there are fewer needs for tillage or cultivation operations [74].

5. Subsoiling/chiseling can be used for eliminating soil compaction, destroying hard pans and ameliorating hard setting soils developed due to traffic and puddling. Deep ripping of compacted soil may also improve soil health and ability of plants to resist disease [75].

6. Incorporating crops and pasture plants with strong tap roots able to penetrate and break down compacted soils in the rotation is desirable to minimize the risks of subsoil compaction [56].

7. Field operations should not be performed under wet soil conditions. Soil is more compressible at wet soil conditions. Traffic during high moisture conditions may compact soil, whereas the same traffic under dry conditions will not [18].

5. Conclusions

In recent years, the structural and technological development of modern agriculture caused a significant increase in the power, size and the weight of vehicles and machinery used on agricultural fields. This dramatic increase in the weight of agricultural machinery and the necessity to use heavy machines in unfavorable soil conditions have caused a significant increase in the subsoil compaction, which is considered as a serious form of soil degradation and may have serious economics and environmental consequences in world agriculture. The main results of our literature evaluation showed that severe soil compaction might result in a decreased root growth and plant development, and consequently, a reduction in crop yield because it adversely affect key soil hydraulic and aeration properties such as saturated hydraulic conductivity and air movement in soil. Soil compaction is also an environmental problem because it is one of the causes of erosion and flooding. In addition, it directly or indirectly increases nutrient and pesticide leaching to the groundwater and nitrous oxide emissions to the atmosphere. Therefore, prevention of soil compaction and alleviation of existing compaction is one of the most important issues in agricultural production in order to sustain or improve soil fertility and productivity.

Author details

Songül Gürsoy
Department of Agricultural Machinery and Technology Engineering,
Faculty of Agriculture, Dicle University, Sur, Diyarbakir, Turkey

*Address all correspondence to: songulgursoy@hotmail.com

References

[1] O'Sullivana MF, Simota C. Modelling the environmental impacts of soil compaction: A review. Soil and Tillage Research. 1995;35:69-84

[2] COM (Commission of the European Communities). Communication from the Commission to the Council, the European Parliament, the Economic and Social Committee and the Committee of the Regions – Towards a Thematic Strategy for Soil Protection [Internet]. 2002. Available from: https://eur-lex.europa.eu/LexUriServ/LexUriServ.do?uri=COM:2002:0179:FIN:EN:PDF [Accessed: 1921-05-09]

[3] Batey T. Soil compaction and soil management – a review. Soil Use and Management. 2009;25:335-345.

[4] Raghavan GSV, Alvo P, McKyes E. Soil Compaction in Agriculture: A View Toward Managing the Problem. In: Lal R, Stewart BA, editors. Advances in Soil Science. Advances in Soil Science. New York: Springer; 1990. P.1-32.

[5] Voorhees WB. The Effect of soil compaction on crop yield. SAE Transactions, 1986;95(3):1078-1084.

[6] Idowu J, Angadi S. Understanding and Managing Soil Compaction in Agricultural Fields [Internet]. 2013. Available from: https://aces.nmsu.edu/pubs/_circulars/CR672.pdf [Accessed: 2021-05-09]

[7] Light S. Strategies to Avoid and Manage Soil Compaction [Internet]. 2018. Available from: https://ucanr.edu/blogs/blogcore/postdetail.cfm?postnum=27421 [Accessed: 2021-05-09]

[8] Bennie ATP, Krynauw GN. Causes, adverse effects and control of soil compaction. South African Journal of Plant and Soil. 1985;2:109-114.

[9] Petelakau H, Dannowski M. Effect of repeated vehicle traffic in traffic lanes on soil physical properties, nutrient uptake and yield oats. Soil and Tillage Research. 1990;15:217-225.

[10] Soane BD, Van Ouwerkerk C. Soil compaction problems in world agriculture. In: Soane BD, van Ouwerkerk C, editors. Soil Compaction in Crop Production. Amsterdam: Elsevier; 1994. p. 1-26.

[11] Alakkuku L, Weisskopf P, Chamen WCT, Tijink FGJ, Van der Linden JP, Pires S, Sommer C, Spoor G. Prevention strategies for field traffic-induced subsoil compaction: A review. Part 1. Machine/soil interactions. Soil and Tillage Research. 2003;73:145-160.

[12] Shah AN, Tanveer M, Shahzad B,Yang G, Fahad S, Ali S, Bukhari MA, Tung SA, Hafeez A, Souliyanonh B. Soil compaction effects on soil health and crop productivity: an overview. Environmental Science and Pollution Research. 2017;24: 10056-10067.

[13] Nawaz MF, Bourrié G, Trolard F. Soil compaction impact and modelling. A review. Agronomy for Sustainable Development. 2013;33:291-309.

[14] Soane BD, van Ouwerkerk C. The global problem of soil compaction in crop production and its implications for the quality of the environment. In: Proceedings of the International Conference on Protection of the Soil Environment by Avoidance of Compaction and Proper Soil Tillage; 23-27 August 1993; Melitopol. Ukraine: Vol. 2; 1993. p. 3-8.

[15] Horn R, Doma H, Sowiska-Jurkiewicz A, Van Ouwerkerk C. Soil compaction processes and their effects on the structure of arable soils and the

environment. Soil and Tillage Research. 1995;35:23-36.

[16] Ishaq M, Hassan A, Saeed M, Ibrahim M, Lal R. Subsoil compaction effects on crops in Punjab. Pakistan I. Soil physical properties and crop yield. Soil and Tillage Research. 2001;59: 57-65.

[17] Saqib M, Akhtar J, Qureshi R. Pot study on wheat growth in saline and waterlogged compacted soil I. Grain yield and yield components. Soil and Tillage Research. 2004;77:169-177.

[18] Hamza M, Anderson W. Soil compaction in cropping systems A review of the nature, causes and possible solutions. Soil and Tillage Research. 2005;82:121-145.

[19] Batey T, McKenzie DC. Letter to the editor 'Deep subsoil compaction'. Soil Use and Management, 1999;15:136.

[20] Batey T, McKenzie DC. Soil compaction: identification directly in the field. Soil Use and Management. 2006;22(4):414-414.

[21] Tekin Y, Kul B, Okursoy R. Sensing and 3D Mapping of Soil Compaction. Sensors (Basel). 2008;8(5):3447-3459.

[22] Ziyaee A, Roshan MRA. Survey study on soil compaction problems for new methods in agriculture. International Research Journal of Applied and Basic Sciences. 2012;3(9): 1787-1801.

[23] Bilbro JD, Wanjura DF. Soil crusts and cotton emergence relationships. Transactions of the ASAE. 1982;25(6):1484-1487.

[24] Raper RL, Schwab EB, Dabney SM. Measurements and variation of site-specific hardpans for silty upland soils in the Southeastern United States. Soil and Tillage Research. 2005;84:7-17.

[25] Biriş SŞt, Ungureanu N, Cujbescu D. Modelling of soil compaction under heavy duty tractors. INMATEH – Agricultural Engineering. 2019;57(1): 261-270.

[26] Silva S, Barros N, Costa L, Leite F. Soil compaction and eucalyptus growth in response to forwarder traffic intensity and load. The Revista Brasileira de Ciência do Solo. 2008;32:921-932.

[27] Sakai H, Nordfjell T, Suadicani K, Talbot B, Bøllehuus E. Soil compaction on forest soils from different kinds of tires and tracks and possibility of accurate estimate. Croatian Journal of Forest Engineering. 2008;29:15-27.

[28] Jimenez KJ, Rolim MM, Gomes IF, de Lima RP, Berrío LLA, Ortiz PFS. Numerical analysis applied to the study of soil stress and compaction due to mechanised sugarcane harvest. Soil and Tillage Research. 2021;206:104847.

[29] Botta GF, Becerra T, Tourn B. Effect of the number of tractor passes on soil rut depth and compaction in two tillage regimes. Soil and Tillage Research. 2009;103:381-386.

[30] Smith H, Dickson L. The contributions of vehicle weight and ground pressure to soil compaction. Journal of Agricultural Engineering. 1992;46:13-29.

[31] Raper L, Bailey A, Burt E, Way T, Liberati P. Inflation pressure and dynamic load effects on soil deformation and soil tyre interface stress. Transactions of ASAE. 1994;38(3):685-689.

[32] Håkansson I, Reeder RC. Subsoil compaction by vehicles with high axle load—extent, persistence and crop response. Soil and Tillage Research. 1994;29(2-3):277-304

[33] DeJong-Hughes J, Moncrief JF. Voorhees WB, Swan JB. Soil

Compaction: Causes, Effects and Control [Internet]. 2001. Available from: https://conservancy.umn.edu/handle/11299/55483 [Accessed: 2021-05-08]

[34] Da Silva RB, Iori Pi, de Souza ZM, Pereira DMG, Filho OJV, de Melo Silva FA. Contact pressures and the impact of farm equipment on Latosol with the presence and absence of sugarcane straw. Ciência e Agrotecnologia. 2016;40(3):265-278.

[35] Porterfield JW, Carpenter TG. Soil compaction: an index of potential compaction for agricultural tires. Transactions of ASAE. 1986;29: 917-922.

[36] Chehaibi S, Khelifi M, Boujelban A, Abrougui K. Effects of tire inflation pressure and field traffic on compaction of a sandy clay soil as measured by cone index and permeability. Canadian Biosystems Engineering. 2012;54: 2.9-2.15.

[37] Zhang XY, Cruse RM, Sui YY, Jhao Z. Soil compaction ınduced by small tractor traffic in Northeast China. Soil Science Society of America Journal. 2006;70:613-619.

[38] Rusanov VA. USSR standards for agricultural mobile machinery: permissible influences on soils and methods to estimate contact pressure and stress at a depth of 0.5 m. Soil and Tillage Research. 1994;29:249-252.

[39] Botta G, Jorajuria D, Rosatto H. Light tractor frequency on soil compaction in the rolling Pampa region of Argentina. Soil and Tillage Research. 2006;86:14-19.

[40] Keller T, Sandin M, Colombi T, Horn R, Or D. Historical increase in agricultural machinery weights enhanced soil stress levels and adversely affected soil functioning. Soil and Tillage Research. 2019;194:104293

[41] Gysi M, Ott A, Flühler H. Influence of single passes with high wheel load on a structured, unploughed sandy loam soil. Soil and Tillage Research. 1999;52:141-151.

[42] Horn R, Lebert M. Soil compactability and compressibility. In: Soane BD, van Ouwerkerk C, editors. Soil Compaction in Crop Production. Amsterdam: Elsevier; 1994. p. 45-69.

[43] Baumgartl Th, Horn R. Effect of aggregate stability on soil compaction. Soil and Tillage Research. 1991;19:203-213

[44] Ellies Sch A, Smith RR, Jose Dorner FJ, Proschle TA. Effect of moisture and transit frequency on stress distribution on different soils. Agro Sur. 2000;28:60-68.

[45] Ekwue EI, Stone RJ. Organic matter effects on the strength properties of compacted agricultural soils. Transaction of the ASAE. 1995;38(2): 358-365.

[46] Barzegar AR, Assodar MA, Ansari M. Effectiveness of sugarcane residue incorporation at different water contents and the Proctor compaction loads in reducing soil compactilibility. Soil and Tillage Research. 2000;57:167-172.

[47] Mosaddeghi MR, Hajabbasi MA, Hemmat A, Afyuni M. Soil compactibility as affected by soil moisture content and farmyard manure in central Iran. Soil and Tillage Research. 2000. 55, 87-97.

[48] Balbuena R, MacDonagh P, Marquina J, Jorajuria D, Terminiello A, Claverie J. Wheel traffic influence on poplar regeneration and grass yield. Biosystems Engineering. 2002;81: 379-384.

[49] Hargreaves PR, Baker KL, Graceson A, Bonnett S, Ball BC,

Cloy JM. Soil compaction effects on grassland silage yields and soil structure under different levels of compaction over three years. European Journal of Agronomy. 2019;109:125916.

[50] Sivarajan S, Maharlooei M, Bajwa SG, Nowatzki J. Impact of soil compaction due to wheel traffic on corn and soybean growth, development and yield. Soil and Tillage Research. 2018;175:234-243.

[51] Patel SK, Mani I, Sundaram PK. Effect of subsoil compaction on rooting behavior and yields of wheat. Journal of Terramechanics. 2020;92:43-50.

[52] Pulido-Moncada M, Munkholm LJ, Schjønning P. Wheel load, repeated wheeling, and traction effects on subsoil compaction in northern Europe. Soil and Tillage Research. 2019;186: 300-309.

[53] Arvidsson J. Subsoil compaction caused by heavy sugarbeet harvesters in southern Sweden. I. Soil physical properties and crop yield in six field experiments. Soil and Tillage Research. 2001;60:67-78.

[54] Hassan FU, Ahmad M, Ahmad N, Kaleem Abbasi M. Effects of subsoil compaction on yield and yield attributes of wheat in the sub-humid region of Pakistan. Soil and Tillage Research. 2007;96(1-2):361-366.

[55] Abu-Hamdeh NH. Effect of compaction and deep tillage on soil hydraulic and aeration properties and wheat yield. Communications in Soil Science and Plant Analysis. 2003;34:2277-2290.

[56] de Moraes MT, Debiasi H , Franchini JC, Mastroberti AA, Levien R, Leitner D, Schnepf A. Soil compaction impacts soybean root growth in an Oxisol from subtropical Brazil. Soil and Tillage Research. 2020;200:104611.

[57] Botta GF, Tolón-Becerra A, Rivero D, Laureda D, Ramírez-Roman M, Lastra-Bravo X, Agnes D, Flores-Parra IM, Pelizzari F, Martiren V. Compactión produced by combine harvest traffic: Effect on soil and soybean (*Glycine max* l.) yields under direct sowing in Argentinean Pampas. European Journal of Agronomy. 2016;74:155-163.

[58] Lipiec J, Hatano R. Quantification of compaction effects on soil physical properties and crop growth. Geoderma. 2003;116:107-136.

[59] Gürsoy S, Türk Z. Effects of land rolling on soil properties and plant growth in chickpea production. Soil and Tillage Research. 2019:195;104425.

[60] United Nations. Threats to soils: global trends and perspectives. A Contribution from the Intergovern mental Technical Panel on Soils, Global Soil Partnership Food and Agriculture Organization of the United Nations [Internet]. 2017. Available from: https://knowledge.unccd.int/sites/default/files/201806/17.%20Threats%2Bto%2BSoils__Pierzynski_Brajendra.pdf [Accessed: 2021-05-08]

[61] Willatt ST. Root growth of winter barley in a soil compacted by the passage of tractors. Soil and Tillage Research. 1986;7:41-50.

[62] Dexter AR. Soil physical quality - Part I. Theory, effects of soil texture, density, and organic matter, and effects on root growth. Geoderma, 2004;120:201-214.

[63] Bengough AG, McKenzie BM, Hallett PD, Valentine TA. Root elongation, water stress, and mechanical impedance: a review of limiting stresses and beneficial root tip traits. Journal of Experimental Botany. 2011;62:59-68

[64] Schjønning P, Lamandé M, Munkholm LJ, Lyngvig HS, Nielsen JA.

Soil precompression stress, penetration resistance and crop yields in relation to differently-trafficked, temperate-region sandy loam soils. Soil and Tillage Research. 2016;163:298-308.

[65] Colombi T, Keller T. Developing strategies to recover crop productivity after soil compaction – a plant eco-physiological perspective. Soil and Tillage Research. 2019;191:156-161.

[66] Ishaq M, Ibrahim M, Hassan A, Saeed M, Lal R. Subsoil compaction effects on crops in Punjab, Pakistan: II. Root growth and nutrient uptake of wheat and sorghum. Soil and Tillage Research. 2001;60(3-4):153-161.

[67] Obour PB, Ugarte CM. A meta-analysis of the impact of traffic-induced compaction on soil physical properties and grain yield. Soil and Tillage Research. 2021;211:105019.

[68] Day SD, Bassuk NL. A review of the effects of soil compaction and amelioration treatments on landscape trees. Journal of Arboriculture. 1994;20(1):9-17.

[69] Chamen WCT, Alakukku L, Pires S, Sommer C, Spoor G, Tijink F, Weisskopf P. Prevention strategies for field traffic-induced subsoil compaction. Part 2. Equipment and field practices. Soil and Tillage Research. 2003;73:161-174.

[70] ten Damme L, Stettler M, Pinet F, Vervaet P, Kellerd T, Munkholm LJ, Lamandé M. The contribution of tyre evolution to the reduction of soil compaction risks. Soil and Tillage Research. 2019;194:104283

[71] Van den Akker, JJH. SOCOMO: A soil compaction model to calculate soil stresses and the subsoil carrying capacity. Soil and Tillage Research. 2004;79(1):113-127.

[72] Kumar V, Butter TS, Samanta A, Singh G, Kumar M, Dhotra B, Yadav NK, Choudhary RS. Soil compaction and their management in farming systems: A review. International Journal of Chemical Studies. 2018;6(3):2302-2313.

[73] Tullberg JN, Yule DF, McGarry D. Controlled traffic farming – from research to adoption in Australia. Soil and Tillage Research. 2007;97:272-281.

[74] Thomas GW, Haszler GR, Blevins RL. The effects of organic matter and tillage on maximum compactability of soils using the Proctor test. Soil Science. 1996;161:502- 508.

[75] Raper RL, Reeves DW, Burt EC. Using in-row subsoiling to minimize soil compaction caused by traffic. Journal of Cotton Science. 1998;2:130-135.

Chapter 6

Potentials of Wood, Bamboo and Natural Fibre-Reinforced Composite Products as Substitute Materials for Fabricating Affordable Agricultural Equipment and Processing Machines in Africa

Abel Olajide Olorunnisola

Abstract

Modern agriculture depends heavily on technology. Land clearing, irrigation, drainage, crop storage and processing all require technological input. By modernising her agriculture, through wise application of science and technology, Africa can make significant headway in economic growth. However, an agricultural technology that is too sophisticated for a particular country/region is beyond its absorptive capacity. Hence, to achieve the objectives of agricultural mechanisation in Africa, it is imperative to take into account prevailing socio-economic conditions and the level of mechanisation necessary for optimal productivity. One major constraint to agricultural mechanisation in sub-Saharan Africa is the relatively high cost of imported metallic machine and equipment fabrication materials. Taking full advantage of substitute non-metallic materials may lower the cost of production and concomitantly empower rural fabricators with limited access to electricity and welding facilities to engage in local manufacturing of sundry agricultural machines and equipment. This Chapter presents illustrative examples of full and partial substitution of metallic with non-metallic materials in the fabrication of affordable machines and equipment for agricultural production, agro-processing, irrigation and drainage, crop drying and storage. Ways of addressing identified critical challenges of technology diffusion are also discussed.

Keywords: Wood, Bamboo, Fibre-reinforced composites, Agricultural machinery, Modern agriculture

1. Introduction

Africa covers about 6% of the earth's total surface and about 20.4% of the land area. It is the second most populous continent after Asia constituting around 14.72% of the world's total population. All the countries in Africa can at best be referred to as developing countries, with the exception of South Africa, Egypt,

Tunisia, Libya and Algeria that take the lead in the stated order as the top five developed countries on the continent. A developing country or less-developed country (LDC) is, in this context, defined as a nation with a low standard of living, underdeveloped industrial base as contrasted with a "More Developed Country" (MDC) that has more a highly developed economy and advanced technological infrastructure [1].

The main engine of economic growth is agriculture. Agricultural development led to the rise of human civilization and a rapidly modernising agriculture produces food without which an economy cannot possibly grow. It also creates demands for many new industries, from fertilisers to farm equipment, from repair shops to farm credit, from transportation and roads to food processing. In the last fifty years, technological developments in agriculture have dramatically changed the performance of farming. For instance, irrigation and drainage have ensured the use of otherwise unusable land for agriculture, while mechanisation has minimised drudgery, improved productivity and decreased farm labour requirement significantly in the more developed countries.

Agriculture has a massive social and economic footprint in Africa where it is practised both for subsistence as well as commercial reasons. It is by far the single most important economic activity, providing employment for about two-thirds of the continent's working population, contributing an average of 30–60% of gross domestic product and about 30% of the value of exports for many African countries. In the last 30 years, Africa's population has doubled overall and tripled in urban areas. The most direct consequence of this exponential population growth rate is that the continent now has more mouths to feed. However, moving from being self-sufficient in the 1960s, Africa has become a net importer of many food items. Indeed, African agricultural exports have fallen by half since the mid-1990s with imports accounting for 1.7 times the value of exports. This is partly due to the fact that over 80% of all farms in Africa are rather small which makes large-scale mechanisation unrealistic [2, 3].

For the foreseeable future, heavy dependence on agriculture is likely to continue being the norm rather than the exception in sub-Saharan Africa (SSA). By modernising her agriculture through wise application of science and technology, the region can make significant headway in economic growth. The modernised agricultural sector can contribute towards major regional and continental priorities, including poverty alleviation, a boost in intra-African trade and investments, rapid industrialisation and economic diversification, sustainable resource and environmental management, job creation, human security, and shared prosperity. The region in particular and the continent in general can thus fulfil the enormous potential of becoming a major player in the global food market.

It is obvious that mechanisation is a necessity in modernising agriculture. To mechanise means to use machines to accomplish tasks, reduce human efforts, and improve timelines and quality of various farm operations. A machine may be as simple as a wedge or as complex as a combine harvester. Hence, there are presently three levels of mechanisation distinguished in agricultural engineering literature: *Hand Tool Technology* (HTT), *Draught Animal Technology* (DAT), and *Engine Power Technology* (EPT) [4]. Agricultural mechanisation cannot, therefore, be restricted to the use of motorised equipment alone. Rather, the term covers the development, maintenance, repair, management, and utilisation of agricultural hand tools, implements, and machines and also applies to agricultural land development, crop production, water control, harvesting, material handling and preparation for storage, on-farm processing and rural transport.

Attempts at agricultural mechanisation in many African countries over the years seem to have failed because of the false notion that mechanisation only implies tractorisation- the use of tractors in agricultural activities which is a mere aspect and level of mechanisation. The performance history of imported tractors and other agricultural machinery has been largely characterised by a chronic chorus of lack of spare parts, repair facilities, capital, skilled operators and mechanics, as well as incompatibility with fragile local soil conditions and farmers' cropping techniques. Adoption of '*appropriate selective mechanisation*', a situation in which HTT, DAT AND EPT are appropriately combined, has therefore been recommended. For example, ploughing may be done with an appropriate type of tractor or draught animal power, while seeding, weeding and harvesting are accomplished with hand or simple mechanical tools [5]. Appropriate selective mechanisation is considered a workable strategy because multiple factors play a role in mechanisation patterns adopted in different countries over time. These include farming systems; agro-climatic conditions, such as soil, terrain and rainfall; institutional environments; and social objectives of societies, such as nation building and modernisation. Since each of these factors differs within and between countries, it is difficult to specify a blueprint of technological change, for all African countries to follow [6]. Other scholars have also recommended the promotion of (i) the use of locally available materials as substitution to high-carbon steel in the manufacture of agricultural machines and equipment and (ii) village level manufacturing of affordable agricultural machines and equipment by blacksmiths, tinsmiths and carpenters [4, 7–10].

In implementing some of the foregoing recommendations, numerous researchers have, over the years, engaged in innovative research culminating in the development of different agricultural equipment and machines.The aim of this chapter, therefore, is to highlight some of the research outputs on the uses of wood, bamboo and natural fibre-reinforced composites as substitute materials in the local fabrication of affordable agricultural machines and equipment in sub-Saharan Africa. However, research results, inventions and innovations have value only when they serve useful purposes in the society. The diffusion of technological improvements, within a country and across international borders, is critical for long run growth. It is most unfortunate that most of the equipment and machines to be discussed are yet to be in common use. Hence, this chapter also discusses ways of addressing the critical challenges of technology diffusion Africa.

2. Rationale for the use of agro-forestry materials and fibre-reinforced composites in agricultural machine and equipment fabrication

The benefits of using agro-forestry materials and fibre-reinforced composites for agricultural machine and equipment fabrication are many and varied. They include the following:

Wood Products: The major wood products used in machine and equipment fabrication include lumber, plywood, particleboard and fibreboard. Wood remains one of the most versatile materials whose natural structure can be retained, as in lumber and plywood, or can be reduced to its basic fibres and reconstituted to a more uniform product such as fibreboard and particleboard. Besides, wood is renewable, available in various sizes, shapes and colours, affordable, easy to machine and join, durable (depending on the species), and aesthetically appealing [10, 11]. Other advantages derivable from the use of wood and wood products in machine and equipment fabrication include easy replacement of damaged machine/equipment parts and reduction in the weight to enhance portability.

Bamboos: Bamboos grow and reach maturity more rapidly than trees and start to yield within three or four years of planting. Unlike most timbers, bamboo is self-regenerating; new shoots that appear annually ensure future raw material after mature culms are harvested. The ease with which bamboo can be worked, its versatility, strength, and availability recommend it for industrial utilisation. Besides, laminated bamboo products do not retain the characteristic shapes of the bamboo raw material, thus offering more versatility to the machine/equipment designer [12]. The cylindrical hollow structure of many bamboo species with the rigid cross walls gives it resistance to collapse from bending. The jointed culm typically has a very hard external surface which contributes to its strength and impermeability to water - characteristics that satisfy many of the requirements of irrigation and drainage pipes.

Cement-bonded fibre-reinforced composites: These are low-cost materials made from a mixture of cement, water, particles of different sizes (strands, flakes, chips, fibres) obtained from agricultural and forestry products. The incorporation of fibrous materials in the composite improves the fracture toughness of the cement [10, 13]. Some of the admirable properties of cement-bonded composites of significant advantage in equipment fabrication include relatively high strength to weight ratio, durability; high resistance to moisture uptake; ease of sawing; excellent insulation against noise and heat; ability to absorb and dissipate mechanical energy, and high resistance against fire, insect and fungus attack [14]. Being environment friendly, natural fibre-reinforced composite pipes are beginning to attract great attention as substitutes to synthetic fibre-reinforced composite irrigation pipes which are difficult to recycle after their designed service life.

3. Selective examples on the of the use of substitute non-metallic materials in agricultural machine and equipment fabrication

Some of the various aspects of agricultural production in which the foregoing materials have been experimentally used for fabricating affordable machines and equipment are the following:

3.1 Tillage and crop production equipment

Tillage and crop production activities performed on a farm include ploughing, harrowing, seed bed preparation, cultivation, weeding, and harvesting. All these activities require power sources, which on large scale farms are derived from tractors of different sizes. However, the average level of tractorisation in SSA is about 28 tractors per 1 000 ha in contrast to 241 tractors per 1000 ha in other regions [3]. While relevant data are scarce and at times out-dated, there is a general consensus that the level of tractorisation in particular and farm mechanisation in general is still very low in Africa. This is in spite of the efforts made by governments in many SSA countries over the years to promote tractorisation in particular and farm mechanisation in general. Such recent interventions include importation and provision of tractors and farm machinery at subsidised rates to farmers as shown in **Table 1**, and setting up state- owned tractor assembly plants and tractor hiring schemes, e.g., the *Nigerian Tractor Hiring Units*, the *Ghanaian Agricultural Mechanisation Service Centres* and the *Mozambican Agricultural Service Centres* [15]. Again, as earlier mentioned, the failure of many of these interventions is largely attributable to a strong focus on machinery importation and the neglect of knowledge and skills development at the local level, among other factors. Hence, the principal power source for 50 to 80% of the land area under cultivation in the region is still human power.

For centuries, wood has played a prominent role in land clearing, tillage and crop production equipment manufacture in SSA, where it is used as handles for hand tools such as hoes, axes and cutlasses (**Figure 1**). Over the years, a number of other

Country	Time period	Number of machines imported	Number of persons trained in 2017 on mechanisation in regular or project-funded programs
Mali	2016–2018	1500 tractors + implements as well as water pumps, motocultors, threshers, dehullers	None
Nigeria	2010–2018	950 tractors +150 implements as well as groundnut and melon threshers, water pumps, power tillers	560
Kenya	2016–2020	Unspecified number of tractors, implements and other machinery worth US$ 100 million	Unspecified number of persons trained on ad hoc basis (no regular training courses)
Benin	2008–2018	1040 tractors as well as 360 rotor tillers	100
Burkina Faso	2015–2018	800 tractors + implements	300

Source: [6].

Table 1.
State-led mechanisation and training in selected countries in sub-Saharan Africa.

Figure 1.
Hoes, cutlasses and axes fabricated with wooden handles.

simple tillage tools made up of metals and wooden handles have been locally developed. Examples include weeders with wooden boards fitted with sharp metal blades, harrows (made of wooden plank to which wood/iron pegs, handle and bamboo shaft are fitted, typically used for breaking soil crust after rain and also for uprooting weeds), mallots (wooden blocks with attached handles, used for the breaking of clods), and levellers with shafts generally made of bamboo sticks used for land levelling. These and numerous other hand tools are typically inexpensive, easy to manufacture, use, maintain and repair. Besides, they are often times multi-purpose tools employed in several crop production operations and are culturally accepted.

Hand hoes, in particular, are still extensively used for land clearing, ridging, weeding and root crop harvesting across sub-Saharan Africa. It has been noted that '*the peasant farmer and his hoe and cutlass are efficient companions in crop production at the subsistence level where he operates*' [4]. At such subsistence level, the farm sizes are usually about 1.0–3.0 hectares, the farmer's income is typically low, and the farmers practice intercropping which discourages the use of tractors but encourages

Figure 2.
A planter in which wood is used for fabricating selected machine parts.

the use of hand tools, which can reduce drudgery, if not area of cultivated land. Researchers have worked on improving the performance efficiency of hand hoes, focusing attention on the angle of inclination of the metal blade (the soil shearing member) to the handle, length of the handle and the weight ratio between the handle and the blade. These efforts have shown potential improvements in its scooping efficiency and field capacity [2, 5–7]. Introducing improved hand hoes can, therefore, be of tremendous benefit to peasant farmers.

However, wood is also relevant in the production of modern crop production equipment, especially, seed planters. Metering mechanism is the heart of every planter. Its function is to distribute seeds uniformly at the desired application rate and control seed spacing in a row. Wood products including lumber and plywood are suitable and very highly recommended for the fabrication of not only metering mechanisms but also handles, hoppers of manually operated seed planters. **Figure 2** shows a planter in which lumber and plywood were used for the fabrication of the handle and the metering device with the associated advantages of portability, low cost, and ease of fabrication, while **Figure 3** shows a maize planter with a wooden roller type seed metering device. A rough estimation showed that substitution of steel with wood in the fabrication of the component parts of the planters shown in **Figures 2** and **3** could reduce the cost of fabrication (i.e., material and labour costs) by 30–45%.

3.2 Crop processing machines

Africa produces numerous crops including legumes and cereals. Groundnut is one of such important leguminous cash crops and a major raw material for several industries especially in the food processing and poultry sectors. It is also processed at small- to-medium scale levels for domestic consumption as snacks in roasted and fried forms. Traditionally, groundnut shelling is a manual operation, a slow process with a maximum throughput per person of 2 Kg/hr. and a shelling capacity 15–20 Kg/day [17]. Groundnut shellers are typically fabricated using steel [18]. In a departure from this norm, a wood-steel manually operated groundnut sheller was developed. The hopper, main frame and collection tray of

Figure 3.
Component parts of the maize planter. Source: [16].

the sheller were fabricated with the sawn wood of *Cordia platythrsa*. The shelling unit was made up of a combination of a lumber casing, metallic pipe and wooden rasp bars, while the turning handle was made of a hollow steel pipe [19]. Wood was selected for fabricating the casing for the shelling unit in particular because of its exceptionally good acoustic properties, i.e., wood absorbs large amounts sound energy before it resonates. Hence, minimal noise would be experienced during machine operation. When tested, the sheller (**Figure 4**) gave a maximum

Figure 4.
Wooden/metallic Sheller. Source: [19].

throughput capacity of 11 Kg/hr., more than five times the shelling capacity of a person, shelling efficiency of 98.6% and kernel damage of 16.6%.

To upgrade the sheller performance, an improved version (**Figure 5**) was developed that incorporated a cleaning device for separating shelled nuts from the chaffs [20]. The cleaner had a wooden housing and a steel sieve cleaner that gave a maximum cleaning efficiency of 84%. In another improvement of the sheller (**Figure 6**), wood products were used in fabricating all the machine component parts [21]. It is apposite to note that the costs of fabricating one-off units of the three versions of the groundnut sheller decreased with increase in metal substitution with wood from approximately 55% (for version one) to 35% (for version two) of the cost of producing an equivalent metallic sheller.

Another major crop produced and consumed in Africa is maize (corn). *Maize* occupies approximately 24% of farmland in *Africa* and the average *yield* is around 2 tons/hectare/year. The largest *African* producer is Nigeria with over 33 million tons, followed by South *Africa*, Egypt, and Ethiopia [22]. The steps involved in maize processing include harvesting, drying, de-husking, shelling, and (often times) milling. Many of these processes, particularly shelling which is the separation of the grains from the cobs, are labour-intensive and time-consuming for rural farmers who engage in hand shelling. The easiest hand shelling method is to press the thumbs on the grains in order to detach them from the ears. Another simple shelling method is to rub two ears of maize against each other. Other methods include beating with stick, crushing with mortar and pestle, et c. Small tools are also sometimes used.. A worker can hand-shell about 2 kg of maize per hour [19, 23]. With the use of hand tools, the output per worker increases to between 8 and 15 kg/hr. [24].

Figure 5.
Modified wooden/metallic Sheller. Source: [20].

Figure 6.
A completely wooden groundnut Sheller. Source: [21].

Figure 7.
Wood-metal hybrid maize Sheller. Source: [27].

Different types of mechanical maize shellers are in existence in forms of handheld, portable, motorised and large commercial sized units and are almost invariably fabricated using metals including mild steel, stainless steel and cast iron for the various components [24–26]. However, a low-cost wood-steel hybrid motorised maize sheller shown in **Figure 7** was developed [27]. The wooden components of the sheller fabricated with the sawn wood of *Cordia millenni* included the hopper, the main frame, the shelling drum housing, and the cleaning unit housing. The cost of a one-off version of the sheller, driven by a 5 hp. electric motor was US $150 compared to the market price of about US $400 - US $500 for the metallic version in Nigeria. The performance of the sheller was evaluated using yellow maize (*Zea saccharata*) variety at 13% moisture content. Its output capacity (118.9 Kg/hr), shelling efficiency, cleaning efficiency, grain recovery, and total grain losses of about 79%, 96%, 91% and 2.8% respectively at a shelling speed of 536 rpm are comparable to similar shellers made entirely of metallic parts. The findings confirmed that wood is an acceptable and relatively cheaper substitute material in the fabrication of critical parts of a maize sheller.

Cassava is another major crop produced in many Sub-Saharan African countries, with Nigeria being the largest producer in the world. It is a consumed as a major source of carbohydrates in human diet and as starch for industrial applications. The tubers of cassava cannot be stored for long after harvest hence processing tends to follow immediately after harvesting. Cassava processing activities include peeling, grating (i.e., transformation of cassava tubers into pulp), dehydrating, milling and sieving. The traditional method of grating involves placing the grater, typically made of perforated metal sheet on the table where it is convenient for effective use and brushes sheet metal. The cassava turns into pulp and drops into container that is being used to collect the grated pulp cassava. It has been shown that cassava graters can be fabricated with wood products. For example, a cassava grater was fabricated with the use of hardwood for constructing the frame, grating chamber, hopper, grating roller and the outlet [28]. The grater had grating capacity of 102.9 kg/h and a grating efficiency of 90.91%. The cost saving associated with substituting metal with wooden component parts was estimated at about 30%.

3.3 Poultry production equipment

There are two options for poultry development in Africa. One option is to attempt to increase large scale intensive poultry production in order to respond to the urban demand. The other option is to explore new channels for developing small and medium scale semi-intensive poultry production to serve both the urban and rural populations. Where possible, the two options should be pursued simultaneously. One of the core aspects of poultry production is egg incubation- the management of fertilised eggs to ensure satisfactory development of embryos into normal chicks. For very small number of eggs, say 6–12, the easiest and usual way of hatching chicks is the natural method, whereby the broody hen sits on the nest to provide the required warmth. However, for larger quantities of eggs, the most cost effective practice is to use artificial incubators- closed heat-insulated chambers in which temperature and relative humidity are strictly monitored and controlled.

Most small-scale poultry farmers in sub-Saharan Africa still are unable to produce day old chicks by artificial incubation due to the relatively high cost of procuring imported incubators. To demonstrate the use of wood in poultry equipment production, a flat-type wooden incubator was developed (**Figure 8**) which was used to successfully hatch chicken eggs [29]. The component parts of the incubator included a cabinet, fabricated using 6.4 mm thick interior grade plywood for the floor, side walls and the lid; and the sawn wood of *Terminalia superba* for the beams and columns; a transparent glass inspection panel; improvised heaters and humidifiers required to achieve mean ambient incubator temperature of 37-39°C and relative humidity of 58%, vents, instrumentation and egg trays. The choice of wood for fabricating the various component parts was based not only on local availability and relatively low cost, but more importantly on effectiveness in performing the desired functions of insulation and structural stability.

Figure 8.
The loaded wooden incubator. Source: [29].

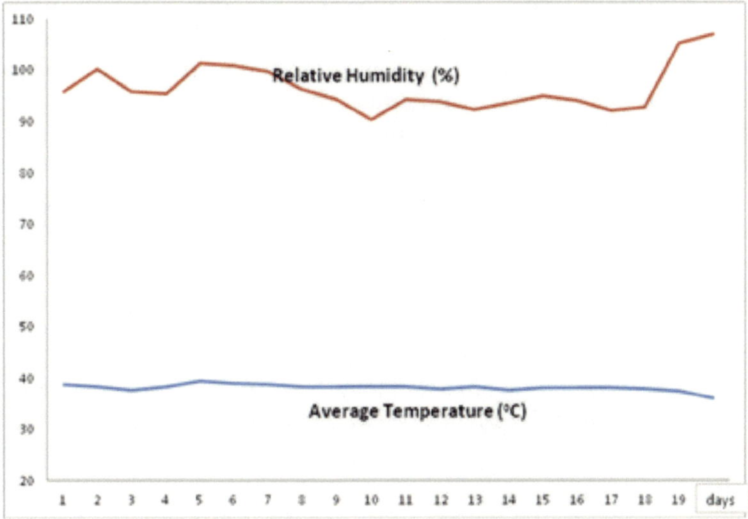

Figure 9.
Temperature and relative humidity variations in the wooden egg incubator. Source: [29].

When tested with 30 chicken eggs, the expected mean internal temperature and relative of 37°C and 58% respectively were achieved by the incubator. It was able to hatch chicks in 20 days with 76% hatchability and 18.5% mortality rates. The temperature and relative humidity variations observed during the incubator loading test are shown in **Figure 9**. The incubation duration was in conformity with observations of Ref. [30] who reported a range of 20 to 21 days for hatching chickens naturally and in artificial incubators. The percentage hatchability and mortality rates were also within acceptable limits. The current cost of producing the incubator is approximately N20,000–25,000 (in Nigerian currency), which is equivalent to US$ 40 – US$ 50.

3.4 Farm irrigation and drainage equipment

Crop production requires large quantities of water. African agriculture, characterised by low levels of productivity relative to population growth, and frequently accompanied by human induced degradation and drought presents special challenges not encountered in other regions. There is draught in many arid or semi-arid parts of Africa, many of which cannot support rain-fed agriculture and hence require irrigation.

Efficient use of water in African farms does not necessary require large scale, energy-intensive irrigation schemes. Small pumps have had an important beneficial effect on irrigation in small-scale farms in a number of African countries. Where surface water is available, this technology represents a well-distributed and energy efficient option. Also, in recent times, the concept of "affordable micro-irrigation" systems has been identified as a corresponding drip irrigation technology for low-income farmers [31, 32]. Bamboo drip irrigation system (**Figure 10**) is a very old but relevant system of tapping stream and spring water by using bamboo pipe and transporting water from higher to lower regions. The advantages of using bamboo are two-fold: it prevents leakage, increasing crop yield with less water, and makes use of natural, local, and inexpensive material.

It has now been scientifically proven that bamboo pipe can be used in both gravity and pressurised conditions for irrigation and drainage pipes provided the transmission pressure do not exceed 13.5×10^5, 13.1×10^5 and 12.9×10^5 N/m^2 for base, middle and top portions respectively. Values of head losses obtained are generally high for a bamboo pipe length of 6 meters. However, head losses can be reduced by proper node removal, increasing velocity of flow; and increasing the pressure head among others. It has also been recommended that node removal mechanism be improved upon so as to reduce the overall roughness size of the bamboo pipe [33].

Some of the hydraulic properties of *Oxytenanthera abyssinica* bamboo species have been investigated with a view to determining its potentials as irrigation piping material [34]. The properties tested included burst strength, head loss and friction factor. These properties were found to vary, some of them significantly, along the culm height. The bursting pressure was found to be about 13×10^5N/m^2, which is higher than that of PVC pipes (11.2×10^5 N/m^2) which are currently in common use as irrigation and drainage pipes. The mean friction factor, determined within the turbulent range (N_{Re} <2000) in a 6 m length, 2.6 mm internal diameter bamboo pipe discharging at about 5.3×10^{-4} m^3/s was 0.020 giving a mean head loss of

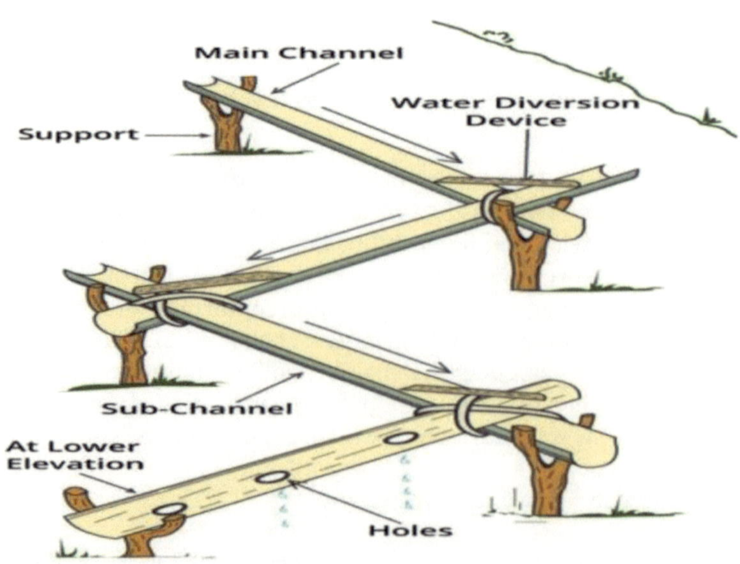

Figure 10.
Bamboo drip irrigation system.

Figure 11.
Bamboo winding composite pipes.

0.14 m/m. However, bamboo has a few limitations to its use in micro irrigation water piping. These include its non-straightness, roughness, existence of the nodal plates that make water flow naturally impossible, non-uniformity of the stalk diameter, jointing problems, strength and durability. These are problem areas requiring further studies.

In a recent development, bamboo winding composite pipe, i.e., ordinary bamboo cut into thin bamboo strips winded by machines and turned into a strong composite pipe, has been developed in China for water conveyance as a green alternative to traditional pipeline materials. Such pipes shown in **Figure 11** have characteristic lightweight, high axial tensile strength and good flexibility which make them suitable for application for urban water supply, farm irrigation and drainage, etc.

Cement-bonded composite pipes are another potential alternative. A study was reported on the possibility of using 6 mm and 8 mm thick cement-bonded sawdust-reinforced composite pipes for water conveyance [35]. The maximum burst strength of the composite pipes, 1.0×10^5 N/m^2, was, however, lower than those of polyvinyl chloride ($8.6–13.8 \times 10^5$ N/m^2) and aluminium pipes (13.8 and 32.4×10^5 N/m^2). The composite pipes shown in **Figure 12** were, therefore, recommended for use in low pressure water drainage.

Figure 12.
Cement-bonded composite pipes. Source: [35].

3.5 Drying equipment

Drying which is an important preservation process for many agricultural crops and food products. It is the phase of the post-harvest system during which an agricultural product is rapidly dried until it reaches the safe-moisture level to guarantee conditions favourable for storage or for further processing of the product. Traditional methods of drying as practised in many SSA countries include direct exposure of agricultural products to sun radiation by spreading the products on ground, polythene sheets, mats, tarred surfaces including roads, cement courts or hanging on eaves. These methods suffer from numerous inadequacies such as infestation by insects, contamination by dirt, loss through rodent attack, and spoilage due to exposure to rain, among others. Modern artificial dryers are generally relatively costly and un-affordable to small-scale farmers. The use indirect solar dryers, where practicable, would be comparatively cheaper and equally efficient.

In a typical indirect solar dryer, a black surface heats the incoming air instead of directly heating the substance to be dried. The heated air is then passed over the substance to be dried and exits upwards often through a chimney, taking moisture released from the substance with it. Indirect solar dryers can enhance the effect of insolation and minimise loss of collected energy to the surroundings. They can also generate higher temperatures and lower air relative humidity than in direct sun drying, both of which are conducive to improved drying rates and lower final moisture contents of dried products. This reduces the risk of spoilage during the drying process and in storage. The higher the temperatures attainable in these devices are deterrent to insects, and microbiological infestation. Also protection against dust, insects and animals are enhanced by drying in an enclosed structure. The use of lumber and other wood products for fabricating solar dryers has been explored by the author, culminating in the development of an indirect solar dryer for seeds, fruits and vegetables shown in **Figure 13**. It is an absorber-type collector device that comprises a double-walled wooden box with a double-glazed tight-fitting glass lid.

Figure 13.
A passive solar dryer for seeds, fruits and vegetables developed by the author.

The gap between the walls of the box was stuffed with dry sawdust. A soot-coated, metal plate was attached to the bottom of the box as the heat absorber.

Another type of solar device that can serve dual purpose of drying and cooking has been developed [36]. For the dying test, beef samples initially at 73% moisture content dried to 17% in 5 hours and cassava samples at 56% moisture content dried to 14% in 5 hours all in bright sunshine.

3.6 Grain storage equipment

Despite the fact that many African countries are blessed with arable land and suitable climate for the production of a myriad of food and cash crops, the major problems with food supply involve not only relatively low agricultural production but also considerable poor storage-induced post-harvest losses incurred in the food supply chain from the farm gate to the final consumer. The major food crops– maize, millet, sorghum, cowpea, et c- are seasonal and require storage if they will be available all year round or as seed until the next planting season. Grain losses of up to 50% have been reported in some sub-Saharan African countries where farmers store their farm produce in rhombus, local cribs, bags, pots, calabashes, baskets, or earthen pots [37, 38].

In reducing the considerable losses associated with traditional grain storage techniques, grain silos are indispensable. However, metallic silos, the most common type of silo employed today, are unsuitable for long-term grain storage in sub-Saharan Africa for several reasons including cost of acquisition and maintenance as well as over-sized capacities [38, 39]. Besides, metallic silos tend to promote moisture condensation, caking and insect infestation of stored grains, as well as the development of hot spots under the prevailing warm and humid climatic conditions in the region [40–42]. The heat flow into the silo may in some cases be sufficiently high to roast the grains directly in contact with the silo wall

Figure 14.
A 1.4 m^3 capacity wooden grain silo. Source [43].

surfaces [39]. To address the afore-mentioned challenges of affordability, inappropriate capacity and material suitability, several interventions have been made by various researchers. For example, a double-walled metallic silo was developed using wood sawdust as insulating material which lowered the interior temperature of the silo [43]. In another series of interventions, 1.4–7 m^3 capacity grain silos that are more efficient in reducing moisture condensation and hot spots and suitable for small-and medium-scale farmers were fabricated with wooden beams and columns and plywood sheathing [43, 44].

An example of the grain silo, shown in **Figure 14**, and erected in Minna, Niger State, Nigeria, retained its structural integrity after four years of erection except for mild peelings of sheathing materials, nail slip and colour change [43]. Nutritional quality of the maize (*Zea mays*) stored in the silo for a period of nine months was also preserved with minimal reduction in crude protein, crude fibre and lipid contents. The use of wood products in grain silo construction has the potential of reducing construction cost by at least 40%. The simplicity of construction and maintenance, and the possibility of small unit capacity recommend such wooden silos for small- and medium-scale farmers in Africa.

4. Addressing the factors militating against agricultural technology diffusion in Africa

The factors responsible for non-adoption of a wide variety of innovative agricultural technologies such as those discussed above are many and varied. Some of the factors and ways of mitigating them are highlighted below:

- **Lack of standardisation**: It is common knowledge that standardisation is a viable tool for achieving high quality, reliability and international competitiveness. However, Africa is still lagging behind in the development of standards for locally developed agricultural equipment and machines. To facilitate mass production and adoption, industry, national and regional standards must be promoted to codify terminologies, basic production techniques and safety rules, dimensions and permissible tolerances, functional requirements and test methods.

- **Dearth of engineering extension workers:** There is a dearth of engineering extension workers in Africa, thereby hampering technology transfer to the society. Hence, many farmers are not aware of the research and development outcomes. A few research institutions and non-governmental organisations have blazed the trail through circulation of research briefs, regular exhibitions, as well as newspaper, radio and television advertisements to promote public awareness about new inventions and innovations. The radio and the television, in particular, are major strategic instruments for extension services. Mounting educational programmes on radio and television is a proven means of reaching out to adults in both rural and urban settings.

- **Illiteracy**: The literacy level in many African societies is generally low. This makes technology diffusion rather difficult. Rapid technology diffusion requires eradication of illiteracy among the adults engaged at various levels in the agricultural production value chain who may be rather too old to participate in the conventional schooling system. One way of doing this is by promoting adult literacy programmes to facilitate the adoption of modern agricultural technologies.

- **Institutional and policy barriers:** In many countries, institutional and policy barriers limit the adoption of technically superior technologies. There has to be a radical departure from the conventional approach which focuses on household characteristics that predispose farmers to technology adoption without taking into account the value chain level or institutional factors and national or regional policy issues that hinder technology adoption, entrepreneurship and commercialisation [45].

- **Sociocultural and Lifestyle factors**: Adoption of new technologies is often limited because interventions fail to adequately account for sociocultural and lifestyle factors that are important to farmers. While they share some common characteristics, of which size is the most obvious, not all smallholder farmers are the same There is considerable diversity across and within regions and countries with regard to their backgrounds, histories and environments. Farms also have different assets, needs and objectives. Furthermore, biophysical, institutional, social and economic drivers also differ between contexts, resulting in different responses from farmers. Hence, the conceptualisation of small farms in theory and policy with emphasis on the dichotomy between market orientation and subsistence has to be modified in line with current realities. It should be understood that some of the small-scale farmers that engage in a high level of self-provisioning do so not necessarily out of poverty but as a lifestyle choice; some small farmers are successful commercial entrepreneurs, while for others farming is a household coping strategy to reduce the risk and poverty [46].

- **Financial Constraints**: The foregoing point notwithstanding, in many instances, small older farmers lack the resources and sometimes the incentives to adopt new technologies. Despite the availability and validation of various agricultural technologies, financial constraint continues to be a major cause of low adoption of new technologies by numerous small-scale farmers in SSA. Subsidising the procurement of selective appropriate technologies is a way out.

- **Dearth of Entrepreneurs**: Another key ingredient in the transformation of innovative ideas to innovative products is entrepreneurship. However, there are very few individuals, organisations, strategic investors and venture capitalists willing to invest in the commercialisation of the numerous agricultural technologies developed in Africa. One major way of promoting entrepreneurship is by the establishing business incubation programmes and facilities. Business incubation has been proven to be a dynamic process of business enterprise development for the purpose of nurturing young firms, new products and technologies as well as helping SMEs upgrade and undergo transformation [47].

5. Conclusion

Africa has all it takes to develop its agriculture and achieve self-sufficiency in food and agro-industrial raw material production. The missing link in sub-Saharan Africa in particular is the non-adoption of appropriate technologies that are compatible with multiple farming systems in use, the environment, socioeconomic status of the proposed users, level of maintenance skill available, as well as construction materials and facilities available for fabrication. Many African countries will for some time to come continue to have a mixture of small, medium and large scale farmers, with perhaps small-scale farmers constituting the majority. While prices have dropped due to the entrance of companies from China and India in the African market, a wide range of agricultural machinery still remains expensive relative to the incomes earned in African agriculture [6].

The use of locally available non-metallic materials as partial or full substitution in manufacturing is, therefore, advocated as a means of making available to small- and medium- scale farmers, suitable and affordable machines for crop production, irrigation systems to supplement natural rainfall, drying and storage facilities to minimise post-harvest losses. This is imperative because smallholder farmers, in particular, are not only key to agricultural development in Africa, excluding them from mechanisation would also result in unequal land and wealth distribution [6, 48, 49]. It is instructive to note also that while '*small farms are typical of the rural landscape in the Global South, small-scale farming continues to exist –and even thrive—in the Global North, including Europe. Small farms are crucial for global food security, producing between 50% and 75% of food calories consumed globally. Small-scale farming also provides key opportunities for employment and livelihoods, is a crucial part of rural communities and landscapes and plays an important role in environmental sustainability and supporting agricultural biodiversity*' [46]. To promote widespread adoption of appropriate agricultural mechanisation, the critical factors militating against technology diffusion have to be addressed. Technology development has to be participatory and coupled with extension efforts that recognise agro-ecological and socio-economic contexts and incorporates knowledge from various sources (e.g., sociologists, economists, historians, etc), rather than from scientists or researchers alone. An enabling environment that supports and/or rewards technology adoption by farmers is also an important prerequisite for success.

Conflict of interest

The author declares no conflict of interest.

Author details

Abel Olajide Olorunnisola
University of Ibadan, Ibadan, Nigeria

*Address all correspondence to: abelolorunnisola@yahoo.com; ao.olorunnisola@ui.edu.ng

References

[1] Anonymous. Africa [Internet]. 2021. Available from: https://en.wikipedia.org/wiki/Africa, [Accessed: 2021-01-02]

[2] Okatahi S.S, Yusuf D.D.: Development and performance evaluation of selected tillage tools for farmers in the middle belt zone of Nigeria. In: Proceedings of the Nigerian Society of Agricultural Engineers Annual Conference on the challenges facing Agricultural Engineering in a depressed economy; 19-22 November 1996; Obafemi Awolowo University, Ile-Ife, Nigeria; 1996. p. 77-86

[3] Lowder SK, Skoet J, Raney T: The Number, Size, and Distribution of Farms, Smallholder Farms, and Family Farms Worldwide. World Development. 2016:87, 16-29.

[4] Odigbo EU. Level of agricultural mechanization in Nigeria. In: Ajav EA, Raji AO, Ewemoje TE, editors. Agricultural Engineering in Nigeria: 30 Years of University of Ibadan Experience. 2nd ed. Aluelemhegbe Publishers; 2002. p. 42-56.

[5] Igbeka JC. Agricultural Engineering and the society. In: Ajav EA, Raji AO, Ewemoje TE, editors. Agricultural Engineering in Nigeria: 30 Years of University of Ibadan Experience. 2nd ed. Aluelemhegbe Publishers; 2002. p. 34-41.

[6] Daum T, Birner R. Agricultural mechanization in Africa: Myths, realities and an emerging research agenda. Elsevier Journal of Global Food Security. 2020: 26: 1-10.

[7] Ige MT. Indigenous machinery development: the challenges before the Agricultural Engineers. In: Proceedings of the Nigerian Society of Agricultural Engineers Annual Conference on Mechanized agricultural production for sustainable development; 9-11 September 1998; Lagos Airport Hotel, Ikeja, Lagos, Nigeria; 1998. p. 50-56

[8] Odigbo EU. Confronting the challenges of agricultural mechanization in Nigeria in the next decade: some notes, some options. In: Proceedings of the Nigerian Society of Agricultural Engineers Annual Conference on The challenges of Nigerian Agricultural Engineers towards Vision 2010; 2-6 September 1997; Federal University of Technology, Owerri, Nigeria; 1997. p. 7-17

[9] Igbeka JC. Review of the indigenous machinery manufacturing industry in Nigeria. In: Proceedings of the Nigerian Society of Agricultural Engineers Annual Conference on Mechanized agricultural production for sustainable development; 9-11 September 1998; Lagos Airport Hotel, Ikeja, Lagos, Nigeria; 1998. p. 37-49

[10] Olorunnisola AO. Harnessing the forester's harvest for sustainable development. An Inaugural Lecture published by the University of Ibadan Publishing House, Ibadan, Nigeria; 2013, 102 p.

[11] Forest Products Society, Wood Handbook. 2010 ed., Forest Products Society; 2011. 600 p.

[12] Anonymous. Transfer of technology model, laminated bamboo furniture manufacturing. International Network For Bamboo And Rattan [Internet]. 2013. Available from: www.anancy.net/documents/file_en/BambooFurniture.ppt, [Accessed: 2013-04-05]

[13] Wolf RW, Gjinolli A. Durability and strength of cement-bonded wood particle made from construction waste. Forest Prod. J. 1999; 4924):24-31

[14] Ramirez-Coretti AC, Eckelman CC, Wolf RW. Inorganic bonded composite

wood panel systems for low-cost housing: A Central American perspective. Forest Prod. J. 1998: 48(4):62-68

[15] FAO. Agricultural mechanization:a key input for sub-Saharan Africa small holders.Int.Crop Manag. 2016:23.

[16] Ikechukwu IB, Gbabo A, Ugwuoke IC. Design and fabrication of a single row maize planter for garden use. J. of Advancement in Engineering and Technology. 2014: 1(2):1-7

[17] Musa HL Peanut Sheller Design Parameters. [Thesis] Department of Agricultural Engineering, Kansas State University, Manhattan, Kansas, USA, 1975

[18] Ademiluyi YS. Performance test for improving of IAR groundnut sheller/ decorticator [Thesis], Department of Agricultural Engineering, University of Ibadan, Nigeria, 1997

[19] Olorunnisola AO, Hamilton OA. Performance evaluation of a manually operated groundnut decorticator. International Journal of Applied Tropical Agriculture.1998:3(2): 109 – 113.

[20] Olorunnisola AO. Improvement of an existing low cost wooden groundnut sheller Journal of Applied Science, Engineering & Technology. 2005: 5(1 & 2): 71 – 74

[21] Kharashy A. Design modification and evaluation of a manual groundnut sheller [Thesis] Department of Wood Products Engineering, University of Ibadan, Nigeria, 2021

[22] IITA. Maize. 2013. Available from: iita.org/cropsnew/maize/, [Internet] 2021. [Accessed: 2021-01-02]

[23] Aremu AK, Olorunnisola AO. Ergonomic evaluation of a maize sheller. Ife Journal of Technology. 2001: 10(1): 25 – 31.

[24] Nwakaire JN, Ugwuishiwu BO, Ohagwu CJ. Design, construction and performance analysis of a maize thresher for rural dweller. Nigerian Journal of Technology. 2011: 30(2):49-54

[25] Ghadi, AJ, Arunkumar P. Design, Development and fabrication of a low cost corn deseeding machine. International Journal of Research in Engineering and Technology. 2014: 3(8):242-248

[26] Aremu DO, Adewumi IO, Ijadunola JA. Design, fabrication and performance evaluation of a motorized maize shelling machine. Journal of Biology, Agriculture and Healthcare. 2015: 5(5): 154-164

[27] Olorunnisola AO, Ogbu BE. Development and performance evaluation of a low-cost motorised wooden maize sheller. Journal of Applied Science, Engineering & Technology. 2017: 17: (1):1-6.

[28] Yusuf KO, Akpenpuun TD, Iyanda MO. Design and fabrication of a simple pedal operated cassava grater suitable for rural dwellers. J. Appl. Sci. Environ. Manage. 2019: 23 (6) 1007-1011

[29] Olorunnisola AO, Ewemoje TA. The development and evaluation of a flat-type wooden incubator. International Journal of Applied Tropical Agriculture. 1998: 3(2): 14 – 19.

[30] Oluyemi JA, Roberts FA. Poultry production in warm wet climates. Macmillan Publishers Limited, London; 1988. 300p.

[31] Masinbe B. Drip irrigation helps small holders. In: Phelan J. editor. African Farming and Food Processing. Alain Charles Publishing Ltd., London. July/August 2003 ed.

[32] Mofoke ALE, Adewumi JK, Mudiare OJ, Ramalan AA. Design,

construction and evaluation of an affordable continuous-flow drip irrigation system. J. Applied Irrig. Sci. 2004: 39:253-269

[33] Baqui MA, Angeles HL. Construction, operation and test of a Bamboo Drip Irrigation System. Journal of Agric. Mech. in Asia, Africa and Latin America. 1994: **2**: 41-44.

[34] Umara BG, Audu I, Ogedengbe K, Omobowale MO. Development of bamboo (*oxytenanthera abyssinica)* stem as irrigation pipe. Journal of Applied Science, Engineering & Technology. 2010: 10: 18 – 24

[35] Nta SA, Olorunnisola AO. Experimental production and evaluation of cement-bonded composite pipes for water conveyance. International Journal of Composite Materials. 2016: 6 (1):9-14.

[36] Akinoso R. The development of a solar device for crop drying and cooking. LAUTECH Journal of Engineering and Technology. 2009: 5(1):75-79.

[37] Alabadan BA. Housing and food security: Now and in the future. Lessons from termites. An Inaugural Lecture, published by the Federal University of Technology, Minna, Nigeria; 2013, 64 p.

[38] Mijinyawa Y. Farm structures and environment: The bedrock of food security. An Inaugural Lecture published by the Publishing House, University of Ibadan; 2019, 81 p.

[39] FAO. Technical cooperation programme: Assistance to the strategic grain reserve scheme in Nigeria. United Nations, Rome, Italy. 1995. 9p.

[40] FAO. Rome declaration on food security and world food summit plan of action. November 13-17, 1996. United Nations, Rome, Italy.

[41] Talabi AE. Implementation of national food security programme: experience so far. In: Proceedings of the National Workshop on strategic grain reserve storge programme: Nucleus for national food security; 19-22 November 1996; Abuja, Nigeria; 1996. p. 67-73

[42] Adejumo BA, Ola FA, Odubiyi AA, Oyefeso AO, Aje JO. The potential of sawdust as an insulator in a double walled metallic silo. J. Agric. Engrg. & Technology. 2010: 18(2):1-6

[43] Alabadan BA. Modelling the performance of a hexagonal wooden silo during storage of maize (*Zea mays*). [Thesis], Department of Agricultural Engineering, University of Ibadan, Nigeria, 2002

[44] Mijinyawa Y. The use of wood products in the design and construction of a grain silo for the humid tropics. [Thesis], Department of Agricultural Engineering, University of Ibadan, Nigeria, 1989.

[45] Balehegn B, Duncan A, Tolera A, Ayantunde AA, Issa S, Karimou M, Zampaligré N, André K, Gnanda I, Varijakshapanicker P, Kebreab E, Dubeux J, Boote K, Minta M, Feyissa F, Adesogan, AT. Improving adoption of technologies and interventions for increasing supply of quality livestock feed in low- and middle-income countries. Elsevier Journal of Global Food Security 2020:26:1-11.

[46] Guarín A, Rivera M, Pinto-Correia T, Guiomar N, Šūmane S, Moreno-Pérez OM. A new typology of small farms in Europe. Elsevier Journal of Global Food Security 2020:26:1-9.

[47] Olorunnisola AO. Challenges of commercialising engineering research outputs from Nigerian universities. Acta Academica Supplementum 2011: (2): 217 - 238.

[48] Birner R, Resnick D. The political economy of policies for smallholder

agriculture. Elsevier Journal of World Development 2010:38(10):1442-1452.

[49] Hazell P, Poultton C, Wiggins S, Dorward A. The future of small farms: Trajectories and policy priorities. Elsevier Journal of World Development 2010:38(10):1349-1361.

Chapter 7

Promoting Domestic Production of Fish Using Recirculating Aquaculture System (RAS)

Shadrack Kwadwo Amponsah and Luiz Guilherme

Abstract

With the increasing global population and its associated high demand for fish protein, engineers are under pressure to develop systems that can maximise and intensify production of fish in an environmentally friendly and sustainable manner. This demand is escalated in the face of pandemics like the novel Covid-19, which have had serious toll on global food production and availability. The increasing fish demand over the years has caused the emergence of new aquaculture technologies such as the recirculating aquaculture systems (RAS). These fishponds are constructed in a way to ensure the efficient use of water. A technology extensively researched and developed by Brazilian researchers; the RAS technology has now been widely adapted to some developing countries in the sub-Saharan African sub region. Learning from the Brazilian and Ghanaian experiences, this chapter provides valuable information on these aquaculture production technologies and offers useful guidelines on their operation and management. The chapter also gives some highlight on available opportunities to better harness the RAS technology to promote sustainable food and nutritional security whiles improving on the general livelihood of adopters.

Keywords: aquaculture, recirculating, fish production, technology, Ghana, sustainable

1. Introduction

While the world's population and its appetite for fish is growing, wild fish stocks are shrinking. Over the last four decades, global consumption of fish has doubled, and the developing world has been responsible for 90% of this growth [1]. Globally, fish has maintained a prime position as the favourite source of animal protein for most people. Aquaculture is the farming of aquatic organisms including fish, molluscs, crustaceans, and aquatic plants. Aquaculture can be freshwater or marine and it ranges from simple ponds using naturally occurring food sources to highly intensive systems with water control, aeration, and supplemental feeding. It is practiced inland, along the coast in brackish water systems, and in marine cages and net pens. Out of an estimated 179 million tonnes global fish production in 2018, aquaculture accounted for 46% of the total production and 52% of fish for human consumption equivalent to an estimated annual supply of 20.5 kg per capita [2].

Harvesting of aquatic resources and production is done either in the wild (capture fisheries) or in controlled environments (aquaculture). Unlike freshwater aquaculture, only a handful of the world population practice capture fisheries probably due to the high resource requirement in terms of skill and equipment. Aquaculture presents huge economic and environmental importance to the world economy. Fisheries and aquaculture are a vital source of jobs, nutritious food, and economic opportunities, especially for small-scale fishing communities. Aquaculture is also important for food security through the significant production of some low-value freshwater species (also through integrated farming) destined mainly for domestic consumption. Some economic importance of aquaculture includes alternative food and fuel source, increased jobs in the market, reduction in seafood trade deficit, increased production of food for human consumption, opening of commercially viable business opportunities, increased national exports and the substitution of imports by local production [3].

Recently, there is heightened interest by most people in developing countries to grow their own food to improve food and nutrition security whilst generating extra income. This move has led to the development of integrated food production systems which incorporates crop/animal production with aquaculture. In parts of Africa and South America, these integrated aquaculture systems consist of fish tanks or ponds that are used to culture Tilapia (*Oreochromis niloticus*), African Catfish (*Clarias gariepinus*) or *Heterobranchus* species whiles the pond effluent is utilised as manure to grow commonly consumed vegetables [4]. Pond aquaculture clearly has a huge potential for growth through sustainable intensification [5]. Similarly, Amponsah et al. [6] assessed the effect of rectangular, ellipsoid and circular tanks on production of Catfish using a recirculating aquaculture setup in an effort to promote the technology.

The increasing fish demand over the years has caused the emergence of new aquaculture technologies. An example of such systems is the water re-circulating pond. This type of fish tank is constructed in a way to ensure the efficient use of water. The effluent is not thrown away but simply emptied out into a reservoir where it undergoes treatment and is then pumped back into the pond [4]. In the phase of pandemics like the novel Covid-19, extra awareness is being created on the need for people in especially the developing countries to grow their own food in the comfort of their homes or communities. This practice, extensively adopted, is expected to create a sense of security, and improve the general livelihoods of the populace. Promoting home production of fish will greatly enhance the contribution of the aquaculture sector to sustainable economic growth. This chapter presents types, operation principles and management guidelines of tank aquaculture systems, focusing on experiences from some developing countries.

2. History of aquaculture

The term aquaculture is not new as it has been around for thousands of years [2]. Ancient Chinese manuscripts from the 5th century B.C. indicate that the Chinese practiced the fish culture of Carp in rice patties. The ancient Hawaiians furthered these methods by building pens to breed and raise Carp and Tilapia. According to Ling [7], the practice of aquaculture started in Asia (around 5000 BC), ancient Egypt (around 2500 BC) and in Central Europe. The earliest species of fish cultured was the common carp (*Cyprinus carpio*), by a native of China. In addition, Indian carp culture existed in the 11th Century AD [8]. Similarly, aquaculture started in Europe with the introduction of common carp culture in monastic ponds. During the 14th century, the propagation of trout was introduced in France as well as the method of artificial impregnation of trout eggs [9]. Subsequently, commercial trout culture in freshwater was developed in France, Denmark, Japan, Italy and Norway [8]. Specifically, the British introduced trout as sport fisheries in their Asian and African Colonies. Moreover, the development of fish culture in North

America became possible through the propagation of trout Salmon and Black bass. In the Czech Republic, these fishes were cultured in large ponds, which were built from around 1650, some of which are still in use [10]. Aquaculture in Brazil probably started in the 17th century, during the Dutch occupation of the north eastern region [11]. Originally, other culture facilities such as pens, and cages were used to grow Catfish in Cambodia. While the earliest brackish-water farming originated in Indonesia during the 15th Century AD. Atlantic Salmon also were cultured in cages in Norwegian fjords [8]. In West Africa, The Gambia started aquaculture in the 1970s in the form of trials using Tilapia culture in rice fields [12, 13]. Later on, in 1982, a company known as West African Aquaculture limited started the culture of Peneaus monodon in the coastal region [12]. This company became well established in The Gambia in 2000. Similarly, in 1988, two fish farms were operated in Western Region by Scan Gambia limited. Fish farming started in Ghana in 1953 by the former Department of Fisheries. Thus, it served as hatcheries to support the then culture-based reservoir fishery development programme of the colonial administration. In 1957, the government of Ghana adopted a policy to develop fishponds for farming within all irrigation schemes in the country [14, 15]. The modern form of aquaculture was started with pond culture, and this type of aquaculture is still prevalent in Ghana presently [16]. Production or holding systems used in Ghana are floating cages, earthen ponds, and concrete tanks. Majority of farmed Tilapia (approx. 90%) are from cage culture system with the remaining percentage from ponds and tanks [17]. Tank culture systems, though in the minority, offer great prospects to change the narrative in Ghana's fish sector if properly harnessed.

3. Overview of production systems

There are a wide range of aquaculture production systems used in the rearing of aquatic species. Production systems are diverse in terms of culture methods, practices, facilities and integration with other agricultural activities [2]. Few among them are cage and pen culture systems, earthen (dug out) ponds and tank culture systems. Earthen ponds remain the most commonly used type of facility for inland

Figure 1.
Earthen ponds or dug out systems.

Figure 2.
Cage/pen aquaculture systems.

Figure 3.
Tank culture systems.

aquaculture production, although raceway tanks, aboveground tanks, pens and cages are also widely used where local conditions allow [2]. Each of these aquaculture production systems/units are unique in its construction (**Figures 1–3**) and presents some pros and cons for consideration by potential adopters (**Table 1**).

4. Recirculating aquaculture system

Technological advances in tank culture systems led to the development of continuous recirculating aquaculture systems (RAS). This type of production system or holding facility is constructed in a way to ensure the efficient use of

Production system	Pros	Cons
Earthen pond/dug out	• Relatively cheaper to construct	• Requires physical security against theft
	• Large stocking capacity	• Fish could escape during flooding
	• Fish live in their natural environment	• Sorting may be difficult in earthen
	• Fish growth is rapid	• Pond water pollution and total pond collapse are common
	• High profit margin may be envisaged	• Expensive to manage and maintain over time
Cage/pen	• Fishes live in their natural habitat	• The construction process is cumbersome
	• Fish growth is rapid	• Risk of total fish kill during water pollution
	• Large stocking capacity	• High skilled labour requirement
	• Oxygen is always available	
Tank	• Lower risk of losing investment	• Higher initial investment
	• Lower skilled-labour requirement	• High dependency on reliable electricity and water
	• Smaller space requirement	• Overcrowding of fishes
	• Ease of construction	• Low stocking capacity
	• Ease of adoption and management at home	

Table 1
Pros and cons of aquaculture production systems.

water. The used water is not thrown away but simply emptied out into a reservoir where it undergoes treatment and is then re-used. A major advantage of the water re-circulating fishpond is that it offers the most efficient water usage, a higher dissolved oxygen percentage as well as a balanced pH level. The main challenge is the fact that electricity is required to power the system to be able to manage the waste effectively and efficiently in the pond. A typical recirculating aquaculture system essentially consists of a tank (pond water holding facility) and a filtration unit (pond waste management). **Figure 4** illustrates a well-labelled recirculatory tank culture system.

These production systems are suitable for both home and commercial culture of fish and comes in different sizes and shapes. Standard size for the circular geometry tanks is 4.2 m diameter at 0.9 m height, whereas the rectangular geometry is 4 m length by 3 m width at a height of 0.9 m. The advent of these production systems has restored hope for aquaculture enthusiasts, making it possible to grow fish in an environmentally friendly manner in the comfort of your home [4]. The tank frames can be constructed with locally available materials such as bamboo, tree branches, stones, clay (earth), empty paper cartons and cement blocks. These tank frames are usually lined with UV-resistant polyethene sheets or high-grade canvas material to make them water-resistant. Waste is managed using biological or mechanical filtration units installed on the tanks.

The innovation is an aquaculture system composed of a raised tank structure (rectangular, circular, or ellipsoid) made of either a cement block or wooden frame and lined with a canvas material to make it watertight. Water in the tank is set

Figure 4.
Parts of a typical recirculating aquaculture setup.

in motion using a submersible water pump, placed at one point along the inner periphery of the tank. The pump also helps with aeration by constantly pumping atmospheric air through the water. The pump inlet is fixed on a perforated PVC pipe and placed on a concrete stand underneath the water. Once the pump is turned on, water is drawn through the perforations and by the centrifugal force generated, suspended solids are pushed along the inner periphery of the tank to the central area of the pond. Water hose is placed at the central bottom area to draw suspended solids by capillary action into a sedimentation tank placed outside the pond. The tank has at its bottom, a nylon thread mesh which filters solid waste from the incoming water for safe disposal through a valve opening at the lower side of tank. Filtered water from the sedimentation tank is brought back into the pond using a second submersible pump (i.e., to achieve water recirculation) through a bucket (stuffed with nylon thread mesh) placed at a height above the pond. Potential energy generated by the falling water produces bubbles aimed at providing extra oxygen for the fishes. The activity bacteria, pumps and mesh together help to remove ammonia from pond through the process of nitrification, making the water safe for fishes. This process is the referred to as biological filtration (biofiltration) which employs sets of biofilters to perform its function.

4.1 Biofilters

Biological filter (biofilter) setup is the single most important component of the pond system as far as waste management is concerned. Biofilters use natural processes and organisms (bacteria) through the Nitrogen cycle to break down Ammonia into less harmful components. It also helps to remove faecal waste and leftover feed from the pond water to ensure that the water is conducive for fish culture.

A standard biofilter setup is composed of submersible pumps (a and f), water hoses (e), nylon mesh-stuffed bucket on a stand (d), nylon mesh-stuffed basket inside the waste tank (b), sedimentation/waste tank with its accompanied plumbing components (c) as shown in **Figure 5**. In each biofiltration setup, there are two submersible pumps, recirculating/aeration (a) and return (f).

DOI: http://dx.doi.org/10.5772/intechopen.99697

Figure 5.
The components of the biofiltration setup.

4.1.1 Biofilter installation procedure

The step-by-step installation procedure for the biofiltration setup is described as follows:

1. Fill your pond with potable water. Note that for chlorinated water, it is important to allow 3–5 days prior to stocking to allow the chemical to breakdown to tolerable levels for the fingerlings.

2. Gradually lower or place your biofilter unit (labelled *d* in **Figure 5**) into the pond – **Figure 6**.

3. Position your waste tank unit (labelled *c* in **Figure 5**) a little below the pond height, making sure it is well levelled and the outlet does not impede movement around the pond – **Figure 6**.

4. Gently fix the nylon mesh-stuffed basket (labelled *b* in **Figure 5**) inside the waste tank – **Figure 6**.

5. Cut the water hoses for the different connections on the biofilter, ensuring that they are well straightened before use – **Figure 7**. Cut at least four sets of 3-meter length and one piece of 1-meter length.

6. Fix the shorter hose inside the component labelled d and secure it with an end cap – **Figure 7**.

7. Place the return pump (labelled *f* in **Figure 5**) inside the waste tank and on top of the nylon mesh-stuffed bucket on a stand (labelled *d* in **Figure 5**) – **Figure 7** (insert).

8. Carefully place the longer hoses in the central area of the pond and gently siphon pond water into the waste tank – **Figure 8**.

9. Position the recirculating/aeration pump (labelled *a* in **Figure 5**) inside the pond on a firmly grounded stand, ensuring that it is fully submerged in water – **Figure 8**.

4.2 Pond management

After the pond has been constructed and biofilter installation is completed, all other production activities from pond stocking to harvesting of matured fish fall under pond management. It is important to note that within the production cycle, management of the pond is critical to ensuring good yield at harvest and nothing should therefore be taken for granted.

Figure 6.
Illustration for steps 2, 3 and 4 of the biofilter installation process.

Figure 7.
Illustration for steps 5, 6 (insert) and 7 of the biofilter installation process.

Figure 8.
Illustration for steps 8 and 9 of the biofilter installation process.

4.3 Pond stocking

The process of putting fish into the pond is termed stocking. This marks the beginning of a production cycle and is among the most stressful processes the fish go through during production. The process of stocking starts with the collection of fingerlings from the hatchery, transporting them to the pond location and finally putting them into the pond.

Poor stocking procedures are among the major causes of low survival in grow-out ponds and usually result in disease outbreaks, reduced growth and mortality. Successful stocking depends on the quality of fingerling, how they are stocked and when they are stocked. Stocking should be done in a manner that minimises stress on the fish. It is therefore advisable to get fingerlings from an approved source or outlet. Apart from the various hatcheries under the Ministry of Fisheries and Aquaculture Development, several private hatcheries are available across the country where quality stock can be procured. It is advisable to stock a pond before sunrise or at sunset to avoid heat stress on the fingerlings. As a precautionary measure to arrest any bacterial/fungal infection, salt treatment (using plain rock salt) of pond water is required prior to stocking. Salt treatment may be repeated from time to time, especially when fungal or bacterial infection is suspected. **Figure 9** illustrates the salt treatment process of pond.

As a caution, fish should not be poured straight into the ponds. Always allow fish to be acclimatised to the new environment for the first 15 to 30 minutes before gradually releasing them into the pond (**Figure 10**). This helps them adapt to the differences in water quality between the transport container and the pond without shocking the fish. When fish are shocked by the sudden changes in water quality, they become stressed or may die. It is important to stock fish of uniform size; otherwise, the larger fish (especially for catfish) will cannibalise the smaller ones. They will also dominate the feeding area which will result in them growing bigger and the smaller fish remaining small. Such a situation can negatively affect fish survival rate and total yield at harvest.

Ensure to keep a record of all the stocking information including the source of the fingerlings, average size stocked, total number and kilogrammes stocked as well as any notable observations such as mortalities and disease outbreaks during the stocking process and all through the production cycle. The recirculating pond systems could take a maximum of 400 tilapia fingerlings or 1000 catfish fingerlings (≥10 g) to be fattened to a size of at least 0.3 kg and 1.0 kg respectively within a period of 4–6 months.

Figure 9.
Salt treatment procedure.

Figure 10.
Pond stocking procedure.

4.4 Feeding and nutrition

After stocking, fish requires the right nutrition to grow and promote good yields at harvest. It is also important to feed your fish with the right quantity of feed at the right time and growth stage. It is advisable to buy commercial floating pelleted feed to minimise feed wastage. These commercial feeds come in various sizes and nutritional contents to meet the demand of the fish at the different stages of growth.

In administering the feed, it is required to use the response feeding method while ensuring that the fishes are fed to satiation (**Figure 11**). Response feeding is proven to be the best feeding method. Not only does it help minimise feed wastage, but it also ensures that the pond water quality is well maintained. The most important rule in fish nutrition is to avoid overfeeding. Overfeeding is a waste of expensive feed. It also results in water pollution, low dissolved oxygen levels, increased biological oxygen demand, and increased bacterial loads.

Figure 11.
Response feeding of fingerlings.

Figure 12.
Removal of waste from waste tank.

4.5 Water quality and maintenance

Water quality is the first most important limiting factor in pond fish production. It is also the most difficult production factor to understand, predict and manage. Water is not just where the fish live. Its quality directly affects feed efficiency, growth rates, the fish's health, and survival. Most fish mortalities, disease outbreaks, poor growth, poor feed conversion efficiency and similar management problems are related to poor water quality. Water quality refers to anything in the water (physical, chemical, or biological) that affects the production of fish. The objective of pond management is to manage the water quality, to provide a relatively stress-free environment that meets the physical, chemical, and biological standards for the fishes' normal health and production performance.

The following are some general pond managements and troubleshooting tips/practices for improved pond water quality and fish production:

- Ensure waste in sedimentation tank is safely removed daily by opening the outlet valve whilst stirring water in the bucket continuously to dislodge the waste and allow it to freely move out (**Figure 12**).

- Ensure to check and service the submersible water pumps at least once every week.

- Check and top up pond water to optimal level from time to time or where necessary, especially after every waste removal activity.

- In extreme cases, flush out the water from the pond entirely and replace with fresh one. This becomes necessary when there is more waste than the biofilter system can handle or when the biofilter pumps are not functioning due to power outage.

- Reduce the feeding frequency to about half in situations where the biofiltration system is not functioning for more than 3 days due to power outage. This is done to reduce the quantity of waste produced and maintain the water quality till the problem is addressed.

4.6 Diseases and management

Disease is a condition in living organisms in which normal physiological functions are being impaired due to alteration in the body systems and typically manifested by distinguishing signs and symptoms. Sustainable aquaculture production can only be feasible when fish are healthy and free from disease. Disease is a significant source of constraint to aquaculture development and sustainability from both social and economic viewpoints. Production costs are increased through investment lost in dead cultured fish, cost of treatment and decreased quality and quantity of yields. Again, livelihood and standard of living may be affected due to reduced products availability, loss of income and employments.

Fish diseases and infections in ponds may occur when pathogen load increases above what the natural resistance of the fish can cope due to external factors like poor water supply. Such external factors may cause drastic changes in water quality and lower fish resistance, making them susceptible to diseases and even the risks of 100% fish mortality. Fish disease management is a combination of both preventive and curative measures required to avoid and treat possible disease, respectively. Healthy fish can adapt to reasonable environmental changes and in turn resist diseases.

Generally, two forms of diseases affect fish: infectious and non-infectious. Infectious diseases are caused by living/pathogenic organisms (viruses, bacteria, fungi, or parasites) present in the aquatic environment or carried by other fish. Fish become vulnerable to pathogenic infections when there are stressors (environmental abnormalities, water quality deterioration, unbalanced nutrition, or bodily injuries) which weaken fish natural resistance (immune system). Infections can occur internally and externally affecting tissues, organs, and other fish body parts. They are mostly contagious diseases, and some types of treatments may be

Figure 13.
Tail rot bacterial infection on an African catfish juvenile.

necessary to control the disease outbreaks. Infectious diseases can be further grouped into parasitic, bacterial, fungal, and viral diseases based on the causative agent. Bacterial infections are considered the major cause of mortality in aquaculture. Antibiotics are commonly used in the treatment of bacterial infection of fish. In some cases, salt bath may be done for fish in combination with antibiotics to remedy the situation. **Figure 13** depicts a tail rot bacterial infection on an African Catfish juvenile.

Non-infectious diseases (systemic diseases) are caused by non-living factors. The diseases are either congenital (such as genetic anomalies or neo plastic conditions) or iatrogenic (induced by external conditions such as environmental or nutritional problems). Non-infectious diseases are not contagious, and medications are generally not required. However, iatrogenic condition can usually be reversed by removing (or adjusting) the cause. Non-infectious diseases of fish include diseases due to environmental, nutritional, and genetic anomalies.

Apart from the application of appropriate (approved) remedial actions and/or treatments for fish diseases, a change in management is necessary in most cases. However, as a rule of thumb, it is advised to consult the nearest aquaculture/fisheries extension office for technical assistance in remedying the situation before things get out of hand. As a precautionary measure, it is advisable to not feed fish with bare hands, as that could be a point for introducing infection into the pond. Either wear gloves or use a small cup to fetch and throw fish feed when feeding.

4.7 Fish harvesting and processing

Harvesting is one of the most important and labour-intensive activities in aquaculture but is often overlooked. In raised ponds, harvesting may be achieved either by using a two-man drag net (**Figure 14**) or by fully draining the pond water (**Figure 15**) and scooping the fish out with a basket or a harvesting scoop net (**Figure 16**). The scoop net and the two-man drag net can be helpful in fish sampling purposes. Harvest fish during the cool hours of the day; preferably, early morning or at sunset, where possible. To keep the harvested fish fresh, especially where long distance transportation is involved, place iced blocks on the fish.

Figure 14.
Fish harvest in pond using a two-man drag net.

Figure 15.
A fully drained pond prior to harvest.

4.8 Handling and processing

The handling, processing, and marketing of fish are essential complementary functions. The marketability of fish is an important constraint in the development of the fish industry. Processing and marketing offer the greatest opportunities for employment within the sector. A rule of thumb is that the timing of harvest must be tailored to meet the local supply and demand patterns. This is because fresh fish cannot be held for long periods of time without serious losses. The transport of fresh or live fish requires: (a) location of ponds close to the market to minimise handling and to limit transportation time (b) early morning harvests to transport fish at cool temperatures and (c) markets equipped with ice facilities or water tanks (cement or small tin containers) with aeration devices and a drainage system. For

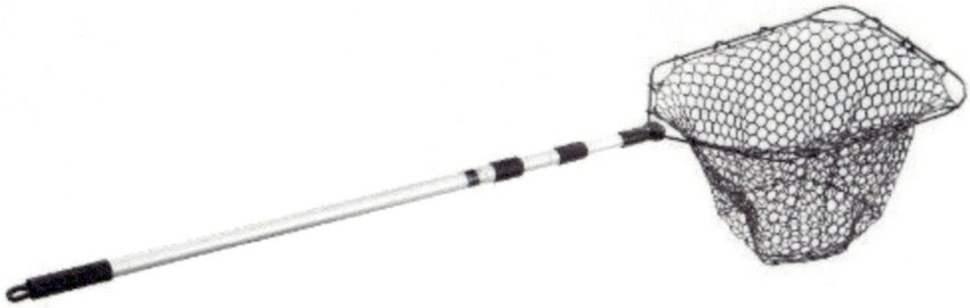

Figure 16.
Fishing scoop net.

Figure 17.
African catfish preservation by smoke drying.

these reasons, simple but efficient preservation and processing methods have been employed over the years by fish farmers.

The processing methods used can vary greatly and are dependent on consumer taste, availability and costs of the processing material, technical knowledge, time needed for processing, price of the final product, storage facilities, marketability and seasonal fluctuations [18]. The most predominant fish preservation method is smoking, which comes in varied forms. In smoke drying of fish, the use of barrels, roofing sheets, cement blocks and kilns are common (**Figure 17**).

5. The Brazilian Sisteminha story

In 2002, researchers from the Federal University of Uberlandia, Minas Gerais State of Brazil developed an integrated food production system. The Brazilian Agricultural Research Cooperation (EMBRAPA) have, since its successful patent in 2006, disseminated the integrated food production system, popularly known as *Sisteminha*, across almost all the regions of Brazil. The project helps low-income families to meet their nutritional needs by producing their own quality foods, with low levels of pesticides and with the use of all available resources around their homes. The project also assisted in fostering entrepreneurship and financial

Figure 18.
Sisteminha fishpond systems using different materials.

independence in regions with low family income. Currently, various aspects of the Sisteminha have been adopted in some African countries including Ghana, Uganda, Ethiopia, Cameroon, Angola, Mozambique through donor-funded projects or own initiative.

The technique of constructing the fish tank in the fish farming module presented itself as the most versatile and therefore underwent more modifications. Initially the tank was designed to have its structure supported by wooden stacks tied with PET bottle wires, cardboard sides, coated with plastic film of 150–200 micron thick [19]. Currently, the tank can even be designed with prefabricated concrete slabs. These EMBRAPA Sisteminha fishponds have been well adopted in Brazil and can be adapted to meet existing local conditions. **Figure 18** provides an illustration of the various Sisteminha tanks.

6. Building on success stories: Wontesty Ventures, Ghana

The recirculating tank culture system was first introduced to Ghana in 2014 through a joint donor-funded project with Scientists from Embrapa Mid-North, Brazil [20]. The closure of the Ghana-Brazil collaborative research and dissemination project on aquaponics-based food production systems in 2016 sparked an interest for recirculating aquaculture ponds, which led to the establishment of a private agri-business venture to meet this demand.

Wontesty Ventures is a start-up business at Kumasi in the Ashanti Region of Ghana specialised in the construction of recirculating aquaculture systems, provision of general aquaculture services including training, production, processing and marketing of Catfish, Tilapia, and other farm produce. With a vision to pond every household in Ghana by 2040, Wontesty Ventures has been working hard through the provision of tailor-made research-based solutions to fish farmers, families and aqua-preneurs across the country. Currently with over 150 established ponds across the country and more than 50 clients since 2017, the business strives to erase the notion that fish farming is possible only with the traditional earthen ponds and cages. The system has since been adapted to Ghanaian conditions through research

Figure 19.
The evolution of the recirculating tank culture system.

and currently available on the market for consumers. These tanks have evolved over the years and now there are even collapsible (mobile) tanks made of industrial-grade canvas tarpaulin material. The mobile tanks come in handy for potential adopters who have plans to relocate after a couple of years. **Figure 19** shows the evolution of the recirculating tank culture system since 2014.

Author details

Shadrack Kwadwo Amponsah[1*] and Luiz Guilherme[2]

1 CSIR-Crops Research Institute, Kumasi, Ghana

2 EMBRAPA Cocais, Sao Luis, Brazil

*Address all correspondence to: skamponsah@hotmail.com

References

[1] Delgado, C., Wada, N., Rosegrant, M. W., Meijer, S., and Ahmed, M. (2003). The Future of Fish Issues and Trends to 2020.

[2] FAO. (2020). The state of world fisheries and aquaculture 2020. Sustainability in action. DOI:10.4060/ca9229en

[3] World Bank. (2014). Fish to 2030-prospects for fisheries and aquaculture. Agriculture and Environmental Services Discussion Paper 03, Washington, D.C. http://www.fao.org/3/i3640e/i3640e.pdf

[4] Amponsah, S. K. (2018). Yali Voices: Making Aquaculture Possible with Simple Fish Tanks. Young African Leaders Initiative (YALI) Network. https://yali.state.gov/yali-voices-making-aquaculture-possible-using-simple-raised-fish-tanks/

[5] Frimpong, E., Ansah, Y., Amisah, S., Adjei-Boateng, D., Agbo, N., and Egna, H. (2014). Effects of two environmental best management practices on pond water and effluent quality and growth of Nile Tilapia, *Oreochromis niloticus*. Sustainability, 6, 652-675. DOI:10.3390/su6020652

[6] Amponsah, S. K., Agodzo, S., Agbeko, E., and Osei, E. A. (2021). Impact of tank geometry on production of African catfish (*Clarias gariepinus*). African Journal of Agricultural Research, 17(1), 165-172.

[7] Ling, S. W. (1977). Aquaculture in South East Asia. A Historical Review. University of Washington Press, Seattle.

[8] Pillay, T. V. R. (1990). Basic of Aquaculture and History of Aquaculture and its Present State. In: Aquaculture Principles and Practice. Fishing News Books.

[9] Davies, H. S. (1956). Culture and Disease of Game Fishes. University of California Press, Berkeley.

[10] Mboge, A. (2010). Opportunities and Constraints of Fish Farming in Ghana, a Case Study of Ashanti Region. KNUST, Kumasi.

[11] Valenti, W. C., Barros, H. P., Moraes-Valenti, P., Bueno, G. W., and Cavalli, R. O. (2021). Aquaculture in Brazil: Past, present and future. Aquaculture Reports, 19, 100611. DOI:10.1016/j.aqrep.2021.100611

[12] Jallow, A. (2009). History of Aquaculture in the Gambia. In: Aquaculture in The Gambia. Daily Observer Forward with The Gambia.

[13] Jawo, S. L. (2007). Fisheries Development and Aquaculture. The Gambia Daily Observer Press. The Gambia.

[14] FAO. (2000). National aquaculture sector overview in Ghana: Fisheries and aquaculture 2:1.

[15] Kassam, L. (2014). Aquaculture, and Food Security, Poverty Alleviation and Nutrition in Ghana: Case Study Prepared for the Aquaculture for Food Security, Poverty Alleviation, and Nutrition Project.

[16] Amenyogbe, E., Chen, G., Wang, Z., Lin, M., and Lu, X. (2018). A review of Ghana's aquaculture industry. J Aquac Res Developmen, 9(545), 1-6. DOI:10.4172/2155-9546.1000545

[17] MOFAD. (2016). Annual Report 2015: Ministry of Fisheries and Aquaculture Development.

[18] Spliethoff, P. C. (1987). Handling, processing and marketing of aquaculture products. In: C. E. Nash, C. R. Engle, and D. Crosetti (Eds.),

Proceedings of the ADCP/NORAD Workshop on Women in Aquaculture. FAO, Rome.

[19] Guilherme, L. C., Sobriera, R. S., and Oliviera, V. Q. (2019). Sisteminha Embrapa – UFU – FAPEMIG Integrated Food Production System – Module 1: Fish Tank. ISSN 0104-866x, August, 2019. Series Documents 259 Embrapa, 63 p.

[20] PAEPARD. (2014). Africa-Brazil Agricultural Innovation Marketplace. http://paepard.blogspot.com/2014/04/africa-brazil-agricultural-innovation.html

Chapter 8

Precision Vegetable Farming Technologies: An Update

Prashant Kaushik

Abstract

The efficiency of precision vegetable farming cannot deny in the current of climate change. As compared to west adoption of reliability precision agriculture approaches in developing world is a gradual procedure. The small scale of farms limits economic benefits from currently accessible precision farming technologies. Nevertheless, horticulture interventions like geographical positioning system (GPS), geographical information system (GIS), artificial intelligence (AI), robotics, sensor technologies, etc., are being utilized for precision vegetable farming to improve production and quality of vegetables. This retains excellent promise for developing vegetable crops within the present farming scenario when climate change makes the whole rethink agriculture practices. Overall, this chapter will provide useful information about precision vegetable farming technologies for vegetable growers, enthusiasts, farmers, and researchers.

Keywords: precision, vegetable farming, GIS, GPS, robotics, AI, sensors

1. Introduction

The demand for vegetable crops is continuously increasing. Sustaining this demand requires the infusion of technologies for efficient utilization of sources for deriving higher result per product of inputs with a superb quality of produce [1]. It will be possible only through the deployment of hi-tech applications and precision farming. Precision agriculture is among the most modern technologies for sustainable agriculture, which has gained impetus in the 21st Century [2, 3]. Precision farming consists of using products and principles to control temporal and spatial variability regarding all factors of gardening generation for bettering crop performance and setting quality [4]. Precision farming calls for effective management of resources through location unique hi-tech interventions. Precision vegetable farming provides a new solution using a systems treatment for today's farming issues, such as the necessity to balance efficiency with ecological problems [5]. It offers to describe and model variation in soils and plant species combining farming methods to meet up with website specific needs. It seeks to improve economic return shipping and delivery and lessen the power feedback and the environmentally friendly impact of farming. Precision agriculture approaches can help both large scale farmers as well as small farm holders. Moreover, precision vegetable farming offers the promise of enhancing productivity while reducing production costs [6].

With the precision vegetable farming describing the variability is going to be the primary key concept. Primarily, it is affected by variation within each region [7]. In comparison, hi-tech horticulture will be the deployment of today's technology

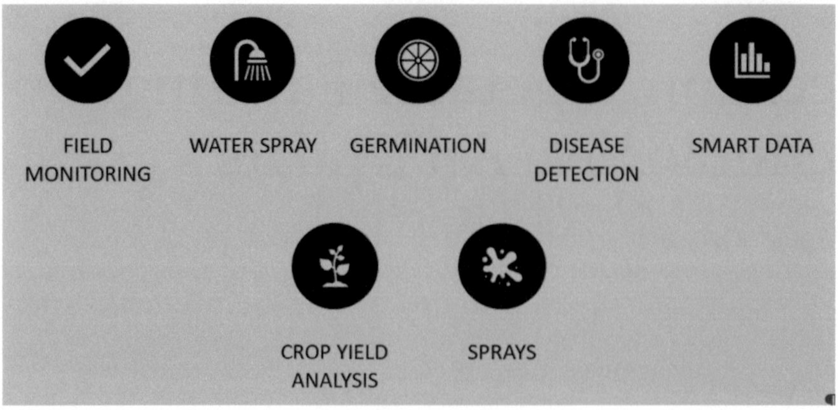

Figure 1.
Application of precision vegetable farming technologies to different aspects of vegetable production.

that is capital intensive, much less environment-dependent, capable of enhancing the output and quality of produce. Utilization of these interventions orchestrated together, keeping the goal of obtaining higher output in granted time results in precision vegetable farming, which is knowledge-driven. Precision vegetable farming aims to enhance crop performance and setting quality [8]. The actual benefit of precision vegetable farming is that the farmer can conduct much more regular tillage, adjust sowing rates, fertilizer application based on soil conditions, plan even more crop protection programmes with increased precision, and understand the yield variation area. These benefits may improve the real cost-effectiveness of harvest production; nonetheless, the grower should be prepared to make changes in his management types to work [9, 10]. Precision vegetable farming is a modern-day procedure utilized to improve crop productivity using the newest technologies, cloud computing, information technology (IoT), WSN, i.e., artificial intelligence (AI) machine learning (ML). Thus far, the majority of research suggests that PA-based methods possess a good impact on productivity and sustainability [11]. Nevertheless, a few obstacles are associated with the growth and deployment stage of these methods. Since the primary goal of precision agriculture is producing surplus yield by optimizing the materials such as for instance water, pesticide sprays, plant foods, etc., for source, prescription maps play a crucial role, that allows farm owners to quantify information needed for useful plants at any specific development stage. The application of precision farming technology to various aspects of the vegetable farming process is presented in **Figure 1**. In general, this chapter is going to provide information that is useful about precision vegetable farming solutions for vegetable growers, farmers, enthusiasts, and scientists.

2. Role of global positioning system (GPS) in precision vegetable farming

Along with its simplicity of using and worldwide all the weather procedure, GPS owes its acceptance on the superior, dependable precision with what place, time, and direction might be determined. This information is provided in healthy time, meaning that continuous place information is supplied while in motion [12]. Having exact location information at any time permits harvest, soil and liquid dimensions being mapped. GPS receivers perhaps have with the spot or possibly mounted on implements enable owners to return to places to sample or perhaps deal with those areas. GPS component in precision vegetable farming is implementing the most significant factor [13]. At the correct spot, at the proper time, understanding the most significant thing to perform may involve all kinds of hi-tech equipment and fancy statistics or other analysis. However, making the perfect point starts with excellent supervisors and new operators doing an excellent job of employing everyday resources as planters, fertilizer applicators, harvesters no matter what else is needed [14, 15]. In this specific context, GPS is going to become part and parcel of perfection agriculture. For analysis and processing of remote Sensed images must have ground-truth information, collected in the region, at many sites and often at several times throughout the crop paper season [16]. Precision farming is an incorporated harvest management method that attempts to enhance inputs' kind and quantity with the actual crop requirements for small regions within a farm region. We have produced an interactive, portable telephone system to gain field info right away into a digital database containing yield, soil, road, water alongside contour maps [17]. The system additionally allows anyone to save GPS info from widening and tracking field exercise at a later working day. Data captured in the field, like the location of weed patches, region boundaries, and harvest quality pointed out, are moved readily to others. Spreadsheet along with image processing techniques to enhance datasets plus much better analysis [18]. GPS-based precision farming applications have been used for farm preparation, region mapping, soil sampling, tractor guidance, harvest scouting, flexible rate programs, and yield mapping. GPS permits farm owners to function during low visibility field problems, for instance, rainfall, particles, fog, and darkness. Growers can achieve additional benefits by pairing better utilization of some other soil amendments and fertilizers, identifying the economic threshold for combating pest and weed infestations and protecting the organic energy sources for future use. Nowadays, GPS derived items are utilized by many growers to enhance organizations within their farming companies. GPS receivers gather location information for mapping region borders, roads, irrigation systems, and problem areas in plants as weeds or disease. The accuracy of growers is permitted by GPS to create farm maps with extremely accurate acreage for subject areas, street places, and ranges between tourist attractions. Farm owners are allowed by GPS to successfully review to particular spots in the area, season after season, to collect soil samples or perhaps monitor crop conditions. GPS likewise allows pilots to provide farmers with precise maps. Growers, together with farming service providers, are in a position to foresee more changes as GPS regularly modernize. With the present civilian service provided by GPS, the nation is focused on using a second together with a third municipal signal on GPS satellites.

3. Role of sensor technologies in precision vegetable farming

Optical and thermal sensors will be the most favored sources in plant phenotyping. Sensors are classified based on the amount and wavelength selection of measured wavelengths, the taller the detection ability [19]. The more comprehensive the band range assessed around a specific wavelength, the diminished the measurement reliability because of the overlap of different wavelengths, even if several indices will most likely be steadier when estimated in broader bands. For a clear understanding of how sensors fixes are positioned to complement breeders' needs in vegetable crop phenotyping, it looks like far more valuable to describe the different PRSs, contemplating their working principles distant relative to how plant life speaks with the electromagnetic light [20]. The use of the evaluation and stereo system camera rigs by computer programs of photographs taken by several angulations enable drawing sophisticated versions because of the reconstruction of growing constructions in 3D. Nevertheless, sensors reliant on reactance measurement can provide a lot more valuable information to develop physiology studies. Multispectral imaging consists of computing leaf reactance at numerous wavelengths, thus providing details for calculating important vegetation indices [21]. In Solanaceae, numerous scientific studies say hyperspectral sensing items for early detection of various symptoms relating to biotic stresses. Results indicated how spectral imaging is a lot more suitable for classifying ripeness stage, lessening the error as an outcome of variations which are easy in ripeness. These studies demonstrate hyperspectral items' power to choose parameters and plant diseases related to quality, thus enhancing synthetic substances in horticulture in the various phases of the supply chain [4]. Spectrophotometry was also utilized to assess chlorophyll concentration. These two phenomena are related to the development of H20, which is free in the tissue and mild penetration. Results evidenced how lettuce has a drop-in plant growth triggered by severe salinity, while no improvement rate decrease was confirmed under temperature stresses [22, 23]. Data gathered from sensors will better offer smallholder farmers, various other stakeholders, and extension workers with updated information about their plant life to enhance productivity. Improved data and information will optimize farming inputs and time invested by farming extension employees along with different actors on the floor, ultimately leading to great utilization of electrical power and cost savings. Finally, information is necessary to allow the private sector and various other intermediaries to make decision-making methods that might get smallholder farmers like financial and insurance applications. Consequently, leaders, technology vendors, farming businesses, academics and funders should commit to coming also and together finding the potential for these technologies. Agricultural receptors gather a significant quantity of info, but they usually must run in shallow connectivity environments when

Vegetable	Character	Instrument with Sensors
Solanum lycopersicum	Leaves damaged by leaf miner	Nexus FT-NIR spectrometer
Solanum lycopersicum	Ripeness	ImSpector V9
Leafy vegetables	Chlorophyll content	ASD Fieldspec FR spectroradiometer
Spinacia oleracea	Crop canopy under water	Specim V10 spectrometer
Citrullus lanatus	Lycopene, -Carotene, and Total Soluble Solids	NIR On-Line® X-One
Lactuca sativa	Plant traits under extreme temperature and salinity stress treatments	Series VNIR Micro-Hyperspec Sensor; Fluor Cam 800 MF
Solanum lycopersicum	Harvest time	AgroSpec VIS–NIR spectrophotometer
Cichorium intybus	Cold stress	CF Imager
Phaseolus vulgaris	Photosyntetic traits, morphological parameters and shoot architecture	Growscreen Fluoro
Brassica rapa subspecies pekinensis and chinensis	Quality	FluorPenFP 100 fluorimeter
Cucumis melo	Grafting compatibility	Imaging-PAM fluorometer
Leafy vegetables	quality in post-harvest storage	SPAD-502; Agriexpert CCN 6000
Solanum melongena	Fruit morphology and shape	Scanner imaging

Table 1.
Non-destructive compact sensors for field use and data procuration in vegetables [22].

applied to the developing world. Agricultural details are being collected through satellites, weather stations and ground sensors. Yet these options do not consult one another, and therefore there is just no centralized method to level data. For the info being genuinely useful in decision making, know-how vendors need to work toward interoperability and discover efficient ways to integrate it. Examples of sensor technologies used for the precision vegetable farming are presented in **Table 1**.

4. Role of geographic information system (GIS) in precision vegetable farming

The advance GIS solutions enable checking, picture, and analysing geographical contents to facilitate information consumption. Furthermore, the mobile version will allow farmers to keep informed about all their plots through information on soil analysis, the crop's problem, and the managing of evapotranspiration [24, 25]. With using remote sensing, GPS and GIS, farmers might be prepared to see site-specific demands of the farms. With this information, they can formulate and implement control techniques that will ensure the perfect use of inputs to enhance their output and profits. Therefore, geospatial technologies give you a farmer with an information resource that he/she can use generating educated decisions that guarantee effective and efficient managing of the farm to enhance its efficiency. Thus, growers need to understand and implement these ways together with their expertise and experience getting the best advantages of their farms [26]. GIS has a hardware-software repository unit used to capture, store, enter, influence, analyze, screen, map as sort, and spatially referenced geographical information. GIS maps are energetic. On the computer display screen, map computer users can check out a GIS chart in almost any program, zoom in or out, and modify the characteristics of the information discovered within the chart. The capability of GIS to evaluate and imagine farming environments and work has shown to be genuinely beneficial to each of those connected with the agriculture business [27]. GIS is a crucial element of automated field operations, known as precision agriculture or possibly satellite farming. Using info collected from remote receptors and sensors installed on farm machinery, farmers have raised decision-making abilities to cook their growing to enhance yields. Last crop yields, surfaces particulars, natural material content, pH, moisture, and nutrient quantities of the grime most help in appropriate preparation for real farming. Combine harvesters constructed with GPS tracking devices are able to evaluate crop yields along with crop quality values as grow water content and chlorophyll levels in time which is genuine and at the particular place in the region from which they are harvested. Adjustable velocity engineering (VRT) will be the element of accuracy agriculture, which really allows the info being placed right to apply. It joins farm machinery, management methods, and application products to don precise amounts of cultivating inputs at specific times or locations. Precision farming with VRT has both economic and environmental benefits. Applying seed, fertilizer, nutrients, and perhaps insecticides, only where and when they are needed, might have sizable cost savings due to the farmer and boost revenues. Furthermore, harmful environmental impacts from overusing of numerous synthetic materials are relieved, and the use of specific chemical compounds might probably be eliminated wholly based on data analysis. Persistent dilemmas as nitrogen program may additionally be dealt with, helping the farmer obtain the appropriate length between excessive and insufficient.

Public, individual, together with non-profit sectors, use GIS to handle public utilities to control the movement and the dispersion of goods and services. GIS is very purposeful in typical map making, to plot items as fire hydrants along a freeway, and maybe to sketch boundaries, like the area of different crop parts on a farm. Real examples of this within the realm of agriculture will be a map showing the range of farm crashes by county, or perhaps the number of crop acres fallen to flood by tax chart parcel. The best benefits of remote sensing are that it is non-invasive and does not negatively affect the spot that is now being observed.

5. Role of grid soil sampling and variable-rate fertilizer (VRT) application in precision vegetable farming

The usual soil test P or K of control strips frequently was Less or perhaps optimum in responsive areas. When GIS strategies were used to analyze yield responses on the consistent P or K program, an incredibly high yield reply variation became obvious. Approximately these outcomes show an exceptionally high potential in numerous fields for dense soil sampling to identify areas with contrastingly several soil tests values. The outcomes also demonstrate an excellent chance for VRT, because this technology place on the strips, hardly ever there's undoubtedly a statistically significant [28]. The standard amount of P or K fertilizer applied per acre by each method varied considerably among fields but generally was a lot less for just about any variable-rate method. Many factors might explain infrequent, small, in addition to inconsistent differences observed between uniform and variable fertilization methods, very possibly for minimal testing field areas [29]. Nevertheless, it might just explain an absence of distinction between application methods for the 1st harvest (because extra P or K likely was utilized by every method) however, not for the following harvest. Therefore, disparities in small field places will be diluted by no response or random differences in larger area areas. Hence, although varying rate program hardly ever improved harvest yield in contrast to a consistent application, it did offer with P application significantly better [30]. The results endorse that adjustable-rate P method will reduce P loss from places compared to a regular application over low-testing or high-testing field areas and may wind up in much better water quality. Moreover, these on-farm trials' results say that the most significant problem to utilize variable-rate fertilization effectively is the dirt sampling method and the soil test chart exactly where it should be seated. The results suggest that a major issue is if the significant small-scale P and K variation can really be calculated cost-efficiently [31, 32]. The convenience of increased cost sampling and fertilizer application methods increases profitability advances only if the view of fertilization is converted to a much more demanding effect-based philosophy. Nevertheless, outcomes indicating VRT does reduce both fertilizer application.

6. Recent developments and tools in precision vegetable farming

Newest solutions in precision vegetable farming include various technologies. In this direction, self-steering tractors have existed for very a while. These tractor types do a great majority of the works; these are establishing toward driverless gadget customized by GPS to spread manure or maybe to furrow land. Novel developments feature a solar fueled scanner which distinguishes weeds and executes them with a portion of herbicide or lasers.

Robots: Farming robots, otherwise referred to as AgBots, as of today, are existing. However, advanced harvesting robots now are being developed to realize fresh fruit, which is ready, comply with their shape and size, and cautiously pluck them from limbs. Recent light aircraft models can take aerial photographs with information from satellite records to foresee coming yields based on the current subject biomass level [33, 34]. Collected pictures can produce maps to follow where water stations, decide adjustable rate seeding and produce yield maps of nearly productive regions.

Smartphone applications: Tablet and cellular phone apps are starting to be steadily well known in precision farming. Cell phones accompany numerous helpful applications earlier published, GPS, like accelerometer, etc. Personal computer only at that information is created by that time and sends linking activities to these items [35]. This involves considering robots to impart the perfect way of computing manure or IoT products to help make the ideal amount of water legally to the environment. The fate of farming pushes a lot more the last machine learning methods each year. It is seen as progressively proficient and exact cultivating with less human labour [36]. Greenhouses are among the most effective techniques for precision agriculture. Sensors, actuators, drones, robots each are IoT items that are meant to improve the effectiveness and precision of greenhouses. With the aid of IoT and a few AI, reliable, unique, innovative farming methods are created right from irrigating to harvesting, which involves forecasting the lifetime of the crop [37]. These days, the farmer can have complete control over his crop worldwide by utilizing the internet finally, which improves the financial system of the nation and is going to save a huge amount of time.

7. Conclusions

In vegetable farming, probably the latest quick adoption of precision agriculture, approaches provide growers with quality solutions and notice food quality and security issues. With automation products recording parameters concerning product quality such as color, size, shape, external defects, sugar content, acidity, and other inner characteristics. Additionally, checking field functions as synthetic substances sprayed and fertilizers can easily provide complete vegetable and fruit processing methods. This information might be disclosed to clients for risk management and food traceability and makers for precision agriculture being much better quality and larger yields with enhanced inputs. Recently many new methods were developed that consider the specific size of the tree, the crop issue, and the ecological variables. Vegetable quality and yield maps are of excellent worth during crop to avoid mixing grapes of several potential wine qualities. Precision vegetable farming will continue to be a concept in many developing nations, and strategic guidance from the public and private sectors is essential to promote its quick adoption. Successful adoption, nevertheless, comprises at least three phases as execution, evaluation, and exploration. Precision vegetable farming can focus on both economic and environmental issues that surround modern farming practices nowadays. Queries remain about cost-effectiveness and the most remarkable techniques to use the technical resources we today have. However, the thought of "doing the best thing in the correct area on the appropriate time" offers an excellent intuitive appeal. Ultimately, precision agriculture's success hinges largely on how properly and how easily the information should point the new solutions might be found. Precision vegetable farming provides a new choice utilizing systems utilized for current-day agricultural issues, such as the necessity to balance efficiency with ecological problems. It is reliant on sophisticated information technologies. The polygons that stand for different ownership or municipalities can impart values in various approaches, most regular being a changing style ramp. GIS can help a farmer change to these different variables, monitor specific plants' health, appraisal yields originating from a particular region, and enhance crop production. There are many choices for GIS info at no cost as well as for a rate. Colleges, federal organizations, and individual business owners tend to be repositories of spatial information.

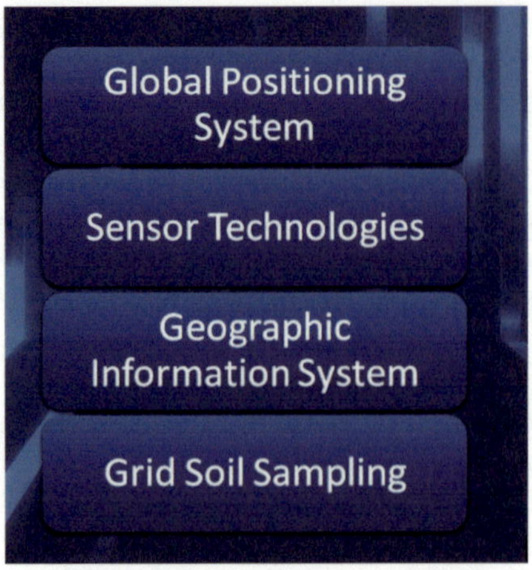

Figure 2.
Most frequently used precision vegetable farming approaches.

But, based on the development of a country resources and reach some precision farming approaches are more frequently used than others (**Figure 2**). Using land-use and primary food crop information, together with info collected by mobile devices and satellites to identify locations in need root causes of food insecurity, GIS is essential to end around the world hunger. Satellites, drones, plus human-crewed aircraft are used for remote sensing, in addition to that's the gathering of information about the planet's surface area by scanning it from high altitudes. Based on the exterior temperature, the intensity of the wavelengths created by different vegetation types and various manmade and natural landscapes differs. The captured info is transformed into explicit electronic imagery and maybe set on to popular objectives like managing water for watering consumption or planting disease detection. It's competent to besides, be placed onto objectives like analyzing the maturity of fresh fruit. It offers to describe and model variation in soils and plant species and mix farming methods to meet up with website specific needs. It seeks to increase monetary return delivery and lessen the considerable feedback and the environmentally friendly impact of farming.

Author details

Prashant Kaushik[1,2]

1 Instituto de Conservación y Mejora de la Agrodiversidad Valenciana, Universitat Politècnica de València, Valencia, Spain

2 Nagano University, Nagano, Japan

*Address all correspondence to: prakau@doctor.upv.es

References

[1] Brar NS, Saini DK, Kaushik P, Chauhan J, Kamboj NK (2019) Directing for Higher Seed Production in Vegetables. Agronomy

[2] Fraser A (2019) Land grab/data grab: precision agriculture and its new horizons. The Journal of Peasant Studies 46:893-912

[3] Council NR (2010) Toward sustainable agricultural systems in the 21st century. National Academies Press

[4] Lee W-S, Alchanatis V, Yang C, Hirafuji M, Moshou D, Li C (2010) Sensing technologies for precision specialty crop production. Computers and electronics in agriculture 74:2-33

[5] Khan AM (2013) Precision farming technique for water resource management. PhD Thesis, Aligarh Muslim University

[6] Groher T, Heitkämper K, Walter A, Liebisch F, Umstätter C (2020) Status quo of adoption of precision agriculture enabling technologies in Swiss plant production. Precision Agriculture (eingereicht)

[7] Goetz SJ, Prince SD, Small J, Gleason AC (2000) Interannual variability of global terrestrial primary production: Results of a model driven with satellite observations. Journal of Geophysical Research: Atmospheres 105:20077-20091

[8] Ray PK (2015) Hi-Tech Horticulture and Climate Change. Climate Dinamics in Horticultural Science, Principles and Applications; Choudhary, ML, Patel, VB, Siddiqui, MW, Mahdi, SS, Eds 1-22

[9] Brar NS, Kumar T, Kaushik P (2020) Integration of Technologies Under Climate Change for Profitability in Vegetable Cultivation: An Outlook. https://doi.org/10.20944/preprints202009.0542.v1

[10] Balafoutis A, Beck B, Fountas S, Vangeyte J, Wal TV der, Soto I, Gómez-Barbero M, Barnes A, Eory V (2017) Precision agriculture technologies positively contributing to GHG emissions mitigation, farm productivity and economics. Sustainability 9:1339

[11] Limpisathian P (2011) Geographic Information System in Agriculture and Precision Farming. The Pennsylvania State University. State College: The College of Earth and …

[12] Edwards PN (2010) A vast machine: Computer models, climate data, and the politics of global warming. Mit Press

[13] Hildreth JC (2003) The Use of Short-Interval GPS Data for Construction Operations Analysis. PhD Thesis, Virginia Tech

[14] Matolcsi T, Matolcsi M (2008) Coordinate time and proper time in the GPS. European journal of physics 29:1147

[15] Rovelli C (2002) GPS observables in general relativity. Physical Review D 65:044017

[16] Shi Y, Thomasson JA, Murray SC, Pugh NA, Rooney WL, Shafian S, Rajan N, Rouze G, Morgan CL, Neely HL (2016) Unmanned aerial vehicles for high-throughput phenotyping and agronomic research. PloS one 11:e0159781

[17] Zijp W (1994) Improving the transfer and use of agricultural information: A guide to information technology. The World Bank

[18] Addicott JE (2020) The Precision Farming Revolution. In: The Precision Farming Revolution. Springer, pp 1-35

[19] Wang H (2017) Crop assessment and monitoring using optical sensors. PhD Thesis, Kansas State University

[20] Yuan W (2019) A Multi-Sensor Phenotyping System: Applications on Wheat Height Estimation and Soybean Trait Early Prediction.

[21] Shum H-Y, Chan S-C, Kang SB (2008) Image-based rendering. Springer Science & Business Media

[22] Tripodi P, Massa D, Venezia A, Cardi T (2018) Sensing technologies for precision phenotyping in vegetable crops: current status and future challenges. Agronomy 8:57

[23] Benson EE (1990) Free radical damage in stored plant germplasm. Bioversity International

[24] Burrough PA, McDonnell R, McDonnell RA, Lloyd CD (2015) Principles of geographical information systems. Oxford university press

[25] Kennedy MD (2013) Introducing geographic information systems with ARCGIS: a workbook approach to learning GIS. John Wiley & Sons

[26] Nath SS, Bolte JP, Ross LG, Aguilar-Manjarrez J (2000) Applications of geographical information systems (GIS) for spatial decision support in aquaculture. Aquacultural Engineering 23:233-278

[27] Afrooz AE, Lowe R, Leao SZ, Pettit C (2018) 3D and virtual reality for supporting redevelopment assessment. Real Estate and GIS: The Application of Mapping Technologies 151

[28] Sawyer JE (1994) Concepts of variable rate technology with considerations for fertilizer application. Journal of Production Agriculture 7:195-201

[29] Lobry de Bruyn L, Andrews S (2016) Are Australian and United States farmers using soil information for soil health management? Sustainability 8:304

[30] Ng CWW, Ni JJ, Leung AK, Zhou C, Wang ZJ (2016) Effects of planting density on tree growth and induced soil suction. Géotechnique 66:711-724

[31] Sharpley A, Sharpley A (2016) Managing agricultural phosphorus to minimize water quality impacts. Scientia Agricola 73:1-8

[32] Kaushik P, Kumar S, Dhaliwal MS (2018) Usefulness of Three under Exploited Vegetable Legumes—A Review. IntJCurrMicrobiolAppSci 7:861-870

[33] Yaghoubi S, Akbarzadeh NA, Bazargani SS, Bazargani SS, Bamizan M, Asl MI (2013) Autonomous robots for agricultural tasks and farm assignment and future trends in agro robots. International Journal of Mechanical and Mechatronics Engineering 13:1-6

[34] R. Shamshiri R, Weltzien C, Hameed IA, J. Yule I, E. Grift T, Balasundram SK, Pitonakova L, Ahmad D, Chowdhary G (2018) Research and development in agricultural robotics: A perspective of digital farming. 1-14. https://doi.org/10.25165/j.ijabe.20181104.4278

[35] Costopoulou C, Ntaliani M, Karetsos S (2016) Studying mobile apps for agriculture. IOSR J Mob Comput Appl 3:44-49

[36] Panda CK, Bhatnagar R (2020) Social Internet of Things in Agriculture: An Overview and Future Scope. In: Hassanien AE, Bhatnagar R, Khalifa NEM, Taha MHN (eds) Toward Social Internet of Things (SIoT): Enabling Technologies, Architectures and Applications: Emerging

Technologies for Connected and Smart Social Objects. Springer International Publishing, Cham, pp 317-334

[37] Ruan J, Hu X, Huo X, Shi Y, Chan FTS, Wang X, Manogaran G, Mastorakis G, Mavromoustakis CX, Zhao X (2020) An IoT-based E-business model of intelligent vegetable greenhouses and its key operations management issues. Neural Comput & Applic 32:15341-15356

Chapter 9

Technological Solutions and Specific Equipment for Improving the Degraded Grasslands by Total Reseeding

Vasile Mocanu, Tudor Adrian Ene and Vasile Adrian Blaj

Abstract

The scientific basis and the development of research on new solutions for organic production of fodder and their conversion into animal products with high biological value, maintaining biodiversity and environmental protection, have created the premises for promoting new technologies to improve, rehabilitate and enhance the grasslands, which are in accordance with a sustainable and efficient agriculture practice. In this chapter the technological solutions and technical equipment for improving the permanent grasslands by total renovation, specific to each stationary area conditions, are presented. The basis of new technologies or technological sequences for improving the degraded grasslands is composition and utilization of complex aggregates, consisting of specific machines and equipment, using the recent research achievements in the field of grassland farming. It is also intended that the proposed technological solutions eliminate or limit the effect of external restrictive factors so as to ensure a high feed production and a high feed value, meeting the agrotechnical requirements for each agricultural component operation. For highlighting the advantages of using the specific machines, traditional technologies, within are used common farming machines and new technologies, when are used specific machinery for grassland farming, are analyzed in comparison. To improve the degraded grasslands by reseeding method, new technologies and technological sequences use different complex aggregates, which perform in a pass 2 or even 3 operations, such as: clearing of non-valuable vegetation, hillocks and liming; tillage and liming; seedbed preparation and spreading the chemical fertilizers; rolling before sowing, sowing, rolling after sowing; the destruction of the old grass carpet, seedbed preparing, sowing and rolling after sowing and fertilization with chemical fertilizers; fertilization with chemical fertilizers, rolling before sowing, sowing and rolling after sowing. Compared to traditional technologies, new technologies and technological sequences for improving the degraded grasslands, require reduced fuel consumption and labor, with a lower number of aggregate passes.

Keywords: grassland farming, technology, total reseeding, specific equipment, feed quality

1. Introduction

The technological solutions and specific agricultural machinery for improving the permanent grasslands by total renovation presented in this chapter meet the general objective of the Grassland Science to enhance pastoral heritage by increasing total feed production and quality, with an optimal conversion into animal products, in accordance with a good agricultural practice characterized by the harmonization between economic and social development, biodiversity conservation and environmental protection.

The very different ecological conditions in which the grasslands are located, as well as the socio-economic changes that have led to a certain stage of degradation, require an integrated and interdisciplinary approach in order to develop new solutions for rational management of pastoral heritage, to achieve higher quantities of healthy animal products, to conserve biodiversity and with a reduced ecological impact.

These technologies are consistent with achieving the target regarding the adaptation of the permanent grassland reseeding technologies, specific to each stationary area conditions and implementation of some specific equipment for mechanization the technological sequences in grassland farming.

It is also intended that the proposed technological solutions eliminate or limit the effect of external restrictive factors, so that on the pastures a high fodder production with a high nutritional value is achieved.

For this purpose, the final decision on establishing the technology for improving a degraded grassland area must be taken after an analysis of the limiting factors, the state of degradation, the stationary conditions, the endowment level of the agricultural holding and the possibility of access for agricultural equipment.

Traditional technologies, within are used in conventional farming machines and new technologies, when are used the specific machinery for grassland farming, are analyzed in comparison.

2. Purpose of technological solutions

The productivity of the grasslands is directly influenced both the environmental conditions: these surfaces are located in very different stationary area conditions, usually occupying areas unsuitable for other crops, the poor physics-chemical properties of the soil, the orography of the land or the insufficient temperature with too short a vegetation period from higher altitudes and other causes, and as well as the activities of man and his animals: abandonment, weed control, improper maintenance and use, water imbalance, pollution, etc.

The implementation of the most appropriate technological sequences for the improvement of permanent grasslands and specific agricultural machinery is based on establishing a favorable interaction on the successful use of natural resources, between grasslands ecosystems improved by total renovation and animal husbandry systems.

It is absolutely necessary that before deciding the appropriate improvement solutions and technologies, the causes of the degradation of permanent grassland must be previously determined, because the application of any solution to improve the sward without removing the degradation causes, leads to some good results, valid only on short time.

In accordance with these requirements, the purpose of these technological solutions consists in:

- adaptation of technologies for reseeding permanent grasslands, specific to each stationary area conditions, to achieve sustainable agricultural systems, with minimal effects caused by climate change and economic optimization of technological sequences for obtaining and utilizing the grassland forage;

- increasing the nutritional value of the grass carpet, which ensures a balanced and efficient feeding of different categories of animals, especially from bovine and ovine species, in order to obtain healthy animal products and welfare;

- promoting some specific agricultural machinery for mechanization of grassland farming in condition of a lower specific fuel consumption, labor force number of aggregate passes and implicitly lower costs.

3. Grasslands suitable for improvement by total reseeding method

This grassland improvement solution consists in a complex of works for the total replacement of old vegetation with valuable species and perennial mixtures of grasses and perennial legumes by reseeding.

On the other hand, creating the grassy carpet on the degraded areas, is the simplest and most economical solution to combat soil erosion and improve its physics-chemical properties [1–5].

The improvement of degraded grasslands by reseeding method is the main technology, which is applied on grasslands with an advanced degree of degradation, as well as the poorly productive ones [6].

Within category of degraded and poorly productive grasslands can be mentioned [1–5]:

- grasslands with degraded vegetal carpet;

- grasslands with small productions (less than 4 t/ha green mass);

- grasslands with a degree of vegetation cover below 60%;

- poor quality grasslands, invaded by hillocks (**Figure 1a**), species with low forage value and weeds (over 25–30% of vegetation cover degree, **Figure 1b**);

a b

Figure 1.
Grasslands invaded by molehills, woody vegetation (a) and species with low forage value and weeds (b).

Figure 2.
Grasslands resulting from the destruction of large hillocks and high density.

- grasslands from which the non-valuable woody vegetation was completely cleared (including removing stumps and roots);
- grasslands resulting from the destruction of large hillocks and high density per unit area (more than 25%, **Figure 2**);

Reseeding with species, grasses and mixtures of valuable grasses and legumes is also done in the following cases:

- intensive grasslands after the end of the economic cycle of use (4–5 years);
- grass and perennial legume seed production;
- the establishment of fodder crops, which are introduced in crop rotation on arable lands;
- the establishment of recreational areas and sports fields, etc.
- a rational, ecological and economic solution to bring back abandoned lands in a sustainable agricultural system (grassland rehabilitation).

4. Factors influencing the solutions for improving degraded grasslands

In determining the most appropriate measures to improve degraded grasslands, special attention should be paid to the major influence of the following factors [2, 3, 7–14]:

- quality, the degree of coverage with old vegetation and the thickness of the old greensward;
- soil characteristics;
- area, altitude, exposure, slope and ground orography: unevenness and erosion stage;
- subsequent destination of the improved grassland (pasture, hayfield or mixed etc.);

- level of logistics equipment of the agricultural holding for mechanization of grassland farming;

- financial possibilities of the agricultural holding.

All of the above factors are important, but some of them still need to be given a higher attention.

The analysis of the soil characteristics is made on the basis of the pedological study and the agrochemical mapping of the grassland. From the pedological study as well as on the basis of the agrochemical mapping of the soil the degraded grassland, which are obligatory in this situation, result: the useful edaphic volume of the soil; the thickness of the fertile soil layer and the level of surface erosion; rate of amendment and fertilization; the type, texture and load capacity of the soil that influence the mechanization variants used.

The location area, the altitude, the slope and the surface exposure have a decisive importance in terms of the species and mixtures of forage plants indicated for improvement, respectively on the optimal working period. It is very important to know that the slope of the land is a restrictive factor in the application of measures to improve degraded grasslands, due on the one hand to the restrictions on anti-erosion protection of these areas and on the other hand to the slope size of the recommended of operation limits of agricultural aggregates [7, 8].

The level of logistics equipment of the agricultural holding is one of the decisive factors for implementing the technological sequences of technology. Depending on it, different mechanization variants are established [7, 8]. Also, the financial possibilities of the agricultural holding are decisive in the application of measures to improve these areas [7, 8]. This is because funds are needed for the purchase or rental of specific machines, for fertilizers, seeds, fuels, for mechanical works, for labor remuneration etc.

For providing optimal conditions for seed germination, for the growth of seedlings and their root system, with the ultimate goal of developing a dense vegetation carpet, the best solution will be decided in accordance with the following: adequate tillage system for the destruction of old vegetation, the period of preparation of the germination bed and sowing, basic fertilization, seed mixture, agricultural machinery and equipment for sowing and post-sowing management.

In conclusion, for the successful establishment of a productive grassland by total reseeding must meet the following minimum requirements:

1. Old sward has as small coverage as possible to reduce competition for sown seeds. If it is necessary to do this by work prior to sowing;

2. Severe destruction of old vegetation before sowing. This can be done by mowing, grazing or chemical treatment;

3. Control of perennial existing weeds before sowing;

4. Improving soil characteristics, such as correction of acidity and/or level of nutrient supply, which limit plant growth;

5. Carefully choose the right soil type; avoid heavy and stony soils that are the most difficult;

6. The characteristics of the soils prepared for sowing correspond to the agro technical requirements of the sowing operation (good seedbed conditions, soil will be well crushed and rolling);

7. The correct use of agricultural machinery and equipment appropriate to the various stationary area conditions;

8. Reseeding should be made at period of the year when soil moisture allows rapid installation of new sward (avoid sowing in mid-summer when soil moisture will be limited); if sowing is made in the late summer-early autumn season, the deadline is mid-August -September 1st (especially for areas with a temperate continental climate, specific to Central Europe);

9. Weed control as soon as possible; pest control if they are a problem - snails, for example;

10. Where possible, perennial mixtures of grasses and perennial legumes should be composed of competitive, fast-growing species and varieties, adapted to different stationary conditions;

11. Establishing the most appropriate solutions for the improvement of degraded pastures in accordance with individual circumstances and with a low impact on the environment;

12. Target objective: the existence of a number of approximately 6,000 seedlings/m^2, in the spring of the year after sowing [15].

5. Farm works and specific agricultural machinery systems necessary for reseeding the grasslands

When determining the type of farming work to be applied on a pasture, the production potential of the grass carpet is a very important criterion, but it is not the only one. The slope of the land limits the degree of mechanization and involves restrictions on some operations that could cause the erosion process. The grasslands located on slopes over 17^0 (30%) no tilled [16]. Also excluded from tillage, no considering their productive level, pastures located on shallow soils with rock fragments on the surface, with groundwater below 50 cm depth and those with a protective role (ravines, gullies etc.) [3, 16].

Improving meadows by reseeding includes the following groups of farming works:

- to create optimal conditions for seed germination, establishment and a good plant growth;

- itself sowing operation;

- grassland farming maintenance after reseeding.

To achieve the optimal conditions for establishment and a good plant growth, the following works are necessary: elimination of excess moisture, combating erosion and landslides, cleaning of hillocks and worthless vegetation, correcting the soil reaction (acidity and alkalinity), fertilizing, preparing the land for sowing. In general, these works are included in pastoral development projects and require agricultural machinery powered by high energy sources, usually used for sward renovation by surface works and other agricultural crops. That is why in this chapter will not be presented the technological sequences and the agricultural equipment used for their realization. In the following this paper will only focus on the farm works and technical systems necessary for seedbed preparing, sowing the grassland fodder plants and grassland maintenance after reseeding.

5.1 Introductory knowledge regarding the choice of the optimal variant for sowing the grassland fodder plants

The degraded grasslands, recommended for improvement by the method of reseeding are located in very different stationary conditions, starting from the plain area to the mountain area. Adding to this diversity of pedoclimatic conditions, the mode of utilization and the degree of grassland productivity, which can also be very different, results the special importance that must be given to the establishment of species and varieties that are components of grass mixtures.

5.1.1 Choice of species and structure of seed mixtures

A mixture of fodder plants consists of grasses and perennial legumes. The percentage of participation in the mixture of perennial grasses is 60–80%, and that of perennial legumes is 20–40% [2, 3].

In the composition of the seed mixtures are chosen those species that correspond to the stationary area conditions (climatic and soil conditions etc.), the method of cultivation and utilization of pasture, with a high capacity of competition and forage value. These conditions can be met only on the basis of knowledge of climate and soil requirements, suitability for different use system and the nutritional value of the main species of grasses and perennial legumes of grasslands.

For choosing perennial grass and grass species and varieties for composition of a seed mixtures, the following should be taken into account:

1. Duration of use of the grass carpet;

2. How to use the reseeded grassland: grazing, mowing or both; environment protection, biodiversity conservation;

3. The intensity degree of sward utilization related to fertilization rate, animal species and stocking rate, cutting and storage type for preservation;

4. Climate and soil conditions; the possibility of special conditions, such as: disease, drought or disappearance due to frost;

5. Compatibility of species and varieties of perennial grasses and legumes in relation to the optimal harvest time for a superior fodder quality (hay, silo, semi-hay);

6. Compatibility of species and varieties of perennial grasses and legumes in order to obtain optimal fodder productions and qualities.

The large number of new varieties, with different qualities, allow the realization of mixtures compatible with the pedoclimatic conditions specific to each location and how to use later.

Recently, multi-species seed mixtures have been recommended, the so-called complex mixtures, especially for grassland used extensively and for the ecological reconstruction of degraded lands.

5.1.2 Season of sowing

In all areas of the country, spring sowing is a complete success, as early as possible, when agricultural machinery can move in the field and when the temperature does not fall below 0°C (during the nights). On the lands and in the years with sufficient humidity, sowing can be done at the end of summer-beginning of autumn, respectively at the end of August-beginning of September.

In areas with irrigation possibilities, the most recommended sowing time is late summer and early autumn.

5.1.3 Sowing distance

In general, for all mixtures of grassland fodder plants and perennial fodder crops, the inter-row distance is 12–15 cm. In areas and years with sufficient rainfall, mixtures of meadow fodder plants and perennial fodder crops such as alfalfa, red clover can also be sown by broadcasting variant. This sowing method has the advantage of a better ground cover and a more uniform distribution of the nutrition space for each sown plant. That is why it is very successfully applied in the case of park areas, sports and landscape beauty etc.

5.1.4 Sowing depth

Sowing depth varies depending on seed size and on water content of the soil. In the case of mixtures, the sowing depth is conditioned by the seed with the smallest dimensions. Thus, white clover and timothy seeds and mixtures containing these species shall be sown to a depth of 1–2 cm, and other seeds and mixtures to a depth of up to 2.0–2.5 cm, except for sainfoin which sown depth is of 3.5–4.0 cm. In the case of broadcasting method, the sowing depth of 1.0–1.5 cm is ensured by the rolling work before and after sowing by special construction rollers, for a better seed contact with the soil moisture.

5.1.5 Seed rate

In general, perennial grasses and legumes, when used as feed, are seeded in a mixture, this leading to a high quantity and quality of fodder. The exception is alfalfa and red clover, which are also grown in pure crops.

The species and mixtures of different species of perennial grasses and legumes are chosen according to the climate and soil conditions, their productive potential, their forage value, the way they respond to fertilization and the suitability for the desired mode of exploitation.

When sowing recreational areas, special mixtures are used, which meet specific requirements such as: appearance, shade resistance, treading resistance, regeneration after mowing etc. For sports fields, perennial grass species resistant to trampling are chosen, with a low high, which are well regrown and resistant to repeated mowing.

The calculation of the sowing rate of each Q_{si} species participating in the mixture is made with the following relation [8]:

$$Q_{si} = k_s \times S \times q_i \times \frac{100}{V_{ui}}, \text{in [kg]} \tag{1}$$

in which:

- S is the size of sowing surface, in [ha];
- q_i is the rate of sowing for i species for each ha, in [kg/ha];
- V_{ui} represents the useful value of the seed, is made with the following relation:

$$V_{ui} = \frac{P_i \times G_i}{100} \text{ in } [\%], \tag{2}$$

where: P_i represents the purity of the seed of species i, in [%], and G_i represents the germination of the seed of species i in [%], according to the quality certificate;

- k_s is a super unit coefficient that takes into account possible seed losses through handling, administration, etc. Usually we take k_s = 1.05–1.10.

The total mixture sowing rate $\boldsymbol{Q_s}$ is determined by the relation:

$$Q_s = \sum_{1}^{n} Q_{si}, \text{ in } [\text{kg}], \tag{3}$$

where n represents the number of species composing the mixture of seeds.

5.2 Farm works and technical systems necessary for seedbed preparing

The reseeding methods depend by local area stationary conditions, especially the circumstances of the land to be cultivated, such as:

a. degraded grasslands with deep layer of fertile topsoil and thin grass sward;

b. degraded grasslands with deep layer of fertile topsoil and deep grass sward;

c. degraded grasslands with thin layer of fertile topsoil and deep grass sward;

d. degraded grasslands with thin layer of fertile topsoil and thin grass sward;

e. grasslands affected by erosion phenomenon;

f. grassland, fodder or grass and forage legume seed crops established in arable land.

5.2.1 Degraded grasslands with deep layer of fertile topsoil and thin grass sward

For destructing the old vegetation, firstly it is recommended harrowing by two perpendicular passages with the help of medium weight disc harrow. After the disc tillage, the plowing is done at a working depth of 18–22 cm. On slopes higher than 9^0 (16%) plowing is performed with reversible plows. The destruction of the old sward can also be performed with heavy weight disc harrows, performing, in this case, two perpendicular passes. Proper soil leveling must be carried out before or at the same time as the germination bed is prepared. A special importance will be given to the leveling of the land in the case of the establishment of the seed production fields. This is performed on the diagonal of the parcel, using either special graders or leveling blades (bars, rails etc.) mounted on the germination bed preparation aggregates. The seedbed preparing is made with light (15–20 daN/disc) and medium weight disc harrows (20–30 daN/disc) or with rotary harrows. The preparation of the germination bed is finalized with total tillage cultivators, unless the rotary harrows were used. The working direction for the preparation of the germination bed is on the diagonal of the surface. Before sowing, as well as after sowing, the soil is rolled with a ring roller, possibly ballasted.

By using complex aggregates or combined agriculture machinery, two or more works can be performed simultaneously, which allows reducing energy consumption and limiting the degree of environmental pollution etc.

5.2.2 Degraded grasslands with deep layer of fertile topsoil and deep grass sward

The first agricultural works consist of cutting at a low height carried out with the rotary cleaning machine (rotary toppers) with a vertical rotor or a horizontal rotor, equipped with articulated knives and a floating linkage for better contour following. Destruction of the old sward and seedbed preparing can be done in several ways, namely:

- by a milling at a depth of 6–8 cm, using total agricultural milling machines with "L" shaped knives. Next, plowing, leveling, preparing the germination bed and rolling are performed the same as in the previous point;

- with heavy disc harrows, making two perpendicular passes. The preparation of the germination bed is done with harrows with medium or light weight discs and with combined tillage cultivator. Leveling is performed by mounting a leveling blade behind the disc harrows. Rolling before and after sowing is performed with ring rollers;

- by rotary milling to a depth of 7–8 cm, using agricultural milling machine with total soil processing. The preparation of the germination bed and the sowing is done with the combined machines of total rotary milling and sowing, in which case this operation is perpendicularly on the first pass, and the working depth is 12–13 cm. It is recommended to allow an interval of 10–14 days between the two passes to dry the processed soil.

Complex aggregates can also be used in this situation, which perform two or more operations in a single pass, allowing the reduction of the number of passes, fuel consumption and the level of environmental pollution.

5.2.3 Degraded grasslands with thin layer of fertile topsoil and deep grass sward

Soil cultivation can be done through two technological variants, respectively with or without the use of herbicides. In the first way, for the destruction of the old degraded grass carpet total herbicide is used, by operating the spraying machines for total herbicide. Herbicide operation is carried out in spring or summer two weeks before tillage. After two weeks, the old vegetation is lightly cutting with the grassland cleaning machines. The destruction of the old sward and the preparation of the germination bed can be made using the heavy weight disc harrow, the light disc harrow, the combine soil cultivator and the roller.

If it is not sprayed, a cutting at a low height of the non-valuable vegetation is made previously using rotary cleaning machines. The control of the old vegetation and seedbed preparing is made in two perpendicular passes of the rotary milling machine equipped with "L" shaped knives for the total processing. The first pass is made at a depth of 6–8 cm with the adjustment of the speed rotation to the maximum. The second passage, recommended to be done after 10–14 days, is performed at a depth of 10–12 cm.

Another variant with low specific consumption consists in destroying the old sward by rotary milling operation to a depth of 6–7 cm, followed by seedbed preparing and sowing, works that are made with the combined rotary tiller-drill machines of total milling and sowing, perpendicular to the first pass, the working depth being of 10–12 cm. Between the first and second pass, to allow the old cultivated soil become dry, an interval of 10–14 days is recommended.

Also if the herbicides are not used, the destruction of the old vegetation and the preparation of the germination bed can be made by two perpendicular passes with heavy disc harrows, followed by medium disc harrows, the combined cultivator and the roller with the ring roller, but the quality of the work, even after the execution of a high number of repeated operations (both with the heavy disk and with the medium one) it is inadequate (degree of incorporation, control of the old sward and soil tillage are unsatisfactory).

5.2.4 Degraded grasslands with thin layer of fertile topsoil and thin grass sward

In the case of these meadow areas, depending on the endowment of the holding, the following variants are recommended:

Option 1: the destruction of the old sward and the preparation of the germination bed can be achieved by two perpendicular passes using heavy disc harrow, followed by 2–3 passes with the medium weight disc harrow, by combined cultivator and the ring roller.

Option 2: control of the old sward is performed with heavy disc harrows by two perpendicular passes. At the second pass, leveling is done by mounting a leveling blade behind the disc harrows. Simultaneously rolling before and after sowing and sowing are operated with special machines for sowing grassland fodder plants.

Option 3: control of the old sward, preparation of the germination bed and sowing are performed in a single pass, in this variant grassland rotary tiller-drill machines are used.

5.2.5 Grasslands affected by erosion phenomenon

In variant of grasslands with a low slope, up to 10–12^0 (17,5-21,0%), and with a colluvium horizon, the soil is mobilized by superficial plowing to a depth of 14–16 cm with reversible plows or by repeated operations with light or medium weight disc harrow [3, 16].

After plowing, the land is cultivated with light disc harrows with the combined seedbed preparation devices equipped with a leveling blade and with the ring roller.

In the case of lands with slopes between 12 and 22^0 (21 and 40%) the control of the old vegetation can be done in the following ways:

- repeated passes with medium disc harrows;
- superficial plowing with reversible plows;
- rotary milling to 6–8 cm depth with rotary cultivator equipped with knives for total processing.

The preparation of the germination bed on the plowed lands is further achieved with rotary harrows or with medium disc harrows, with the combined seedbed preparation devices with leveling blade and with the ring roller.

The preparation of the germination bed on the rotary milling lands is done by a second passage at a depth of 8–10 cm with the rotary cultivator with "L" shaped knives, for total processing, followed by the ring roller.

Rolling before and after sowing and sowing can be carried out simultaneously with special machines for sowing grassland fodder plants.

The use of combined rotary tiller-drill machines allows the realization of the old vegetable carpet control, the seedbed preparation, the sowing and the rolling after sowing by one pass.

On slopes greater than 7^0 (12%), the farm works must be carried out on the level curves, strictly observing the following technological requirements: on long slopes, where soil erosion process is favored, the operations necessary for grassing must be done in alternative lanes parallel to the level curves. Uncultivated strips are to be operated on next year when the first series is already established. The width of the lanes varies depending on the slope as follows: on a slope of $7–9^0$ (12–16%) between 30 and 40 m; on a slope of $9–14^0$ (16–25%) between 20 and 30 m; on a slope of $14–18^0$ (25–32%) between 12 and 20 m; respectively on a slope of $18–22^0$ (32–40%) between 7 and 12 m [1, 7, 8, 16, 17].

The machines and equipment are operated with special slope tractors (4WD tractor, equipped with double wheels or caterpillars).

5.2.6 Grassland, fodder or grass and forage legume seed crops established in arable land

The land is plowed to a depth of 20–22 cm with the plows, then the germination bed is prepared with rotary harrows or with light or medium weight discs harrow followed by the combined cultivator with leveling blade and rolling with the ring roller. The preparation of the germination bed has to be executed on the diagonals of the parcel. Particular attention will be paid to leveling the ground, which can be realized either before the preparation of the germination bed, or at the same time. For leveling before the seedbed preparation, the trailed graders are used, and for the leveling during the seedbed preparation, leveling blades mounted behind the germination bed preparation aggregates are used.

By using complex or combined farming aggregates, two or more operations can be performed simultaneously, which allows reducing energy consumption and limiting the degree of environmental pollution etc. [1, 8, 17, 18]. Thus, by using the combined rotary tiller-drill machines, the soil processing, the preparation of the seedbed, sowing and rolling after sowing are performed in a single pass.

On the other hand the use of the special sowing machines for grassland fodder plants, the seeding, and the rolling before and after sowing is simultaneously performed. Also, by forming and using compatible complex farming aggregates, which can perform in a single pass several works for example: plowing simultaneously with the fertilizer; old sward control simultaneously with the seedbed preparation etc., considerably reduces fuel consumption, the level of environmental pollution and production costs.

5.3 Technologies and specific agricultural machinery systems necessary for reseeding the grassland

In this subsection are presented the different reseeding methods for improving the degraded grasslands, located in different stationary area conditions.

For improving the degraded grasslands it is necessary to adopt optimal technological solutions so as to obtain the desired results, avoiding technical errors. Traditional technologies, within are used in conventional farming machines and new technologies, when are used in specific machinery for grassland farming, are analyzed in comparison.

The differences of necessary of specific fuel consumption, labor force and number of aggregate passes between usual and new technologies have been determined.

According to the stationary area conditions, are recommended to apply the most appropriate alternatives to improve the degraded grassland by reseeding method, taking into account also the endowment of farm. Therefore there are situations when for mechanization of grassland farming is used conventional agricultural machinery.

Classic technology for reseeding the degraded grasslands use aggregates providing the realization of one operation by one pass machine, such as: liming; spreading the fertilizers; destruction of the old grass sward; seedbed preparing; rolling before sowing; sowing; rolling after sowing [8, 17].

The new technological solutions are based on utilization of complex farming aggregates, using specific equipment and machines.

Depending on work conditions, to improve the degraded grassland by reseeding method, new solutions of mechanization technology using complex aggregates which provide the realization of 2, 3 or 4 operations by one pass machine are used.

Thereby specific fuel consumption, labor force and involving the number of passes are lower.

New technology for reseeding the degraded grasslands use complex aggregates providing the realization of two or more operation by one pass machine, as following:

- destruction of the old sward and liming;
- seedbed preparation and fertilization with chemical fertilizers;
- rolling before and after sowing, sowing of fodder grass plants;
- destruction of the old grass sward, seedbed preparing, sowing of fodder grass plants, rolling after sowing and fertilization with chemical fertilizers;
- clearing of non-value vegetation, of mole-hills and fertilization with chemical fertilizers;
- clearing of non-value vegetation, of mole-hills and lime spreading;
- destruction of the old grass sward and seedbed preparing;
- fertilization with chemical fertilizers, rolling before/after sowing, sowing of fodder grass plants.

For a guaranteed success, it is important that each technological sequence corresponds to the agronomic requirements and has a beneficial effect on the environment.

In **Table 1** there are presented the new technology for reseeding the degraded grassland.

In **Figure 3** are presented the total fuel consumption, necessary labor force and the number of machine passes, both for usual variants and new technological solutions. The data are given for each situation in which the degraded grasslands are located (a, b, c, d, e and f).

The graphs were drawn based on the data obtained during the experimental results for the variants of using the usual agricultural equipment and using the specific agricultural equipment for mechanization of grassland farming [8, 9].

Operation	**Recommended aggregate***
0	1
a. Degraded grassland with deep layer of fertile topsoil and thin grass sward	
a.1. Variant 1	
Destruction of the old sward and liming/ fertilization	 Wheel tractor of 74 kW + Heavy disc harrow GDG 2,7 type +Equipment for chemical fertilization EF 2,5 type**
Seedbed preparation and fertilization with chemical fertilizers	 Wheel tractor of 74 kW + Rotary harrow 2,5 m with + Equipment for chemical fertilization EF 2,5 type
Rolling before/after sowing, sowing of fodder grass plants	 Wheel tractor of 74 kW + Special machine for sowing fodder grass plants MSPFP 2,5 type**
a.2. Variant 2	
Destruction of the old grass sward, seedbed preparing, sowing of fodder grass plants, rolling after sowing and fertilization with chemical fertilizers	 Wheel tractor of 74 kW + Rotary tiller-drill machine improved MCT 2,5 M type** + Equipment for chemical fertilization EF 2,5 type

Operation	Recommended aggregate*
0	1
b. Degraded grassland with deep layer of fertile topsoil and deep grass sward	
b.1. Variant 1	
Cleaning of non-value vegetation, of mole-hills and fertilization with chemical fertilizer	 Wheel tractor of 74 kW+ Machine for cleaning the grassland MCP 2,5 type + Equipment for chemical fertilization EF 2,5 type
Destruction of the old sward and liming/ fertilization	 Wheel tractor of 74 ... 88 kW + Heavy disc harrow + Equipment for chemical fertilization EF 2,5 type
Seedbed preparation and fertilization with chemical fertilizers	 Wheel tractor of 74 kW + Rotary harrow with + Equipment for chemical fertilization EF 2,5 type
Rolling before/after sowing, sowing of fodder grass plants	 Wheel tractor of 74 kW + Special machine for sowing fodder grass plants MSPFP 2,5 type
b.2. Variant 2	
Cleaning of no value vegetation, of molehill and administration of the amendments	 Wheel tractor of 74 kW + Machine for cleaning the grasslands MCP 2,5 + Equipment for chemical fertilization EF 2,5 type

Operation	Recommended aggregate*
0	1
Destruction of the old grass sward, seedbed preparing, sowing of fodder grass plants, rolling after sowing and fertilization with chemical fertilizers	 Wheel tractor of 74 kW + Rotary tiller-drill machine improved MCT 2,5 M type + Equipment for chemical fertilization EF 2,5 type
c.Degraded grassland with thin layer of fertile topsoil and deep grass sward	
Cleaning of no value vegetation, of molehill and administration of the amendments	 Wheel tractor of 74 kW + Machine for cleaning the grasslands MCP 2,5 + Equipment for amendment administration EF 2,5 type
Destruction of the old grass sward, seedbed preparing, sowing of fodder grass plants, rolling after sowing and fertilization with chemical fertilizers	 Wheel tractor of 74 ... 88 kW + Rotary tiller-drill machine improved MCT 2,5 M type + Equipment for chemical fertilization EF 2,5 type
d. Degraded grassland with thin layer of fertile topsoil and thin grass sward	
Destruction of the old grass sward, seedbed preparing, sowing of fodder grass plants, rolling after sowing and fertilization with chemical fertilizers	 Wheel tractor of 74 kW + Rotary tiller-drill machine improved MCT 2,5 M type + Equipment for chemical fertilization EF 2,5 type
e. Grassland affected by erosion	
Destruction of old grass sward and seedbed preparation	 Wheel tractor of 59 kW + Medium or light weight disc harrows in aggregate with fixed toothed harrows

Operation	Recommended aggregate[*]
0	**1**
Sowing of fodder grassland plants and rolling before and after sowing	 Wheel tractor of 74 kW (80 HP) + Special machine for sowing fodder grass plants MSPFP 2,5 type
f. Grassland, fodder or grass and forage legume seed crops established in arable land	
f1. Variant 1	
Tillage for preparing the seedbed	 Wheel tractor of 88 kW + Medium weight disc harrow
Rolling before/after sowing, sowing of fodder grass plants	 Wheel tractor of 74 kW + Special machine for sowing fodder grass plants MSPFP 2,5 type
f2. Variant 2	
Destruction of the old grass sward, seedbed preparing, sowing of fodder grass plants, rolling after sowing and fertilization with chemical fertilizers	 Wheel tractor of 74 kW + Rotary tiller-drill machine improved MCT 2,5 M type + Equipment for chemical fertilization EF 2,5 type

[*] *When the operations are carried out on the grasslands affected by erosion and also located on slope conditions are required following the indications provided in chapter 5.2, letter e.*
[**] *Specific agricultural equipment for grassland farming designed, made and tested in different stationary area conditions by the Grassland Research Institute from Brasov, Romania.*

Table 1.
New technological variants for mechanization of grassland improvement by reseeding method [8, 9, 19].

In according with working conditions and degradation stage of grassland, the data presented in **Figure 3** demonstrates the following:

- total fuel consumption for usual alternatives varies between 27,0 and 86,8 l ha^{-1};
- total fuel consumption for new mechanization solutions range between 17,4 and 65,4 l ha^{-1};
- the fuel consumption economy range between 9,6 and 21,4 l ha^{-1} in favor of the new technologies;
- consumption of labor force for usual variants of mechanization varies between 3,9 and 12,1 man hour ha^{-1};
- consumption of labor force for new mechanization technologies range between 1,7 and 5,7, man hour ha^{-1};
- the labor force economy is between 2,2 and 6,4 man hour per ha, in favor of the new technologies;
- the number of aggregate passes for usual solutions varies between 5 and 10;
- the number of aggregate passes for new mechanization solutions range between 1 and 4.

6. Grassland maintenance after reseeding

The maintenance farming works are different depending on the sowing season, respectively spring or summer-autumn [1, 3, 8, 17].

6.1 Grassland reseeded in spring

At an interval of 3–4 weeks after sowing it is necessary to control weeds. This can be made in two ways, namely: chemically by selective herbicide, respectively mechanically by mowing and evacuation of mowed plants from the field. The choice of herbicides is made according to the weed species that need to be controlled based on the catalogs of plant protection products. For administration, the recommended doses and the application phase indicated by the manufacturer must be within the rules, strictly observing the environmental protection norms.

Mowing and removing from the field must be done in a good weather condition and in a short time so as not to damage the new established sward.

The following cycles are mowed or grazed, depending on the destination for which the respective grassland was reseeded.

In autumn it is necessary to remove excess grass by grazing sheep for a short time.

6.2 Grassland reseeded in end of summer-early autumn

In this situation, weed control by mowing operation is no necessary, as weeds are usually annual plants that disappear with the arrival of winter and do not have the necessary time for seed production. In the spring, the state of sown field is

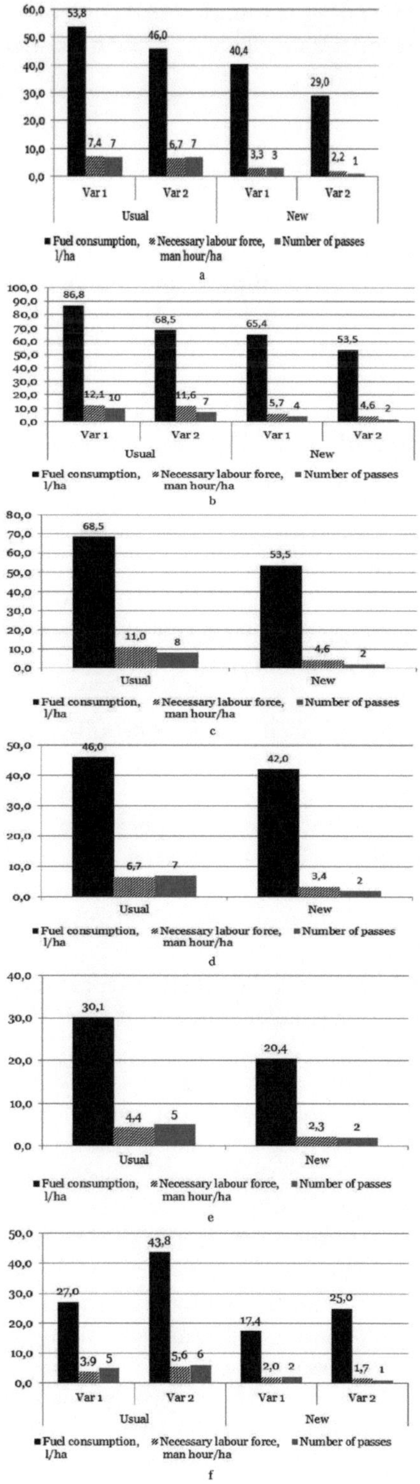

Figure 3.
Comparative of fuel consumption, necessary labor force and number of passes between new and usual technology, depending on local stationary area conditions (a, b, c, d, e and f): a-degraded grasslands with deep layer of fertile topsoil and thin grass sward; b-degraded grasslands with deep layer of fertile topsoil and deep grass sward; c- degraded grasslands with thin layer of fertile topsoil and deep grass sward; d-degraded grasslands with thin layer of fertile topsoil and thin grass sward; e-grasslands affected by erosion phenomenon; f-grassland, fodder or grass and forage legume seed crops established in arable land.

monitored and only if its weeding is found, weed control is used, proceeding as in the same way of the spring season. The other farm works recommended for grassland sown in spring remain valid.

7. Conclusion

Permanent grasslands cover a very large percentage of the world's surface and are of great importance to the environment, so their sustainable management is a widespread problem of interest and is not limited to those who exploit them.

For this purpose, the elaboration of technological solutions and the promotion of specific agricultural machines for improving the degraded pastures through total renovation, was based taking into account the causes of degradation, the diversity of stationary area conditions, destination (economic function, protection, landscape beauty etc.) and the specific equipment used for mechanization of these farming works.

It was also intended that the proposed technological solutions eliminate or limit the effect of external restrictive factors, so as to ensure a high feed production and a high quality, based on the causes of degradation, the diversity of seasonal conditions in which these areas are located, destination (economic function, protection, landscape etc.) and the specific equipment used for the mechanized execution of the works.

To establish these technologies, the results of multidisciplinary research were combined (agrochemistry, pedology, pratotechnics, soil processing works, fertilization, weed control, disease and pest control, soil analysis etc.). Based on the use of experimental results, zonal technologies and specific technical systems were promoted, focusing a sustainable development of agriculture in conditions of multifunctionality, biodiversity and stability of environmentally friendly yields.

In comparison to usual variants, the new mechanization technologies for improving the degraded grasslands by reseeding method, for different stationary area conditions, involve a reduced fuel comsumption of 10–46% and labour forces of 49–67% with a smaller number of agricultural aggregate passes of 3–8 units.

By lowering fuel consumption, necessary labour force and the number of machine passes, new technological solutions of mechanization of work for improving degraded grasslands by reseeding method have a reduced environmental impact, environment pollution (air, water, soil) is less, inputs are lower and costs decrease proportionally, within 15–30% [9].

The promotion of the most suitable technological sequences for improving the grasslands must be based on establishing a favorable interaction between the grassland ecosystems improved by total renovation and the animal breeding systems. To be successful these initiatives, with good results, avoiding technical errors, optimal technological solutions are necessary to adopt.

Therefore, in general, the human factor, through his activities, has a great responsibility for rehabiliting these degraded areas, by practicing the sustainable agricultural systems, to protect the environment.

8. Final conclusion

Using science and technology it can make grassland farming more profitable, agreeable and sustainable.

Author details

Vasile Mocanu*, Tudor Adrian Ene and Vasile Adrian Blaj
Research-Development Institute for Grasslands, Brasov, Romania

*Address all correspondence to: vasmocanu@yahoo.com

References

[1] Mocanu V, Hermenean I, Tehnologii de mecanizare pentru îmbunătățirea pajiştilor prin reînsămânțare. Revista Mecanizarea Agriculturii 2002;5 9-12.

[2] Cardaşol V, Simtea N, Proca M, Razec I, Hermenean I. Măsuri tehnico-organizatorice pentru sporirea producției pajiştilor. București (România): Redacția de propagandă tehnică agricolă; 1989.

[3] Maruşca T, Mocanu V, Cardaşol V, Hermenean I, Blaj VA, Oprea G, et all. Ghid de producere ecologică a furajelor de pajişti montane. Brașov (România): Editura Universității Transilvania; 2010.

[4] Soucek R, Pippig G. Maschinen und Geräte für Bodenbearbeitung, Düngung und Aussaat. Berlin (Germania): Verlag Technick GmbH; 1990.

[5] Mocanu V, Hermenean I. Mecanization of grassland farming - The final Meeting for FAO Grassland Ecosystem from mountain and Related Farm Products. Braşov (România): Editura ICDP Bv; 2003.

[6] Mocanu V, Hermenean I. MECANIZAREA LUCRĂRILOR AGRICOLE PE PAJIȘTI-Tehnologii, mașini și echipamente. Brașov (România): Editura Universității Transilvania; 2013.

[7] Hermenean I, Mocanu V, Gîțan D. Mecanizarea lucrărilor agricole în zona montană. Vatra Dornei (România): Centrul de Formare şi Inovație pentru Dezvoltare în Carpați CEFIDEC; 2005.

[8] Mocanu V, Hermenean I. Tehnologii pentru mecanizarea cu inputuri minime a lucrărilor de întreținere şi îmbunătățire a pajiştilor. Brașov (România): Editura Universității Transilvania; 2008.

[9] Mocanu V, Ene TA, Blaj VA, Silistru D. Solutii tehnologice și mijloace tehnice de îmbunătățire a pajiștilor permanente prin renovare totală. Brașov (România): Editura Universității Transilvania; 2018.

[10] Săplăcan L, Îndrumătorul mecanicului agricol. București (România): Editura Ceres; 1970.

[11] Wyss U, Jans F. Rundballensilage: Konservierung und Fütterung, Eidgenössiche Forschungsanstalt für viehwirtschafliche Produktion, Posieus, 1992.

[12] Moser D. Leistungs - Qualitäts - und Komfortsteigerung durch den Einsatz leistungsfähiger Zweiachsmäher und Tansporter, 1. Tagung ,,Landtechnik in Alpenraum" Österreich; 1992.

[13] Ott A. Anforderungen an die Mechanisierung von Hang und Bergbetrieben aus schweizerischer Sicht. 1. Tagung ,,Landtechnik in Alpenraum", Österreich; 1992.

[14] Rinaldi M, Naef E. Motorauslastung von Landwirtschaftstraktoren. FAT-Berichte. 426, Schweizerland. 1992.

[15] Frame J, Laidlaw A.S. Improved grassland management-New Edition. Ramsbury, Marlborough (United Kingdom): The Crowood Press; 2011.

[16] Dumitrescu N. Pajişti degradate de eroziune şi ameliorarea lor. Bucureşti (România): Editura Ceres; 1979.

[17] Mocanu V, Hermenean I, Maruşca T. Ecological and economical technology for mechanization of grassland improvement works. Uppsala (Suedia): Proceding of the 22nd General Meeting of the European Grassland Federation; 2008.

[18] Hermenean I, Necșulescu V. Încercări ale remorcii autoîncărcătoare

de furaje modernizate RAF -4 M. Lucrãri ştiinţifice ale ICPCP Braşov, 1992; 15:173-186.

[19] Mocanu V, Hermenean I. New mechanization alternatives with low inputs for reseeding degraded grasslands. Timişoara (România): Proceeding of Symposium Trends in European Agriculture Development, 2009; 41(2):462-467.

Section 2

Biosystems Engineering and Applications

Chapter 10

Use of Biofertilizers in Agricultural Production

José L. Zambrano-Mendoza, Carlos A. Sangoquiza-Caiza, Diego F. Campaña-Cruz and Carlos F. Yánez-Guzmán

Abstract

Most of agricultural production in Latin America consists of smallholder farmers who need the development of sustainable technologies, with costs according to their economic condition. Biofertilizers composed of free-living bacteria promote plant's growth, improve productivity through the strengthening of its roots, and reduce the amount of synthetic fertilizer applied to the crops. The aims of this chapter are to highlight the microorganisms commonly used in agriculture as biofertilizers and the main researches carried out in several countries of Latin America, and to describe the development of an experimental biofertilizer for maize, based on strains of *Azospirillum* spp. and *Pseudomonas fluorescens*, in the highlands (Sierra Region) of Ecuador. Seven phylum and 95 genera o microorganism used as biofertilizers or Plant Growth-Promoting Rhizobacteria (PGPR) are summarized, along with the benefits, challenges and future prospect of their use. The effectiveness of the experimental biofertilizer developed in Ecuador was demonstrated through several experiments at the green house and field, in which it was evident the increase of root's size, the amount of crocket, the percentage of dry matter, and the crops´ yield. The evaluations, accomplished on farmers' fields showed 30% of increase in yield and 21% of decrease in the cost of production per kilogram; as a consequence of the use of biofertilizer plus 50% of the recommended chemical fertilization, in comparison with standard farming techniques. Farmers can reduce the application of synthetic fertilizers and sustainably increase crop yield through the use of this technology.

Keywords: bioinoculant, corn, family farming, maize, microbiology, plant growth-promoting rhizobacteria, PGPR, sustainability

1. Introduction

In order to supply the growing demand for food, the expansion of the agricultural frontier has been one of anthropogenic activities with greater impact on ecosystems mainly due to: monocultures, watering systems, and use of pesticides and fertilizers. It's probably that the world's population will become 8000 billion people in 2030 and 9 billion in 2050, it means that food production should increase in between 60 and 70% [1–3].

There are two ways in order to achieve the challenge of growing agricultural production for supplying the growing population's food needs, such as:

- Increase the crops areas, with little possibility because it would affect large forest masses and nature reserves, especially in developed countries. In Latin America, the main source of new land for agriculture has been native forests or forests which were already intervened [4]. One of the most important challenges of Latin American countries is to achieve develop with a suitable level of environmental sustainability [5].

or

- Enhance crop yield through seed improvement (plant breeding) and use of environmentally safe and more assimilable fertilizers for plants. This option is the most feasible to choose because there are several ways to nourish plants, either by the development of synthetic or inorganic fertilizers (e.g. urea, ammonium nitrate, among others), organic fertilizers (e.g. humus, compost, leachates, among others), or biofertilizers (e.g. bacteria, fungi, algae), also known in a broad sense as bioinoculants or microbial inoculants.

Since 2005 to 2018, Latin America and the Caribbean increased the consumption of synthetic fertilizers from 15.2 million to 26.3 million tons, which means an increase of 73.59% (**Figure 1**). The use of inorganic fertilizers can pollute (e.g. heavy metals), damage water sources (e.g. nitrates and phosphates), and increase production cost. It means a great difficulty for those production systems based on peasant family farming, which has faced a hard situation in the last decades [7, 8]. Synthetic fertilizers represent in between 30 to 50% of production costs, depending on features of each crop.

This chapter summarizes the microorganisms commonly used in agriculture as biofertilizers or Plant Growth-Promoting Rhizobacteria (PGPR) and the main researches carried out in several countries of Latin America. The literature review was conducting using Google Scholar, identifying research literature published after 2000 using the term "biofertilizer". Then, only the most recent papers whose study was conducted in Latin America was kept to give examples of their use.

In addition, this chapter describes the development of an experimental biofertilizer for maize in the highlands (Sierra Region) of Ecuador. The biofertilizer was developed with native strains of *Azospirillum* spp. and *Pseudomonas fluorescens* collected in the highlands of Ecuador. The strains were first evaluated in greenhouse and only those isolates that showed a beneficial effect on plant growth

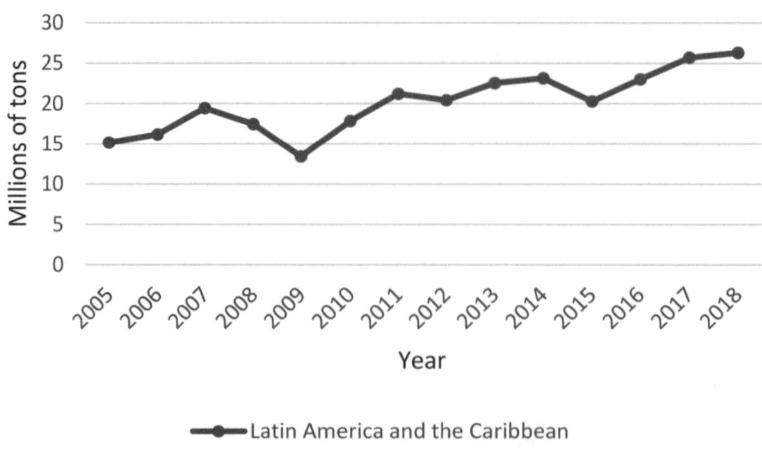

Figure 1.
Annual consumption of synthetic fertilizers in Latin America and the Caribbean. Source: Economic Commission for Latin America and the Caribbean (ECLAC), 2021 [6]. Made by: Authors.

were evaluated in the field. This chapter describes the entire protocol used for the development and production of the biofertilizer.

2. Biofertilizers in agriculture

Biofertilizers are made in laboratory with live or latent cells of organisms, either nitrogen fixers, solubilizers of phosphates, cellulites microorganisms, growth promoters, among others, which are applied to seeds or plants in order to boost their growth. In contrast to synthetic fertilizers, biofertilizers have microorganisms that are not a source of nutrients by themselves, but allow the access of available nutrients in the rhizosphere [9, 10].

In the last two decades, the use of biofertilizers or microbial inoculants has increased notably in several parts of the world [11]. Biofertilizers are considered as a feasible and sustainable attractive biotechnological alternative to increase crop yield, improve and restore soil fertility, stimulate plant growth, reduce production costs and the environmental impact associated with chemical fertilization [12–14]. Several microorganisms are commonly used as biofertilizers, including nitrogen-fixing soil bacteria (e.g. *Azotobacter*, *Rhizobium*), nitrogen-fixing cyanobacteria (e.g. *Anabaena*), solubilizing phosphate bacteria (e.g. *Pseudomonas*), and arbuscular mycorrhicical fungi (**Table 1**). Similarly, the producer bacteria of phytohormones (e.g. auxins) and those cellulite microorganisms are also used as a biofertilizers [10, 15]. In addition, the use of plant growth promoting bacteria can be useful in developing strategies to facilitate plant growth under normal and abiotic stress conditions [21].

It is likely that the market of biofertilizers in Latin America would grow at an annual rate of 10% during the period 2020-2025. Argentina is the largest and fastest growing market, followed by Brazil. The main incentive factors of market in these two countries are: a favorable government policy, easy registration process, and the increase of organic farms [22]. On the other hand, Peru, Colombia, Bolivia, Ecuador, and Venezuela's growth in the biofertilizer market would be slow, because they are ruled by ambiguous regulations.

2.1 Use of biofertilizers in major crops

Biofertilizers and PGPR have been evaluated in a wide variety of crops, including: rice, cucumber, wheat, sugarcane, oats, sunflower, corn, flax, beet, tobacco, tea, coffee, coconut, potato, fan cypress, grass sudan, eggplant, pepper, peanut, alfalfa, tomato, alder, sorghum, pine, black pepper, strawberries, green soybeans, cotton, beans, lettuce, carrots, neem, among others [23].

The most important example of the use and importance of biofertilizers in crop production is in soybean. Soybean production is mainly carried out by inoculating the seed with selected strains of *Bradyrhizobium japonicum*, *Bradyrhizobium diazoefficiens* or *Bradyrhizobium elkanii* (jointly referred to as *Bradyrhizobium* spp.). In Argentina, one of the main soybean producers in the world, there are around 70 companies producing and marketing biofertilizers for this crop [24].

Castro et al. [25] evaluated the benefits of the rice-*Azolla* association for rice cultivation in Cuba. The results showed that this cereal was positively influenced by the use of *Azolla*, which allowed a rise in the number of grains per panicle, panicle per m^2, and consequently, a significant increase in yields. In addition, it was observed that the association regulated the temperature and pH of the water.

Grageda-Cabrera et al. [26] evaluated the effect of the inoculation of bacterial and fungal isolates on nitrogen use efficiency (NUE) in wheat, using the ^{15}N

Phylum	Genus	Host	Benefit
Actinobacteria	*Arthrobacter, Brevibacterium, Cellulomonas, Corynebacterium, Kocuria, Microbacterium, Micrococcus, Mycobacterium, Rhodococcus, Streptomyces, Streptomonospora, Haloglycomyces, Haloactinospora, Actinopolyspora, Amycolatopsis, Prauserella*	Maize, pea, rice, soybean, sugarcane, sunflower, wheat.	Increase plant vigor (growth promoters), and tolerance to biotic and abiotic stress. Improve nutrients use efficiency.
Firmicutes	*Bacillus, Paenibacillus, Alicyclobacillus, Aneurinibacillus, Virgibacillus, Salibacillus, Gracilibacillus, Brevibacillus, Amphibacillus, Paraliobacillus, Oceanobacillus, Salimicrobium, Halobacillus, Pontibacillus, Thalassobacillus, Sediminibacillus, Alkalibacillus, Tenuibacillus, Ammoniphilus, Salinibacillus, Exiguobacterium, Marinilactibacillus, Alkalinebacterium, Sporosarcina, Planomicrobium, Lysinibacillus, Planococcus*	Amaranth, apple, barley, buckwheat, maize, mustard, oat, pepper, rice, sorghum, sunflower, tomato, wheat	Solubilization of phosphorus, potassium, zinc; production of indole acetic acids, hydrogen cyanide, gibberellic acid, and siderophore. Nitrogen fixation, biocontrol.
Proteobacteria	*Allidiomarina, Marinobacter, Aquisalimonas, Microbulbifer, Marinobacterium, Pseudomonas, Salicola, Deleya, Halomonas, Marinospirillum, Methylophaga, Achromobacter, Alcaligenes, Rhizobium, Albimonas, Paracoccus, Pantoea, Enterobacter, Kluyvera, Azospirillum, Methylobacterium, Arcobacter, Oceanibaculum, Fodinicurvata, Altererythrobacter, Glycocaulis, Xanthobacter, Bradyrhizobium, Amorphus, Sinorhizobium*	Amaranth, barley, bean, buckwheat, cotton, cowpea, maize, millet, mustard, oat, pea, rice, soybean, sunflower, tomato, wheat.	Increase plant vigor (growth promoters), nitrogen fixation, solubilization of nutrients, and biocontrol.
Bacteroidetes	*Flavobacterium, Shingobacterium*	Barley, Millet, Wheat	Plant growth promoting attributes
Ascomycota	*Trichoderma, Penicillium, Fusarium, Phoma, Aspergillus, Phomatropica, Acremonium*	Horticultural, fruit and forest crops	Biocontrol, biodegradation

Phylum	Genus	Host	Benefit
Glomeromycota	*Glomus, Gigaspora, Acaulospora, Scutellospora, Sclerocystis, Laccaria, Pisolithus, Boletus, Amanita, Pezizella.*	Horticultural, fruit and forest crops	Phosphate-mobilizing
Cyanobacterias	*Asterocapsa, Chroococcus, Aphanothece, Gloeocapsa, Microcystis, Synechococcus, Rhabdoderma, Merismopedia, Aphanocapsa, Coelosphaerium, Leptolyngbya, Pseudanabaena, Komvophoron, Oscillatoria, Lyngbya, Phormidium, Nostoc, Anabaena, Scytonema*	Bean, maize, rice.	Fixation of nitrogen, bioremediation, biocontrol.

Source: Modified by the authors from [15–20].

Table 1.
Groups of microbial inoculants for agriculture.

isotopic dilution technique. The inoculation of wheat with arbuscular fungi significantly increased grain yield up to 1 291 kg ha^{-1}, and the NUE up to 11%, in relation with the non-inoculated treatment.

The solubilizing phosphate bacteria *Pseudomonas putida*, *Microbacterium laevaniformans* and *Pantoea agglomerans*, were evaluated in potato to determine the effect of inoculation on growth and crop yield. The mixture of *P. agglomerans* or *M. laevaniformans* and *P. putida* substantially increased biomass and improved tuber growth. The yield was possibly due to the higher supply of phosphorus (P) from the bacteria to the growing plants. Among the microorganisms, *P. agglomerans* significantly improved potato growth and yield by approximately 20-25% [27].

Like several horticultural crops, tomato is influenced by the application of growth promoters. Bernabeu et al. [28] showed that the inoculation of seedlings with *Burkholderia tropica* had an effective colonization of the roots that spread to aerial tissues. This effective colonization led to an increase in tomato production in two growing seasons. Trials carried out by Mirik et al. [29] with pepper and *Bacillus* strains increased yield up to 23.5%.

Garza et al. [30] evaluated the response of annual and perennial crops to the application of biofertilizers in the central region of Mexico, in a series of experiments and plots in which the bacteria *Azospirillum brasilense* and *Rhizobium etli* were tested, as well as the fungus *Glomus intraradices* in cereals, legumes, and citrus. In most of the test locations, increases in production were recorded, that were up to 60% in maize, 85% in wheat, 74% in barley, 25% in oats (biomass), 36% in beans, and 111% in orange, in relation to the non-inoculated treatment.

Biofertilizers are also used in forest species. Inoculum prepared with the ectomycorrhizal fungi *Suillus luteus* and *Rhizopogon luteolus*, and the saprobes *Coriolopsis rigida* and *Trichoderma harzianum*, alone and combined, were evaluated as potential biofertilizers for the growth of *Pinus radiata* seedlings in greenhouse. At the end of the test (after 10 months), it was determined that the inoculants stimulated the growth of the plants compared to the control without inoculation. The inoculant formulated with the mixture *C. rigida* and *R. luteolus* produced *P. radiata* plants with the highest quality indices, being a viable alternative for their use in the production of *Pinus* spp. [31].

3. Mechanism of action of the biofertilizers

Biofertilizers increase the growth and yield of crops in an eco-friendly manner. They show synergistic and antagonistic interactions with the soil native microbiota and participate in many process of ecological importance. Biofertilizers promote plant growth by enhancing biotic and abiotic plant stress tolerance and supporting its nutrition by fixing atmospheric nitrogen and solubilizing soil nutrients [32, 33].

The detailed mechanisms of biofertilizers and PGPR and their specific contribution to plant growth have been reviewed comprehensively [34–37]. The action modes that PGPRs use to benefit plant growth can be classified into direct and indirect mechanisms, which occur inside and outside the plant, respectively. PGPRs directly promote plant growth by enhancing nutrient acquisition and by regulating phytohormones. The indirect effects of PGPRs on plant growth are caused by the induction of systemic resistance of plants against a wide range of pathogenic microbes. Direct action modes include an improvement in plant nutrition by providing nutrients, such as nitrogen, or solubilized minerals from the soil (e.g. P, K, Fe, Zn, among others) and/or stimulating plant growth by regulating the levels of phytohormones (e.g. gibberellins, auxins, ethylene, cytokinins, and abscisic acid). Indirect effects on plant development are given by the suppression of pathogens and other harmful microorganisms through parasitism, competing for nutrients and niches within the rhizosphere, producing antagonistic substances (e.g. antibiotics, hydrogen cyanide and siderophores) and enzymes. lytic (e.g. glucanases, proteases and chitinases), and the induction of plant systemic resistance against a wide spectrum of pathogens [33].

4. Biofertilizers for maize in Latin America

During the last 30 years, several researchers have discovered many species of nitrogen-fixing bacteria associated with cereals and other crops which no generate nodules. Virtually all of these bacteria are microaerophilic, they fix nitrogen only when there is a low oxygen pressure. One of the most important is *Azospirillum* spp. which are associated with the rhizosphere of plants nourish them [38]. Döbereiner et al. [39], mentioned that these bacteria, in tropical and subtropical regions, occur naturally in numbers between 103 to 106 per gram of soil, and in even higher numbers on the root surface of cereals and forage grasses.

Maize is the most cultivated cereal in the region and its planted area is bigger than other economically important ones, such as rice or wheat [40]. Maize requires intensive use of nitrogen fertilizers for suitable production, which leads to increased costs and possible environmental pollution [41]. Nowadays, it is necessary to preserve the soil's productive capacity. Plant nutrition and soil improvement practices must be integrated to enable an adequate management of nutrients, giving importance to the biodiversity found in soils [42]. An appropriate management of the association *Azospirillum*-maize can result in productivity gains and lower production costs [43].

In Latin America, several cases of use of biofertilizers in maize cultivation have been reported with promising results. Girón Molina and Llallahui Isasi [44], in an experiment with purple maize, conducted in Peru's Ayacucho region, determined substantial performance improvements: treatments with a biofertilizer got better yields, that reached 6376 kg ha^{-1}, while the control achieved just 2225 kg ha^{-1}. Ccente Gaspar [45] in his study "Identification of Azospirillum spp. associated with the roots of maize (*Zea mays* L.) in Pomacocha-Acobamba-Huancavelica" demonstrated that the inoculation with Azospirillum spp. increased plant height, dry matter, dry root weight, and root length of inoculated plants, evaluated at 60 days after planting.

Studies in Ecuador reported that the use of 50% of the recommended dose of synthetic fertilizer and *Azospirillum* spp, increased fresh corn harvesting (choclo) by 81% [46], in comparison with the control. Sangoquiza-Caiza [47] in his study reported strains of *Azospirillum* spp. capable of mitigating damage caused by salinity for maize. Subsequently, Sangoquiza et al., [48], reported that inoculation with *Azospirillum* spp. and *Pseudomonas fluorescens*, alone or in combination, promoted greater assimilation of N and P content in the plant tissue of maize plants.

In Venezuela, López et al., [49] studied the effect of nitrogen-fixing bacteria and phosphorus solubilizers (*Azotobacter* sp. and *Bacillus megatherium*) on maize in two contrasting soils (high and low fertility). There was an evident higher plant height, stem diameter, leaf length and width, biomass of roots and aerial parts, and N and P content in the inoculated plants; the greatest effect was evident in the low fertility soil. Later, in Táchira, Valery & Reyes [50] demonstrated that inoculation with two isolations of diazotropic bacteria, individually and in consortium, increased the relative agronomic efficiency of the dry weight of maize grain by 130 and 403%, the N content by 463 and 116%, and the P content by 152 and 376%, respectively. They achieved up to 18% of increase in maize grain production.

Mexico is one of the most advanced countries in research and use of biofertilizers in the Region. Biofertilizers are found in the market with diverse trademarks such as: VOP, Bactiva, Endospor, and Bioraíz [12]. The National Institute of Agricultural and Livestock Forestry Research (INIFAP) has developed biofertilizers for maize with the use of bacterial genera *Azospirillum*, *Azotobacter*, *Pseudomonas,* and *Bacillus*, and arbuscular mycorrhiza fungi. The most widespread products are: "Biofertilizante" (with *Azospirillum* and mycorrhiza) and " Biofertilizante Bacteriano INI2709" (with several strains of *Pseudomonas fluorescens*). The use of these biofertilizers allow the replacement of 20 to 60 kg of N per hectare, depending on the strain and the way of application [51]. In Chiapas, seed inoculation with *Azospirillum* increased grain yield by 28% over the absolute control, with a greater economic benefit [52]. Recently, Hernández-Reyes et al., [53] showed that the use of a cyanobacterias consortium in maize produced a higher plant height, number of leaves, and amount of fodder; while yield was similar to those treated with chemical fertilization. In addition, an increase in the amount of protein in the grain was observed.

For more than 40 years, Brazil has studied the effect of *Azospirillum* spp. on maize's development, not only in terms of crop yield, but also in relation to the physiological causes responsible for it. In addition, there are technology packages that use multiple and efficient bacterial maize varieties and strains. The bacteria can supply more than 50% of the nitrogen needed by the plant, due to the greater ease of assimilation of nitrogen [43]. The answer of several maize genotypes to the inoculation of four strains of *Azospirillum* spp. was evaluated by Salomone and Döbereiner [54], who detected an increase in grain weight of up to 7300 kg ha^{-1}, which was greatly influenced by soil conditions, environment, and the genotypes used. In another study lead by EMBRAPA in the Amazon Region, the inoculation with *A. brasilense* increased maize yield by about 50% compared to plants not fertilized with nitrogen [55]. Recently, Pereira et al., [56] reported increases in grain yield until 39.5% and 34.7%, when maize seed was inoculated with *Bacillus subtillis* and *A. brasilense*, respectively.

Argentina is one of the countries that has developed more biofertilizers containing *Azospirillum* worldwide. There is a great variability in the response of maize yield to this bacterium, which fluctuates in between negative values to more than 100%, regarding to the non-inoculated control [57]. For example, experiments in the province of Buenos Aires have shown a 34% average increase in maize yield during 15 years, while studies carried out in other regions got 11.5% increase related to the non-inoculated control [58].

5. Development of a biofertilizer for maize in the highlands of Ecuador

In Ecuador, maize is a crop of paramount importance because of the significant role it plays in the food security of the population. Most of the cultivated area in the highlands is managed in small fields by farmers with few economic resources, who characterize by the low use of technology that causes low productivity of the crop (1.1 t ha^{-1}) and low profit [59]. For these reasons, it is advisable to exploit to the maximum the mechanisms of the so-called "Biological Fertilization" through the use of beneficial microorganisms that has the ability to fix nitrogen and solubilize phosphates, as a natural fertilization alternative, which preserves the environment and improves soil quality, with costs available to farmers.

During almost two decades, the National Agricultural Research Institute (INIAP) has leaded research with diazotrophic bacteria through the collection of soil samples and maize roots in the main cultivated areas of the Sierra. Several strains were isolated and characterized, whose effect have been evaluated in greenhouse and field with liquid and solid inoculation media (**Table 2**).

Year	Subject	Main Author, Reference
2003	Collection, isolation and characterization of diazotrophic bacteria *Azospirillum* sp., associated with maize.	Espinosa, [60]
2005	Development of a biofertilizer from *strains of Azospirillum* sp., for the maize *(Zea mays* L.), INIAP-102 variety with two chemical fertilizations and two organic fertilizers in the province of Chimborazo.	Molina, [61]
2009	Evaluation of bio-fertilizer based on *strains of Azospirillum* sp. in maize *(Zea mays* L.) INIAP 101, in the sector Ainche, province of Chimborazo.	Cool, [46]
2009	"Evaluation of the effect of four inoculation methods of two strains *of Azospirillum* sp., in maize crops *(Zea mays* L) INIAP-122 and INIAP-102 in the provinces of Imbabura and Pichincha.	Ortiz, [62]
2009	Evaluation of solid and liquid supports, for the production of a biofertilizer based on *Azospirillum* sp. applicable to maize cultivation *(Zea mays* L.)	Pallo, [63]
2010	Selection of strains of *Azospirillum* sp., such as biofertilizer *of Zea mays,* L. under saline stress.	Sangoquiza, [47]
2011	Evaluation of biofertilizer based on *strains of Azospirillum* sp. in maize (Zea *mays* L) INIAP-111 Guagal, in addition to three types of fertilization and two methods of inoculation, on Laguacoto II farm, in the province of Bolivar.	Rivadeneira, [64]
2012	Response of maize (*Zea mays* L.) INIAP 111 to biofertilizer and nitrogen fertilization, on Laguacoto, Guaranda, Bolivar.	Changoluisa, [65]
2013	Characterization and evaluation of *Pseudomonas* sp. bacteria, phosphate solubizers found in the maize rhizosphere (*Zea mays*. L) of the provinces of Imbabura, Bolívar, Chimborazo, and Pichincha.	Pincay, [66]
2016	Response of nitrogen and phosphorus absorption of a maize variety when inoculating *Azospirillum* sp. And *Pseudomonas fluorescens*.	Sangoquiza, [48]

Table 2.
Studies carried out with Azospirillum *sp. and Pseudomonas fluorescens and maize in the highlands of Ecuador.*

Ecuador started the research with *Azospirillum* through collection, isolation, and identification of the bacteria in 2003. In this year, a collection of 19 strains of the maize crop rhizosphere was obtained from the highlands of Ecuador [60]. From this collection, four strains were selected because they showed in vitro a greater ability to fix nitrogen. They were evaluated in the field and the C2 strain was selected as the most efficient in promoting higher plant growth. Strain C2, increased plant height by 11.92% relative to the control [61]. A subsequent study verified the beneficial effect of the C2 strain on plant growth [46].

Pallo [63] evaluated solid and liquid supports (carriers) for the production of the biofertilizer. The best solid medium for biofertilizer production was vermiculite peat, evaluated at 180 days of storage at room temperature (average of 20°C), with a production cost of USD 2.11 per 30 ml. For the production of the liquid biofertilizer, the most suitable support was the 2% molasses solution, with a production cost of USD 1.10 per 30 ml. The liquid supports maintained a higher concentration of Colony Forming Units (CFUs) than solids. The liquid inoculum applied to the seed increased the percentage of emergence, plant height, and ear insertion height, showing a more vigorous maize plants with less pest attack, and increases in yield between 14.34 and 26.60%, depending on the variety used [62].

Rivadeneira [64] evaluated several doses of synthetic fertilizers in combination with the biofertilizer, finding that the biofertilizer and 50% of the recommended synthetic fertilization, (55 Kg of N ha^{-1}), produced the highest yield, obtaining 11.21 t ha^{-1} of fresh corn (choclo), and 4.05 t ha^{-1} of grain in a floury open pollinate maize variety. This study also determined that the biofertilizer supplemented with 50% of the recommended nitrogen fertilization, obtained the highest marginal rate of return, indicating that for each dollar invested, 9.6 dollars returned. Later, similar results were observed by Changoluisa [65].

In order to complement the biofertilizer with other beneficial species, the isolation, characterization and evaluation of 21 isolates of bacteria of the genus *Pseudomonas* and three of the genus *Acinetobacter*, were accomplished. The isolates were also collected from highlands of Ecuador. With the isolates, tests were carried out to evaluate their ability to solubilize phosphorus in vitro and under greenhouse conditions. In the in vitro evaluation, the aI3 strain (*P. fluorescens*) showed a higher solubility index with an average of 4.8; followed by nP2 (*P. putida*) with 4.5. In the greenhouse evaluation, the strains that demonstrated the greatest capacity to produce a beneficial effect in the maize plants were cnC2, cnI5, caB1, and cnP3, showing higher values in terms of root length (**Figure 2**), percentage of dry matter, foliar area, and phosphorus accumulation in the tissues, with respect to the control treatment not inoculated [66]. A subsequent evaluation selected the *P. fluorescens* strain nl5 to be included in the biofertilizer.

Once the best isolates of *Azopirillum* sp. and *Pseudomonas fluorescens* were identified, field experiments were conducted to test the efficacy of the biofertilizer. Sangoquiza et al. [48] determined that seed inoculated with *Azospirillum* sp. and/or *P. fluorescens*, promoted a significant increase in morphological and agronomic traits, compared to the absolute control (no inoculated and without synthetic fertilizers). Additionally, the inoculated plots showed similar trait values to the plots treated with the conventional doses of synthetic fertilizers. The biofertilized treatments with *Azopirillum* sp. and *P. fluorescens* showed foliar contents of N-total and P-total higher than the controls, evidencing the effect of these rhizospheric microorganisms in the absorption and translocation of nutrients. Seeds inoculated with *Azospirillum* sp. showed the highest foliar absorption of N with 24.49 g $plant^{-1}$, while seeds inoculated with *Azospirillum* sp. + *P. fluorescens* showed the greatest foliar absorption of P, with 10.86 g $plant^{-1}$. An economic analysis of the application of the biofertilizer showed that the use of *Azospirillum* sp. + *P. fluorescens* allowed

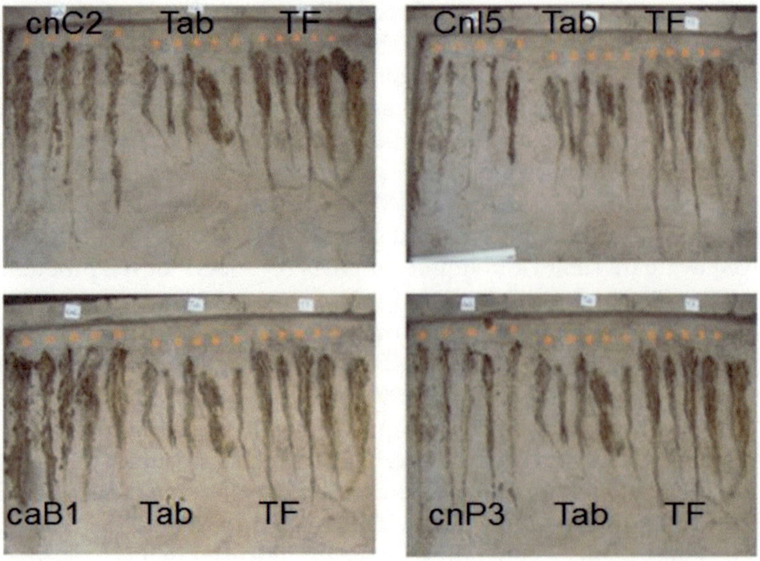

Figure 2.
Root growth of sixty-day-old maize seedlings whose seeds were inoculated with isolates of Pseudomonas fluorescens*: cnC2, isolate from Chimborazo; cnl5, isolate from Imbabura; caB1, isolate from Bolívar; cnP3, isolate from Pichincha; tab, control not inoculated; TF, control with synthetic fertilizers. Source: [66].*

a saving of USD 281.12 ha^{-1}, in relation to the treatment that used conventional synthetic fertilization, which represents a decrease of approximately 16% in the cost of production, without significantly affecting yield.

The validation of the biofertilizer (*Azospirillum* sp. and *P. fluorescens*) with farmers in the provinces of Imbabura, Chimborazo, and Bolívar started in 2019, financed by the Korean Program on International Agriculture (KOPIA). Results of nine locations have shown that the use of the biofertilizer plus 50% of the recommended synthetic fertilization increased on average 30% of the yield, compared to the farmer's plot, and reduced the cost of production per kg of grain produced by 21% [67]. Proposals are currently being made to disseminate and spread the use of this biofertilizer in the highlands of Ecuador and transfer its production and distribution to the private sector.

5.1 Procedure for the production and use of the biofertilizer

The isolates are conserved lyophilized at 4°C in the laboratory of the Maize Program at the Experimental Station Santa Catalina. For reactivation of the isolates, 1000 micro liters (μL) of 1% peptone are added to the Eppendorf tubes containing the lyophilized strains. Then, the tubes are shaken until homogenizing the mixture with the help of a vortex. After that, 50 μL of the strains were taken and placed in Petri dishes containing: solid medium Malic Acid - Congo Red for *Azospirillum* sp., and King B for *P. fluorescens*, with the help of a sterile glass triangle. The isolate was dispersed until dryness, and the Petri dish was placed in incubation at 30°C for seven days. After these days, several colonies of *Azospirillum* sp. and *P. fluorescens* are chosen with the aid of a platinum beam. Pure sections of these strains are transferred to a glass flask containing broth media culture. Subsequently, the strains are placed on a rotary shaking (120 rpm) at 19° C, for 48 hours. Finally, the inoculum was introduced into a liquid support (2% molasses) with a concentration of $1x10^9$ CFU ml^{-1}, and it is incubated for seven days at 30°C. The inoculant is placed in aluminum bags and sealed (**Figure 3**).

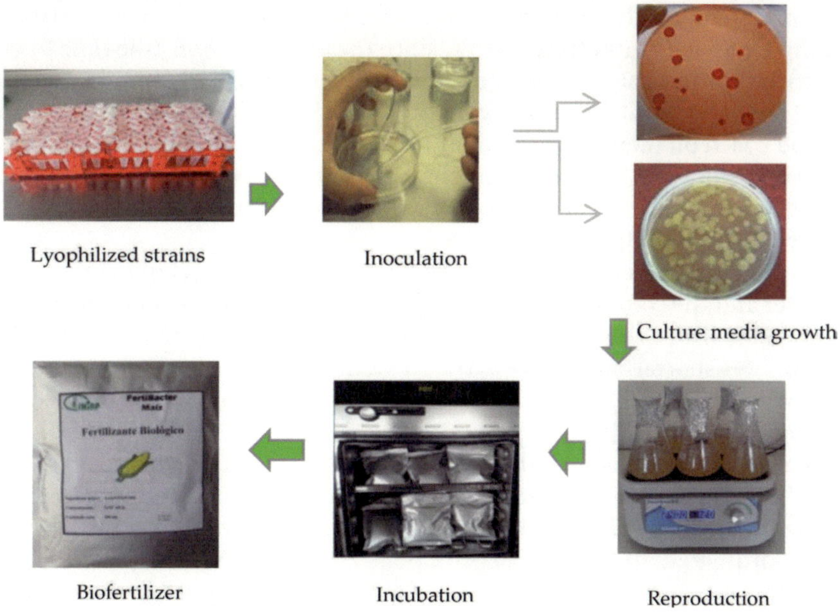

Figure 3.
Production flow of a biofertilizer with Azospirillum *sp. and* Pseudomonas fluorescens *for maize used by the maize program at INIAP, Ecuador. Photo credits: Authors.*

Before planting, the inoculant is applied to the seed and homogenized, trying to ensure that the solution comes into contact evenly with the seed. Then, the seed is allowed to stand for 10 to 15 minutes before it is ready to be planted. It is not recommended to store the inoculated seed for more than two days. The bags can be conserved sealed at room temperature (20-26°C) for 180 days.

5.2 Regulations for the registration of biofertilizers in Ecuador

In Ecuador, the Agencia de Regulación y Control Fito y Zoosanitario (AGROCALIDAD) is in charge of regulating the procedure, registration, control, and surveillance of agricultural inputs. To register the biofertilizer the developer has to be considered as an "operator", which can manufacture, formulate, package, export, import, and distribute fertilizers, soil amendments, and related products for agricultural use. After that, it is required to specify and comply with a series of requirements related to the product to be marketed, such as: type of inoculant, microbial composition, heavy metal content, physical-chemical properties, application methods, security information, among others [53].

6. Challenges of biofertilizers development and use

The use of biofertilizers is a biological approach toward the sustainable intensification of agriculture. However, their application for increasing agricultural yields have several challenges that have to be solved yet. Biofertilizers tend to be susceptible to biotic and abiotic stress and those that perform well in laboratory and greenhouse, often do not perform the same way in the field. Crops are grown under diverse environmental conditions, including diverse ranges of temperature, rainfall, soil type, soil biodiversity, and crop variety. Therefore, such variations cause inconsistency in the efficacy of the biofertilizers. In addition, biofertilizers act slow compared with synthetic fertilizer, since the inoculum will take time to build its concentration and colonize root [68]. These responses could affect the adoption of biofertilizers by the farmers. To avoid these challenges, potential isolates should be selected based on their performance under field conditions, with several crops, across diverse soil types and environmental conditions [33]. In addition, biofertilizers should not complete replace the other fertilizers, but they can complement them and reduce their use [68].

Another significant challenges encountered during the development of a biofertilizer and the commercialization is its shelf life. Biofertilizers contain live microbial cells, with a short shelf life (approx. 6 months, under 20-25°C.) and their storage and transportation require extra care and precaution, that increase the cost of the product [69]. This situation also causes that the product is not often available in remote rural areas [68].

Regulatory constraints include the challenges in product registration and patent filling. The lack of a standardized legal and regulatory definition for "plant biofertilizer" or "plant biostimulant" is the primary reason behind the lack of a globally coordinated uniform regulatory policy. The process to register a biofertilizers is quite ambiguous or complex, extensive and complicated in most parts of the world [33].

7. Future prospects of biofertilizers

There is a growing demand for agricultural goods produced in a sustainable manner; therefore, the use of eco-friendly inputs for food production, such as biofertilizers, will have a significant increase in the coming years. The global market of biofertilizers is expected to increase from 2.3 billion US dollar in 2020 to 3.9 billion in 2025 [70]. The increase in the market value will be supported by government agencies and industry to create awareness among farmers and consumers about the benefits of the use of biofertilizers, in concordance with the Development Sustainable Goal 12 proposed by FAO: "ensure sustainable consumption and production patterns" [71].

The role of metagenomics to have a better understanding of microbial communities associated with rhizosphere is a growing field of research, and it will continue to increase. Plant rhizosphere is colonized by a large number of microorganisms and this results in a huge number of microbial genes that interact with plant genes. However, more elaborated research has to be conducted, such as meta transcriptomic and meta proteomics, and their relation with plant growth [72]. The results of the omics must be translated into the field to have better agronomic practices and new biofertilizer formulations.

An alternative way to improve the growth promoting traits of rhizobacteria is by genetic modification. On this basis, it could be likely that genes of PGPR might be identified by their particular plant-beneficial functions and used for gene editing or transgenic approaches [73]. For instance, certain tumor-inducing Agrobacterium strains have the potential to promote plant growth on non-susceptible plant hosts [74], and bacteria genes directly conferring plant-beneficial properties, such as nif (nitrogen fixation) or phl (phloroglucinol synthesis), have been identified [75].

The use of nano fertilizers in agriculture is growing. Nano fertilizers are non-toxic, minimize production costs, and increase the nutrients use efficiency. Encapsulation of nano biofertilizers will contribute to extend the release of PGPR to target cell by a conjugation of gold, aluminum, and silver nanoparticles [75].

8. Conclusions

The use of biofertilizers have proved to be an efficient way to produce food in a sustainable manner. Many scientific reports assure the benefits of PGPR in the growth and yield of several crops, including: corn, rice, cucumber, wheat, sugar cane, oats, sunflower, flax, beet, tobacco, tea, coffee, coconut, potato, cypress, sudan grass, pepper, peanuts, alfalfa, tomato, sorghum, pine, black pepper, strawberries, soybeans, cotton, beans, lettuce, carrots, among others. The most important microorganisms that have been used in biofertilizer formulations are Rhizobium, Azotobacter, Azospirillum, Pseudomonas, Bacillus, and vesiculo-arbuscular mycorrhizae.

In Ecuador, the development of the biofertilizer took almost 20 years. The process could have been much faster; however, there were no permanent resources to develop this product. The technological development of the biofertilizer depended on sporadic national founding for science and technology; and currently, it depends on the international cooperation. The development of a biofertilizer for maize containing *Azospirillum* spp. and *P. fluorescens* increased yield and reduced the use of synthetic fertilizers by around 50%, lowering the cost of production. This demonstrated that it is feasible to substitute the use of synthetic fertilizers with biofertilizers, opening the opportunity for a sustainable and environmentally friendly maize production system in the highlands of Ecuador.

Acknowledgements

The development of the biofertilizer in Ecuador was financed by the Programa de Modernización de los Servicios Agropecuarios (PROMSA), and the Secretaría de Educación Superior, Ciencia, Tecnología e Innovación (SENESCYT). The validation and scaling up of the biofertilizer is being funded by KOPIA.

Conflict of interest

We declare no conflict of interest.

Author details

José L. Zambrano-Mendoza[1]*, Carlos A. Sangoquiza-Caiza[2], Diego F. Campaña-Cruz[1] and Carlos F. Yánez-Guzmán[1]

1 Instituto Nacional de Investigaciones Agropecuarias – INIAP, Estación Experimental Santa Catalina, Mejía, Ecuador

2 Korean Program on International Agriculture – KOPIA, Estación Experimental Santa Catalina, Mejía, Ecuador

*Address all correspondence to: jose.zambrano@iniap.gob.ec

References

[1] Lee DR, Edmeades S, De Nys E, McDonald A, Janssen W. Developing local adaptation strategies for climate change in agriculture: A priority-setting approach with application to Latin America. Global Environmental Change 2014;29:78-91.

[2] Reyes G, Cortés-Sánchez J. Intensidad En El Uso De Fertilizantes En América Latina Y El Caribe (2006-2012) (Intensity of Fertilizer Use in Latin America and the Caribbean (2006-2012)). Rochester, NY: Social Science Research Network; 2017.

[3] Sosa Baldivia A, Ruíz Ibarra G. La disponibilidad de alimentos en México: un análisis de la producción agrícola de 35 años y su proyección para 2050. Papeles de Población 2017;23:207-30.

[4] Gibbs HK, Ruesch AS, Achard F, Clayton MK, Holmgren P, Ramankutty N, et al. Tropical forests were the primary sources of new agricultural land in the 1980s and 1990s. Proc Natl Acad Sci USA 2010;107:16732.

[5] Gligo N. La dimensión ambiental en el desarrollo de América Latina. Cepal; 2001.

[6] Comisión Económica para América Latina y el Caribe (CEPAL). CEPALSTAT Bases de datos para publicaciones estadísticas 2021. https://cepalstat-prod.cepal.org/cepalstat/tabulador/ConsultaIntegrada.asp?IdAplicacion=22&idTema=696&idIndicador=2038&idioma=e (accessed January 19, 2021).

[7] Cadahia C. Fertirrigación: cultivos hortícolas, frutales y ornamentales. Mundi-prensa Libros; 2005.

[8] Reyes LM, Jiménez CEA, Montiel MGC, Galdámez JG, Cabrera JAM, Aguilar FBM, et al. Biofertilización y fertilización química en maíz (*Zea mays* l.) en Villaflores, Chiapas, México. Siembra 2018;5:26-37.

[9] Itelima J, Bang W, Onyimba I, Sila M, Egbere O. Bio-fertilizers as key player in enhancing soil fertility and crop productivity: A Review 2018.

[10] Umesha S, K. Singh P, P. Singh R. Chapter 6 - Microbial Biotechnology and Sustainable Agriculture. In: Singh RL, Mondal S, editors. Biotechnology for Sustainable Agriculture, Woodhead Publishing; 2018, p. 185-205.

[11] Yadav AN, Verma P, Singh B, Chauhan VS, Suman A, Saxena AK. Plant growth promoting bacteria: biodiversity and multifunctional attributes for sustainable agriculture. Adv Biotechnol Microbiol 2017;5:1-16.

[12] Aguado-Santacruz G. Introducción al Uso y Manejo de los Biofertilizantes en la Agricultura. 2012.

[13] Vassilev N, Vassileva M, Lopez A, Martos V, Reyes A, Maksimovic I, et al. Unexploited potential of some bio-technological techniques for biofertilizer production and formulation. Applied Microbiology and Biotechnology 2015;99:4983-96.

[14] Ronga D, Biazzi E, Parati K, Carminati D, Carminati E, Tava A. Microalgal Biostimulants and Bio-fertilisers in Crop Productions. Agronomy 2019;9.

[15] Thomas L, Singh I. Microbial biofertilizers: types and applications. Biofertilizers for Sustainable Agriculture and Environment, Springer; 2019, p. 1-19.

[16] Camarena-Gutiérrez G. Interacción planta-hongos micorrízicos arbusculares. Revista Chapingo Serie Ciencias Forestales y Del Ambiente 2012;18: 409-21.

[17] Morales MC, Villalobos K, Rodríguez AR, Simón NM, Umaña-Castro R. Identificación y

caracterización molecular de cianobacterias tropicales de los géneros *Nostoc, Calothrix, Tolypothrix* y *Scytonema* (Nostocales: *Nostocaceae*), con posible potencial biotecnológico. Cuadernos de Investigación UNED 2017;9:280-8.

[18] Yadav AN, Verma P, Sachan SG, Kaushik R, Saxena AK. Psychrotrophic microbiomes: molecular diversity and beneficial role in plant growth promotion and soil health. Microorganisms for green revolution, Springer; 2018, p. 197-240.

[19] Hernández-Melchor DJ, Ferrera-Cerrato R, Alarcón A. *Trichoderma*: Importancia agrícola, biotecnológica, y sistemas de fermentación para producir biomasa y enzimas de interés industrial. Chilean Journal of Agricultural & Animal Sciences 2019;35:98-112.

[20] Kumar B, Jacob J. Plant growth promoting rhizobacteria as a biological tool for augmenting productivity and controlling disease in agriculturally important crops-A review. Journal of Spices & Aromatic Crops 2019;28.

[21] Singh VK, Singh AK, Singh PP, Kumar A. Interaction of plant growth promoting bacteria with tomato under abiotic stress: a review. Agriculture, Ecosystems & Environment 2018;267: 129-40.

[22] Mordor Intelligence. South America Biofertilizers Market | Growth, Trends, and Forecast (2020-2025) 2019. https://www.mordorintelligence.com/industry-reports/south-america-biofertilizers-market (accessed January 26, 2021).

[23] García-Fraile P, Menéndez E, Rivas R. Role of bacterial biofertilizers in agriculture and forestry. AIMS Bioeng 2015;2:183-205.

[24] Lodeiro AR. Interrogantes en la tecnología de la inoculación de semillas de soja con Bradyrhizobium spp. Revista Argentina de Microbiología 2015;47:261-73.

[25] Castro R, Novo R, Castro R. Uso del género Azolla como biofertilizante en el cultivo del arroz (*Oryza sativa* L.). Cultivos Tropicales 2002;23:5-10.

[26] Grageda-Cabrera OA, González-Figueroa SS, Vera-Nuñez JA, Aguirre-Medina JF, Peña-Cabriales JJ. Efecto de los biofertilizantes sobre la asimilación de nitrógeno por el cultivo de trigo. Revista Mexicana de Ciencias Agrícolas 2018;9:281-9.

[27] Malboobi MA, Behbahani M, Madani H, Owlia P, Deljou A, Yakhchali B, et al. Performance evaluation of potent phosphate solubilizing bacteria in potato rhizosphere. World Journal of Microbiology and Biotechnology 2009;25:1479-84.

[28] Bernabeu PR, Pistorio M, Torres-Tejerizo G, Estrada-De los Santos P, Galar ML, Boiardi JL, et al. Colonization and plant growth-promotion of tomato by Burkholderia tropica. Scientia Horticulturae 2015;191: 113-20.

[29] Mirik M, Aysan Y, Çinar Ö. Biological control of bacterial spot disease of pepper with Bacillus strains. Turkish Journal of Agriculture and Forestry 2008;32:381-90.

[30] Garza MBI, Vázquez PV, García DG, Tut C, Martínez IR, Campos AT, et al. Respuesta de cultivos agrícolas a los biofertilizantes en la región central de México. Agricultura Técnica En México 2003;29:213-25.

[31] Chávez D, Pereira G, Machuca Á. Estimulación del crecimiento en plántulas de Pinus radiata utilizando hongos ectomicorrícicos y saprobios como biofertilizantes. Bosque (Valdivia) 2014;35:57-63.

[32] Dineshkumar R, Kumaravel R, Gopalsamy J, Sikder MNA, Sampathkumar P. Microalgae as Bio-fertilizers for Rice Growth and Seed Yield Productivity. Waste and Biomass Valorization 2018;9:793-800.

[33] Basu A, Prasad P, Das SN, Kalam S, Sayyed RZ, Reddy MS, et al. Plant Growth Promoting Rhizobacteria (PGPR) as Green Bioinoculants: Recent Developments, Constraints, and Prospects. Sustainability 2021;13.

[34] Glick BR. Plant Growth-Promoting Bacteria: Mechanisms and Applications. Scientifica 2012;2012:963401.

[35] Basu S, Rabara R, Negi S. Towards a better greener future - an alternative strategy using biofertilizers. I: Plant growth promoting bacteria. Plant Gene 2017;12:43-9.

[36] Backer R, Rokem JS, Ilangumaran G, Lamont J, Praslickova D, Ricci E, et al. Plant Growth-Promoting Rhizobacteria: Context, Mechanisms of Action, and Roadmap to Commercialization of Biostimulants for Sustainable Agriculture. Frontiers in Plant Science 2018;9:1473.

[37] Singh I. Plant Growth Promoting Rhizobacteria (PGPR) and their various mechanisms for plant growth enhancement in stressful conditions: a review. European Journal of Biological Research 2018;8:191-213.

[38] Boddey RM, Dobereiner J. Nitrogen fixation associated with grasses and cereals: recent progress and perspectives for the future. Nitrogen economy in tropical soils, Springer; 1995, p. 241-50.

[39] Döbereiner J, Paula M de, Monteiro E. A pesquisa em microbiologia do solo no Brasil. Revista Brasileira de Biologia, Rio de Janeiro 1990;50:841-54.

[40] FAO. FAOSTAT Database 2020. http://www.fao.org/faostat/en/#data/QC (accessed January 30, 2021).

[41] Kouchebagh S, Mirshekari B, Farahvash F. Improvement of Corn Yield by Seed Biofertilization and Urea Application. World Applied Sciences Journal 2012;16.

[42] Ávarez-Solís JD, Gómez-Velasco D, León-Martínez NS, Gutiérrez-Miceli FA. Manejo integrado de fertilizantes y abonos orgánicos en el cultivo de maíz. Agrociencia 2010;44:575-86.

[43] Bárbaro IM, Brancalião SR, Ticelli M. possível a fixação biológica de nitrogênio no milho. Pesquisa e Tecnologia 2008;5.

[44] Girón Molina JB, Llallahui Isasi C. Abonamiento orgánico y micro-organismos eficientes en la absorción de fósforo por maíz morado (Zea mays L.) - Ayacucho. RevInvestigacion 2018;26:11-6.

[45] Ccente Gaspar RC. Identificación de *Azospirillum* spp. asociada a las raíces del maíz amiláceo (*Zea mays* L.) en Pomacocha-Acobamba-Huancavelica 2019.

[46] Cool Zambrano CA. Evaluación de bio-fertilizante a base de cepas de *Azospirillum* spp. en el cultivo de maíz (Zea mays L.) variedad INIAP 101, en el sector Ainche, provincia de Chimborazo. Tesis Ingeniero Agrónomo. Universidad Estatal de Bolívar, 2010.

[47] Sangoquiza Caiza CA. Selección de cepas de Azospirillum spp. como biofertilizante de Zea mays, L. bajo estrés salino. 2011.

[48] Sangoquiza C, Yanez C, Borges M. Respuesta de la absorción de nitrógeno y fósforo de una variedad de maíz al inocular Azospirillum sp. y Pseudomonas fluorescens. ACI Avances En Ciencias e Ingenierías 2019;11.

[49] López M, Martínez Viera R, Brossard Fabré M, Bolívar A, Alfonso N, Alba A, et al. Efecto de biofertilizantes

bacterianos sobre el crecimiento de un cultivar de maíz en dos suelos contrastantes venezolanos. Agronomía Tropical 2008;58:391-401.

[50] Valery A, Reyes I. Evaluación de rizobacterias promotoras del crecimiento bajo diferentes esquemas de fertilización en el cultivo de maíz variedad HIMECA-95. Revista Colombiana de Biotecnologia 2013;15:81-8.

[51] Aguado-Santacruz G, Moreno-Gómez B. Biofertilizantes Bacterianos Desarrollados por el INIFAP. Introducción al Uso y Manejo de los Biofertilizantes en la Agricultura, México: INIFAP/SAGARPA; n.d., p. 151-70.

[52] Martínez Reyes L, Aguilar Jiménez CE, Carcaño Montiel MG, Galdámez Galdámez J, Gutiérrez Martínez A, Morales Cabrera JA, et al. Biofertilización y fertilización química en maíz (Zea mays l.) en Villaflores, Chiapas, México. Siembra 2018;5:001-12.

[53] Hernández-Reyes B, Rodríguez-Palacio M, Castilla-Hernández P, Sánchez-Robles J, Vela-Correa G, Schettino-Bermúdez B. Uso potencial de cianobacterias como biofertilizante para el cultivo de maíz azul en la Ciudad de México. Revista Latinoamericana de Biotecnología Ambiental y Algal 2019;10:13-27.

[54] de Salomone IG, Döbereiner J. Maize genotype effects on the response to Azospirillum inoculation. Biology and Fertility of Soils 1996;21:193-6.

[55] Oliveira IJ, Fontes JRA, Pereira BFF, Muniz AW. Inoculation with Azospirillum brasiliense increases maize yield. Chemical and Biological Technologies in Agriculture 2018;5:6.

[56] Pereira NCM, Galindo FS, Gazola RPD, Dupas E, Rosa PAL, Mortinho ES, et al. Corn Yield and Phosphorus Use Efficiency Response to Phosphorus Rates Associated With Plant Growth Promoting Bacteria. Frontiers in Environmental Science 2020;8:40.

[57] Mendoza-Herrera A, Cruz-Hernández MA. Empleo de Azospirillum como biofertilizante. Introducción al Uso y Manejo de los Biofertilizantes en la Agricultura, México: INIFAP/SAGARPA; n.d., p. 171-94.

[58] Anriquez AL, Silberman JE, Nuñez JAD, Albanesi AS. Biofertilizers in Argentina. Biofertilizers for Sustainable Agriculture and Environment, Springer; 2019, p. 225-44.

[59] Zambrano Mendoza JL, Yánez C, Sangoquiza Caiza CA, Limongi Andrade R, Alarcón Cobeña F, Zambrano E, et al. Situación del cultivo de maíz en Ecuador: investigación y desarrollo de tecnologías en el Iniap 2019.

[60] Espinosa L. Caracterización y selección de la bacteria diazotrófica Azospirillum spp., asociado con el maíz de altura (Zea mays L). Tesis Ingeniero Agrónomo. Universidad Central del Ecuador, 2004.

[61] Molina Herrera S. Desarrollo de un biofertilizante a partir de cepas de Azospirillum spp. para el cultivo de maíz (*Zea mays* L.), variedad INIAP 102 con dos fertilizaciones químicas y dos fertilizaciones orgánicas Sibambe-Chimborazo. Tesis Ingeniero Agrónomo. Universidad Técnica de Cotopaxi, 2006.

[62] Ortíz Bustos GM. Evaluación del efecto de cuatro métodos de inoculación de dos cepas de *Azospirillum* spp., en el cultivo de maíz (*Zea mays* L.), variedades Iniap 122 y 102, en las provincias de Imbabura y Pichincha. Tesis Ingeniera en Bioquímica. Universidad Técnica de Ambato, 2010.

[63] Pallo Y, Velasteguí R. Evaluación de soportes sólidos y líquidos, para la producción de un biofertilizante a base de Azospirillum spp. aplicable al cultivo de maíz (*Zea mays*, L.). Alimentos, Ciencia e Ingeniería 2013;22:32-9.

[64] Rivadeneira Aguay MJ. Evaluación del biofertilizante a base de cepas de Azospirillum Spp. en el cultivo de maíz (*Zea mays* L.) Iniap-111 guagal mejorado, en complemento con tres tipos de fertilización y dos métodos de inoculación, en la granja Laguacoto II, provincia Bolívar. Tesis Ingeniero Agrónomo. Universidad Estatal de Bolívar. Facultad de Ciencias Agropecuarias. 2012.

[65] Changoluisa Gavi GF. Respuesta del maíz (*Zea mays* L.) Iniap 111 al biofertilizante y fertilización nitrogenada, en la granja Laguacoto III, cantón Guaranda, provincia Bolívar. Tesis Ingeniero Agrónomo. Universidad Estatal de Bolívar, 2013.

[66] Pincay Verdezoto AK. Caracterización y evaluación de bacterias *Pseudomonas* sp. solubilizadoras de fósforo, presentes en la rizósfera de maíz (*Zea mays* L) de los ensayos experimentales del INIAP de las provincias de Imbabura, Bolívar, Chimborazo y Pichincha. Tesis Ingeniería en biotecnología. Universidad de las Fuerzas Armadas, 2014.

[67] Zambrano J, Sangoquiza C, Campaña D. KOPIA Annual Report. Quito, Ecuador: 2020.

[68] Roy A. Biofertilizers for Agricultural Sustainability: Current Status and Future Challenges. Current Trends in Microbial Biotechnology for Sustainable Agriculture, Springer; 2021, p. 525-53.

[69] Mitter EK, Tosi M, Obregón D, Dunfield KE, Germida JJ. Rethinking Crop Nutrition in Times of Modern Microbiology: Innovative Biofertilizer Technologies. Frontiers in Sustainable Food Systems 2021;5:29.

[70] Marketandmarkets. Biofertilizers Market by Form (Liquid, Carrier-Based), Mode of Application (Soil Treatment, Seed Treatment), Crop Type, Type (Nitrogen-Fixing, Phosphate Solubilizing and Mobilizing, Potash Solubilizing and Mobilizing). Region – Global Forecast to 2025 2020.

[71] Food and Agriculture Organization (FAO). Food and agriculture in the 2030 Agenda for Sustainable Development 2016.

[72] Jha P, Kumar V. Role of Metagenomics in Deciphering the Microbial Communities Associated with Rhizosphere of Economically Important Plants. Current Trends in Microbial Biotechnology for Sustainable Agriculture, Springer; 2021, p. 79-94.

[73] Bruto M, Prigent-Combaret C, Muller D, Moënne-Loccoz Y. Analysis of genes contributing to plant-beneficial functions in plant growth-promoting rhizobacteria and related Proteobacteria. Scientific Reports 2014;4:6261.

[74] Walker V, Bruto M, Bellvert F, Bally R, Muller D, Prigent-Combaret C, et al. Unexpected Phytostimulatory Behavior for *Escherichia coli* and Agrobacterium tumefaciens Model Strains. MPMI 2013;26:495-502.

[75] Gupta K, Dubey NK, Singh SP, Kheni JK, Gupta S, Varshney A. Plant Growth-Promoting Rhizobacteria (PGPR): Current and Future Prospects for Crop Improvement. Current Trends in Microbial Biotechnology for Sustainable Agriculture, Springer; 2021, p. 203-26.

Chapter 11

Nanotechnology: Past, Present and Future Prospects in Crop Protection

Kallol Das, Pijush Kanti Jhan, Srijan Chandra Das, F.M. Aminuzzaman and Benjamin Yaw Ayim

Abstract

Nanotechnology is an advanced and evolving discipline in the field of science and technology with various applications in other fields such as the life sciences, and is increasingly important in the plant sciences as well. It is estimated that 20–40% of crops are lost each year due to plant pests and pathogens. The current plant disease management, which primarily relies on toxic pesticides that may be harmful to humans and the environment, has the benefit of utilizing nanotechnology. It has capabilities in determining the outbreak of an epidemic as well as diagnosing different types of diseases. It can also distinguish between similar microbes like bacteria, fungi, viruses, complex genomic portions, and how two versions of genes on an individual chromosome differ. This chapter will cover the plant disease management implementation of this technology.

Keywords: Crop Protection, Nanotechnology, Nanoparticles, Pathogens, Plant Disease Management

1. Introduction

Nanotechnology is a versatile discipline which embraces information from natural science, chemistry, physics and other fields. Sustainable crop production experiences approximately 20–30% of its total annual loss due to plant infection [1]. Food sustainability has been identified as one of the biggest problems of humanity faces. Conflict has plagued nations, societies, and administrations for a long time. Attack by plant pathogens on either cash crops or food crops results in decreased yields, economic loss, and possible crop damage [2, 3]. There is a constant increase in the global population therefore the main challenge is how to meet the growing population's needs while reducing the stress on the environment. Controlling the spread of disease appears to be a main task at times. Nanotechnology is described as the operation or assemblies of discrete atoms, molecule, or molecular collections into structures in order to generate novel or extremely diverse assets [4]. The application of nanotechnology in agriculture can modify the agricultural science with advanced apparatuses for quick infection recognition, directed dealing, improved plant nutrient absorption, microbial infection resistance, and ecological stress resistance. Agronomic production will benefit from smart sensors and smart delivery systems to combat viruses and boost the harvest.

Moreover, the early and efficient pathogen detection is critical for infection resistance and disease management, thus helping to reduce yield damage. Magnetic nanoparticles, quantum dots, and gold nanoparticles are all commonly used nanoparticles for molecular detection. Gold nanoparticles, because of their unique characteristics such as small size, catalytic and surface effects, are frequently utilized for rapid insusceptible identification. Gold nanoparticles are capable of exhibiting covalent bonding with DNA, which allows it to be used for DNA assessment and reorganization. At the preliminary phase, nano-based applications are used for detection, diagnosis, and controlling of plant pathogens due to the application of safeguards for plant harvesting.

2. Defining nanotechnology in agriculture

The US Environmental Protection Agency has defined nanotechnology as the science of understanding and control of matter at dimensions of roughly 1–100 nm, where unique physical properties make novel applications possible [5]. This definition is somewhat inflexible when it comes to size dimensions. Materials could have received a greater emphasis on their problem-solving abilities. The particles referred to in this definition have "particulate between 10 and 1,000 nm in size dimensions that are simultaneously colloidal particulate" [6, 7]. Ultimately, When the research into nanotechnology has been completely explored, it can be said to be the science of designing and building machines in which every atom and chemical bond is precisely specified. Rather, it is not a particular collection of techniques, devices, or products, but rather the full capabilities that we will be able to attain when our technology nears the theoretical limits set by atomic physics [8]. The "assembler" is Drexler's ultimate goal for nanomachine technology. The assembler is a nano-machined machine that is used to conduct routine operations at the atomic level [9]. The various types of nanoparticles are described below, along with their definitions and potential applications in plant pathology (**Table 1**).

2.1 Nanoparticle for plant fungal disease management

Nanotechnology may play a key role in providing healthier food by promoting precision farming [11], in particular when facing huge losses due to a wide range of phytopathogens [12]. Agrochemicals use in farming are seeing a dramatic reduction due to the increasing popularity of metallic NPs. With the potential to eliminate targeted microbes from plants, soil, and hydroponics, they have outstanding capabilities [13, 14]. NPs can be used to combat phytopathogens directly, as well as in developing fungicides [15]. Foliar sprays can be used to eliminate pathogens by NP application and also helps in plants growth [16]. Coefficient of variation is increased in low concentrations to control pathogens [13, 17]. Metallic NPs cause hyphal plasmolysis, damaging fungal cell walls, resulting in cell death [18]. Some scientists believe that each agent has a different mechanism of action, whereas others claim the different mechanisms of action vary [19–21]. They include:

1. Permeability of plasma membrane, which is disturbed preventing a proper functioning due to attachment of the NPs proteins sulfur groups.

2. DNA damage.

3. Disturbance of electron transport chain and protein oxidation.

Type	Definition	Potential Use in Plant Pathology
Metalloids, metallic oxides, nonmetals, and their composites	Engineered metals at nanoscale in cubes, spheres, bars, and sheets	Bactericides/Fungicides Nanofertilizers Delivery vehicle for antimicrobials and genetic materials
Carbon nanomaterials	Allotropes of carbon designed at the nanoscale	Multiple uses
Single-walled or multiwalled nanotubes	Graphene sheets rolled into single or multiple tubes	Antimicrobial agents Delivery vehicle for antimicrobials and genetic materials
Fullerenes (bucky balls)	60 carbon atoms in a specific soccer-ball arrangement	Antimicrobial agents Delivery vehicle for antimicrobials and genetic materials
Graphene oxide sheet (reduced or oxide forms)	Graphene oxide sheet	Antimicrobial agents Delivery vehicle for antimicrobials and genetic materials
Liposomes	A lipid enclosing a water core	Delivery vehicle for genetic or antimicrobial products
Dendrimers	Nanomaterial with tree-like appendages that radiate from a central core	Delivery vehicle for genetic or antimicrobial products
Nanobiosensor	A nanoparticle that combines a biological component for detection	Diagnostics, research tool
Nanoshell	Nanoparticles composed of a gold shell surrounding a semiconductor	Diagnostics, research tool
Quantum dots	Inorganic fluorescent, crystalline semiconductor nanoparticles used in biosensors	Diagnostics, research tool

Sources of description [10].

Table 1.
Types of nanoparticles and their definitions and potential uses in plant pathology.

4. Reactive oxygen species (ROS) may be generated, which cause cellular damage.

5. Hindrance in nutrients uptake.

All of the mechanisms mentioned above have interconnected and exert interdependent effects on pathogens [22, 23]. As part of a new technology that attempts to synthesize metallic nanoparticles, bio-reduction reactions have been evaluated for bio-reduction of different metals (e.g. iron, silver, zinc, gold, copper). In vitro and in vivo tests have also been done to see if metallic NPs have antimicrobial potential.

2.2 Nanotechnology based detection of plant diseases

Several plant species have been designated as hyper-accumulators, meaning that they have a high capacity to concentrate trace metals and then use those NPs [24]. Silver, silica, gold, zinc, and copper are frequently used metallic nanoparticles as antimicrobial agents. Silver nanoparticles were shown to be antimicrobial in both ionic and nano forms, and when tested and studied, the particles were shown to be capable of killing plant pathogens [25]. Silver has a strong antifungal and antimicrobial mode of action against bacterial and fungal pathogens [26–28]. Fungicides' hydroxyl radicals-like degradation of fungal and bacterial cell material is achievable with copper NPs [29, 30]. Since the size of these nanoscale copper particles helped control bacterial blight of rice and mung bean leaf spot disease, they were considered useful [31]. This is something you could try if you have plants that are ill or weak. Silica NPs would assist in the acquisition of resistance to diseases and in the activation of plant physiological mechanisms [32]. Iron NPs come into direct contact with fungal cell membranes and disturb the cell's permeability, reducing the cell's growth and eventually causing death through the development of oxidative stress [33]. In solution, zinc NPs release hydroxyl and superoxide radicals, destroying fungal cell walls, hyphae, and preventing conidiogenesis, all of which leads to cellular death [34, 35]. The gold NPs had greater toxic effects on *Salmonella* than on its macroform [4]. As of now, there is no evidence to suggest that silica itself has any antimicrobial capability, but it can indirectly promote plant disease resistance and other stress resistance [36, 37]. For some plant diseases, nano-silver was found to be highly effective [13]. Nano-sized mesoporous silica particles feature a regular pattern of pores with increased surface area. It's improved delivery, efficiency, and effectiveness for site-specific chemicals.

2.3 A Brief Guide to Fungicides, Herbicides and Insecticides

2.3.1 Fungicides (Nano fungicides)

Nano-scale materials development has seen considerable progress in the recent years, with distinct characteristics from corresponding bulk materials. Nanotechnology promises a bright future while delivering pesticides in a safer manner [38]. Agrochemicals have enhanced solubility when polymeric Nano fungicide formulations deliver them at a slow rate, thus increasing their bioavailability [39, 40]. Nanofungicide developed and tested so far proved effective in plant protection strategies [41, 42]. For developing Nano fungicides, we should opt for Nano emulsions (NEs) with smaller size, lower viscosity, and higher stability [43]. The active fungicide ingredient is placed within a core surrounded by a membrane in a nanocapsule. Additionally, nanoencapsulation could be utilized in nanopesticide formulations. Polymers and inorganic compounds have been tested for their possible use in nanopesticide formulations for crop protection [44]. For these formulations to be as potent and stable as possible, while still meeting the safety criteria of the systems to the environment and human beings, it is critical that more work be done. Nanotechnology has a large capacity to develop completely new systems and formulations [45].

2.3.2 Herbicides

They aimed to improve low water solubility, decrease volatility and to deliver the active molecules slowly. Organic solvents are needed to help solubilize insecticides that have low water-solubility, which increases the cost and toxicity. To decrease toxicity, nanoparticles may be used to increase solubility. Another problem that occurs following insecticide application is the evaporation or volatilization of the active ingredient. Even though essential oils have a reputation for causing insecticidal effects, the unstable chemical nature of the substances in the presence of air, light, moisture, and heat causes them to rapidly evaporate. Another motivation for the development of Nano-insecticides is to make the active molecules more stable, which would help ensure that they release at consistent rates over time, resulting in a reduction in insecticide usage while also improving the safety of humans and animals. Whereas fipronil, a commercial insecticide, extended the 100 percent mortality window by 3 days, compared to that commercial product, fipronil-loaded silica extended the 100% mortality window by 3 days, allowing better control of the colony. In groundnut bruchid storage conditions, utilized azadirachtin-loaded zinc oxide or chitosan nanoparticles, and examined the effectiveness over 180 days. The groundnut bruchid, 54.61% of the total weight, was found in the Neem seed kernel extract, loaded into zinc oxide nanoparticles, when compared to the other formulations tested [46]. If the gradual release of active molecules decreases toxicity, then this could also potentially lower the toxicity of insecticides. The results of these studies are promising, but additional studies are required to fully confirm the reductions in insecticide toxicity.

2.3.3 Insecticides

The use of nanosized preparations or nano-materials based herbicide formulations offers unique opportunities for delivering chemical or biological pesticides. Herbicide formulations based on nanomaterials may be more effective, have better solubility, and reduce toxicity than conventional herbicides. If early-stage weed control is employed with the use of nanoparticle-based herbicide release systems, the resistance potential will be minimized while the active ingredient's effectiveness will be maintained and prolonged release will be prolonged [47]. Herbicides can be selective, killing only target weeds, or non-selective and potentially hazardous to people and the environment if they kill all vegetation. Before seedlings sprout up from the soil, you can apply a pre-plant, pre-emergence, or post-emergence herbicide (weed seedlings already emerged from the soil).

2.4 Types of nanoparticles for plant disease management

2.4.1 Nano-phytopathology

New techniques such as nanotechnology will remain novel for this century, but as time passes and extensive research is done in the field, they will be more widely used in the treatment of plant infection. Plant pathology is set to offer an exciting future of research with regard to their antibacterial and antifungal properties in the context of nanotechnology. Using nanoparticles to protect the seeds and foliage from pathogens that would otherwise intrude can be the most effective method. Therefore, it can be summarized here that nanoparticles can play their role in plant disease management in following ways-.

- Nanoparticles being used as pesticides themselves and being applied to plants directly for the control of disease.

- Nanoparticles as carriers of other pesticides/nanoparticles for their controlled and targeted release and to increase their effect.

- Nanodevices to detect diseases at early stages.

Nanoparticles can be successfully applied to disease affected plants with enhanced and effective results due to their extremely reactivity/affectability. This increased reactivity can be attributed to their extremely small size and large surface area.

2.4.2 Effect of nanoparticles on the pathogens/microorganisms

After scientific testing, scientists have confirmed that the chemical and physical properties of nanoforms of materials actually shift between their macroform and nanoform. All of these properties transformations ultimately end up with useful real-world applications in plant defense and plant protection. Since nanoparticles have the advantages of both small size and surface area, they influence plant pathogens in a more precise way. Larger macroscopic entities are less likely to interact with microorganisms because they are farther away from them. There is an abundant amount of evidence available to bolster this claim.

2.4.3 Effect of nanoparticles on Bacteria

New research shows that nanoparticles have antibacterial properties, which is most likely the result of the cell wall of the bacteria being broken or of high levels of reactive oxygen species (ROS) being produced [48–50]. Bacterial infection is a leading cause of contamination and mortality due to prolonged presence of pathogens. Due to low cost and positive results, antibiotics have been selected for the treatment of bacterial contamination. Although multiple studies have demonstrated that overuse of antibiotics leads to multidrug-resistant bacteria strains, this has not yet been confirmed though. A very powerful strain of antibiotic-resistant bacteria has emerged. Previous researches states that, these kinds of bacteria consist of gene, which are super resistant [51]. The mode of action of nanoparticles is directly linked with the bacterial cell wall, which enables the nanoparticles to control super-resistant bacteria. Bacteria may become resistant to nanoparticles if they are exposed to them. If these new NP-based materials are able to perform antibacterial activity, it's entirely possible [52–54].

2.4.4 Nanotechnology for plant viral diseases

The various types of nanoparticles were investigated and used to control *Sitophilus oryzae* and baculovirus BmNPV (*Bombyx mori* nuclear polyhedrosis virus) in silkworm (*B. mori*) disease caused by *S. oryzae* and BmNPV [55]. To complete their investigation, the researchers performed bioassay, which involved preparing the nanoparticles' solid and liquid formulations; after which, they applied these formulations to rice and stored it in a plastic box with 20 adult *S. oryzae* for seven days. On the first day, the researchers reported that hydrophilic silver nanoparticles were most effective. More than 90% of mortality was gained on day 2, and silver and aluminum nanoparticles were the primary source. Seventy-three percent of the insects died after seven days of exposure to lipophilic silver nanoparticles. But with respect to nanoparticles of aluminum, 100% mortality was observed. Thus, a significant decrease in viral load was found in a hydrophobic aluminosilicate nanoparticle suspension when *B. mori* leaves were used in an ethanolic treatment of grasserie disease.

2.4.5 Metallic nanoparticles: effective tool for plant disease management

There are different nanotechnology approaches for the detection of plant diseases (**Figure 1**).

2.4.6 Bio-nano materials

There are some bio-nano materials which are categorized through X-ray diffraction (XRD) technique, X-ray photoelectron spectroscopy (XPS), Energy-dispersive X-ray spectroscopy (EDS), UV visible spectroscopy, scanning electron microscopy (SEM), Fourier transforms infrared spectroscopy (FTIR), Coupled plasma spectrometry (ICP), Transmissions electron microscope (TEM) and Atomic force

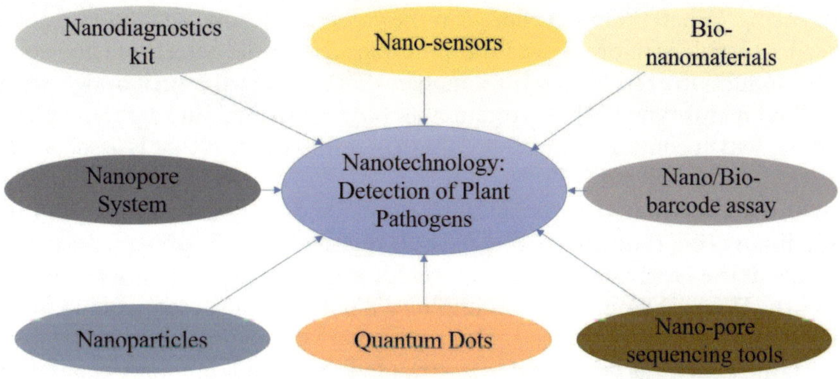

Figure 1.
Different nanotechnology approaches for the detection of plant diseases.

microscopy (AFM) techniques. These bio-nano materials played noteworthy role in field of agriculture, medicine and biology. The increasing usage of bio- nano materials in many areas will enhance their productivity in the atmosphere by developing more analytical tools in nanotechnology for controlled environmental risk management [56]. Bio-synthesis of bio-nanomaterials can be also achieved by using plant extracts [57], microbial cultures or their enzymes and proteins.

2.4.7 Nano bio-barcode assay

The technology of biological barcodes is improving in nanotechnology and incorporating new advancements in nanotechnology to aid in the identification of non-enzyme-containing ultra-sensitive proteins and DNAs. Instead of using an orthodox ELISA, a protein barcode assay could be employed, which is both more complicated and far more sensitive and profound. Because of its dependence on nanoparticles, the nanoparticle-based biobarcode assay is more sensitive to finding pathogens than traditional techniques like ELISA, Real-time PCR, etc. [58, 59]. It can also help in early detection of plant diseases. The biobarcode technique consists of two probes.

1. Magnetic micro beads (MMB): Target recognition and carry an antibody or DNA as a biological probe.

2. Gold nanoparticles (Au-NP): It has a polyclonal antibody or an oligonucleotide (Bio-barcode) Bio-barcode is a developing technique with the help of advancements in nanotechnology. It is an enzyme free, PCR free technique and highly sensitive for protein and DNA detection.

Moreover, the advancement has been made on different aspects of Bio-barcodes to make it more applicable in fields. DNA barcoding has been suggested for fungal identification [60]. It has been reported to be reliable and rapid method of detection. A DNA barcode should be standardized and scalable. And also, the similar techniques can be developed for speedy and onsite detection of plant pathogens particularly viruses to decrease the losses to crops.

2.4.8 Nanopore system

Nano pore used in nano-pore systems can be used to examine genetic information at a low cost, have low sample preparation requirements, and operate quickly [61]. In fact, nanopore is a nano-sized pore through which a flow of nanoparticle ions is flowing. When there is a change in the current, it shows that an analysis of the biological molecule is in progress. Nanopore-based systems determine nucleotides through conductivity changes, which enables them to identify nucleotides because of their lipid membrane [62]. A protein nanopore is injected into a polymer bilayer membrane, which contains a sensory chip to measure current that is associated with the identity of the molecule [63].

Newly, nanopore based sequencing (Nano-SBS) distinguished four DNA bases through discovering four different sized tags released from 5′-phosphate-modified nucleotides at the particular molecule level for sequence determination [64].

Recently, UK-based nanopore technology, a portable DNA sequencing machine (MinION) has been released. It enables researchers to sequence a 10 kb sample of single-stranded DNA as well as double-stranded DNA, making next-generation sequencing easily approachable [65].

The discovery of this new technique offers both the ability to detect and track the spread of an epidemic, as well as the ability to differentiate between various bacteria, fungi, viruses, complex genomic components, and difference between two different gene sequences located on the same chromosome. In other words, a nanopore platform that is already inside a current diagnostic machine can conduct whole genome analysis in minutes. It could be used for analyzing plant and pathogen genomics for the purpose of increasing agricultural crops.

2.4.9 Nanodiagnostic kit

Nanodiagnostic kit also called "lab in a box" is used as a small box for measuring important tasks in plants which can be done in small space [66]. A smart kit helps to detect the plant pathogens and can help the farmers in prevention of wide spread diseases [1, 67]. Nanodiagnostic kit contained four myco-sensors which can detect the of ZEA, T-2/HT-2, DON and FB1/FB2 myco-toxins on only one strip used for cash crops like wheat, barley and corn [68]. This method is fast, convenient, and less expensive for finding out if crops have a fungal infection. Antigen and antibodies, the nucleotide sequence in which nano kit can be used, all have multiple additional purposes. Moreover, it can also detect particular gene target, isolation and purification of specific genes. But nano kit has not fully checked practically for the plant pathogen detection in field conditions. More extensive research works are still needed in this field.

2.4.10 Quantum dot (QDs)

This is another level of nanocrystals that release specific wavelengths of light: Quantum Dots (QDs). There are three-dimensional nanoparticles that have a broad excitation spectrum [69]. Narrow-tunable emission peak, extensive fluorescence lifetime, resistance to photo bleaching, and molar extinction coefficient ten to one hundred times higher than for most QDs. However, one of the findings revealed multiple enzymatic properties of QD-based nanosensors [70] and stated CdTe quantum dots usage as biosensors with specific antibiotic coatings against *Polymyxa betae* specific glutathione-S-transferase (GST) protein [71]. Because of the interaction of CdTe quantum dots and rhodamine, which cause a resonance dipole–dipole coupling, the resonance dipole–dipole coupler, which is required for fluorescence resonance energy transfer, is created (FRET). For more efficient results, this device can be used to test plants in under 30 minutes. Although it was concluded that there was a highly sensitive detection of diseased lime trees by showing 100% specificity with a sensitive detection limit for *P. aurantifolia*, the opposite appears to be true in this instance, according to the findings of this study what [72].

In the near future, quantum dots will be used in almost every form of diagnostics and medical testing. For example, fluorescent QDs can be used for a variety of molecular diagnostics and genotyping procedures. Additionally, these studies contribute to the complex diagnosis and combination with therapies which help lead to possible cancer diagnosis applications. QD bio conjugates enable the visualization of living cancer cells in animals and the visual differentiation of cancer cells in the context of a fluorescence microscope.

3. Future prospects of nanotechnology in plant disease management

To find out more about using NMs in plant sciences, we have covered all relevant aspects here. According to recent researches, the use of nanotechnology has resulted in considerable advancements in the development of NMs and their implementation in medicine for the detection and treatment of illnesses. This is where NMs (naturally occurring mini-molecules) for plants has fallen short. Concluding remarks: Additional studies are likely needed to optimize this synthesis and biofunctionalization direction for plant applications, but also to investigate further the underlying mechanisms of plant uptake, especially with regards to sustainable agriculture. The significant aspects of plant physiology that have thus far gone undocumented are something that the applications will need to support. Nanobiosensors for monitoring plant development and interactions with the environment, especially in controlling growth conditions, could be a means of improvement. Polymeric and hydrogel-based NPs, because of their safety profile, high loading capacity, and resistance to degradation, offer undeniable advantages for drug delivery. This unique NMs strategy is elegant in that it allows for the spatial and temporal cargo release from cell and animal models based on environmental cues (e.g., UV, NIR, ultrasound, etc.) [73–75]. Finally, as a final remark, we encourage multidisciplinary approaches for designing and synthesizing smart nanomaterials, in order to help expedite plant nanotechnology. The point of such a project is to open new horizons in phytonanotechnology by uniting complementary professional competencies, including those of plant biologists, geneticists, chemists, biochemists, and engineers.

Every sanitation/quarantine strategy requires that infested or infected material be discovered ahead of time, because any greenhouse, field, state, or country where this treatment is used is inevitably where you find infested or infected material. In order for this operation to succeed, prompt detection of pathogens is critical. There are large advances in the speed and sensitivity of pathogen probes with the introduction of nanotechnology. Nanoparticles can be used as rapid diagnostics for the detection of bacterial, fungal, nematode, and viral pathogens, which can help in the early detection of diseases [76, 77]. Pathogen detection has only recently been explored with super paramagnetic iron oxide nanoparticles, though the techniques have been pursued in other applications for nearly a decade [78, 79]. Nanoanalytical devices employ a biologically-integrated sensing element with a transducer built into it to form an electronic signal when they come into contact with the analyte of interest (pathogen). However, Nanotechnology innovations in the last decade have greatly reduced the technical barriers of making biosensors with a variety of nanoparticles and nanostructures [80].

Moreover, this novel use of nano-enabled biosensors to monitor, map, and treat specific areas in a field prior to or during the onset of symptoms can be coupled with robotics and GPS systems to produce smart delivery systems that monitor, map, and treat locations prior to or during symptoms appearing. This new technology could help farmers reduce their use of agrochemicals while also increasing yield and profits [81, 82]. Biosensors, being capable of detecting pathogens with a higher level of sensitivity, can also be used in ports of entry,

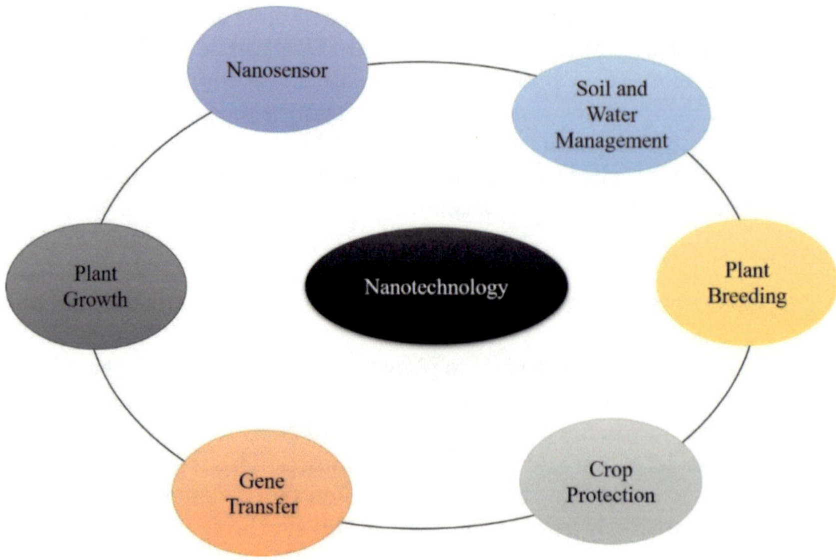

Figure 2.
Applications of nanotechnology in plant disease management and crop improvement.

where quarantined pathogens can be apprehended with greater efficiency. When food pathogens and mycotoxins are detected quickly, the value of rapid analysis is obvious [83–85]. Applications of nanotechnology in plant disease management and crop improvement are shown in **Figure 2**.

4. Limitations and future directions of nanotechnology

Nanotechnology with promising results in the agricultural sector, such as its unique method of applying pesticides, fertilizers, and so on, may finally allow the human population to visualize the dream of achieving sustainable and ecofriendly agricultural technology. Recent findings have shown that nanomaterials could potentially harm a beneficial soil organism, the earthworm [86]. Xu et al. [87] summarized increased safety concerns regarding nanomaterials in food and agriculture. They concentrated on the most common exposure routes and factors involved in nanotoxicity. More and more engineered nanomaterials are reaching the environment, due to new technologies. Production scale and cost currently limit the application of nanocarriers in agriculture. Large-scale manufacturing of nanomaterials and its effective application to agriculture will greatly reduce the cost. The difficult commercialization of nanomaterials for agricultural applications necessitates well-protected materials, superior testing priorities, a clear-cut risk assessment, and international regulatory guidance [88]. However, a concern has been raised on the impact of increased nanomaterial production on ecosystem health. Many commercial nanomaterials are more toxic than their bulk-form counterparts, even though bulk-form nanomaterials are legally allowed for sale. There still needs to be further research on the various applications of nanomaterials, such as synthesis, toxicity, and its use at the field level. Even the dream of utilizing nanotechnological methods in agriculture is still in its infancy. As a result, the focus areas that require further investigation are the development of systems that would improve the release profile of herbicides without altering their characteristics and novel carriers with enriched activity without significant environmental damage. There are some examples of recent breakthroughs in nanotechnology in agriculture (**Table 2**) [89].

Product	Application	Institution
Nanocides	Pesticides encapsulated in nanoparticles for controlled release Nanoemulsions for greater efficiency	BASF, Ludwigshafen, Germany Syngenta, Greensboro, NC, USA
Buckyball fertilizer	Ammonia from buckyballs	Kyoto University, Kyoto, Japan
Nanoparticles	Adhesion-specific nanoparticles for removal of Campylobacter jejuni from poultry	Clemson University, Clemson, SC, USA
Food packaging	Airtight plastic packaging with silicate nanoparticles	Bayer AG, Leverkusen, Germany
Use of agricultural waste	Nanofibers from cotton waste for improved strength of clothing	Cornell University, Ithaca, NY, USA
Nanosensors	Contamination of packaged food Pathogen detection	Nestle, Kraft, Chicago, USA Cornell University, Vevey, Switzerland
Precision farming	Nanosensors linked to a global positioning system tracking unit for real-time monitoring of soil conditions and crop growth	US Department of Agriculture, Washington, DC, USA
Livestock and fisheries	Nanoveterinary medicine (nanoparticles, buckyballs, dendrimers, nanocapsules for drug delivery, nanovaccines; smart herds, cleaning fish ponds (Nanocheck [Nano-Ditech Corp., Cranbury, NJ, USA]), and feed (iron nanoparticles)).	Cornell University NanoVic, Dingley, Australia

Sources of description [89].

Table 2.
Some examples of recent breakthroughs in nanotechnology in agriculture.

5. Conclusion

Nanotechnology has the potential to improve plant growth, disease resistance and nutrient use with controlled delivery of agrochemicals. More efficient and targeted use of fungicides, herbicides and insecticides may be seen through using environmentally friendly nanocapsules. In order to preserve the freshness and quality and help in disease prevention, post-harvest nanotechnology research and development is needed. As the use of nanotechnology progresses, applications of green chemistry have reduced the use of toxic solvents, allowing for crop protection. Through use of biotechnology and nanotechnology, crop protection and production are now available to a much larger portion of the population. Despite being unverified, the effects of nanomaterials on the environment are clearly noticeable because of their unique physical and chemical properties. Nanomaterials application in agriculture is new, and further research is required. Nanomaterials are expected to have a huge impact on the pricing and environmental friendliness of crop protection techniques. With the application of nanotechnology, new methods will be developed for managing disease in greenhouses and fields, while advancements in disease diagnostics and the construction of molecular manipulations of plants and pathogens will be made possible. However, only a few laboratories are currently exploring the incorporation of nanotechnology into phytopathology, but we expect that as new research is geared toward discovering, adapting and applying nanotechnology, the barriers to global food production will be lessened.

Acknowledgements

We would like to express our gratitude to the management of INTECHOPEN LIMITED for inviting us to write the book chapter and for providing the assistance during the chapter's preparation.

Conflict of interest

The authors declare no conflict of interest.

Author details

Kallol Das[1*], Pijush Kanti Jhan[2], Srijan Chandra Das[3], F.M. Aminuzzaman[4] and Benjamin Yaw Ayim[5]

1 College of Agriculture and Life Sciences, Kyungpook National University, Daegu, Korea

2 Department of Agriculture, Noakhali Science and Technology University, Noakhali, Bangladesh

3 Breeder Seed Production Center, Bangladesh Agricultural Research Institute, Panchagarh, Bangladesh

4 Department of Plant Pathology, Faculty of Agriculture, Sher-e-Bangla Agricultural University, Dhaka, Bangladesh

5 Ministry of Food and Agriculture, Plant Protection and Regulatory Services Directorate, Ashanti, Ghana

*Address all correspondence to: kalloldas91@gmail.com

References

[1] Nezhad AS. Future of portable devices for plant pathogen diagnosis. Lab Chip. 2014;14:2887-2904.

[2] Pan L, Qiu H, Dou C, Li Y, Pu L, Xu J, Shi Y. Conducting polymer nanostructures: template synthesis and applications in energy storage. International Journal of Molecular Sciences. 2010;11:2636-2657.

[3] Thind TS. Phytopathogenic Procaryotes and plant diseases. Research in Plant Disease. 2012;27:125-125.

[4] Joseph T, Morrison M. Nanotechnology in agriculture and food. Institute of Nanotechnology: London; 2006.

[5] US Environmental Protection Agency. Nanotechnology White Paper. Report EPA 100/B-07/001, Washington, DC, USA; 2007. Available from: http://www.epa.gov/osainter/pdfs/nanotech/epananotechnologywhitepaper-0207.pdf. Accessed June 9, 2014.

[6] United States Department of Agriculture. Nanoscale science and engineering for agriculture and food systems. Report submitted to Cooperative State Research, Education and Extension Service, United States Department of Agriculture, National Planning Workshop; 18-19 November 2002; Washington, DC, USA. 2002.

[7] Nakache E, Poulain N, Candau F, Orecchioni AM, Irache JM. Biopolymer and polymer nanoparticles and their biomedical applications. In: Nalwa HS, editor. Handbook of Nanostructured Materials and Nanotechnology. Academic Press; New York, NY, USA. 1999.

[8] Storrs Hall J. Nanofuture: What's Next for Nanotechnology? Amherst, NY, USA: Prometheus Books; 2006.

[9] Drexler KE. Engines of creation: the coming era of nanotechnology. 1986. Available from: http://e-drexler.com/p/06/00/EOC_Cover.html. Accessed June 9, 2014.

[10] Elmer W, Jason CW. The Future of Nanotechnology in Plant Pathology. Annual Review of Phytopathology. 2018; 56:111-33.

[11] Gruère GP. Implications of nanotechnology growth in food and agriculture in OECD countries. Food Policy. 2012;37:191-198.

[12] Tournas VH. Spoilage of vegetable crops by bacteria and fungi and related health hazards. Critical Reviews in Microbiology. 2005; 31:33-44.

[13] Park HJ, Kim SH, Kim HJ, Choi SH. A new composition of nanosized silicasilver for control of various plant diseases. The Plant Pathology Journal. 2006;22: 295-302.

[14] Sharma K, Sharma R, Shit S, Gupata S. Nanotechnological application on diagnosis of a plant disease. International Conferences on Advances in Biological and Medical Sciences (ICABMS'2012); July 15-16, 2012; Singapore. 2012.

[15] Khan MR, Rizvi TF. Nanotechnology: Scope and application in plant disease management. Plant Pathology Journal. 2014;13:214-231.

[16] Agrawal S, Rathore P. Nanotechnology pros and cons to agriculture: A review. International Journal of Current Microbiology and Applied Sciences. 2014;3:43-55.

[17] Nel A, Xia T, Madler L, Li N. Toxic potential of materials at the nano level. Science. 2003;311: 622-627.

[18] Min JS, Kim KS, Kim SW, Jung JH, Lamsal K, Kim SB, Lee YS. Effects of colloidal silver nanoparticles on sclerotium-forming phytopathogenic

fungi. Plant Pathology Journal. 2009;25: 376-380.

[19] Zeng H, Li XF, Zhang GY, Dong JF. Preparation and characterization of beta cypermethrin nanosuspensions by diluting O/W microemulsions. Journal of Dispersion Science and Technology. 2008;29:358-361.

[20] Prabhu S, Poulose EK. Silver nanoparticles: Mechanism of antimicrobial action, synthesis, medical applications, and toxicity effects. International Nano Letters. 2012;2:32.

[21] Lemire JA, Harrison JJ, Turner RJ. Antimicrobial activity of metals: Mechanisms, molecular targets and applications. Nature Reviews. Microbiology. 2013;11:371.

[22] Alghuthaymi MA, Almoammar H, Rai M, Said-Galiev E, Abd-Elsalam KA. Myconanoparticles: Synthesis and their role in phytopathogens management. Biotechnology and Biotechnological Equipment 2015;29:221-236.

[23] Abd-Elsalam, K. A., & Prasad, R. (2018). Nanobiotechnology applications in plant protection. Dordrecht: Springer. https://www.springer.com/us/book/9783319911601.

[24] Dubey M, Bhadauria S, Kushwah BS. Green synthesis of nanosilver particles from extract of Eucalyptus hybrida (safeda) leaf. Digest Journal of Nanomaterials and Biostructures. 2009; 4:537-543.

[25] Sharon M, Choudhary AK, Kumar R. Nanotechnology in agricultural diseases and food safety. Journal of Phytopathology. 2010;2:83-92.

[26] Clement JL, Jarrett PS. Antibacterial silver. Metal-Based Drugs. 1994;1: 467-482.

[27] Kim YH, Lee DK, Cha HG, Kim CW, Kang YC, Kang YS. Preparation and characterization of the antibacterial Cu nanoparticle formed on the surface of SiO2 nanoparticles. The Journal of Physical Chemistry. 2006;110: 24923-24928.

[28] Wei D, Sun W, Qian W, Ye Y, Ma X. The synthesis of chitosan-based silver nanoparticles and their antibacterial activity. Carbohydrate Research. 2009; 344:2375-2382.

[29] Esteban TL, MalpartidaF, Esteban CA, Pecharroman C, Moya JS. Antibacterial and antifungal activity of a soda-lime glass containing copper nanoparticles. Nanotechnology. 2009; 20:505701.

[30] Brunel F, El Gueddari NE, Moerschbacher B M. Complexation of copper (II) with chitosan nanogels: Toward control of microbial growth. Carbohydrate Polymers. 2013;92: 1348-1356.

[31] Gogoi R, Dureja P, Singh PK. Nanoformulations – A safer and effective option for agrochemicals. Indian Farming. 2009;59:7-12.

[32] Brecht MO, Datnoff LE, Kucharek TA, Nagata RT. Influence of silicon and chlorothalonil on the suppression of gray leaf spot and increase plant growth in St. Augustine grass. Plant Disease 2004;88:338-344.

[33] Xie Y, He Y, Irwin PL, Jin T, Shi X. Antibacterial activity and mechanism of action of zinc oxide nanoparticles against *Campylobacter jejuni*. Applied and Environmental Microbiology. 2011;77: 2325-2331.

[34] Borkow G, Gabbay J. Copper as a biocidal tool. Current Medicinal Chemistry. 2005;12:2163-2175.

[35] Patra P, Mitra S, Debnath N, Goswami A. Biochemical-, biophysical-, and microarray-based antifungal evaluation of the buffer-mediated

synthesized nano zinc oxide: An in vivo and in vitro toxicity study. Langmuir. 2012;28:16966-16978.

[36] Wang S, Lawson R, Ray PC, Yu H. Toxic effects of gold nanoparticles on *Salmonella typhimurium* bacteria. Toxicology and Industrial Health. 2011; 27:547-554.

[37] Brecht M, Datnoff L, Nagata R, Kucharek T. The role of silicon in suppressing tray leaf spot development in St. Augustine grass (pp. 1-4). Gainesville: Publication in University of Florida. 2003.

[38] Ghormade V, Deshpande MV, Paknikar KM. Perspectives for nano-biotechnology enabled protection and nutrition of plants. Biotechnology Advances. 2011;29:792-803.

[39] Lauterwasser C. Small sizes that matter: Opportunities and risks of nanotechnologies. Report in cooperation with the OECD International Futures Programme. Munchen: Allianz Center for Technology. 2006. pp. 45.

[40] Kah M, Hofmann T. Nanopesticide research: Current trends and future priorities. Environment International. 2014;63:224-235.

[41] Bordes P, Pollet E, Avérous L. Nano-biocomposites: Biodegradable polyester/nanoclay systems. Progress in Polymer Science. 2009;34:125-155.

[42] Bergeson LL. Nanosilver pesticide products: What does the future hold? Environmental Quality Management. 2010;19:73-82.

[43] Bernardes PC, Nélio JDA, Nilda de Fátima FS. Nanotechnology in the food industry. Bioscience Journal. 2014;30: 1919-1932.

[44] Chuan L, He P, Pampolino MF, Johnston AM, Jin J, Xu X, Zhou W. Establishing a scientific basis for fertilizer recommendations for wheat in China: Yield response and agronomic efficiency. Field Crops Research. 2013;140:1-8.

[45] Kanto T, Miyoshi A, Ogawa T, Maekawa K, Aino M. Suppressive effect of potassium silicate on powdery mildew of strawberry in hydroponics. Journal of General Plant Pathology. 2004;70: 207-211.

[46] Jenne M, Kambham M, Tollamadugu NP, Karanam HP, Tirupati MK, Balam RR, Shameer S, Yagireddy M. The use of slow releasing nanoparticle encapsulated Azadirachtin formulations for the management of *Caryedon serratus* (groundnut bruchid). IET Nanobiotechnology. 2018;12:963-967.

[47] Worrall EA, Hamid A, Mody KT, Mitter N, Pappu HR. Nanotechnology for Plant Disease Management. Agronomy. 2018;8:285.doi:10.3390/agronomy 8120285.

[48] Zhang L, Jiang Y, Ding Y, Daskalakis N, Jeuken L, Povey M, O'Neill AJ, York DW. Mechanism investigation into antibacterial behaviour of suspension of ZnO nanoparticles against *E. coli*. Journal of Nanoparticle Research. 2010;12:1625-1636.

[49] Cabiscol E, Tamarit J, Ros. Oxidative stress in bacteria and protein damage by reactive oxygen species. International Journal of Microbiology. 2007;3:3-8.

[50] Wiesner MR, Lowry GV, Alvarez P, Dionysiou D, Biswas P. Assesing the risk of manufactured nanomaterials. Environmental Science & Technology. 2006;15:4336-4345.

[51] Hsueh PR. New Delhi metallo-β-lactamase-1 (NDM-1): an emerging threat among Enterobacteriaceae. Journal of the Formosan Medical Association. 2010;109:685-687.

[52] Wang C, Zhao M, Li J, Yu J, Sun S, Ge S, Huang Y. Silver nanoparticles/

graphene oxide decorated carbon fiber synergistic reinforcement in epoxy-based composites. Polymer. 2017a;131:263-271.

[53] Wang J, Qiufang Y, Chengmin S, Chunde J, Qingfeng S. One-step preparation of graphene oxide/cellulose nanofibril hybrid aerogel for adsorptive removal of four kinds of antibiotics. Journal of Nanomaterials. 2017b:5150613. https://doi.org/10.1155/2017/5150613.

[54] Wang L, Hu C, Shao L. The antimicrobial activity of nanoparticles: present situation and prospects for the future. International Journal of Nanomedicine. 2017c;12:1227-1249.

[55] Goswami A, Roy I, Sengupta S, Debnath N. Novel applications of solid and liquid formulations of nanoparticles against insect pests and pathogens. Thin Solid Films. 2010;519:1252-1257.

[56] Sahayaraj K. Bionanomaterials: synthesis and applications. In: Proceedings of the first national seminar on new materials research and nanotechnology (NSNMRN'2012); 12-14 September 2012; Government Arts College, Ooty, Tamil Nadu, 2012.

[57] Sahayaraj K, Roobadevi M, Rajesh S, Azizi S. Vernonia cinerea (L.) Less. silver nanocomposite and its antibacterial activity against a cotton pathogen. Research on Chemical Intermediates. 2015;41:5495-5507.

[58] Bao YP, Wei TF, Lefebvre PA, An H, He L, Kunkel GT, Müller UR. Detection of protein analytes via nanoparticle-based bio bar code technology. Analytical Chemistry. 2006;78:2055-2059.

[59] Nam JM, Thaxton CS, Mirkin CA. Nanoparticle-based bio-bar codes for the ultrasensitive detection of proteins. Science. 2003;301:1884-1886.

[60] Xu J. Fungal DNA barcoding. Genome. 2016;59:913-932.

[61] Branton D, Deamer DW, Marziali A, Bayley H, Benner SA, Butler T, Di Ventra M, Garaj S, Hibbs A, Huang X, Jovanovich SB et al. The potential and challenges of nanopore sequencing. Nature Biotechnology. 200;6:1146-1153.

[62] Egan AN, Schlueter J, Spooner DM. Applications of next-generation sequencing in plant biology. American Journal of Botany. 2012;99:175-185.

[63] Clarke J, Wu HC, Jayasinghe L, Patel A, Reid S, Bayley H. Continuous base identification for single-molecule nanopore DNA sequencing. National Nanotechnology Initiative. 2009;4:265.

[64] Kumar S, Tao C, Chien M, Hellner B, Balijepalli A, Robertson JW, Ju J. PEG-labeled nucleotides and nanopore detection for single molecule DNA sequencing by synthesis. Scientific Reports. 2012;2:684.

[65] Hayden CE. Pint-sized DNA sequencer impresses first users. Natural News. 2015;521(7550):15.

[66] Khiyami MA, Hassan A, Yasser MA, Mousa AA, Kamel AAE. Plant pathogen diagnostic techniques: forthcoming changes? Biotechnology & Bio-technological Equipment. 2014;28: 775-785.

[67] Pimentel D. Invasive plants: their role in species extinctions and economic losses to agriculture in the USA. In: Management of invasive weeds, invading nature – Springer Series in invasion ecology. Dordrecht: Springer; 2009.pp 1-7.

[68] Lattanzio VMT, Nivarlet N, Lippolis V, Gatta SD, Huet AC, Delahaut P, Granier B, Visconti *A. multiplex* dipstick immunoassay for semi-quantitative determination of Fusarium mycotoxins in cereals. Analytica Chimica Acta. 2012;718:99-108.

[69] Edmundson MC, Capeness M, Horsfall L. Exploring the potential of

metallic nanoparticles within synthetic biology. New Biotechnology. 2015;31: 572-578.

[70] Knudsen BR, Jepsen ML, Ho YP. Quantum dot-based nanosensors for diagnosis via enzyme activity measurement. Expert Review of Molecular Diagnostics. 2013;13:367-375.

[71] Safarpour H, Safarnejad MR, Tabatabaei M, Mohsenifar A, Rad F, Basirat M, Shahryari F, Hasanzadeh F. Development of a quantum dots FRET-based biosensor for efficient detection of *Polymyxa betae*. Canadian Journal of Plant Pathology. 2012;34:507-515.

[72] Rad F, Mohsenifar A, Tabatabaei M, Safarnejad MR, Shahryari F, Safarpour H, Foroutan A, Mardi M, Davoudi D, Fotokian M. Detection of Candidatus Phytoplasma aurantifolia with a quantum dots FRET-based biosensor. Journal of Plant Pathology. 2012;94:525-534.

[73] Ma J, Du LF, Chen M, Wang HH, Xing LX, Jing LF, Li YH. Drug-loaded nano-microcapsules delivery system mediated by ultrasoundtargeted microbubble destruction: a promising therapy method. Biomedical Reports. 2013;1:506-510. doi: 10.3892/br.2013.110.

[74] Ambrosone A, Marchesano V, Carregal-Romero S, Intartaglia D, Parak WJ, Tortiglione C. Control of Wnt/β-catenin signaling pathway in vivo via light responsive capsules. ACS Nano. 2016;10: 4828-4834. doi: 10.1021/acsnano.5b07817

[75] Linsley CS, Wu BM. Recent advances in light responsive on-demand drug-delivery systems. Therapeutic Delivery. 2017;8:89-107. doi: 10.4155/tde-2016-0060

[76] Boonham N, Glover R, Tomlinson J, Mumford R. Exploiting generic platform technologies for the detection and identification of plant pathogens. In Sustainable Disease Management in a European Context, ed. DB Collinge, L Munk, BM Cooke. Berlin: Springer; 2008. pp.355-63.

[77] Yao KS, Li SJ, Tzeng KC, Cheng TC, Chang CY, Chiu CY, Liao CY, Hsu JJ, Lin ZP. Fluorescence silica nanoprobe as a biomarker for rapid detection of plant pathogens. Advanced Materials Research. 2009;79:513-16.

[78] Li XM, Xu G, Liu Y, He T. Magnetic Fe3O4 nanoparticles: synthesis and application in water treatment. Journal of Nanoscience and Nanotechnology. 2011;1:14-24.

[79] Tartaj P, del Puerto Morales M, Veintemillas-Verdaguer S, Gonzalez-Carre ́ no T, Serna CJ. The preparation of magnetic nanoparticles for applications in biomedicine. Journal of Physics D. 2003;36:182.

[80] Fang Y, Ramasamy RP. Current and prospective methods for plant disease detection. Biosensors. 2015; 5:537-61.

[81] Bergeson LL. Nano silver pesticide products: What does the future hold? Environmental Quality Management. 2010;19:73-82.

[82] Khot LR, Sankaran S, Maja JM, Ehsani R, Schuster E. Application of nanomaterials in agricultural production and crop protection: a review. Crop Protection. 2012;35:64-70.

[83] Alonso-Lomilloa MA, Dom'ınguez-Renedoa O, Ferreira-Gonc¸alves L, Arcos-Mart'ıneza MJ. Sensitive enzyme-biosensor based on screen-printed electrodes for ochratoxin A. Biosensors and Bioelectronics. 2010; 25:1333-37.

[84] Ansari AA, Kaushik A, Solanki PR, Malhotra BD. Nanostructured zinc oxide platform for mycotoxin detection. Bioelectrochemistry. 2010;77:75-81.

[85] Tothill IE. Biosensors and nanomaterials and their application for mycotoxin determination. World Mycotoxin Journal. 2011;4:361-74.

[86] Ruitenberg R. Earthworm Health Hurt by Nanoparticles in Soil in Alterra Study, Bloomberg. 2013.

[87] Xu L, Liu Y, Bai R, Chen C. Applications and toxicological issues surrounding nanotechnology in the food industry. Pure and Applied Chemistry. 2010;82:349-372.

[88] Chen H and Yada R. Nanotechnologies in agriculture: New tools for sustainable development. Trends in Food Science & Technology. 2011;22: 585-594.

[89] Kalpana-Sastry R, Rashmi HB, Rao NH, Ilyas SM. Nanotechnology and Agriculture in India: The Second Green Revolution? Presented at the OECD Conference on "Potential Environmental Benefits of Nanotechnology: Fostering Safe Innovation-Led Growth" Session 7. Agricultural Nanotechnology, Paris, France, July 15-17, 2009.

Chapter 12

Chaetomium Application in Agriculture

Kasem Soytong, Somdej Kahonokmedhakul, Jiaojiao Song and Rujira Tongon

Abstract

Chaetomium species for plant disease control are reported to be antagonize many plant pathogens. It is a new broad spectrum biological fungicide from *Chaetomium* species which firstly discovered and patented No. 6266, International Code: AO 1 N 25/12, and registered as Ketomium® mycofungicide for plant disease control in Thailand, Laos, Vietnam, Cambodia and China. *Chaetoimum* biofungicide and biostimulants are applied to implement integrated plant disease control. It showed protective and curative effects in controlling plant disease and promoting plant growth. It has been successfully applied to the infested soils with integrated cultural control for the long-term protection against rice blast (*Magnaporte oryzae*), durian and black Pepper rot *(Piper nigram* L.) (*Phytophthora palmivora*), citrus rot (*Phytophthora parasitica*) and strawberry rot *(Fragaria* spp.) caused by *Phytophthora cactorum,* wilt of tomato (*Fusarium oxysporum* f. sp. *lycopersici*), basal rot of corn *(Sclerotium rolfsii)* and anthracnose (*Colletotrichum* spp.) etc. Further research is reported on the other bioactive compounds from active strains of *Chaetomium* spp. We have discovered various new compounds from *Ch. globosum*, *Ch. cupreum*, *Ch. elatum*, *Ch. cochliodes*, *Ch. brasiliense*, *Ch. lucknowense*, *Ch. longirostre* and *Ch. siamense.* These new compounds are not only inhibiting human pathogens (anti-malaria, anti-tuberculosis, anti-cancer cell lines and anti-*C. albicans* etc) but also plant pathogens as well. These active natural products from different strains of *Chaetomium* spp. are further developed to be biodegradable nanoparticles from active metabolites as a new discovery of scientific investigation which used to induce plant immunity, namely microbial degradable nano-elicitors for inducing immunity through phytoalexin production in plants e.g. inducing tomato to produce alpha-tomaline against *Fusarium* wilt of tomato, capsidiol against chili anthracnose, sakuranitin and oryzalexin B against rice blast, scopletin and anthrocyaidin against *Phytophthora* or *Pythium* rot Durian and scoparone against *Phytophthora* or *Pythium* rot of citrus. *Chaetomium* biofungicide can be applied instead of toxic chemical fungicides to control plant diseases.

Keywords: *Chaetomium*, Biofungicides, plant diseases, plant immunity, phytoalexin

1. Introduction

Chaetomium Kunze belongs to Ascomycota of the *Chaetomiaceae* which established by Kunze in [1] and it is one of the largest genera of saprophytic ascomycetes which comprise more than 300 species worldwide [2–6]. *Chaetomium* is recorded to be a potent antagonists of various plant pathogens, especially soil-borne and seed borne pathogens [4, 7–9]. *Chaetomium* is also reported to be antagonistic to the growth of bacteria and fungi through competition for nutrients, mycoparasitism, antibiosis, or various combinations [10]. The application of *Chaetomium* as a antagonist to control plant pathogens was first reported in 1954 by Tveit and Moore who found *Ch. globosum* and *Ch. cochliodes* grown on surface of oat seeds provided to control *Helminthosporium victoriae* [11]. *Chaetomium globosum* and *Ch. cochlioides* noted to inhibit the growth of *Fusarium* spp. and *Helminthosporium* spp. [11]. The living spore of Ch. *globosum* significantly controlled the apple scab (*Venturia inequalis)* [12]. *Ch. globosum* produces metabolites inhibiting the growth of *P. ultimum* causing damping-off of sugar beet [13], *Rhizoctonia solani* [14], leafblight of brassicas causing *Alternaria brassicicola* [15] and can reduce the pathogenic inoculum of *Botrytis cinerea* causing deadly lily leaves in the field [16]. *Chaetomium cupreum* is reported to control soybean plant pathogens e.g. *Phomopsis* and *Colletotrichum* spp. [17]. It is reported that isolate of *Ch. globosum* produce antibiotic substances that suppressed the damping – off of sugar beet cause by *P. ultimum* [13]. *Chaetomium globosum* is noted to be a strong cellulose decomposer [18] and showed an effective antagonist of various soil microorganisms [7, 8]. Moreover, *Chaetomium* spp. are recorded to produce their active metabolites for biological activities. Several types of pure compounds are found from *Chaetomium* species e.g. benzoquinone derivatives [19].

In Thailand, *Chaetomium* spp. were studied and screened for abilities as antagonistic effect againist phytopathogens in 1989 by Soytong *et al.* [20]. *Chaetomium cupreum* and *Ch. globosum* were reported to decrease leaf spot disease of corn caused by *Curvularia lunata*, rice blast caused by *Pyricularia oryzae* and sheath blight of rice caused by *Rhizoctonia oryzae* [21, 22]. *Chaetomium* spp. are also strongly recommended in competitive and high ability to make organic compost and to suppress pathogen in the soil. *Chaetomium* spp. are known to be the strictly saprophytic coprophilousn. It can be found in organic materials, plant debris and high organic soil [23]. Some species produce cellulase to decompose cellulose, lignin and other organic materials and serve as a biological control agent against several phytopathogens [20].

Chaetomium cupreum and *Ch. globosum* are noted to decrease leaf spot of corn (*Curvularia lunata*), rice blast (*Pyricularia oryzae*), sheath blight (*Rhizoctonia oryzae*) and tomato wilt (*Fusarium oxysporum* f.sp. *lycopersici*) [22, 24]. Chetomanone, ergosterol, ergosteryl palmitate, chrysophanol, chaetoglobosin C, alternariol monomethyl ether, echinuline and isochaetoglobosin D were produced from *Ch. globosum* KMITL-N0802 [25] which implies antibiosis. *Chaetomium cupreum* was significantly reduced the growth of seed-borne pathogen of rice e.g. *Curvularia lunata, Drechslera oryzae, Fusarium moniliforme* and *Pyricularia oryzae* [26]. *Chaetomium globosum* was exprsssed significantly suppression to tomato wilt caused by *Fusarium oxysporum* f. sp. *lycopersici* and *Pseudomonas solanacearum in Thailand* [23, 27] while *Ch. cupreum* reported to give a good control tomato wilt in the fields [28]. The specific effective strains of *Chaetomium* spp. have been formulated as biological products in the forms of pellet and powder [29] that could effectively control many soil borne plant pathogens. The *Chaetomium* bioproducts were reported to have a good potential in control of *Thielaviopsis* Bud Rot of Bottle palm caused by *Hyophorbe lagenicaulis* in the fields [30]. *Chaetomium globosum* and *Ch. cupreum* are successfully applied to control root rot disease of citrus, black pepper, strawberry and reduce damping off disease of sugar beet [20, 31]. The powder and pelletized formulations of Ketomium®, as a broad spectrum biofungicides has been registered as a biological biofertilizer for degrading organic matter and to induce plant immunity and stimulate plant growth [20]. It was recorded that Ketomium® mycofungicide from Thailand was the most efficient to control raspberry spur blight (*Didymella applanate*) and reduced potato disease (*Rhizoctonia solani*) and increasing potato yield [32]. After 2 years in storage,

Ketomium® biofungicide from Thailand was still capable to inhibit the growth of plant pathogens in higher doses [31].

The biofungicide produced from *Ch. globosum* and *Ch. cuprem* resulted in controlling *Phytophthora* root rot of durian (*Phytophthora palmivora*) in the fields which the *Phytophthora* pathogen reduced after applying *Chaetomium* into soils planted Durian var. Monthong in Thailand [33]. *Chaetomium* biofungicide is also reported to control *Phytophthora parasitica* in fields planted to Citrus in the most serious disease area in Thailand [34]. Soytong *et al.* [20] stated that Ketomium® was applied to *Fusarium*-infested soils where tomatoes were grown and it had sucessfully controlled the pathogen inoculum and reduced disease prevalance. Ketomium® have been successfully controlled as *Fusarium*-suppressive soils. The tomato plants treated with Ketomium® and pentachloronitrobenzen (PCNB) completely controlled *F. oxysporum* f. sp. *lycopersici* (tomato wilt). *P. palmivora* (black pepper rot). It is penciled that the tested Ketomium® biological product has been proved to control several diseases in the fields to control root rot of durian [33], black paper [35] and citrus [36]. Further research has done to formulate those active antagonistic *Chaetomium* species to be biological fungicide to control plant diseases e.g. late blight of potato, citrus rot disease, white rot disease of para rubber. It is also reported that the specific strains of *Ch. globosum* can be inhibited root knot nematode and insect control.

2. Biological control of potato late blight using *Chaetomium* biofungicide

Application of *Chaetomium*-biofungicide is sucessfully controlled late blight late blight of potato caused by *Phytophthora infestans* in northern Thailand. Before experiment, It was found that *P. infestans* spread directly to the potato plants in the field through zoosporangia and resulted in a serious diseaase and lost of yield. It was investigated that *P. infestans* found in the soil planted with potato in high pathogen inocula before experiment (**Figure 1**). Ketomium®-biofungicide decreased the disease incidence in the infested fields when compared with the non-treated

Figure 1.
Symptoms of late blight caused by Phytophthora infestans *on leaves (upper left) and dry rot on potato tuber (upper right). Sporangiophore, sporangium of* Phytophthora infestans *(lower part).*

Figure 2.
Experimental plot testing Chaetomium *biofungicide against the late blight of potato.*

control and the difference between *Chaetomium*-biofungicide and chemical pesticide treated fields were not significantly differed. The field experiment was carried out in Chiang Mai province, Thailand, where the soil was seriously infested with *P. infestan* and wherein disease previously destroyed the potato plants and resulted in total loss of the yield. Ketomium® biofungicide reduced late blight incidence by 38% as seen in **Figure 2** [37].

3. The control of citrus rot disease with *Chaetomium* biofungicide in Cambodia

P. ultimum was firstly recorded by our investigation to cause citrus (*Citrus reticula* Blanco) root rot in Cambodia. The experimental field was carried out in 1 ha of 200 trees of citrus orchard in Battambang province where the infested soil with *P. ultimum* causing root rot disease. It was observed before experiment that *P. ultimum* infected the citrus trees over 90%, and mostly citrus trees were completely destroyed which can be seen the sign of yellowing and small leaves, dieback and root rot. The four-year-old citrus trees were selected for experiment with the same disease level.

The pathogen causes a serious damage almost everywhere planted to citrus in Battambang province. *P. ultimum* infects citrus plants starting from seedlings which show yellow leaves, die back, stem rot, root rot and die. The infected citrus trees slowly decline from the second year planting and gradullay died which starting from six to seven years old, even using the chemical fungicides. The reserch found that the detached leave method was proved that only three days inoculation of *P. ultimum*, the citrus leaves turned to pale yellow and completely dark brown. The re-isolation the aggressive pathogen from lesion confirmed its pathogenicity. In field expriment, chemical fungicide, *Chaetomium* and *Trichoderma* treamenrts were compared and periodically applied every month to four-year old citrus trees in one year. All treatments were sprayed above plants and to rhizosphere soil every month as metalaxyl-10 g/20 L of water in combination with chemical fertilizers, *Chaetomium* treatment (20 g/20 L of water) and *Trichoderma* treatment (20 g/20 L of water). It was noted that all treated citrus trees significantly recovered within 3–4 months of applications. As it was seen new leave flashes and roots emerged, then the citrus trees were recovered from rot root diseaase. The biological products of either *Chaetomium* or *Trichoderma* showed significant disease control along with the metalaxyl chemical fungicide [38].

4. Application of *Chaetomium* biofungicide to control white root disease of Para rubber trees caused by Rigidoporus microporus

White root disease caused by *Rigidoporus microporus* (**Figures 3** and **4**) that was controlled by *Ch. bostrychodes* BN08, *Ch. cupreum* RY202 which inhibited the

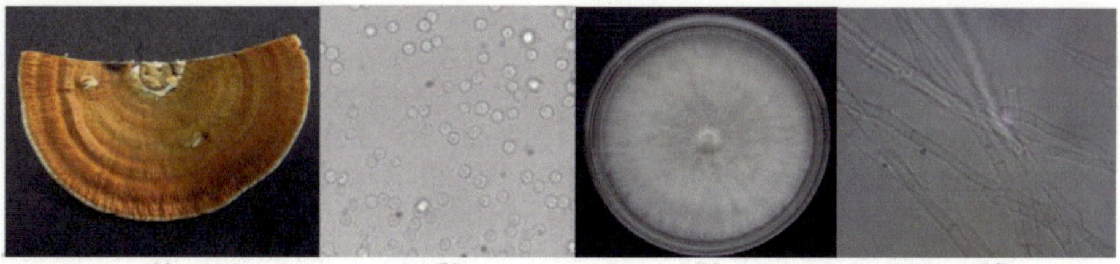

Figure 3.
Rigidoporus microporus: *Colony on PDA at 6 days (a), hypha (b) fruiting body (c), and basidiospores (d).*

Figure 4.
White root disease of Para rubber tree caused by Rigidoporus microporus.

growth of pathogen over 50%. *Chaetomium bostrychodes* BN08 and *Ch. cupreum* RY202 were grown over the colony of *R. microporus* within 30 days. The crude extracts from *Ch. cupreum* RY202 showed the best growth inhibition of *R. microporus* with ED_{50} values of 170, 402, and 1,220 μg/L, respectively. Rotiorinol is a bioactive compound produced from *Ch. cupreum* inhibited the growth of *R. microporus* which the ED_{50} value of 26 μg/L. The bioformulation of *Ch. cupreum* RY202 in the powder and oil form significantly inhibited *R. microporus* to infect the root of the rubber trees as reported by Kaewchai and Soytong [39]. *Chaetomium cupreum* RY202 formulated as biofungicide to control white root disease of para rubber was resulted to disease reduction in the treatments of *Ch. cupreum* RY202 in the powder form and oil form of 75%. This is the first report of our study using *Ch. bostrychodes* and *Ch. cupreum* to control the white root disease of para rubber.

5. Application of *Chaetomium* antagonistic fungi to plant parasitic nematodes and insect control

Diverse research was on *Chaetomium* which carried out for agricultural purpose. Some reserchers demonstrated the endophytic *Chaetomium* exhibited to antibitosis against nematodes e.g. root-knot nematode (*Meloidogyne incognita*) and hatch of soybean cyst nematode (*H. glycines*) and *Globodera pallida* (cyst nematode) or insects e.g. cotton aphids (*A. gossypii*) and beet armyworms (*Spodoptera exigua*).

In [40], Nitao *et al.* reported that the culture broths from *Ch. globosum* inhibited the egg hatch and juvenile mobility of root-knot nematode (*Meloidogyne incognita*) and also inhibited the hatching of soybean cyst nematode (*H. glycines*). It demonstrated that avipin is an active low molecular weight compound responding for antoganistic activity. But the testing on Muskmelon (*C. melo*) plants in steamed and unsteamed soil were inoculated with root-knot nematodes and different concentrations of avipin were not clearly shown. Zhou *et al.* [41] recorded that *Ch. globosum* TAMU 520, is endophyte from cotton (*G. hirsutum*) and systemically colonize the cotton plants through seed treatment. The endophytic *Ch. globosum* suppressed the infection of root-knot nematode (*Meloidogyne incognita*) which reduced female reproduction.

In 2012, *Ch. globosum* with nematicidal activity, metabolite and application is patented in China as a patent No. CN102925369A in 2012 by a group of Chinese Scientists which claimed that *Ch. globosum* NK106 (CGMCC6716) is preserved number. Chaetoglobosin A prevents the plant parasitic nematode due to its high toxicity and nematicidal activity against the juvenile stage of *Meloidogyne incognita*. Chaetoglobosin A at 300 mg/mL showed the lethality is 90%, decreased the nematode eggs by 63%. In [42], Hu *et al*. stated that chaetoglobosin A, the secondary metabolites produced by *Ch globosum* NK102 expressed the nematicidal activity *Meloidogyne incognita. Chaetomium globosum* NK102 also showed repellant second-stage juveniles. Chaetoglobosin A showed strongly adverse effects in secondary stage mortality of nematode with 99.8% at 300 μg /mL (LC_{50} = 77.0 μg/mL) at 72 h. Chaetoglobosin A and filtrates from *Ch. globosum* NK102 were not affect on egg hatching until 72 h. The filtrate treatments inhibited the penetration of second-stage juveniles at 12.5% dilution treatment. Chaetoglobosin A also inhibited the penetration of secondary stage and reduced the number of nematode eggs. Kooliyattil *et al.* [43] found *Ch. globosum* as a fungal parasite which isolated from egg of *Globodera pallida* (cyst nematode) in USA. It showed the greatest reduction of the infection by *G. pallida* in potato of about 76%. In [44], Khan *et al*. reported that endophytic *Ch. globosum* YSC5 showed nematicidal metabolite activities against the second stage juveniles of *Meloidogyne javanica*. Chaetoglobosin A, chaetoglobosin B and flavipin strongly inhibited (91.6, 83.8 and 87.4%, respectively) on mortality of the second stage juveniles at 200 μg/mL with LC_{50} values of 88.4, 107.7 and 99.2 μg/mL after 72 h, respectively, 3-methoxyepicoccone and 4,5,6-trihydroxy-7-methylphthalide moderately inhibited at 78.0 and 75.5%, respectively with LC_{50} values of 124.0 and 131.6 μg/mL, respectively. The promising metabolites Chaetoglobosin A, chaetoglobosin B significantly reduced nematode reproduction in pot experiment.

Chaetomium also showed insect control which was recorded by Zhou *et al.* [41] who stated that endophytic *Ch. globosum* showed negative affect on the fecundity of cotton aphids (*A. gossypii*) and beet armyworms (*Spodoptera exigua*). The beet army worms colovized on *Chaetomium* treated plants were with smally head capsule.

6. Bio-formulation of *Chaetomium* metabolites

Our research and development on *Chaetomium* biological products have been investigated since 1989. The first biological product of *Chaetomium* is contributed as a new broad spectrum biological fungicides from *Chaetomium* (Thailand Patent No. 6266, International Code: AO 1 N 25/12 and registered as Ketomium® biofungicide for plant disease control) which has been developed and improved from 22-strains of *Ch. cupreum* CC01-CC10 and *Ch. globosum* CG01-CG12 in the form of pellets, powder and liquid formulations. The practical integrated biological for plant pathogens is successfully introduced to farmers. Technology of disease control can be demonstrated either alone or integrated with other control measures. The products are scientifically proved not only for protection but also with curative effects as well as promoting plant growth. The formulations have been successfully used in infested field-soils integrated with cultural control for the long-term protection against durian and black pepper *(Piper nigram* L.) rot caused by *Phytophthora palmivora,* citrust rot caused by *Phytophthora parasitica* and strawberry rot caused by *Phytophthora cactorum,* tomato wilt caused by *Fusarium oxysporum* f. sp. *lycopersici* and basal rot of corn caused by *Sclerotium rolfsii*. Further research is undergoing to develop bioactive metabolites from active strains of *Chaetomium* spp. for plant disease control and immunity.

7. Efficacy of *Chaetomium* to control brown leaf spot of rice

The active metabolites of *Chaetomium cochliodes* inhibited spore production of *Drechslera oryzae* causing brown leaf spot of rice var. Pittsanulok 2. *Chaetomium cochliodes* was reported to be a new antagonist to control brown leaf spot of rice var. Pittsanulok 2. Bio-fungicides produced from *Ch. cochliodes* were developed from active strain of *Ch. cochliodes* resulted in crude metabolites, powder, and liquid formulations which significantly inhibited the brown leaf spot of rice and increased plant growth. They significantly reduced leaf spot of rice var. Pitsanulok 2. The biopowder, crude metabolite and benlate-fungicide applied to rice seedlings at 40 days revealed plant height of 13, 13 and 12 cm, respectively when compared to the control (8 cm). The bio-powder, crude metabolite and benlate showed plant height of 50, 50 and 48 cm, respectively when compared to the control (21 cm) for 70 days. It was concluded that crude metabolite produced significantly higher rice growth parameters than the non-treated control [45].

8. Efficacy of *Chaetomium* to control fusarium wilt

The metabolites extracted from *Ch. cupreum* CC3003, *Ch. globosum* CG05, and *Ch. lucknowense* CL01 significantly inhibited growth and spore production of *F. oxysporum* NHP-Fusa-2 (**Figure 5**). The MeOH extract of *Ch. cupreum* CC3003 revealed to be more effective spore inhibition of *F. oxysporum* than the others which the ED_{50} was 85 μg/mL. The ED_{50} of hexane crude extract was 49 μg/mL, and EtOAc crude extract was 62 μg/mL in *Ch. globosum* CG05 and *Ch. lucknowense* CL01. Metabolites extracted from *Ch. cupreum* CC3003, *Ch. globosum* CG05, and *Ch. lucknowense* CL01 can be used to control tea wilt and root rot diseases caused by *Fusarium oxysporum.* It was the first time that *F. oxysporum* reported the causal pathogen of wilt and root-rot disease of tea in Vietnam [46].

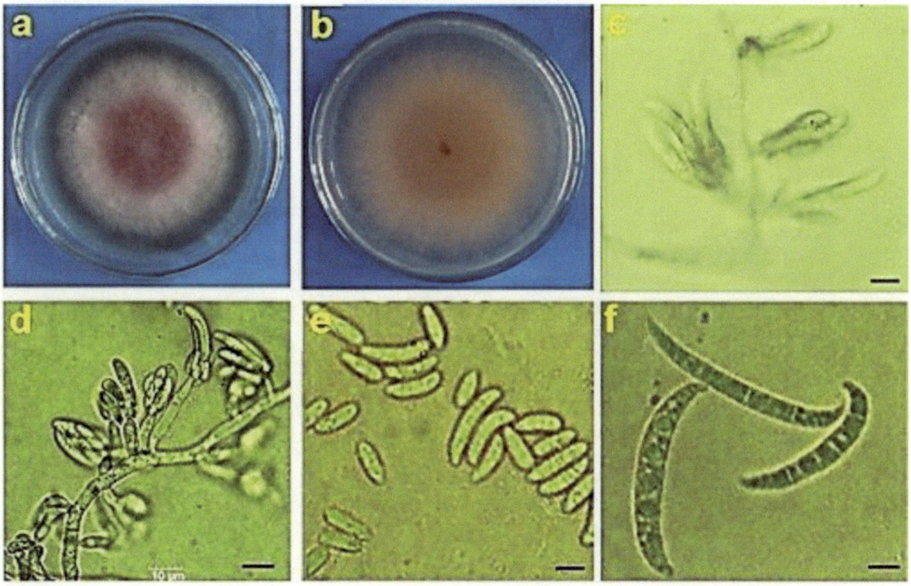

Figure 5.
Fusarium oxysporum *at 7-day-old culture on potato dextrose agar (colony) and water agar (conidia) (scale bar: 10 µm): (a) front surfaced colony; (b) back surfaced colony; (c) macroconidia on phialides; (d) microconidia on sporodochia; (e) microconidia; (f) macroconidia (source: [46]).*

The ED_{50} value of hexane crude extract from *Ch. globosum* N0802 inhibited *Fusarium oxysporum* f. sp. *lycopersici* NKSC01 causing tomato wilt var. Sida at 157 μg/mL. The crude hexane from *Ch. lucknowense* CLT and crude methanol from *Trichoderma harzianum* PC01 showed ED_{50} values of 188 and 192 μg/mL. The developed biofungicides - namely N0802, CLT and PC01 showed significantly higher wilt reduction of 44, 36 and 41%, respectively, than prochoraz fungicide (22%). All biofungicides significantly increased the yield more than prochoraz and inoculated control.

9. Efficacy of *Chaetomium* to control phytophthora root rot in citrus

Hung *et al.* [47, 48] found that pomelo rot caused by *Phythopthora nicotianae* in Thailand (**Figure 6**). Crude metabolites of *Ch. globosum*, *Ch. lucknowense*, *Ch. cupreum* showed antifungal activities against the growth of *P. nicotianae*, with ED_{50} of 3–101 μg/mL (**Figure 7**).

Chaetomium CG05, CL01, and CC3003 were reported to antagonize *P. nicotianae* showing inhibition between *Chaetomium* spp. and *P. nicotianae* before contact was made. The colonies of CG05 and CL01 made contact with *P. nicotianae* colonies without the clear zone of inhibition (**Figure 7**: A3 and B3). CG05, CL01, and CC3003 grew over colony of *P. nicotianae* in biculture plates in 30 days. Hyphae of *Chaetomium* penetrated or coiled around hyphae of *P. nicotianae*, maceration and discoloration of *P. nicotianae* KA1 colonies (from white to light yellowish-brown). Application of living *Chaetomium* spp. and metabolites reduced root rot of pomelo by 66–71% and increased plant weight by 72–85% compared to the non-treated control.

10. Biofungicide from *Chaetomium* elatum ChE01 against banana anthracnose caused by Colletotrichum musae

Colletotrichum musae is seriously caused banana anthracnose. *Chaetomium elatum* ChE01 showed antifungal activity to inhibit the growth of *C. musae* in

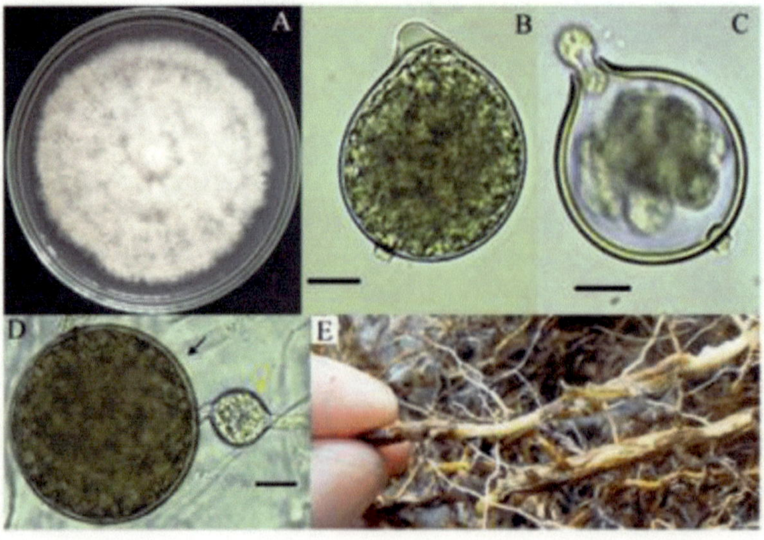

Figure 6.
Phytophthora nicotianae. *A, seven-day-old culture on PDA; B, C, sporangium with a short pedicel; D, Chlamydospore (black arrow) and hypha swelling (yellow arrow) (scale bars: B ~ D = 10 μm); E, root rot symptoms (source: [48, 47]).*

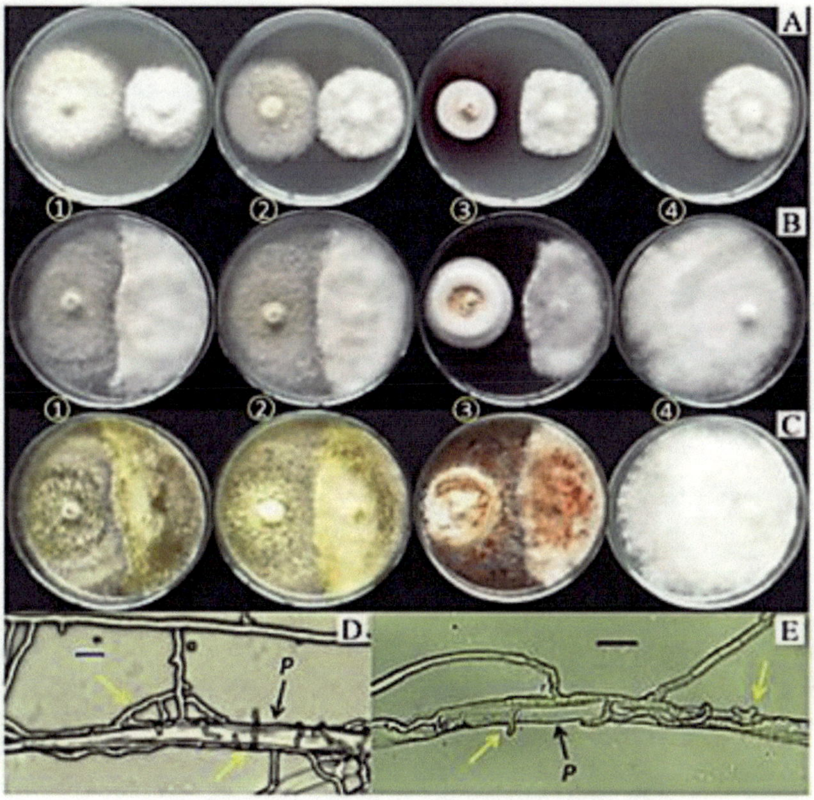

Figure 7.
Phytophthora nicotianae *in dual culture tests (A), 10 (B), and 30 days (C) incubation (1, 2, and 3:* P. nicotianae *[placed on the right side of the plates] against CG05, CL01, and CC3003, respectively; 4:* P. nicotianae *Alone); D, E,* Chaetomium *hyphae (yellow arrows) coiled around and grew inside hypha of* P. nicotianae *KA1 (P) (scale bars: D, E = 10 μm). Source: Hung et al. [48, 47].*

biculture evaluation over 60% and the spore production of *C. musae* was inhibited 57% in biculture trial. Metabolites from *Ch. elatum* ChE01 (crude methanol, crude ethyl acetate and crude hexane extracts) inhibited the spore production of *C. musae* with the ED_{50} values of 5, 7 and 19 μg/mL, respectively. The bio-fungicide produced from *Ch. elatum* ChE01 showed significantly better reduction of disease prevalence than benomyl fungicide. It is reported for the first time that biofungicide produced from *Ch. elatum* ChE01 controlled banana anthracnose caused by *C. musae* for the first time. *Chaetomium elatum* ChE01 produced a new chaetoglobosin V, two natural products, prochaetoglobosin III and prochaetoglobosin IIIed, six known chaetoglobosins B-D, F and G and isochaetoglobosin D [49] implying to a control mechnism of antibiosis. Moreover, crude methanol extract higher of *Ch. elatum* ChE01 showed significantly high spore inhibition of *C. musae* and the ED_{50} value was 5 μg/mL. The crude ethyl acetate and crude hexane extraction showed ED_{50} values 7 and 19 μg/mL, respectively. The higher concentration showed more inhibition of pathogens than the lower ones (**Figure 8**).

The coffee var. arabica plants were inoculated with 1 x 10^6 spores/mL suspension of the anthracnose pathogen (*Colletotrichum gloeosporioides*). Ten wounds were artificially made on leaves/ seedling with the fifth leaf from the top. Wounds were punctured with sterilized needles 10 times. The treatments were usually applied at 15 day intervals: T1 was inoculated with anthracnose pathogen, T2 was treated with a spore suspension of *Ch. cupreum* CC3003 at concentration of 1 x 10^6 spores/mL, T3 was treated with a biofungicide powder of *Ch. cupreum* CC3003 at

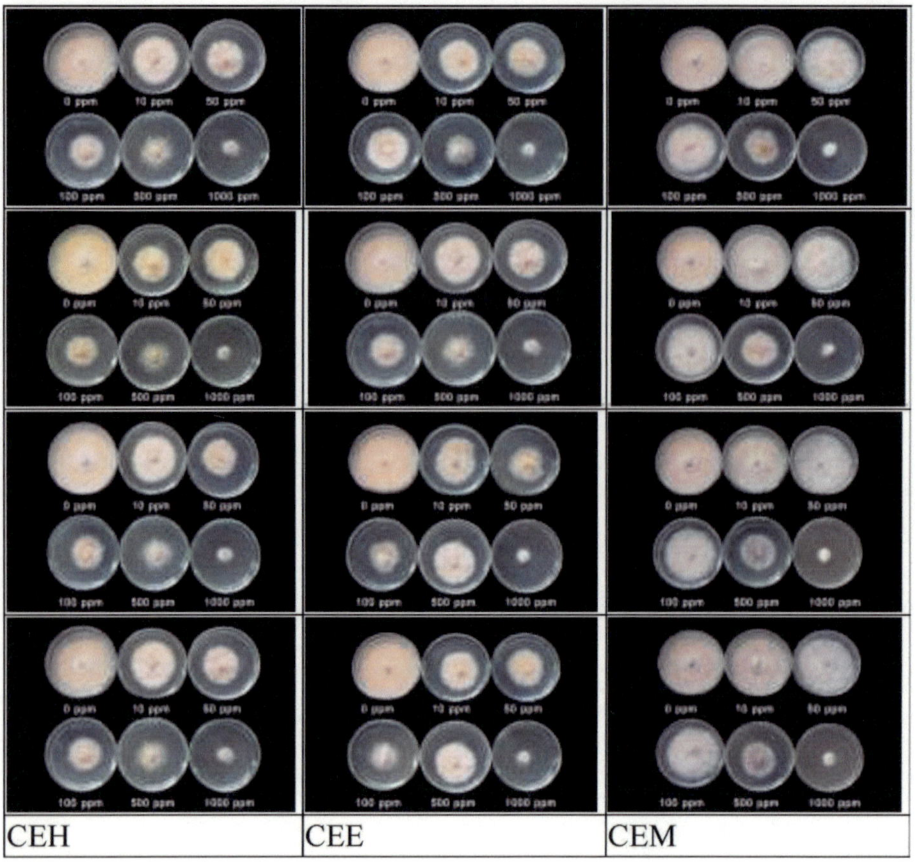

Figure 8.
Crude extracts of Chaetomium elatum *ChE01 testing to inhibit spore production of* Colletotrichum musae.

a concentration of 10 g/20 L of water, T4 was treated with nano- trichotoxin-A50 (made from *Trichoderma harzianum* PC01) and T5 was treated with nano-rotiorinol made from pure compound of *Ch. cupreum* CC3003). Nano- trichotoxin-A50 and nano-rotiorinol were produced by Dr. Kasem Soytong and Joselito Dar at the Biocontrol Research Laboratory, KMITL, Bangkok, Thailand [50]. *T. harzianum* PC01 reported to produce antibiotic polypeptides trichotoxin A50 [51]. Moreover, Vilavong and Soytong [52] reported that crude extracts with hexane, ethyl acetate and methanol from *Ch. cupreum* CC3003 expressed significant inhibition of *C. gloeosporioides* (coffee anthracnose) with ED_{50} values of 13, 11 and 28 μg/mL, respectively. Metabolites of *Ch. cupreum* CC3003 showed antifungal activity against *C. gloeosporioides* with abnormal appearance of spores of tested pathogen. A powder of the biofungicide produced from *Ch. cupreum* reduced anthracnose 54%.

11. Biodegradable nano-particles constructed from active natural products of different species of *Chaetomium* for immunity

Chaetomium species have been reported to degrade cellulolytic plant debris into soil to increase soil fertility as well as organic matter in the soil, and a specific isolate of *Ch. globosum* was reported to inhibit *Pyricularia oryzae* causing rice blast [4]. Soytong *et al.* [20] showed that chaetoglobosin C from *Ch. globosum* KMITL-N0805 actively inhibited several plant pathogens, such as *C. lunata* (leaf spot of corn), *Colletotrichum* sp. (citrus anthracnose), and *Fusarium oxysporum* f.sp. *lycopersici* (tomato wilt).

Kanokmedhakul *et al.* [25] reported that *Ch. globosum* KM ITL-N0802 produces a novel anthraquinone-chromanone compound, named chaetomanone, and known compounds as chaetoglobosin C and echinulin. Chaetomanone and echinulin were reported to be inhibitors of *M. tuberculosis* causing Tuberculosis of human being.

The biodegradable nanoparticles from natural products of active metabolites of *Chaetomium* spp. are further investigated and discovered as a new scientific investigation, namely microbial degradable nano-elicitors for inducing immunity in plants (the authors).

In recent years, scientists have interested nanoparticles possessing biological properties [53, 54]. The nanotechnology for agriculure is interested in various areas [55]. Plant disease control is to decrease or eliminate the nontarget effects either abiotic or biotic factors. Nano-sciences have become a new method to restructure materials at atomic level. Molecular nanotechnology can construct the organic materials into defined structures and atom by atom [56]. Natural products from *Chaetomium* species proved to be functioned for antifungal strategy against several plant pathogens [20]. The alternative disease control is to acheive, safe, effective, and environmentally friendly. The construction and characterization of copolymer nanoparticles loaded with bioactive compounds from *Chaetomium* species have been searched rather than toxic chemical pesticides. The natural products from *Chaetomium* spp. constructed to be fine particles at molecular level as degradable nanoparticles used to control plant disease and induce plant immunity. Biocontrol research unit, Faculty of Agricultural Technology, King Mongkut's Institute of Technology (KMITL), Bangkok, Thailand developed the research on biocontrol technology for years. Degradable nano-CGH, nano-CGE, and nano-CGM constructed from *Ch. globosum* KMITL-N0805 actively inhibited *Curvularia lunata* causing leaf spot disease of rice var. Sen Pidoa in Camobodia. The effective dose of 50% (ED_{50}) of degradable nano-CGH, nano-CGE, and nano-CGM were 1.21, 1.19, and 1.93 μg/mL, respectively at very low concentration was sufficeint to inhibit leaf spot pathogen of rice. These biodegradable nanoparticles actively penetrated to the pathogen cells causing to their disruption and distortion. Those pathogen inocula lost pathogenicity according to preliminary Koch's postulate test. The nano-CGH, nano-CGE, and nano-CGM inhibited spore production by 93%, 93%, and 84%, respectively and resulted antifungal activity against *C. lunata* with ED_{50} values of 1.2, 1.2, and 1.9 μg/mL, respectively.

The applications of degradable nano-CGH, nano-CGE, and nano-CGM to inoculated *C. lunata* on rice seedlings var. Sen Pidoa reduced disease incidence in pot experiments. Degradable nano-CGH and nano-CGM gave higher disease reduction of rice leaf spot caused by *C. lunata* (61%) than nano-CGE (54%). These nanoparticles significantly increased the height and number of tillers of the rice plants 60 days after treatment [57].

Ditta [58] stated that application of chemical pesticides causing climate change, and affected to urbanization, and natural resources. Those chemicals runoff or accumulate toxic pesticides in soil and it is needed to solve these problems immediately. However, there are still a few reports for the use of nanocarrier systems in agriculture [59]. Rai and Ingle [60] suggested that nanotechnology provides efficient to control diseases and insect pests. Ditta [58] added that nanotechnology has a great potential in agriculture due to its enhancing life quality, especially in crop production. Nanotechnology should be carefully evaluated for postive and negative impacts as done with any new technology. Some nanoparticles are formulated to contain chemical pesticides in colloidal suspensions or as powders, at the nano or micro scale that must be noticed. Soutter [56] stated that application of nanotechnology in agriculture should be done in precision farms and become the new "industrial revolution" in agriculture. Ditta [58] suggested that there is a

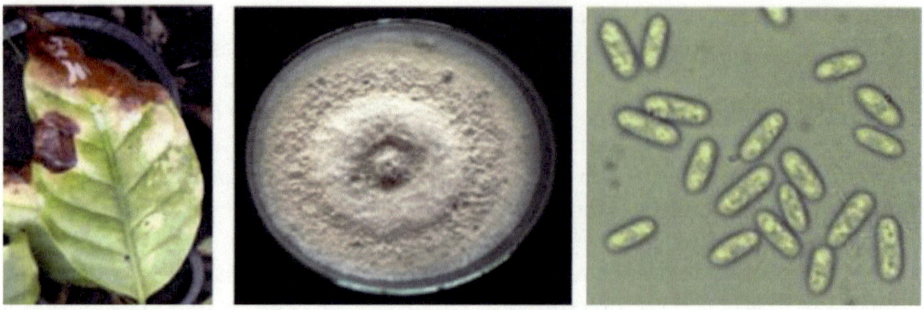

(a) (b) (c)

Figure 9.
Leaf anthracnose of coffee var. Arabica caused by Colletotrichum gloeosporioides *(A); pure culture on PDA at 20 days (B); and conidia, 400 X (C) source: Vilavong and Soytong [52].*

great potential for nanoscience and nanotechnology to face various challenges in agriculture. Perlatti *et al.* [61] expressed that nanoparticles serve as "magic bullets" that contains the bioactive substances from antagonistic fungi enable penetrating through cuticles and tissues, and constantly releasing active substances. The most popular shapes of nanomaterials being used for biocides delivery are nanospheres, nanocapsules, and nanogels. These preparations are advantageous in increasing the stability of active organic compounds, systemic activity, synergism, and specificity, with reducing foliar settling and leaching. The amount of pesticide dosage, number of applications, human exposure to pesticides, and environmental impacts are reduced. The nano formulation is employed in synthetic insecticides and alternative products such as herbal extracts and microorganisms to control insects.

Dar and Soytong [62] first characterized their electrospun materials through visual observation by the naked eye. It was observed that the control has white color. The size of the control nanoparticles ranged from 185 to 218 nm, while the extract-loaded nanoparticles was 241 nm. Tann and Soytong [63] reported that bioformulations and nano product from *Ch. cupreum* CC3003 showed a good control leaf spot of rice var. Sen Pidoa caused by *C. lunata*. Vilavong and Soytong [52] reported that the application of nano-rotiorinol, nano-trichotoxin and a spore suspension of *Ch. cupreum* reduced anthracnose incidence of 46, 43 and 18%, respectively while the inoculated control had high anthracnose disease caused by *Colletotrichum gloeosporioides* (**Figure 9**). The application of bio- formulation of *Ch. cupreum* in a powder form, nano-rotiorinol, and nano-trichotoxin to reduce coffee anthracnose was reported for the first time in Laos.

12. Testing bio-formulations and nano-products against anthracnose of coffee

The powder biofungicide of *Ch. cupreum* CC3003 at 0.5 g/L of water showed effectively control anthracnose of coffee which caused by *C. gloeosporioides,* resulted a disease index (DI) of 0.9 after application. The biodegradable nano-rotiorinol, nano- trichotoxin and living spores of *Ch. cupreum* gave DI of 1.07, 1.14 and 1.62, respectively compared to non-treated control (1.99). However, a biofungicide powder produced by *Ch. cupreum* CC3003 showed the highest disease reduction of 54%. The biodegradable nano-rotiorinol resulted in decrease of anthracnose diseasae 46%. The biodegradable nano-trichotoxin and spores of *Ch. cupreum* decreased the anthracnose disease of 42 and 18%, respectively, when compared to the non-treated control. Biofungicide produced from *Ch. cupreum* CC3003 provided a good control effectively to coffee anthracnose caused by *C. gloeosporioides*. Moreover, it is concluded that biodegradable nano-microbial elicitors from *Chaetomium* and *Trichoderma* PC01 was expressed to induce plant immunity through phytoalexin production against anthracnose of coffee caused by *C. gloeosporioides* [52].

13. Natural product of nano-particles constructed from *Chaetomium* spp. for rice blast disease control

Song *et al.* [64] stated that agricultural nanotechnology is started to be a new tool for plant disease management by restructuring natural active metabolites at the molecular level. Metabolites' nanoparticles contain bioactive compounds from natural products that can rapidly and effectively penetrate through plant cells and can increase the stability of active compounds to control plant diseases. This research finding is a new scientific discovery suggesting that natural products of nano-particles derived from *Ch. cochliodes* isolate CTh05 actively controlled *Magnaporthe oryzae* isolate PO1causing rice blast for the first time.

Chaetomium cochliodes isolate CTh05 expressed actively against the rice blast pathogen, *Magnaporthe oryzae* isolate PO1 on rice var. RD57 in Thailand. The rice blast reduced infection (59%) after applying nano-CCoM at 7 μg/mL, followed by nanoCCoE and nanoCCoH which reduced the blast infection of 57 and 50% respectively. But tricyclazole fungicide reduced blast infection of 55% in 30 days. Later, Song *et al.* [65] further reported on antifungal efficacy of microbial nano-particles constructed from *Ch. elatum*, *Ch. lucknowense* and *Ch. brasiliense* against rice blast pathogen in rice var. PSL 2 in Thailand. Nano particles of *Ch. elatum* (nano-CEE, nano-CEM and nano-CEH) inhibited sporulation of *M. oryzae* which the ED_{50} values of 7, 8 and 16 μg/mL. respectively. The nano-CBH, nano-CBE and nano-CBM constructed from *Ch. brasiliense* suppressed sporulation of the blast pathogen with the ED_{50} values of 6, 9 and 13 μg/mL, respectively. Nano-particles from *Ch. lucknowense* (nano-CLM, nano-CLE and nano-CLH) inhibited sporulation of rice blast pathogen with the ED_{50} values of 5, 7 and 10 μg/mL, respectively. Interestingly, all tested nano-particles deriverd from *Chaetomium* caused pathogenicity lost of rice blast pathogen due to broken down of pathogen cells. The treated rice leaves with nano-CBH from *Ch. brasiliense* showed the Rf values of 0.05 and 0.28 which defined to produce Sakuranertin and Oryzalexin B as phytoalexin against blast disease. Our reserch findings have developed to be a natural product of nanoelicitor for rice blast immunity.

14. Concluding remarks

The scientific investigations have proved that *Chaetomium* biofungicide can be directly applied into rhizosphere soil. *Chaetoimum* biofungicide can be sucessfully applied to control of various diseases, except for downy mildew, powdery mildew and rust as they are obligate parasites. Hence, it can be applied to control plant diseases for good agricultural practices (GAP), pesticide free production (PFP), non agrochemical production (NAP) and organic agriculture (OA). *Chaetoimum* biofungicide and biostimulants are applied to implement integrated plant disease control. It is proved to be a biological agent for plant diseases which has been successfully integrated with other control measures for suitable disease control. *Chaetomium* biofungicide showed protective and curative effects in controlling plant disease and promoting plant growth. It has been successfully applied to the infested soils for the long-term protection against many plant diseases. The new bioactive compounds from *Ch. globosum*, *Ch. cupreum*, *Ch. elatum*, *Ch. cochliodes*, *Ch. brasiliense*, *Ch. lucknowense*, *Ch. longirostre,* and *Ch. siamense* are reported to inhibit plant pathogens. These active natural products from different strains of *Chaetomium* are further developed to be biodegradable nanoparticles from active metabolites as a new discovery of scientific investigation which used to induce plant immunity, namely microbial degradable nano-elicitors for inducing immunity in plants. The biodegradable nano-elicitors are developed to induce plant immunity through phytoalexin production in plants e.g. inducing tomato to produce alpha-tomaline against *Fusarium* wilt of tomato, capsidiol against chili anthracnose, sakuranitin against rice blast, scopletin and anthrocyaidin against *Phytophthora* or *Pythium* rot Durian and scoparone against *Phytophthora* or *Pythium* rot of citrus etc.

Chaetomium is a unique board spectrum biofungicide which is registered in Thailand, China, Laos, Vietnam, Cambodia, and BioAgriCert, IFOAM (International Federation of Organic Agriculture Movements). These countries contributed and tested *Chaetomium* biofungicide. At present, it is registered in Cambodia, Laos, Vietnam and being registered in China and BioAgriCert, IFOAM. Biodegardable nano-elicitors constructed from active metabolites from *Chaetomium* species are the new unique science for plant immunity which have been contributed in Thailand, Finland, Indonesia, India, Laos, Cambodia, Myanmar, Vietnam and China. Our experienced research investigation in this area of specialization would be contributed for sustainable development goals (SDGs) in agriculture.

Author details

Kasem Soytong[1*], Somdej Kahonokmedhakul[2], Jiaojiao Song[1] and Rujira Tongon[1]

1 Faculty of Agricultural Technology, Department of Plant Production Technology, King Mongkut's Institute of Technology Ladkrabang, Bangkok, Thailand

2 Faculty of Science, Natural Products Research Unit, Department of Chemistry and Centre of Excellence for Innovation in Chemistry, Khon Kaen University, Khon Kaen, Thailand

*Address all correspondence to: ajkasem@gmail.com

References

[1] Kunze, G. and Schmidt, J. K. (1817). Chaetomium. Mykologische Hefte nebst einem allgemein-botanischen Anzeiger, Leipzig, 1-2pp.

[2] Decock, C. and Hennebert, G. L. (1997). A new species of *Chaetomium* from Ecuador. Mycological research, 101:309-310.

[3] Rodríguez, K., Stchigel, A. and Guarro, J. (2002). Three new species of *Chaetomium* from soil. Mycologia, 94:116-126.

[4] Soytong, K. and Quimio, T. H. (1989). A taxonomic study on the Philippine species of *Chaetomium*. The Philippine agriculturist , 72:59-72.

[5] Udagawa, S., Muro, I. T., Kurata, H., Sekita, S., Yoshihira, K. and Natori S. (1979). *Chaetomium udagawae*: a new producer of sterigmatocystin. Transactions of the Mycological Society of Japan, 20:475-480.

[6] Von Arx, J. A., Guarro, J. and Figuers, M. J. (1986). The Ascomycete Genus *Chaetomium*. Nova Hedwigia, 84:1-162.

[7] Aggarwal, R., Tewari, A. K., Srivastava, K. D. and Singh, D. V. (2004). Role of antibiosis in the biological control of spot blotch (*Cochliobolussativus*) of wheat by *Chaetomium globosum*. Mycopathologia, 157: 369-377.

[8] Dhingra, O. D., Mizubuti, E. S. G. and Santana, F. M. (2003). *Chaetomium globosum* for reducing primary inoculum of *Diaporthe phaseolorum* f. sp. *meridionalis* in soil- surface soybean stubble in field conditions. Biological Control, 26:302-310.

[9] Park, J. H., Choi, G. J., Jang, K. S., Lim, H. K., Kim, H. T., Cho, K. Y. and Kim, J. C. (2005). Antifungal activity against plant pathogenic fungi of chaetoviridins isolated from *Chaetomium globosum*. FEMS Microbiology Letters, 252:309-313.

[10] Zhang, H. and Yang, Q. (2007). Expressed sequence tags-based identification of genes in the biocontrol agent *Chaetomium cupreum*. Applied Microbiology and Biotechnology, 74:650-658.

[11] Tveit, M. and Moore, M. B. (1954). Isolates of *Chaetomium* that protect oats from *Helminthosporium victoriae*. Phytopathology, 44:686-689.

[12] Cullen, D. and Andrews, J. H. (1984). Evidence for the role of antibiosis in the 0antagonism of *Chaetomium globosum* to the apple scab pathogen, *Venturia inaequalis*. Canadian Journal of Botany, 62:1819-1823.

[13] Di Pietro, A., Kung, R., Gutrella, M. and Schwinn, F. J. (1991). Parameters influencing the efficacy of *Chaetomium globosum* in controlling *Pythium ultimum* damping-off of sugar- beet. Journal of Plant Disease and Protection, 98:565-573.

[14] Walter, D. and Gindrat, D. (1988). Biological control of damping-off of sugarbeet and cotton with *Chaetomium globosum* or a fluorescent *Pseudomonas* sp. Canadian Journal of Microbiology, 34:631-637.

[15] Vannacci, G. and Harman, G. E. (1987). Biocontrol of seed-borne *Alternaria raphani* and *A. brassicicola*. Canadian Journal of Microbiology, 33:850-856.

[16] Kohl, J. W., Molhoek, H. L., van der Plas, C. H. and Fokkema, H. J. (1995). Effect of *Ulocladium atrum* and other antagonists on sporulation of *Botrytis cinerea* on dead lily leaves exposed to field condition. Phytopathology, 85:393-400.

[17] Manandhar, J. B., Thapliyal, N., Cavanaugh, K. J. and Sinclair, J. B. (1986) Interaction between pathogenic and

saprobic fungi isolated from soybean roots and seeds. Mycopathologia, 98:69-75.

[18] Umikalsom, M. S., Ariff, A. B., Hassan, M. A. and Karim, M. I. A. (1998). Kinetics of cellulose production by *Chaetomium globosum* at different levels of dissolved oxygen tension using oil palm empty fruit bunch fibre as substrate. World Journal of Microbiology and Biotechnology 14:491-498.

[19] Brewer, D., Jerram, W. A. and Taylor, A. (1968). The production of cochliodinol and a related metabolite by *Chaetomium* species. Canadian Journal of Microbiology, 14:861-866.

[20] Soytong, K., Kanokmedhakul, S., Kukongviriyapa, V. and Isobe, M. (2001). Application of *Chaetomium* species (Ketomium®) as a new broad spectrum biological fungicide for plant disease control: A review article. Fungal Diversity, 7:1-15.

[21] Soytong, K. (1989). Antagonism of *Chaetomium cupreum* to *Pyricularia oryzae*: a case study to biocontrol of a rice blast disease. Thai Phytopathology, 9:28-33.

[22] Soytong, K. (1992a). Antagonism of *Chaetomium cupreum* to *Pyricularia oryzae*. Journal of Plant Protection in the Tropics, 9:17-24.

[23] Soytong, K. (1990). A taxonomic study of *Chaetomium* spp. In Thailand. Abstract, 4th International Mycological Congress, Regensburg, Germany, August 28-September 3,1990.

[24] Soytong, K. (1992b). Biological control of tomato wilt caused by *Fusarium oxysporum* f.sp. *lycopersici* using *Chaetomium cupreum*. Kasetsart Journal (Natural Science), 26:310-313.

[25] Kanokmedhakul, S., Kanokmedhakul, K., Phonkerd, N., Soytong, K., Kongsaeree, P. and Suksamrarn, A. (2002). Antimycobacterial antraquinone-chromanone compound and diketopiperazine alkaloid from the fungus *Chaetomium globosum* KMITL-N0802. Planta medica, 68:834-836.

[26] Soytong, K. (1992c). Biological control of rice blast disease by seed coating with antagonistic fungi. Songklanakarin Journal of Science and Technology, 14:59-65.

[27] Soytong, K. (1991). Species of *Chaetomium* in Thailand soils. Thai Phytopathology, 11:86-94.

[28] Soytong, K. (1992d). Antagonism of *Chaetomium cupreum* to *Pyricularia oryzae*. Journal of Plant Protection in the Tropics, 9:17-23.

[29] Soytong, K. (1993). Biological control of plant pathogens using *Chaetomium* spp. A paper presented at Kasetsart University - Mycology in Thailand -incorporation with British Mycological Society, March, 3, 1993. And Encapsulation of potential biocontrol agents in sodium alginate. The 6th International Congress of Plant Pathology. Canada. July, 28-August, 6, 1993, 273. 274pp.

[30] Soytong, K., Pongnak, W. and Kasiolarn, H. (2005). Biological control of *Thielaviopsis* bud rot of *Hyophorbe lagenicaulis* in the field. Journal of Agricultural Technology, 1:235-245.

[31] Tomilova, O. G. and Shternshis, M. V. (2006). The effect of a preparation from *Chaetomium* fungi on the growth of phytopathogenic fungi. Applied Biochemistry and Microbiology, 42:76-80.

[32] Shternshis, M., Tomilova, O., Shpatova, T. and Soytong, K. (2005). Evaluation of Ketomium-mycofungicide on Siberian isolates of phytopathogenic fungi. Journal of Agricultural Technology, 1:247-253.

[33] Prechaprome, S. and Soytong, K. (1997). Integrated biological control of durian stem and root rot caused by *Phytophthora palmivora*. Proceedings of the First International Symposium on Biopesticides. Thailand, October 27-31, 1996:228-237.

[34] Usuwan, P. and Soytong, K. (1998). Biological control of *Phytophthora* rot of citrus in the th field. Proc. of 24.

[35] Sodsa-art, P. and Soytong, K. (1999). Biological control of black pepper root and basal stem rot in the field. Proc. os Symposium on Biological Control in Tropics. MARDI Training Centre, 18-19 Malaysia 1999, 68-70 pp.

[36] Soytong, K., Usuwan, P., Kanokmedhakul, S., Kanokmedhakul, K., Kukongviriyapan, V. and Isobe, M. (1999). Integrated biological control of *Phytophthora* rot of sweet orange using mycofungicides in Thailand. Proceedings of the 5th International Conference on Plant Protection in the Tropics. March 15-18. Malaysia. 329-331 pp.

[37] Soytong, K. and Ratancherdchai, K. (2005). Application of mycofungicide to control late blight of potato. Journal of Agricultural Technology, 1:19-32.

[38] Kean, S., Soytong, K. and To-Anun, C. (2010). Application of biofungicide to control citrus root rot under field condition in Cambodia. Journal of Agricultural Technology, 6:219-230.

[39] Kaewchai, S. and Soytong, K. (2010). Application of biofungicides against *Rigidoporu microporus* causing white root disease of rubber trees. Journal of Agricultural Technology, 6:349-363.

[40] Nitao, J. K., Meyer, S. L. F., Oliver, J. E., Schmidt, W. F. and Chitwood, D. J. (2002). Isolation of avipin, a fungus compound antagonistic to plant-parasitic nematodes. Nematology, 4:55-63.

[41] Zhou, W., Starr, J. L., Krumm, J. L. and Sword, G. A. (2016). The fungal endophyte *Chaetomium globosum* negatively affects both above- and belowground herbivores in cotton. FEMS Microbiology Ecology, Volume 92, Issue10,fiw158, https://doi.org/10.1093/femsec/fiw158.

[42] Hu, Y., Zhang, W. P., Zhang, P., Ruan, W. B. and Zhu, X. D. (2013). Nematicidal Activity of Chaetoglobosin A Poduced by *Chaetomium globosum* NK102 against *Meloidogyne incognita*. Journal of Agricultural and Food Chemistry, 61:41-46.

[43] Kooliyattil, R., Dandua, L. M. and Knudsen, G. R. (2017). Prospecting fungal parasites of the potato cyst nematode *Globodera pallida* using a rapid screening technique. Journal of Basic Microbiology, 57:386-392.

[44] Khan, B., Yan, W., Wei, S., Wang, Z. Y., Zhao, S. S., Cao, L. L., Rajput, N. A. and Ye, Y. (2019). Nematicidal metabolites from endophytic fungus *Chaetomium globosum* YSC5. DOI: 10.1093/femsle/fnz169.

[45] Soytong, K. (2014). Biological control of stem rots of oil palms caused by *Ganoderma boninense* using *Chaetomium lucknowense* and *Chaetomium cochiliodes*. Journal of Agricultural Technology, 10:87-92.

[46] Phong, N. H., Pongnak, W. and Soytong, K. (2016). Antifungal activities of *Chaetomium* spp. against *Fusarium* wilt of tea. Plant Protection Science, 52:10-17.

[47] Hung, P. N., Pongnak, W., Soytong, K. and Supatta, P. (2015b). Biological control of *Phytophthora palmivora* causing root rot of pomelo using *Chaetomium* spp. Mycobiology, 43:63-70.

[48] Hung, P. N., Pongnak, W., Soytong, K. and Supattra, P. (2015a). Efficacy of *Chaetomium* species as biological control agents against *Phytophthora nicotianae* root rot in Citrus. Mycobiology, 43:288-296.

[49] Thohinung, S., Kanokmedhakul, S., Kanokmedhakul, K., Kukongviriyapan, V., Tusskorn, O. and Soytong, K. (2010). Cytotoxic 10-(indol-3-yl)-[13] cytochalasans from the fungus *Chaetomium elatum* ChE01. Archives of Pharmacal Research, 33:1135-1141.

[50] Kanokmedhakul, S. K., Nasomjai, P., Loungsysouphanh, S., Soytong, K., Isobe, M., Kongsaeree, K., Prabpai, S. and Suksamran, A. (2006). Antifungal Azaphilones from the fungus, *Chaetomium cupreum* CC3003. Journal of Natural Products, 69:891-895.

[51] Suwan, S., Isobe, M., Kanokmedhakul, S., Lourit, N., Kanokmedhakul, K., Soytong, K. and Koga, K. (2000). Elucidation of high micro-heterogeneity of an acidic-neutral trichotoxin mixture from *Trichoderma harzianum* by electrospray ionization quadrupole time-of-flight mass spectrometry. Journal of Mass Spectrometry, 35:1438-1451.

[52] Vilavong, S. and Soytong, K. (2016) Application of bio-formulation of *Chaetomium* for biocontrol of *Colletotrichum gloeosporioides* causing coffee anthracnose in Arabica variety in Laos. Journal of Agricultural Technology, 12:1239-1245.

[53] Elibol, O. H., Morisette, D. D., Denton, J. P. and Bashir, R. (2003). Integrated nanoscale silicon sensors using top-down fabrication. Applied Physics Letters, 83:4613-4615.

[54] Salata, O. V. (2004). Application of nanoparticles in biology and medicine. Journal of Nano- biotechnology, 2: 3.

[55] Li, M., Huang, Q. and Wu, Y. (2011). A novel chitosan- poly (lactide) copolymer and its submicron particles as imidacloprid carriers. Pest Management Science, 67:831-836.

[56] Soutter, W. (2012). Nanotechnology in agriculture. AZoNano.com Publishers. Available at http:// www.azonano.com/ article.aspx?Article ID=3141#1%204. Accessed on May 25, 2013.

[57] Tann, H., Soytong, K., Makhonpas, C., Adthajadee, A. (2011). Comparison between organ- ic, GAP and chemical methods for cultiva- tion of rice varieties in Cambodia. Journal of Agricultural Technology, 7:2239-2245.

[58] Ditta, A. (2012). How helpful is nanotechnology in agriculture? Advances in Natural Sciences: Nanoscience and Nanotechnology, 3:033002.

[59] Nguyen, H. M., Hwang, I. C., Park, J. W. and Park, H. J. (2012). Enhanced payload and photo-protection for pesticides using nanostructured lipid carriers with corn oil as liquid lipid. Journal of Microencapsulation, 29: 596-604.

[60] Rai, M. and Ingle, A. (2012). Role of nanotechnology in agriculture with special reference to management of insect-pest. Applied Micro- biology and Biotechnology, 94:287-293.

[61] Perlatti, B., de Souza Bergo, P. L., das Graças Fernandes da Silva, M. F., Fernandes, J. B. and Forim, M. R. (2013). Polymeric nanoparticle-based insecticides: a controlled release purpose for agrochemicals. In: Insecticides – Development of Safer and More Effective Technologies. S. Trdan (ed.). InTechOpen Access, Croatia. Ch. 20.

[62] Dar, J. and Soytong, K. (2014). Construction and characterization of copolymer nanomaterials loaded with bioactive compounds from *Chaetomium* species. Journal of Agricultural Technology, 10:823-831.

[63] Tann, H. and Soytong, K. (2016). Effects of nanoparticles loaded with *Chaetomium globosum* KMITL-N0805 Extracts against leaf spot of rice var. Sen Pidoa". Malaysian Applied Biology Journal, 45:1-7.

[64] Song, J. J., Soytong, K., Kanokmedhakul, S., Kanokmedhakul, K and Poeaim, S. (2020a). Natural Product of Nano-particles constructed from *Chaetomium* spp. to control rice blast disease caused by *Magnaporthe oryzae*. International Journal of Agriculture & Biology 23:1013-1020.

[65] Song, J. J., Soytong, K., Kanokmedhakul, S., Kanokmedhakul, K. and Poeaim, S. (2020b). Antifungal activity of microbial nanoparticles derived from *Chaetomium* spp. against *Magnaporthe oryzae* causing rice blast. Plant Protection Science, 56:180-190.

Chapter 13

Synchrotron Based Techniques in Soil Analysis: A Modern Approach

Surabhi Hota

Abstract

Soil is a highly heterogenous system where a number of physical, chemical and biological processes are taking place. The study of these processes requires analytical techniques. The electromagnetic radiations in the form spectroscopy, X-Ray diffraction, magnetic resonance etc. have been used in the field of soil analysis since decades. The study of soil nutrients, mineralogy, organic matter and complex compounds in soils use these techniques and are successful tools till date. But these come with a limitation of lesser spatial and spectral resolution, time consuming sample preparation and destructive methods of study which are mostly ex-situ. In contrast to the conventional spectroscopic techniques, the synchrotron facility is of high precision and enables non-destructive study of the samples to a nano scale. The technique uses the high intensity synchrotron radiation which is produced in a special facility, where the electrons are ejected using very high voltage and accelerated in changing magnetic field, at a speed of light resulting in a very bright radiation that enables a very précised study of the subject. For example, in studying the dynamics of P and N in soils, SR aided XAS are used to study the K-edge spectra of these nutrients, without any matrix interference, which used to be a problem in conventional SEM, IR or NMR spectroscopy. These radiations provide high energy in GeV, which imparts high sensitivity and nanoscale detection. Basically, the SR facility improves the precision of the existing spectroscopic techniques. This chapter discusses how the Synchrotron radiations aid to improve precision in various field of soil analysis such as, carbon chemistry, nutrient dynamics, heavy metal and contaminant speciation and rhizosphere study. However, the technique also come with major limitations of requirement of very high skill for preparation of samples, inadequate availability of references for studies related to absorption spectrum and control of radiation damage. Applications and limitations of the technique thoroughly reviewed in this chapter with an aim to provide a brief idea of this new dimension of soil analysis.

Keywords: Soil, Synchrotron, analysis, energy, resolution

1. Introduction

Soil is a heterogenous system and habitat for numerous flora and fauna, where umpteen chemical, physical and biological processes are operating to make it a living system. These processes are, element mineralization and immobilization, redox reactions, decomposition of organic matters by microbes, movement of nutrient elements towards the plant roots and related soil water movements, the complex processes near the rhizosphere, accumulation and distribution of elements and their chemical transformations in to different complexes and valency, weathering of rocks and minerals

IntechOpen

to form soil and related soil forming processes. To understand these processes, several techniques have been developed and experimented by the soil researchers from time to time. And there are numerous remarkable techniques that have aided to the human understanding of complex matters. Most of these techniques use source of energy and illumination for quantification of the elements and studying the complex structure of a system. Such techniques are Diffraction/scattering for crystallography; Polarimetry for measuring the shape of complex molecules and the properties of magnetic materials; Imaging for highly detailed imaging of small animals, and ultimately humans, to the substructure level of biological and physical material, using light from infrared to hard X-rays; Spectroscopy for analysis of chemical compositions and speciation of elements [1]. However, these techniques have several limitations in terms of their sensitivity, spatial resolution, lengthy sample preparation procedures and most of them are dependent upon destructive sampling and ex-situ analysis. With the development of synchrotron facility around the world, the exploitation and study of these processes have been made possible to the molecular level and in-situ because of the unique qualities of a synchrotron radiation viz.; high brightness, wide energy spectrum, highly polarized radiation and time structured emission. This gives the SR an advantage of due to high proton flux, possibility of beam tunability and weak scattering, high sensitivity resulted from acceleration of heavy ion beams to high energy.

Synchrotron Radiation (SR) is basically an electromagnetic radiation which is emitted when the charged particles are radially accelerated. Natural synchrotron radiation (SR) is as old as the stars produced due to charged particles spiraling around magnetic field lines in space [2], while short-wavelength synchrotron radiation generated artificially by in circular acceleration of relativistic electrons only ways back to early nineties [3, 4]. The synchrotron principle was invented by Vladimir Veksler in 1944 [5]. Edwin McMillan constructed the first electron synchrotron in 1945 [6]. The first ever synchrotron was successfully made in 1947 under the direction of Herbert Pollock at General Electric Research Laboratory, NY. And the first synchrotron radiation was observed by Floyd Haber on 24th April, 1947. This was a remarkable achievement in the field of physics which today has led to large, full-fledged synchrotron facilities around the world. The first-generation synchrotron facilities were only being used to study the fundamental properties of matter and was more of a particle accelerator [2]. Presently, the synchrotron facilities have upgraded to third generation, which enable the use of low energy or soft and high energy or hard X-rays, UV and longer wavelengths. The first ever third-generation hard X-ray source was European Synchrotron Radiation Facility (ESRF) in Grenoble followed by the Advanced Photon Source (APS) at Argonne National Laboratory with an energy of 7 GeV. The synchrotron is one of the first accelerator concepts to operate as large-scale facilities, where bending, beam focusing and acceleration could be separated into different components and where the magnetic field is guided in a time dependent manner, to synchronize with a particle beam of increasing kinetic energy.

There are more than 50 facilities in operation worldwide (http://www.lightsources.org) and are being used by more than 15,000 researchers. Some large facilities are listed in **Table 1**. There are a few large synchrotron facilities that have been operating with high electron energies, as high as 8 GeV (SPring-8 in Harima Science Garden City in Japan). PETRA III, which took up operation in 2009, is the most brilliant storage-ring-based X-ray radiation source in the world. As the most powerful light source of its kind (https://petra3.desy.de/ index _eng.html). These machines are physically large (850 to 1,440 m in circumference) with a capability for 40 or more beamlines. There is are also many smaller facilities (120 to 280 m in circumference) which have come in to existence in recent past, having lower electron energies (from 1.3 to 3 GeV). There are also some fourth-generation synchrotron facilities which have been conceptualized by few countries recently.

S.No.	Name of facility	Country	Energy (GeV)	Circumference (m)	Year of Operation
	Indus II	India	2.5	173	2005
1	National Synchrotron Light Source (NSLS-II)	US	3	792	2015
2	PETRA III	Germany	6	2304	2009
4	SPring-8	Japan	8	1436	1997
5	Advanced Photon Source (APS)	US	7	1104	1995
6	Progetto Utilizzazione Lucedi Sincrotrone (PULS)	Italy	1.5	33.5	1980
7	Canadian light source	Canada	2.9	147	2004
8	Australian Synchrotron	Australia	3	216	2006
9	Shanghai Synchrotron Radiation Facility	China	3.5	432	2007
10	ALBA	Spain	3	270	2010
11	Sirius	Brazil	3	518.2	2018
12	Pohang Light Source II	South Korea	3	281.82	2011
13	Swiss Light Source	Switzerland	2.4	288	2001
14	European Synchrotron Radiation Facility	France	6	844	1992
15	Kurchatov Synchrotron	Russia	2.5	124	1999

Table 1.
Synchrotron facilities around the world.

Synchrotron facilities can generate frequencies over entire range of the electromagnetic spectrum. In addition to having a high intensity, the synchrotron light has other properties useful to researchers. Wavelengths produced by SR, may range from 10 to 11 m hard X-rays) to 106 m (infrared light), emitted in short pulses (~100 ps wide spaced at 2 ns), partially coherent and arrives in parallel rays, highly polarized, isolation of specific wavelengths are possible with the use of diffraction or crystal gratings. And most importantly, light is released as a very narrow cone [7] which is then focused to provide a very narrow beam at the work station.

There are a number of techniques that have been used by the researchers for study of soils, they are: STXM- Scanning Transmission X-ray Microscopy, NEXAFS- Near Edge X-ray Absorption Fine Structure Spectroscopy, FTIR- Fourier Transform InfraRed Spectroscopy, μXRD- Micro X-ray Diffraction, EELS- Electron Energy Loss Structure spectroscopy, EXAFS- Extended Absorption Fine Structure Spectroscopy, XANES- X-Ray Absorption Near Edge Structure Spectroscopy [8], HRTEM-High Resolution Transmission Electron Microscopy, SXRF- Synchrotron Based X-ray Radiation Fluorescence, XFS- X-ray fluorescence spectroscopy, X-ray μ-computed tomography [9–11], SEM- Scanning Electron Microscopy, TEM-Transmission Electron Microscopy.

These synchrotron-based techniques have proved to increase the precision of determination of many soil parameters such as carbon chemistry of soil [12, 13], rhizosphere nutrient transport and other mechanisms [14, 15], heavy metal quantification and remediation, microbe-metal interaction mineral transformation, study of physical parameters of soil, soil fertility study.

Methodology: This chapter reviews the works done in the field of soil analysis for physical, chemical and biological parameters, using SR facility, all over the world and how it has aided to the improvement in precision and accuracy of the study.

2. Components and working of a synchrotron facility

A synchrotron facility in general, consists of:

1. Electron gun: It consists of a transformer which feeds a very high voltage (in hundred thousand Volts, approximately, 200000 V of DC) of electricity through a cathode which causes the electrons to jump off the surface. The cathode is made of tungsten-oxide disk, otherwise called a button. As electricity flows through the disk, the disk gets heated to about 1000° C, the temperature at which the electrons are released. A screen near the button is given a short, strong positive charge 125 times per second, which pulls the negative charge particles, the electrons, away from the disk.

2. Linear Accelerator (LINAC): The electron ejected from the electron gun, are fed into the LINAC. The energy of acceleration is provided by the Microwave radio frequency fields to accelerate the electrons to the speed of light. The speed of the electrons reaches approximately to 3 x 108 m/s. The LINAC produces electron pulses in a duration of approximately up to 132 nano seconds (ns). The electrons which become the photos now, must travel in a vacuum to avoid hinderance and reduction in velocity by colliding with other atoms and molecules. The vacuum chamber pressure is kept around 10–11 torr.

3. Booster Ring: The electrons circulate in the booster ring, to get a boost in energy from approximately 250 MeV (Mega electron volt) to approximately 2.9 to 6 GeV by the radio frequency cavity generated microwaves. The Booster Ring increases the speed of the electrons close to the speed of light (approximately 99.9999985 percent of the speed of light). There are two types of electro magnets in the booster ring. These magnets create the magnetic field, which is used by the synchrotron to guide the electrons around the booster ring. The magnetic field of the quadruple magnets are used to force the bunches of electrons into a fine beam within the vacuum chamber.

4. Storage Ring: After the electrons have gained sufficient energy to produce light, the injection system transfers them from the booster ring to the storage ring. The process of transferring electrons from booster to storage ring occurs approximately once per second up to 600 cycles which may take about 10 minutes. Once entered the storage ring, the electrons keep circulating for four to twelve hours and every time the dipole magnets/bending magnets bend the path of electron flow, photons are released. The bending magnets are aimed to circulate the electrons in a circular path [16]. After each turn of the bending

magnets, there is a photon port to allow the light to travel down the beam lines to the research stations.

5. Beam line: Each beam line consists of an optics cabin, experimental cabin, and control cabin. The optics cabin has optical instruments used to extract the type of radiation required for the experiment. The experimental cabin is for support mechanism, and provides environment for the sample study. Instruments called detectors record the information generated from the sample are recorded by the detectors. The control cabin, as the name suggest, allows the researchers to control the experiments and collect the data.

6. End Station: The final result is obtained at the end station. The components of end station are experiment specific. There is an optic hutch, consisting of mirrors by which the selected wavelengths of synchrotron light are focused onto the sample in an end station. The samples are placed in the experimental hutch, which consists of a sample holder and a detection system, specific to the technique adopted by the researcher. Also, includes computers through which the experiment is controlled by the researcher.

3. Organic carbon chemistry in soils observed by synchrotron based spectroscopy

Soil is believed to be the largest pool of terrestrial organic carbon containing about 1500 Pg C which is higher than the atmospheric and biotic pools combined [17, 18]. Organic C is crucial in maintaining the balance of inorganic form, specially, CO_2 in the atmosphere, which in turn, controls the global climate scenario [19]. The entire carbon cycle revolves around the organic C fraction. The decomposition of organic matter, from animal and plant residues and subsequent humification, leads to the formation of complex organic compound leading to significant structural and composition changes [20]. Organic compounds released from animals, plants, microbial cells, have a range of size and complexity from monomers to mixtures of biopolymers. This complexity makes chemistry of soil organic matters difficult to study. Organic C shows very high chemical and spatial complexity, creating significant analytical problems in the existing methods [21–24]. Soils characteristically have low amounts of C as compared to pure biological samples. They require relatively higher energy to substantially improve the detection limits and data that allow better quantification of C chemistry. The existing techniques are mostly destructive, or the samples are processed ex-situ, for analysis. This sometimes causes the loss of delicate structure of the complex organic matter. SR based spectroscopic techniques have proven to be an advanced tool to study the organic C chemistry of soils, as it enables the non- destructive micro and nano-scale analysis and enables keen study of reactivity, composition, heterogeneity, physical site of organic materials, and mineral-organic matter interactions. The high energy of SR provides very high spatial and spectral resolution.

SR-based scanning transmission Fourier transform infrared (FTIR) spectroscopy, X-ray microscopy (STXM) and C (1 s) near-edge X-ray absorption fine structure (NEXAFS), have proven to be unique and non-invasive in studying the complex processes of C chemistry in soils, in the recent past. These techniques have been used to reveal the Spatial organization of C forms in microstructures, functional group chemistry of C, interaction of C with soil matrix, surface Interactions of organic C and minerals.

The identification of the compound structure of the SOC and DOC involves basically, identifying the type of chemical bonds and functional groups in a molecule. And because of complex nature of the spectrum produced by the organic compounds, the identification and differentiation of the bond become difficult with low energy, conventional spectroscopic techniques. At low spectral resolution, the peaks seem to be overlapping and the fingerprinting becomes only partially useful. This limitation has been overcome by the incorporation of SR in to the spectroscopic techniques. The identification of the structural characteristics with high precision has become possible and even the study of impact of management on SOC cycle and composition could be successfully carried out by aid of SR based techniques in various studies [13, 23, 25–32].

The aided advantage of SR is its brightness, which can be defined as the photon flux or power emitted per source area and solid angle, is 100–1000 times greater than the brightness obtained from a conventional source [33, 34]. FTIR spectroscopy has been considered the most powerful technique for identifying types of chemical bonds and functional groups in a molecule which produces IR absorption spectrum, in the form similar to fingerprint. FTIR combined with microscopy allows probing of small areas in shorter time intervals and with an optimum signal-to-noise ratios. However, SR-FTIR spectroscopy could reveal the quantity, structure composition, and distribution of chemical constituents in humus fractions [13, 23]. In some studies, it has also been used to identify stable soil aggregates and black C particles chemistry [27, 35] and these studies show that, the SR-FTIR spectra has stronger band width compared to conventional FTIR. Studies revealed that, FTIR spectra below 1450 cm-1 normally leads to overlapping bands of indefinite quantity. SR-FTIR has a spectrum ranging from 4000 to 400 cm^{-1} and could measure the molecular level vibration of organic and inorganic functional groups [36, 37]. SR-FTIR can also identify the changes in intensities and vibrational characteristics of bands, shifts in frequencies and shapes of a particular band [38]. These properties have been exploited to study the molecular level changes in SOM composition due to management practices [13, 23, 39]. it was also made possible by this technique to study organo-mineral assemblages [27, 35].

NEXAFS is also such a technique which is being used by many researchers for similar studies in recent past. It is highly element-specific as the different elements show X-ray absorption peaks at different energies [40]. NEXAFS when combined to STXM, information as précised as 30–50 nm spatial resolution could be obtained. This technique is termed as NEXAFS spectromicroscopy. This technique has been used in study of humic fractions mined from highly complex soil matrices [13, 23, 24, 29, 41, 42], study nanostructures of polymers [43] and fingerprinting of C in biological materials [44, 45].

The results obtained from the SR-based techniques are basically an improvement over the pre-existing spectroscopic techniques, where the bond stretch (eg. C-O or O-H), π and σ transitions [23] are taken as the reference for speciation of the C containing species. However, the spectral signatures, or fingerprints of pure biological materials obtained from SR based spectroscopy can serve as a reference for interpretation of the results obtained from conventional spectroscopy. The results can then be quantified using chemometrical methods, multivariate analysis, regression modeling, principal component analysis or predictive modeling [27, 46]. In laboratory analysis of samples, chemical analysis of C samples can only provide the total OC, or different pools of C or fractions of humus, which are mostly quantitative. But the SR based techniques can reveal what is the actual composition of the TOC fraction or humus fraction. These revelations can be related to the stage of weathering of a soil, its pedogenesis and can also correlate the data to predict the responsible chemical, physical and biological processes. Information related to fingerprints of organic compounds such as proteins, nucleic acids, carbohydrates and lipids, predictive models and algorithms can be developed, which can be used by researchers with basic facilities, to predict the organic C chemistry and related processes. However, the data from fingerprints of standard substances is still inadequate to fully identify functional group chemistry for soil organic C and a strenuous effort is required to fill that breach.

4. Synchrotron based techniques in elemental speciation in soils

Soil is a very heterogenous system and the chemical and biological processes happening inside it are even more complex. The important processes, as mentioned in the introductory section, are dependent upon the concentration of different elements involved in process, their distribution and their chemical behavior, which makes their speciation very much crucial. Also, the speciation of environmental contaminants such as heavy metals are important as they lead to degradation of soils [47, 48]. XAS is a powerful technique to provide answer to questions related to the chemistry of an element present in ant system starting from minerals, to biological specimens and solutions [49–52]. The XAS technique basically scans the object of study using a range of X-ray energies. Hence, with the advent of synchrotron, the energy of X-ray could be increased. XANES and EXAFS are the two techniques XAS techniques which have been used in many studies for studying the elements in heterogenous environments, by selecting appropriate energies of the incoming X-ray photon with the aid of SR facilities and thereby generating chemical maps of an element in relation to its oxidation state and chemical bonding [53]. K-edge XANES has been used to test the speciation of Al on the thin surface coatings of quartz and feldspar grains in a loess soil [54], and studying the coordination of Al in imogolite and allophones [55].

X-ray techniques such as low-energy XRF (LEXRF) has been used to quantify Al in plant tissues to determine the toxicity levels. A study was carried out to determine Al accumulation in tea plants, in the soils with high acidity, resulted to high concentration of Al. it revealed that Al accumulated mainly within the cell walls of the leaf epidermal cells, also confirming that the apoplast are the tolerance mechanism of toxicity in tea plants because of results showing Al accumulation in apoplast and traces in symplast. Similar study in soybean showed higher accumulation of Al in roots [56] using LEXRF. Cr speciation has been done by using EXAFS in soils contaminated by leather tanning, reporting the dominant form as Cr (III) with low levels of Cr (VI) [57]. Different industrially contaminated soils have been studied for determination of and speciation of Cr using this technique [58, 59]. XANES has been used to identify associations of metals in contaminated environments. Fandeur et al. [60] reported that Cr (VI) was found in closed association with Mn-oxides concluding the role of Mn-oxides in oxidation of Cr (III) to Cr (VI). Jacobson et al. [61] reported Cu to be present in hotspots that are associated with organic matter within the soil matrix using in-situ XANES. Similarly, a study by [62], revealed the Cu hotspots not to be associated with calcium (Ca) carbonates, Fe oxides, or Cu sulfates. EXAFS has been used for speciation of As, based on the absorption spectra of As-O distances. Grafe et al. [63] used m-XRF and m-XAS to study the speciation of As in a chromated copper-arsenate (CCA) contaminated soil to determine the effects of co-contaminating metal cations (Cu, Zn, Cr) on As speciation. X-ray computed μ-tomography (SR-μCT) is also a powerful technique that enables study of internal structure and composition of sample without sectioning it physically. Other advantage over conventional technique is enables a 3D study of the subject. 2D study of the samples limit the projection of metals present in the sample, only in the direction of beam. SR-μCT has higher sensitivity, resolution and speed. SR-XRF is also a powerful technique devised to study the elements in environment. It has a very précised detection limit for many elements (<0.01 ppm).

5. Synchrotron based techniques in study of mineral structure, their occurrence and abundance in soils

It is known that X-ray diffraction (XRD) is a basic technique which has been used widely for the characterization and identification of crystalline structure of minerals. But it comes with some imitations like it cannot be used to study amorphous minerals, the slide preparation procedures are very much time consuming, requires invasive sampling. Also, the low energy and low flux make the diffractometer geometry very limited. The use of diffracted X-rays, instead of transmitted X-rays restricts the analysis of hydrated and oxygen samples. Compared to the conventional technique, the SR-XRD provides very high sensitivity due to energy tunable nature and high flux. It enables the non-invasive study of trace matters, amorphous minerals, hydrated and oxygen samples, thin films and solution phase. The study of areas as small as 1 mm^2 can be done with a beam spatial resolutions of <5 mm.

synchrotron-based techniques of XRD and XAFS, have greatly improved the understanding about the structural and chemical properties of amorphous minerals, such as iron oxide minerals like lepidocrocite, goethite and hematite. Their transformations of metastable phases to stable phases of iron oxides [64] and chemical reactions occurring on their surfaces have also been possible to study with the aid of these techniques., and chemical reactions occurring on their surfaces. [65] used XANES spectroscopy for identification and quantification of iron compounds in soils. Massive improvements have been achieved in understanding the structure of poorly crystalline and nanosized iron oxides species like ferrihydrite, by synchrotron XRD, EXAFS, and total X-ray scattering data. Studies also have used laboratory XRD to determine changes in the structural properties of goethite and hematite due to substitution by Al^{3+} than Fe^{3+} in mineral structure (up to 33 mol.%). Atomistic and long-range scales data regarding substitution of trace to low amounts of other elements in the structure of iron oxides on the basis of indirect, dissolution data have been predicted using synchrotron XRD and EXAFS [66–70]. The valence state of substituent elements, coordination states and their effects on element substitution in iron oxides nave been revealed by XANES spectroscopy [66].

6. Synchrotron based techniques in study of microbial interactions with contaminants/pollutants in the soil

SR-FTIR spectromicroscopy is a powerful technique which has come in to use in recent years for study of microbial activities with high spatial resolution to μm scale and temporal resolution to seconds scale, enabling keen study of processes at cellular level [71]. It combines three techniques viz.; microscopy, mid IR spectroscopy and SR based light source, which makes it possible to precisely locate the target, obtain its vibrational spectra and obtain very high signal to noise ratio.

Earlier FTIR spectromicroscopy technique used an aperture to take reading of smaller areas within the sample, because to take reading, one had to adjust the incoming light to either eliminate some part of it or to distribute it among detectors, uniformly. This resulted to reduction in signal strength. The thermal emission sources are limited to focus an area of 75–100 mm in diameter only. In contrast to these conventional sources, a synchrotron infrared source can obtain spectra from smaller target areas without using aperture [72]. In SR-FTIR Spectro microscopy, one can use the infrared beam to visualize a spot as small as the diameter of its wavelength resulting to 0.7 times of spatial resolution [73–75].

The study of action of microbes over a geological surface (here, soil) is based on the fact that, the changes carried out by microbial action (mineralization, immobilization, biodegradation) is monitored by the study of the vibrational spectra of the substrate (heavy metal or pollutant). Many studies have been carried out to find out the reference spectra of several contaminants such as, Chromium and molybdenum hexacarbonyls [76], Pentacyanonitrosyl-complexes of chromium and molybdenum [77], Aromatic chromium tricarbonyls [78], Explosive molecule vapors [79], polycyclic aromatic hydrocarbons (PAH) [80–83], Fluoranthene and benzofluoranthenes [84], Organo-arsenic (III), –antimony (III) and -bismuth (III) thiolates [85], PAHs incorporating the peropyrene structure [86], Nonregular PAHs [87], large PAHs [88] Naphthalene and anthracene [89], 1-nitropyrene [72] Toluene-.

3,4-dithiolatoantimony (III) derivatives [90], Trimethylarsine oxide [91] phosphorus tricyanide [92] Oil spill [93] Quinoline and Phenanthridine in Solid Argon and H2O [94] 6-nitrochrysene [95, 96] Arsenate on Fe-Ce bimetal oxide [97].

A general step-by-step scheme of studying the microbe interaction is given as follows:

Visual examination of the specimen surface under stereoscope to study the overall morology and topography.

↓

Application of sterile liquid to keep the surface microbes alive during study.

↓

Removal of uneven fragments over the surface to prevent hinderance in scattering using mini hammer.

↓

Cleaning of specimen surface by sonification and UV-radiation sterilization.

↓

Specimen placed inside the sterile chamber of microscope and identification of surface to be studied.

↓

Microscopic volume of sample collected from field or microbial culture on to the specimen surface selected.

↓

Coupling FTIR-spectroscopic system to SR light source.

↓

Recording of vibrational spectrum of target sample.

Apart from the above-mentioned organic contaminants, metal contaminants speciation and bioavailability has also been widely determined using SR techniques. In a study [98] XAS was used to explore the short-term aging profile and the long-term speciation of dissolved Cu, CuO, and CuS nanoparticles, concluding that the short-term reactions with nanoparticles are dependent on the Cu form and soil chemistry, while the long-term reaction are independent of the form of Cu and transformed into iron oxyhydroxide or natural organic matter bound Cu. Similarly, Colzato et al. [99] used Cadmium L-edge XANES to study short-term temporal changes in Cd speciation and long-term changes in Cd extractability to conclude that highly weathered Oxisols and less weathered Entisols and Mollisols contain Cd bound to organic matter and Fe and Al oxides. Mariet et al. [100] used EXAFS to study a century old Pb leftover from metallurgical activities and reported the presence of Pb mainly as sorbed to Fe oxy(hydr)oxides, which was a significant finding of great consequence to environment. SR based techniques have also been used recently to study real time biodegradation of organic chemicals [101], where it was predicted that humic acid accelerates the degradation of polyaromatic hydrocarbons (PAHs) by enhancing the solubility of the PAH, thereby increasing PAH bioavailability to microorganisms. SR-FTIR was used to study the effect of HA on the degradation kinetics of pyrene by Mycobacterium sp. on a magnetite surface [101].

7. Synchrotron based techniques in studying soil nutrient (nitrogen and phosphorus) dynamics

Phosphorus is a major essential nutrient to plants and directly involved in the metabolic activities of plants, maintain structural integrity and directly affecting the economic yield of crops due to its crucial role in fruit setting and flower development. P found in many forms in soil, organic or inorganic. The form of P in a soil decides its fate of mineralization, fixation, and availability as well as interaction with other compounds in soil. This makes the study of P speciation very much important. But low concentrations of P in chemical and organic species make it difficult to specify the form of occurrence of P due to matrix interferences [102, 103] using conventional techniques like SEM, IR spectroscopy and NMR spectroscopy.

But due to the aided advantages of SR-XAS, a matrix interference free data has been made possible. The technique uses XAS spectra for identification of the species. However, it represents the average of spectra of all the P species present, hence, model analogues of P species are used as reference to extract the individual spectra from the overall spectra. The conventional techniques involve the chemical processing of sample followed by colorimetric analysis, or other extractant methods, which mostly involves the removal of P from its natural environment (soil matrix or colloid) and resulting to possible alteration of original form of occurrence. However, use of SR techniques such as XANES and EXAFS have proved to be promising in soil P speciation. These techniques involve the study of spectrum produced by only one element (i.e., K-edge spectrum of P) of target irrespective of other elements present. Brandes et al. [104] have studied the phosphorus K-edge XANES spectra of potassium phosphate and various organic P species. The K-edge spectrum of P species in XANES can be identified by characteristic line feature, for example, a pre-white line feature indicated the PO_4 tetrahedra in coordination with a transition metal. Similarly, a weak pre-white line showing absorption peak at 2–5 eV indicated Fe (III) phosphate and adsorbed Phosphate on it [16]. Such features are very prominent for P minerals like phosphorsiderite and strengite [105]. Simlar studies have been carried out by few researchers till date, which uses XANES for speciation of Soil P [106–113]. XANES has been used to study adsorbed species of P on Fe and Al oxides [110].

N is also a major essential plant nutrient involved in structural integrity to metabolic activities of plants. The inorganic forms of N have been well explored and its analysis procedure in soil system is well established. But the nature of soil organic nitrogen (SON) has still remained relatively unexplored given its importance in the N cycle [114], as half of the SON have been termed as the unknown N [115]. The dissolved organic N (DON) also has the same fate as SON in terms of its characterization [116]. There are several methods that have been devised to characterize the SON fraction as hydrolysable N, NH3-N and unidentified N, such as, acid hydrolysis [117], pyrolysis, mass spectrometry and NMR spectroscopy [118, 119]. The use of X-ray spectroscopic methods in this field has come in to existence in recent past only [25, 120–124], which come with their obvious limitations of low energy, sensitivity and incapability to determine very small fraction of samples and limited number of reference spectra.

As mentioned in the previous regarding P determination, XANES K-edge N spectra has been used in few recent studies for speciation of SON and DON, which are present in very low concentrations, due to their highly sensitive beamline characteristics. And these works have been focusses in the direction of developing appropriate N reference standards [125]. The N K-edge XANES requires very high flux beamline critical of the study subject, and it occurs at photon energy of around 400 eV. An N-XANES represents the average of spectra of all the N species present, hence, reference spectrum of N species is used to extract the individual spectra from the overall spectra. This is almost like a fingerprinting technique where reference spectra and resultant K-edge spectra are compared to conclude the speciation. The peak resonance spectra of ammonium compounds were found to be around 401 eV [126], for KNO3 and Ca (NO3)2 it is around 412 eV, pyridine-based compounds at 398.7 eV. N-XANES has been used to characterize Maillard reaction-produced humic substances in DOM-related geomaterials [122], to study the aquatic sediment humic acids and their characterization to pyridines and oxidized pyridines [123]. These approaches had the major disadvantages of physical alteration and poor correlation between soil properties and chemical characteristics of extracts [125]. Because organic N fraction is closely associated with the clay size fractions of soil, the analytical technique that did not involve much chemical alteration, came in to popularity in the recent years, with the advent of SR based XANES. N-XANES has been used to study the dynamics of N-compounds in the rhizosphere of field pea (Pisum sativa L.) [127], characterizing heterocyclic (pyridinic) N in pyridines, pyrazines, nitrilic-N, –N$^{1/4}$, amide N and –NH– and nitroaromatic-N. N K-edge XANES was used to derive the spectra of soil to study the changes in SON compound by changes in pedoclimatic properties and cultivation duration [128].

8. Synchrotron based techniques in soil rhizosphere studies

Soil rhizosphere is a very complex yet important part of soil–plant root interaction. The microbes inhabiting the rhizosphere, the root exudates from plants, their interactions with each other and with the surrounding environment is crucial for nutrient cycling, nutrient release from SOM and ultimately plant growth. However, the study of rhizosphere has not been easy due to its complexity. The very fine root structures are not easily identifiable with existing techniques. The visualization of complex processes taking place in the rhizosphere require very high intensity beam as well as nanoscale spatial resolution, which have been met recently by some interventions which use SR for study.

STXM has been used to investigate the rhizosphere in few recent studies, for studying OC distribution within bacterial biofilms at fine scale [129–131] soil micro-aggregates [12, 26, 51] and bacteria-soil mineral interfaces [132]. Along with STXM, NEXAFS has been used to study oxidative damage of lignocellulosic materials [133] in bleaching and if such technique could be used in soil, it may help in quantifying the degradation of plant cells [134]. Synchrotron -IR microscopy (SIRM) is such a technique which has enabled the non-invasive mapping of root zone by using mid-IR spectroscopic signature. For SIRM studies, seedlings are sown in rhizoboxes, with modification of Lexan faces with windows such as Zn Selenide windows [135] and diamond window [136]. These windows allow direct observation of root-soil interface using SIRM.SIRM enables to focus the IR ton area as small as 6–10 mm [74, 75].

9. Limitations

SR based techniques have proven to be a boon in the field of soil analysis. However, it is still unexplored and there is a lot of scope for further research in this field. No technique is perfect and SR based techniques also come with several limitations. For example, μ- XRF uses hard X-ray beamlines for metal and metalloids with Z > 20, and at this beam line the lighter elements (N, P and S) cannot be studied. Similarly, soft X-ray cannot be used for heavy elements. The interpretation of data in K-edge XANES requires very much skill and accuracy, also when a specific element is being detected, there are chances of self-absorption leading to faulty interpretation. Also, the low energy of K-edge shell and low concentrations of the elements in soil pose limitations to this technique. Studies involving preparation of thin sections require very high skill. Detailed mapping of spatial distribution of organic C also poses challenges when information on single locations needs to be interpolated to processes encountered at the Pedon or landscape scale. Also, constraints such as sample preparation, control of radiation damage, or spectral quantification are faced in SR-FTIR studies. Other limitation of SR-FTIR is that it cannot be applied to all geological materials, because it is not possible to obtain IR-spectrum for those elements which are opaque at IR e.g., Iron oxides or sulfides and non-silica bearing minerals. The data from spectral information of standard substances is still insufficient to fully identify functional group chemistry for soil organic C and a concerted effort is required to fill that gap as soon as possible. However, these limitations also indicate the areas where the future research should be focused.

10. Future perspective

The SR based techniques are far more improved in terms of their accuracy compared to the pre-existing techniques that use EM radiation for chemical and physical analysis of soils. Soils are inseparable entity of agriculture production systems and environment as a whole. The advent of SR techniques has provided a better insight into the soil nutrient dynamics, organic C speciation, soil mineralogy, fate of trace elements, metal contaminants, organic pollutant bioavailability and microbial interactions. These results obtained from SR based techniques can be used to make significant predictions for sustainability of an agricultural system. For example, the forms of nutrient in a soil can indicate the underlying biogeochemical process which is occurring in real time and can be a good indicator of soil health. The correlation of the different forms with soil microbial flora and fauna will give insight into the health status of a soil. The forms of OC in soils are very good indicator of soil genesis, aggregate stability and capacity of a soil to provide nutrition for the microbial population. The clay coatings on the ped faces, cutans and other micromorphological features can be accurately visualized to understand the soil development processes and their stage of weathering, which in turn, decides the nutrient holding capacity, permeability, infiltration etc. and these are important parameters to decide cropping systems. The future researches should be focused towards overcoming the short falls of the techniques. For SR absorption techniques, a wide range of spectral libraries should be priority. The different techniques should be combined together to increase the array of analysis of the samples. In addition to this, significant constraints like need sample preparation, control of radiation damage and spectral quantification should be addressed.

11. Conclusion

The SR based techniques are remarkable improvement over the pre-existing spectroscopic techniques. The use of SR in soils is quite new and fairly unexplored dimension. Not a very less but also not a lot of research has been done towards analysis of soil parameters. The techniques such as XAS, FTIR and XRD have been widely used in soil studies, for soil nutrient dynamics, organic C speciation, soil mineralogy, fate of trace elements, metal contaminants, organic pollutant bioavailability and microbial interactions. The exploration of unexplored at nano scales has been possible with the advent of this technique. However, some limitations such as unavailability of adequate spectral references, converting the qualitative data in to quantifiable terms high skill requirements for handling and sample preparations and most importantly, the confinement of SR facilities to select places have restricted their wide adoption. Hence, future researches focused in this direction to popularize and further explore this technique to make it reachable in public domain.

Author details

Surabhi Hota
ICAR—National Bureau of Soil Survey and Land Use Planning, Regional Centre, Jorhat, Assam, India

*Address all correspondence to: surabhi.hota@gmail.com; surabhi.hota@icar.gov.in

References

[1] Nolan AL, Lombi E, McLaughlin MJ. Metal bioaccumulation and toxicity in soils - Why bother with speciation? Aust J Chem. 2003; 56:77-91 doi:10.1071/CH02226

[2] Lombi E, Susini J. Synchrotron-based techniques for plant and soil science: opportunities, challenges and future perspectives. Plant Soil. 2009; 320:1-35. DOI 10.1007/s11104-008-9876-x

[3] Baldwin GC. Origin of synchrotron radiation. Phys Today. 1975; 28(1):9-17

[4] Pollock HC. The discovery of synchrotron radiation Am J Phys. 1983; 51:278-280

[5] Veksler VI. "A new method of accelerating relativistic particles". Comptes Rendus de l'Académie des Sciences de l'URSS. 1944;43 (8): 346-348.

[6] Jackson J. David and W.K.H. Panofsky (1996). "EDWIN MATTISON MCMILLAN: A Biographical Memoir". National Academy of Sciences.

[7] Balerina, A., and S. Mobilio. Introduction to synchrotron radiation. In: S. Mobilio, F. Boscherini, and C. Meneghini, editors, Synchrotron radiation. SpringerVerlag, Berlin, Heidelberg. 2015; doi:10.1007/978-3-642-55315-8_1

[8] Beauchemin S, Hesterberg D, Beauchemin M. Principal component analysis approach for modeling sulfur KXANES spectra of humic acids. Soil Sci Soc Am J 2002; 66:83-91

[9] Al-Raoush RI, Willson CS. Extraction of physically realistic pore network properties from three-dimensional synchrotron X-ray microtomography images of unconsolidated porous media systems. J Hydrol (Amst). 2005; 300:44-64 DOI: 10.1016/j.jhydrol.2004.05.005

[10] Altman SJ, Peplinski WJ, Rivers ML. Evaluation of synchrotron X-ray computerized microtomography for the visualization of transport processes in low-porosity materials. J Contam Hydrol. 2005a; 78:167-183 doi:10.1016/j.jconhyd.2005.05.004

[11] Altman SJ, Rivers ML, Reno MD, Cygan RT, McLain AA. Characterization of adsorption sites on aggregate soil samples using synchrotron X-ray computerized microtomography. Environ Sci Technol. 2005b; 39:2679-2685 DOI:10.1021/es049103y

[12] Wan J, Tyliszczak T, Tokunaga TK. Organic carbon distribution, speciation, and elemental correlations within soil microaggregates: applications of STXM and NEXAFS spectroscopy. Geochim. Cosmochim. Acta. 2007; 71: 5439-5449.

[13] Solomon D, Lehmann J, Kinyangi J, Liang B, Scha¨fer T. Carbon Kedge NEXAFS and FTIRATR spectroscopic investigation of organic carbon speciation in soils. Soil Sci. Soc. Am. J. 2005;69: 107-119.

[14] Hinsinger P, Gobran GR, Gregory PJ, Wenzel WW. Rhizosphere geometry and heterogeneity arising from root-mediated physical and chemical processes. New Phytol. 2005;168: 293 – 303.

[15] Cardon ZG, Gage DJ. Resource exchange in the rhizosphere: molecular tools and the microbial perspective. Annual Review of Ecology, Evolution, and Systematics. 2006; 37:459-488

[16] Toor GS, Hunger S, Peak JD, Sims JT, Sparks DL. Advances in the characterization of phosphorus in organic wastes: environmental and agronomic applications. Adv. Agron. 2006;89: 1-72.

[17] Batjes N H,. Total carbon and nitrogen in the soils of the world. Eur. J. Soil Sci. 1996; 47:151-163.

[18] Schlesinger WH. Biogeochemistry: Analysis of Global Change, second ed. Academic Press, Orlando, FL. 1997

[19] Stevenson FJ, Cole MA. Cycles of Soil, second ed John Wiley and Sons, London. 1999.

[20] Wershaw RL. Membrane-Micelle Model for Humus in Soils and Sediments and Its Relation to Humification. U.S. Geological Survey Water-Supply Paper. 2410. United States Government Printing Office. 1994

[21] Gleixner G, Poirier N, Bol R, Balesdent J. Molecular dynamics of organic matter in a cultivated soil. Org. Geochem. 2002;33: 357-366.

[22] Ko¨gel-Knabner I. Analytical approaches for characterizing soil organic matter. Org. Geochem. 2000;31: 609-625

[23] Solomon D, Lehmann J, Kinyangi J, Amelung W, Lobe I, Ngoze S. Long-term impacts of anthropogenic perturbations on the dynamics and speciation of organic carbon in tropical forest and subtropical grassland ecosystems. Glob. Chang. Biol. 2007a; 13: 511-530.

[24] Solomon D, Lehmann J, Thies J, Schafer T, Liang B, Kinyangi J. Molecular signature and sources of biochemical recalcitrance of organic C in Amazonian Dark Earths. Geochim. Cosmochim. Acta. 2007b;71: 2285-2298.

[25] Jokic A, Cutler JN, Ponomarenko E, van der Kamp G, Anderson D. Organic carbon and sulfur compounds in wetland soils: insights on structure and transformation processes using K-edge XANES and NMR spectroscopy. Geochim. Cosmochim. Acta. 2003; 67: 2585-2597

[26] Kinyangi J, Solomon D, Liang B, Lerotic M, Wirick S, Lehmann J. Nanoscale biogeocomplexity of the organomineral assemblage in soil: application of STX microscopy and C-1s NEXAFS spectroscopy. Soil Sci. Soc. Am. J. 2006;70: 1708-1718.

[27] Lehmann J, Liang B, Solomon D, Lerotic M, Luiza˜o F, Kinyangi J. Near-edge X-ray absorption fine structure (NEXAFS) spectroscopy for mapping nanoscale distribution of organic carbon forms in soil: application to black carbon particles. Global Biogeochem. Cycles. 2005; 19: 1013-1025.

[28] Myneni SCB, Brown JT, Martinez GA, Meyer-Ilse W. Imaging of humic substance molecular structures in water and soil. Science 1999;286: 1335-1337

[29] Scha¨fer T, Hertkorn N, Artinger R, Claret F, Bauer A. Functional group analysis of natural organic colloids and clay association kinetics using C(1s) spectromicroscopy. J. Phys. France. 2003; 104: 409-412.

[30] Scheinost A C, Kretzschmar R, Christl I, Jacobsen C. Carbon group chemistry of humic and fulvic acid: a comparison of C-1s NEXAFS and 13C-NMR spectroscopies. In: Ghabbour, E.A., Davies, G. (Eds.), Humic Substances: Structures, Models and Functions. Royal Society of Chemistry, Gateshead, UK,2001. pp. 39-47.

[31] Schumacher M, Christl I, Scheinost AC, Jacobsen C, Kretzschmar R. Chemical heterogeneity of organic soil colloids investigated by scanning transmission X-ray microscopy and C-1s NEXAFS microspectroscopy. Environ. Sci. Technol. 2005; 39: 9094-9100.

[32] Schumacher M, Christl I, Vogt RD, Barmettler K, Jacobsen C, Kretzschmar R. Chemical composition of aquatic dissolved organic matter in five boreal forest catchments sampled in spring and fall seasons. Biogeochemistry. 2006; 80: 263-275.

[33] Carr GL, Reffner JA, Williams GP. Performance of an infrared

microspectrometer at the NSLS. Rev. Sci. Instrum. 1995; 66: 1490-1492.

[34] Miller LM, Dumas P, Chemical imaging of biological tissue with synchrotron infrared light. Biochim. Biophys. Acta. 2006; 1758: 846-857

[35] Lehmann J, Kinyangi J, Solomon D. Organic matter stabilization in soil microaggregates: implications from spatial heterogeneity of organic carbon contents and carbon forms. Biogeochemistry. 2007;85: 45-57.

[36] Bonetta DT, Facette M, Raab TK, Somerville CR. Genetic dissection of plant cell-wall biosynthesis. Biochem. Soc. Trans. 2002;30: 298-301.

[37] Marinkovic NS, Huang R, Bromberg P, Sullivan M, Toomey J, Miller LM. Center for Synchrotron Biosciences' U2B beamline: an international resource for biological infrared spectroscopy. J. Synchrotron Radiat. 2002;9: 189-197.

[38] Johnston CT, Aochi YO. Fourier transform infrared and Raman spectroscopy. In: Sparks, D.L. (Ed.), Methods of Soil Analysis, Part 3. Soil Sci. Soc. Am., Madison, WI, 1996;pp. 269-321.

[39] Solomon D, Lehmann J, Kinyangi J, Liang B, Hanley K, Heymann K. Carbon (1s) NEXAFS spectroscopy of biogeochemically relevant organic reference compounds. Soil Sci. Soc. Am. J. 2009; 73: 1817-1830

[40] Sto¨hr J. N85EXAFS Spectroscopy. Springer Series in Surface Sciences. Springer, Berlin, Germany. 1992.

[41] Scha¨fer T, Buckau G, Artinger R, Kim JI, Geyer S, Wolf M. Origin and mobility of fulvic acids in the Gorleben aquifer system: implications from isotopic data and carbon/sulfur XANES. Org. Geochem. 2005;36: 567-582.

[42] Scha¨ferT, Chanudet V, Claret F, Filella M. Spectromicroscopy mapping of colloidal/particulate organic matter in Lake Brienz, Switzerland. Environ. Sci. Technol. 2007;41: 7864-7869.

[43] Smith AP, Urquhart SG, Winesett DA, Mithchell G, Ade H. Use of near-edge X-ray absorption fine structure spectromicroscopy to characterize multicomponent polymeric systems. Appl. Spectrosc. 2001; 55: 1676-1681.

[44] Hitchcock AP, Sto¨ver HD, Croll LM, Childs RF. Chemical mapping of polymer microstructure using soft X-ray spectromicroscopy. Aust. J. Chem. 2005;58: 423-432.

[45] Lerotic M, Jacobsen C, Scha¨fer T, Vogt S. Cluster analysis of soft X-ray spectromicroscopy data. Ultramicroscopy. 2004; 100: 35-57.

[46] Rumpel C, Janik L J, Skjemstad J O, Ko¨gel-Knabner I. Quantification of carbon derived from lignite in soils using mid-infrared spectroscopy and partial least squares. Org. Geochem. 2001; 32: 831-839.

[47] Sharma GK, Hota S, Kumar A, Ray P, Fagodiya RK, Malav LC, Yadav KK, Gupta DK, Khan SA, Ray SK. Recent Development in Bioremediation of Soil Pollutants Through Biochar for Environmental Sustainability. In: Singh J., Singh C. (eds) Biochar Applications in Agriculture and Environment Management. Springer, Cham. 2020. DOI: 10.1007/978-3-030-40997-5_6

[48] Hota S, Sharma G K, Subrahmanyam G, Kumar A, Shabnam A A, Baruah P, Kaur T, Yadav A N. Fungal Communities for Bioremediation of Contaminated Soil for Sustainable Environments. In: Yadav A N.Recent Trends in Mycological Research, Fungal Biology. 2021. DOI: 10.1007/978-3-030-68260-6_2

[49] Brown GE, Bargar JR, Towle S.N. Use of X-ray absorption spectroscopy to

study reaction mechanisms at oxide-water interfaces. Abstr. Pap. Am. Chem. Soc. 1997a ;213: 36-Geoc.

[50] Brown G E, Foster A, Ostergren J, Bargar JR. X-ray absorption spectroscopy and molecular environmental science. Abstr. Pap. Am. Chem. Soc. 1997b; 214: 10-Geoc.

[51] Schulze DG, Bertsch PM. Synchrotron X-ray techniques in soil, plant, and environmental research. Adv. Agron. 1995;55: 1-66.

[52] Sparks DS, Borda MJ. Kinetics and mechanisms of sorption-desorption in soils: a multiscale assesment. In: Violante, A., Huang, P.M., Gadd, G.M. (Eds.), Biophysico-Chemical Processes of Heavy Metals and Metalloids in Soil Environments. 2008. Wiley and Sons, Inc, New York.

[53] Ginder-Vogel M, Sparks D L. The Impacts of X-Ray Absorption Spectroscopy on Understanding Soil Processes and Reaction Mechanisms. In: Singh B, Gräfe M, editors. Synchrotron-Based Techniques in Soils and Sediments. Vol.34. Elsevier; 2010.p. 1-26. DOI: 10.1016/S0166-2481(10)34001-3.

[54] Doyle CS, SJ Traina, H Ruppert, T Kendelewicz, JJ Rehr, G E Brown. XANES studies at the Al K-edge of aluminium-rich surface phases in the soil environment. J. Synchrotron Radiat. 1999;6:621-623. DOI:10.1107/ S090904 9598015568

[55] Ildefonse P, RJ Kirkpatrick, B Montez, G Calas, AM Flank, P Lagarde. 27Al MAS NMR and aluminum X-ray absorption near edge structure study of imogolite and allophanes. Clays Clay Miner. 1994;42:276-287. DOI:10.1346/ CCMN.1994.0420306

[56] Kopittke PM, KL Moore, E Lombi, A Gianoncelli, BJ Ferguson, FPC Blamey. Identification of the primary lesion of toxic aluminum (Al) in plant roots. Plant Physiol. 2015;167:1402-1411. DOI:10.1104/pp.114.253229

[57] Landrot G R, Tappero SM, Webb, DL Sparks. Arsenic and chromium speciation in an urban contaminated soil. Chemosphere. 2012; 88:1196-1201. DOI:10.1016/j.chemosphere.2012.03.069

[58] Terzano R, Spagnuolo M, Vekemans B, De Nolf W, Janssens K, Falkenberg G. Assessing the origin and fate of Cr, Ni, Cu, Zn, Pb, and V in industrial polluted soil by combined microspectroscopic techniques and bulk extraction methods. Environ. Sci. Technol. 2007;41: 6762-6769. doi:10.1021/ es070260h

[59] Hopp L, PS Nico, MA Marcus, S Peiffer. Arsenic and chromium partitioning in a Podzolic soil contaminated by chromated copper arsenate. Environ. Sci. Technol. 2008;42:6481-6486. DOI:10.1021/ es800615f

[60] Fandeur DF, Juillot G Morin, L Olivi, A Cognigni, SM Webb. XANES evidence for oxidation of Cr(III) to Cr(VI) by Mn-oxides in a lateritic regolith developed on serpentinized ultramafic rocks of New Caledonia. Environ. Sci. Technol. 2009;43:7384-7390. doi:10.1021/es900498r

[61] Jacobson A R, S Dousset, F Andreux, P.C. Baveye. Electron microprobe and synchrotron X-ray fluorescence mapping of the heterogeneous distribution of copper in high-copper vineyard soils. Environ. Sci. Technol. 2007; 41:6343-6349. DOI:10.1021/es070707m

[62] Strawn DG, and Baker LL. Molecular characterization of copper in soils using X-ray absorption spectroscopy. Environ. Pollut. 2009; 157:2813-2821. doi: 10.1016/j. envpol.2009.04.018

[63] Grafe M, Tappero RV, Marcus MA, Sparks DL. Arsenic speciation in multiple metal environments - II.

Micro-spectroscopic investigation of a CCA contaminated soil. J. Colloid Interface Sci. 2008;321: 1-2.

[64] Gustafsson J P, Visual M.I.N.T.E.Q. KTH, Dep Land and Water Resource Engineering, Stockholm. 2006

[65] Pritzel J, Thieme J, Eusterhues K, Eichert D. Iron speciation in soils and soil aggregates by synchrotron-based X-ray microspectroscopy (XANES, m-XANES). Eur. J. Soil Sci. 2007; 58, 1027-1041.

[66] Singh B, Sherman D M, Gilkes RJ, Wells MA, Mosselmans JFW. Structural chemistry of Fe, Mn, and Ni in synthetic hematites as determined by extended X-ray absorption fine structure spectroscopy. Clays Clay Miner. 2000;48, 521-527.

[67] Huynh T, Tong AR, Singh B, Kennedy BJ. Cd substituted goethites: a structural investigation by synchrotron X-ray diffraction. Clays Clay Miner. 2003; 51, 397-402

[68] Kaur N, Singh B, Kennedy BJ, Gra¨fe M. The preparation and characterization of vanadium substituted goethite: the importance of temperature. Geochim. Cosmochim. Acta. 2009a; 73, 582-593.

[69] Kaur N, Gra¨fe M, Singh B, Kennedy B. Simultaneous incorporation of Cr, Zn, Cd, and Pb in the goethite structure. Clays Clay Miner. 2009b ;57, 234-250.

[70] Kaur N, Singh B, Kennedy B. Copper substitution alone and in the presence of chromium, zinc, cadmium and lead in goethite (a-FeOOH). Clay Miner. 2009c; 44, 293-309.

[71] Holman Hoi-Ying N. Synchrotron Infrared Spectromicroscopy for Studying Chemistry of Microbial Activity in Geologic Materials. In: Singh B, Gräfe M, editors. Synchrotron-Based Techniques in Soils and Sediments. Vol.34. Elsevier; 2010.p. 103-130. DOI:10.1016/S0166-2481(10)34004-9.

[72] Carr GL. High-resolution micro-spectroscopy and sub-nanosecond time-resolved spectroscopy with the synchrotron infrared source. Vib. Spectrosc. 1999; 19: 53-60.

[73] Levenson E, Lerch P, Martin MC. Infrared imaging: synchrotrons vs. arrays, resolution vs. speed. Infrared Phys. Technol. 2006; 49: 45-52.

[74] Levenson E, Lerch P, Martin MC. Spatial resolution limits for synchrotron-based infrared spectromicroscopy. Infrared Phys. Technol. 2008a;51: 413-416.

[75] Levenson E, Lerch P, Martin MC. Spatial resolution limits for synchrotron-based spectromicroscopy in the mid- and near-infrared. J. Synchrotron Radiat. 2008b; 15 :323-328

[76] Hawkins NJ, Mattraw HC, Sabol WW, Carpenter DR. Spectroscopy of gaseous carbonyls. 1. Infrared spectra and thermodynamic properties of chromium and molybdenum hexacarbonyls. J. Chem. Phys. 1955;23: 2422-2427

[77] Griffith WP, Lewis J, Wilkinson G. Infrared spectra of transition metal-nitric oxide complexes. 4. The penta-cyanonitrosyl-complexes of chromium and molybdenum. J. Chem. Soc. 1959; 872-875.

[78] Humphrey RE. The infrared spectra of aromatic chromium tricarbonyls. Spectrochim. Acta. 1961; 17: 93-100.

[79] Janni J, Gilbert BD, Field RW, Steinfeld JI. Infrared absorption of explosive molecule vapors. Spectrochim. Acta A Mol. Biomol. Spectrosc. 1997; 53: 1375-1381.

[80] Bauschlicher C W. Infrared spectra of polycyclic aromatic hydrocarbons: nitrogen substitution. Chem. Phys. 1998;234; 87-94.

[81] Bauschlicher C W, Langhoff SR. Infrared spectra of polycyclic aromatic

hydrocarbons: methyl substitution and loss of H. Chem. Phys. 1998; 234: 79-86.

[82] Hudgins DM, Sandford SA. Infrared spectroscopy of matrix isolated polycyclic aromatic hydrocarbons. 1. PAHs containing two to four rings. J. Phys. Chem. A 1998a;102: 329-343.

[83] Hudgins DM, Sandford SA. Infrared spectroscopy of matrix isolated polycyclic aromatic hydrocarbons. 2. PAHs containing five or more rings. J. Phys. Chem. A 1998b;102: 344-352.

[84] Hudgins DM, Sandford SA. Infrared spectroscopy of matrix isolated polycyclic aromatic hydrocarbons. 3. Fluoranthene and the benzofluoranthenes. J. Phys. Chem. A. 1998c; 102: 353-360.

[85] Ludwig C, Dolny M, Gotze HJ. Fourier transform Raman and infrared spectra and normal coordinate analysis of organo-arsenic(III), –antimony(III) and -bismuth(III) thiolates. Spectrochim. Acta A Mol. Biomol. Spectrosc. 2000;56: 547-555.

[86] Mattioda AL, Hudgins DM, Bauschlicher C, Allamandola, LJ Infrared spectra of matrixisolated polycyclic aromatic hydrocarbons and their ions: PAHs incorporating the peropyrene structure. Abstr. Pap. Am. Chem. Soc. 2002;224: U323.

[87] Pauzat F, Ellinger Y. The infrared signatures of nonregular PAHs. Chem. Phys. 2002;280: 267-282.

[88] Ruiter kamp R, Halasinski T, Salama F, Foing BH, Allamandola LJ, Schmidt W, .Spectroscopy of large PAHs: laboratory studies and comparison to the diffuse interstellar bands. Astron. Astrophys. 2002;390: 1153-1170.

[89] Chauhan HPS, Shaik NM, Kori K. Synthesis and characterization of some toluene-3,4-dithiolatobismuth (III) alkyl dithiocarbonates. Synth. React. Inorg. Met. Org. Chem. 2004;34: 323-333.

[90] Jensen JO, Jensen JL. Vibrational frequencies and structural determination of trimethylarsine oxide. Spectrochim. Acta A Mol. Biomol. Spectrosc. 2004;60: 3065-3070.

[91] Jensen JO, Jensen JL. Vibrational frequencies and structural determination of trimethylarsine oxide. Spectrochim. Acta A Mol. Biomol. Spectrosc. 2004;60: 3065-3070.

[92] Jensen JO. Vibrational frequencies and structural determination of phosphorus tricyanide. Spectrochim. Acta A Mol. Biomol. Spectrosc. 2004;60: 2537-2540.

[93] Li J F, Fuller S, Cattle J, Way CP, Hibbert DB. Matching fluorescence spectra of oil spills with spectra from suspect sources. Anal. Chim. Acta. 2004; 514: 51-56.

[94] Bernstein MP, Mattioda AL, Sandford SA, Hudgins DM. Laboratory infrared spectra of polycyclic aromatic nitrogen heterocycles: quinoline and phenanthridine in solid argon and H2O. Astrophys. J. 2005;626: 909-918.

[95] Carrasco-Flores EA, Clavijo RE, Campos-Vallette MM, Aroca RF. Vibrational spectra and surface enhanced vibrational spectra of 1-nitropyrene. Appl. Spectrosc. 2004;58: 555-561.

[96] Carrasco-Flores EA, Clavijo RE, Campos-Vallette MM, Aroca RF. Vibrational and surface-enhanced vibrational spectra of 6-nitrochrysene. Spectrochim. Acta A Mol. Biomol. Spectrosc. 2005; 61: 509-514.

[97] Zhang Y, Yang M, Dou X M, He H, Wang DS. Arsenate adsorption on an Fe-Ce bimetal oxide adsorbent: role of surface properties. Environ. Sci. Technol. 2005;39: 7246-7253.

[98] Sekine R, Marzouk E, Khaksar M, Scheckel K, Stegemeier J, Lowry G. Aging of dissolved copper and copper-based

nanoparticles in five different soils: Short-term kinetics vs. long-term fate. J. Environ. Qual. 2017; 46:1198-1205. doi:10.2134/jeq2016.12.0485

[99] Colzato M, Kamogawa M, Carvalho H, Alleoni L, and Hesterberg D. 2017. Temporal changes in cadmium speciation in Brazilian soils evaluated using Cd LIII– edge XANES and chemical fractionation. J. Environ. Qual. 46:1206-1214. doi:10.2134/jeq2016.08.0316

[100] Mariet A, Sarret G, Bégeot C, Walter-Simonnet A, and Gimbert F. 2017. Lead highly available in soils centuries after metallurgical activities. J. Environ. Qual. 46:1236-1242. doi:10.2134/jeq2016.12.0469

[101] Holman HYN, Nieman K, Sorensen DL, Miller CD, Martin MC, Borch T, McKinney WR, Sims RC (2002) Catalysis of PAH biodegradation by humic acid shown in synchrotron infrared studies. Environ Sci Technol 36:1276-1280 doi:10.1021/es0157200

[102] Giesler R, Andersson T, Lovgren L, Persson P. Phosphate sorption in aluminum- and iron-rich humus soils. Soil Sci. Soc. Am. J. 2005; 69: 77-86.

[103] Hinedi Z R, Chang A C. Solubility and P-31 magic angle spinning nuclear magnetic-resonance of phosphorus in sludge-amended soils. Soil Sci. Soc. Am. J. 1989;53: 1057-1061.

[104] Brandes JA, Ingall E, Paterson D. Characterization of minerals and organic phosphorus species in marine sediments using soft X-ray fluorescence spectromicroscopy. Mar. Chem. 2007; 103: 250-265

[105] Khare N, Martin JD, Hesterberg D. Phosphate bonding configuration on ferrihydrite based on molecular orbital calculations and XANES fingerprinting. Geochim. Cosmochim. Acta .2007; 71: 4405-4415.

[106] Kelly SD, Hesterberg D, Ravel B. Analysis of soils and minerals using X-ray absorption spectroscopy. In: Ulery, A.L., Drees, L.R. (Eds.), Methods of Soil Analysis Part 5: Mineralogical Methods, Soil Science Society of America, Inc., Madison, WI, 2008; 387-463.

[107] Manceau A, Marcus MA, Tamura N. Quantitative speciation of heavy metals in soils and sediments by synchrotron X-ray techniques. In: Fenter, P., Rivers, M., Sturchio, N.C., Sutton, S. (Eds.), Reviews in Mineralogy and Geochemistry, Applications of Synchrotron Radiation in Low-Temperature Geochemistry and Environmental Science. Mineralogical Society of America, Washington, D.C., 2002a pp. 341-428.

[108] Ajiboye B, Akinremi O O, Hu Y, Jurgensen A. XANES speciation of phosphorus in organically amended and fertilized Vertisol and Mollisol. Soil Sci. Soc. Am. J. 2008; 72: 1256-1262.

[109] Beauchemin, S., Hesterberg, D., Chou, J., Beauchemin, M., Simard, R.R., Sayers, D.E. Speciation of phosphorus in phosphorus-enriched agricultural soils using Xray absorption near-edge structure spectroscopy and chemical fractionation. J. Environ. Qual. 2003;32: 1809-1819.

[110] Hesterberg D, Zhou WQ, Hutchison KJ, Beauchemin S, Sayers DE. XAFS study of adsorbed and mineral forms of phosphate. J. Synchrotron Radiat. 1999;6: 636-638.

[111] Lombi E, Scheckel KG, Armstrong RD, Forrester S, Cutler JN, Paterson D. Speciation and distribution of phosphorus in a fertilized soil: a synchrotronbased investigation. Soil Sci. Soc. Am. J. 2006;70: 2038-2204

[112] Myneni SCB. Soft X-ray spectroscopy and spectromicroscopy studies of organic molecules in the environment. In: Fenter, M.L.R.P.A., Sturchio, N.C., Sutton, S.R. (Eds.),

Applications of Synchrotron Radiation in Low-Temperature Geochemistry and Environmental Sciences, vol. 49. The Mineralogical Society of America, Washington, DC, 2002. pp. 485-579.

[113] Sato S, Solomon D, Hyland C, Ketterings QM, Lehmann J. Phosphorus speciation in manure and manure-amended soils using XANES spectroscopy. Environ. Sci. Technol. 2005;39: 7485-7491.

[114] Schulten HR, Schnitzer M. The chemistry of soil organic nitrogen: a review. Biol. Fertil. Soils 1998;26: 1-15.

[115] Stevenson FJ. Humus Chemistry, Genesis, Composition, Reactions, second ed. 1994. John Wiley and Sons, New York.

[116] Jeff JC, Chappin FS, Vitousek PM. Breaks in the cycle: dissolved organic nitrogen in terrestrial ecosystems. Front. Ecol. Environ. 2003;1: 205-211

[117] Leinweber P, Schulten HR. Nonhydrolyzable organic nitrogen in soil size separates from long-term agricultural experiments. Soil Sci. Soc. Am. J. 1998;62: 383-393.

[118] Catroux G, Schnitzer M. Chemical, spectroscopic, and biological characteristics of the organic matter in particle-size fractions separated from an Aquoll. Soil Sci. Soc. Am. J. 1987;51: 1200-1207.

[119] Knicker H. Biogenic nitrogen in soils as revealed by solid-state 13C and 15N NMR spectroscopy. J. Environ. Qual. 2000;29: 715-723.

[120] Abe T, Watanabe A. X-ray photoelectron spectroscopy of nitrogen functional groups in soil humic acids. Soil Sci. 2004; 169: 35-43

[121] Jokic A, Cutler JN, Anderson DW, Walley FL. Detection of heterocyclic N compounds in whole soils using N-XANES spectroscopy. Can. J. Soil Sci. 2004b ;84: 291-293.

[122] Jokic A, Schulten HR, Cutler JN, Schnitzer M, Huang PM. A significant abiotic pathway for the formation of unknown nitrogen in nature. Geophys. Res. Lett. 2004a ; L05502 doi:10.1029/ 2003GL018520.

[123] Vairavamurthy A, Wang S. Organic nitrogen in geomacromolecules: insights on speciation and transformation with K-edge XANES spectroscopy. Environ. Sci. Technol. 2002;36: 3050-3056.

[124] Gustafsson J P, Visual M.I.N.T.E.Q. KTH, Dep Land and Water Resource Engineering, Stockholm. 2006

[125] Leinweber P, Kruse J, Walley F, Gillespie A, Eckhardt K, Blyth R I R, Regier T Z. Nitrogen Compounds in Dissolved and Solid Environmental Samples. In: Singh B, Gräfe M, editors. Synchrotron-Based Techniques in Soils and Sediments. Vol.34. Elsevier; 2010.p. 255-287.DOI:10.1016/S0166-2481(10) 34009-8.

[126] Leinweber P, Kruse J, Walley FL, Gillespie A, Eckhardt KU, Blyth RIR. Nitrogen K-edge XANES: an overview of reference compounds used to identify 'unknown' organic nitrogen in environmental samples. J. Synchrotron Radiat. 2007;15: 500-551

[127] Gillespie AW, Walley FL, Farrell RE, Leinweber P, Schlichting A, Eckhardt KU. Profiling rhizosphere chemistry: evidence from C & N K-edge XANES and Py-FIMS. Soil Sci. Soc. Am. J. 2009;73: 2002– 2012.

[128] Leinweber P, Walley FL, Kruse J, Jandl G, Eckhardt KU, Blyth R. Cultivation affects soil organic nitrogen: pyrolysis-mass spectrometry and N K-edge XANES evidence. Soil Sci. Soc. Am. J. 2009;73: 82-92

[129] Dynes JJ, Tyliszczak T, Araki T, Lawrence JR, Swerhone GDW,

Leppard GG. Speciation and quantitative mapping of metal species in microbial biofilms using scanning transmission X-ray microscopy. Environ. Sci. Technol.2006; 40: 1556-1565.

[130] Lawrence JR, Swerhone GDW, Leppard GG, Araki T, Zhang X, West MM. Scanning transmission X-ray, laser scanning, and transmission electron microscopy mapping of the exopolymeric matrix of microbial biofilms. Appl. Environ. Microbiol. 2003;69: 5543-5554.

[131] Toner B, Fakra S, Villalobos M, Warwick T, Sposito G. Spatially-resolved characterization of biogenic manganese Oxide production within a bacterial biofilm. Appl. Environ. Microbiol. 2005;71: 1300-1310.

[132] Benzerara K, Yoon TH, Menguy N, Tyliszczak T, Brown GE. Nanoscale environments associated with bioweathering of a Mg-Fe-pyroxene. Proc. Natl. Acad. Sci. 2005;102: 979-982.

[133] Mancosky DG, Lucia LA, Nanko H, Wirick S, Rudie AW, Braun R. Novel visualization studies of lignocellulosic oxidation chemistry by application of C-near edge X-ray absorption fine structure spectroscopy. Cellulose 2005; 12: 35-41.

[134] Raab T K, Lipson D A. The Rhizosphere: A Synchrotron-Based View of Nutrient Flow in the Root Zone. In: Singh B, Gräfe M, editors. Synchrotron-Based Techniques in Soils and Sediments. Vol.34. Elsevier; 2010.p. 171-198.DOI: 10.1016/S0166-2481(10)34007-4.

[135] Raab TK, Vogel JP. Ecological and agricultural applications of synchrotron IR microscopy. IR Phys. Technol. 2004;45: 393-402.

[136] Wetzel DL. A new approach to the problem of dispersive windows in IR microspectroscopy. Vib. Spectrosc. 2002; 29: 291-297.

Chapter 14

Bioinoculants in Technological Alleviation of Climatic Stress in Plants

Rafia Younas, Shiza Gul, Rehan Ahmad, Ali Raza Khan, Mumtaz Khan, Tauseef Anwar and Huma Qureshi

Abstract

Global climate change is leading to a series of frequent onset of environmental stresses such as prolonged drought periods, dynamic precipitation patterns, heat stress, and cold stress on plants and commercial crops. The increasing severity of such stresses is not only making agriculture and related economic sector vulnerable but also negatively influences plant diversity patterns. The global temperature of planet Earth has risen to 1.1°C since the last 19th century. An increase in surface temperature leads to an increase in soil temperature which ultimately reduces water content in the soil, thereby, reducing crop growth and yield. Moreover, this situation is becoming more intense for agricultural practices in arid and semi-arid regions. To overcome climatically induced stresses, acclimatization of plant species via bioinoculation with Plant Growth Promoting Rhizobacteria (PGPR) is becoming an effective approach. The PGPR are capable of colonizing rhizosphere (exophytes) as well as plant organs (endophytes), where they trigger an accumulation of osmolytes for osmoregulation or improving gene expression of heat or cold stress proteins, or by signaling the synthesis of phytohormones, metabolites, proteins, and antioxidants to scavenge reactive oxygen species. Thus, PGPR exhibiting multiple plant growth-promoting traits can be employed via bioinoculants to improve the plant's tolerance against unfavorable stress conditions.

Keywords: Plant growth promoting rhizobacteria, plant growth-regulating hormones, ACC deaminase, osmoregulation, psychrotrophic microbes

1. Introduction

The agriculture sector has been considered as the main channel through which irreparable climatic change will influence the global economy [1]. The commercial plantation of crops has been severely influenced by various environmental factors mainly being drought, salinity, and intense temperature variation. These factors have become more adverse with continuing fluctuation in climate change. Globally, only 10% of arable land is reported to be free from environmental stress [2]. In arable land, more than 50% is vulnerable to drought and salinization which is expected to inflict an almost 10% rise in global water consumption by 2050. Moreover, the consistent rise in the global population will be succeeded by an increase in food demand double fold in 2050 [3].

Plants adapt and develop their own mechanism to cope the environmental stress. The most effective means of combating environmental stress is the release and accumulation of plant growth-regulating hormones such as abscisic acid (ABA) and salicylic acid (SA). ABA, commonly known as stress hormone is released at the root tip in response to lower water potential, followed by its translocation to leaves where it alters osmotic pressure of the stomatal guard cells, leading to the closure of stomata openings to conserve moisture, thereby, preventing water loss via transpiration. Moreover, accumulated carbohydrates and amino acids being low molecular weight compounds play an active role in osmoregulation. SA, on the other hand, protects the plant against pathogenic attack [4].

PGPR/Treated Plant	Climatic stress	Alleviation of climatic stress
Bacillus cereu/Solanum lycopersicum	Heat stress	Increased production of exo-polysaccharide, cleavage of ACC-deaminase [9]
Bacillus amyloliquefaciens, Agrobacterium fabrum/Triticum aestivum	Water stress	Increase in biomass and grain yield [10]
Aneurinibacillus aneurinilyticus, Paenibacillus sp./Phaseolus vulgaris	Salinity stress	Facilitating ACC-deaminase activity, increased production of IAA, hydrogen cyanide and siderophore [11]
Stenostrophomonas maltophilia/Triticum aestivum	Salinity stress	Improved growth and yield, elevated antioxidative enzymatic activity, increased K^+ uptake [12]
Paenibacillus polymyxa/Brassica napus	Soil temperature	Production of antimicrobial and plant growth promoting volatile organic compounds (VOCs) [13]
Bacillus subtilis, Bacillus thuringiensis, Bacillus megaterium/Cicer arietinum	Climatically induced salinity stress	Regularized photosynthesis via assimilation of soluble sugars, chlorophyll and proline [14]

Table 1.
Alleviation of climatic stress via bioinoculants.

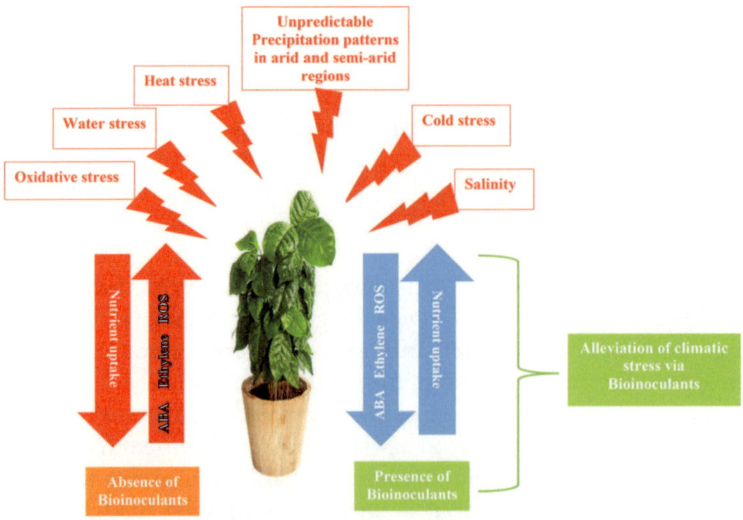

Figure 1.
Plant's responses to various stresses in presence and absence of bioinoculants.

Along with plant internal defense mechanism, the supportive role of PGPR has been widely acknowledged as an effective tool for sustaining plant growth and yield [5–7]. The plant root system serves as a habitat for numerous microorganisms that also interact with soil, thereby developing a complex ecosystem [8]. Certain PGPR inhabiting the rhizosphere penetrates roots and migrates through plant tissues; thereby impact the physiological and biochemical traits of plant cells (**Table 1**). This ecological aspect of the microbial community can be employed to strengthen the stress tolerance in plants, hence increasing their adaptation towards invincible climatic changes (**Figure 1**) [15].

The escalating demand and supply of chemical fertilizers and pesticides along with the persistence of their residual particles has posed an ultimate threat to humans as well as to the ecosystem. As an alternative, the acceptability of bio-fertilizers bearing bioinoculants has been increasing day by day. These bioinoculants act as a potential source of phytohormones which can rectify the climatic stress faced by plants. This chapter will explore various aspects of the utilization of bioinoculants for combating the climatic stress faced by plants.

2. Amelioration of heat stress via thermotolerant microbes

The major climatic stress faced by plants due to the rise in global temperature is heat stress which leads to water stress with counter effect on photosynthetic rate and flowering and fruiting in both tropical and sub-tropical crop systems [16]. The steady increase in surface temperatures is followed by an increase in soil temperature which can influence the root elongation process. Although, root elongation requires optimum soil temperature (specific to each plant), beyond which root elongation is stopped resulting in a stunted root system. Moreover, the activity of rhizobacteria is also relying on optimum soil temperature. In the forest ecosystem, N being a macronutrient acts as a growth-limiting factor. The flow of N cycling is deliberately subjected to temperature variation. The relatively increased temperature triggers N mineralization by microbes and thereby, prompting N uptake by plants as nitrates. However, in this process, the net nitrification is decreased [17]. The inoculation of PGPR is found to be more convenient than hsfs (heat stress transcription factors) genes, transgenic varieties, and breeding of heat-tolerant cultivators [18, 19].

Thermotolerance of certain PGPR has been evaluated to decrease the effect of heat stress on plants. The soybean plants inoculated with *Bacillus cereus* strain SA1 were exposed to heat stress for 5–10 days. The inoculated plants exhibited improved physiological (biomass) and biochemical (chlorophyll content and chlorophyll fluorescence) characteristics. Moreover, SA1 inoculation reduced the synthesis of abscisic acid and increased salicylic acid (a phenolic phytohormone involved in signaling and defense against pathogens) along with remarkable production of antioxidants (ascorbic acid peroxidase, superoxide dismutase, and glutathione) for sequestration of reactive oxygen species (ROS). A continuous increase in the synthesis and assimilation of Heat Shock Protein (HSP) was also claimed throughout the stress period of 5 to10 days, mainly due to the perpetual onset of GmHSP gene expression. Also, over gene expression of GmLAX3 and GmAKT2 were associated with enhanced potassium gradient and altered auxin and ABA stimuli (Aaqil et al. 2020). The bacterial strain *Bacillus cereus* capable of producing ACC-deaminase (0.76–C0.9 μM/mg protein/h) can cleavage ACC to α-ketobutyrate and ammonia to increase heat tolerance in tomato (*Solanum lycopersicum L.)*. Moreover, increase in synthesis of exopolysaccharide (0.66–C0.91 mg/mL) showed promising plant growth in tomato [9] (**Table 2**). Moreover, the presence of enzymatic antioxidants

Treated Plant/PGPR	Stress Alleviation Effect
Potato/*Paraburkholderia phytofirmans*	ACC deaminase production [20]
Tomato/Mycorrhizae	Reduction in lipid peroxidation and H_2O_2, higher ROS scavenging activity [21]
Bacillus amyloliquefaciens, Azospirillum brasilense Pantoea dispersa, *Serratia marcescens*, *Pseudomonas* spp.	ROS reduction, pre-activation of heat shock proteins [22]
	ACC deaminase production [23]
Apple and pear/*Ps. fluorescens*	Competition with INA+ bacteria [24]
Grapevine/*Paraburkholderia phytofirmans*	ACC deaminase production [25]
Soy/*Bacillus aryabhatthai*	ABA production [26]

Table 2.
Alleviation of heat stress via bioinoculants.

such as superoxide dismutase (SOD), peroxidase (POD) and catalase (CAT), and accumulation of metabolites such as amino acids and proline had boosted the scavenging activity against ROS under heat stress.

Among different agro-ecological zones, the arid and semi-arid regions are highly vulnerable to climatic changes mainly due to dynamical precipitation patterns. Being a staple cereal crop, the wheat and its associated endophytic and epiphytic bacterial diversity was investigated for thermotolerance at a temperature range of 30–40°C, pH 3–11, and 3–20% NaCl concentration [27]. Among 32 bacterial strains of 15 genera, 10 strains were reported to exhibit six different plant growth-promoting traits under heat stress. *Arthrobacter sp.* (66.0 ± 0.7 mg L^{-1}) possessed high solubilization of phosphorus followed by *Pseudomonas japonica* (64.6 ± 0.9 mgL^{-1}). Whereas, the potassium solubilization was highest for *Methylobacterium mesophilicum*. *Pseudomonas putida* exhibited the highest IAA production (70.8 ± 1.5 μg mg^{-1} protein day^{-1}) followed by *Rhodobacter capsulatus* (69.1 ± 0.5 μg mg^{-1} protein day^{-1}). The highest production of siderophore had been shown by *Alcaligenes faecalis* (4.9 ± 0.1 mm). Being thermotolerant, *Alcaligenes faecalis* and *Pseudomonas poae* were also characterized as alkali tolerant with 5% NaCl and 10% NaCl tolerance respectively. Besides the production of IAA, siderophore, and ammonia, these bacterial strains were also involved in the solubilization of phosphorus and zinc under heat stress. The chief bacterial groups characterized for nitrogen fixation at high temperatures were *Acromobacter*, *Alcaligenes*, *Bacillus*, *Delftia*, *Providencia*, *Pseudomonas*, *Rhodobacter*, and *Salmonella*.

Similarly, Sorghum being native to arid habitat was inoculated with a thermotolerant bacterial strain of *Psuedomonas aeruginosa* AKM-P6 which was isolated from the rhizosphere of Pigeon Pea plant grown under semi-arid conditions [28]. The inoculated Sorghum seedlings survived the temperature range of 47–50°C up to 15 days. The thermotolerance was imparted by the biosynthesis of high molecular weight proteins in leaves which protected cellular membranes from injury and improved metabolite production mainly proline, chlorophyll, sugars, amino acids, and proteins along with plant biomass. Another strain of *Pseudomonas sp.* PsJN was used to evaluate the heat-stress tolerance level in 18 clones of potatoes via both *in vitro* and *ex-vitro* inoculation [29]. The inoculated potato nodal cuttings were exposed to the temperature range of 20/15°C or 33/15°C day and night for six weeks. The increase in temperature had drastically affected the root system and tuber number and tuber fresh weight. An average root to shoot ratio decreased from 3.7 at 20/15°C to 1.7 at 33/15°C in non-inoculated plantlets and respectively, from 4.3

to 1.5 for inoculated ones. ABA deficient plantlets lack the tuber formation whereas tuberization in inoculated plantlets was significant at 33/15°C. As compared to in vitro, the ex-vitro performance of potato clones (LT-7) exhibited the effectiveness of rhizobacteria in colonizing certain potato clones under heat stress. Hence, the thermotolerant bacterial strains may serve as inoculants and bio-control agents for improving crop productivity with accelerating climatic temperature stress.

3. Amelioration of water stress via 1-aminocyclopropane-1-carboxylate deaminase/exopolysaccharide production

Due to the changes in minimum and maximum temperature, high water requirement and reduction in yield was observed in plants. Although the production of 'Ethylene' as plant stress hormone occurs in response to climatic stress encountered by plants; its accumulation in high amounts could be deleterious. The accumulation occurs under the influence of ACC oxidase that prompts exudation of ACC, a precursor to ethylene synthesis. The PGPR capable of producing ACC deaminase degrades ACC into its intermediates i.e., ammonia and α-ketobutyrate, thereby, declining ethylene level and restoring plant development [30]. The inoculation of *Mucuna pruriens L.* (velvet bean), a well-adapted plant to arid and semi-arid regions with *Bacillus spp.* and *Enterobacter spp.* had shown significantly reduced ACC accumulation by 41% and 21% in leaves respectively, and in the roots by 46% and 15%, respectively. Thereby, the ethylene synthesis was reduced by 45% with *Bacillus spp.* (G9) and 65% with *Enterobacter spp.* (HS9) [29].

In addition to ACC deaminase activity, certain bacterial strains were found to be efficient in plant growth. For instance, *Achromobacter piechaudii* ARV8, a bacterial strain isolated from the arid region, experiencing frequent episodes of water stress was found to be better plant growth promoter as compared to *Pseudomonas putida* GR12–2, a bacterial strain inhabiting area with surplus water supply [31]. The bioinoculation of tomato (*Lycopersicum esculentum* Mill) and pepper (*Capsicum annuum L.*) seedlings with *A. piechaudii* strain improved relative water content (RWC) which subsequently succored plants in maintaining their fresh weight in water stress conditions. The maintenance of the fresh weight in bioinoculated seedling can be justified by the declined production of ethylene to 6.1 nlh^{-1} as compared to stressed seedling with the rise in ethylene level up to 23 nlh^{-1}. Overall, the plant biomass was enhanced four times as compared to uninoculated controls. Similarly bacterial strains mainly *Variovorax paradoxus, Pseudomonas spp. Achromobacter spp.* and *Ochrobactrum anthropi* islolated from rain-fed agricultural soil were reported to possess growth-promoting traits such as N_2 fixation, siderophore, and phosphate solubilization along with sufficient production of ACC deaminase. These bacterial strains both as single inoculation and consortium imparted significant growth, foliar concentration, and antioxidant activity to the wheat (*Triticum aestivum L.*) production [32].

The exopolysaccharide (EPS) released by PGPR tends to form rhizosheath or biofilms around the surface of the roots containing sufficient moisture content, thereby, protect them from prolonged desiccation. Moreover, plants inoculated with EPS are capable of accumulating sugars, amino acids, and proline. The survival of PGPR in dried sandy soils with low moisture content is conceivable due to EPS, which may serve as a biological tool to combat climatic water stress and ultimately providing a path towards global food security [33]. The coalition of ACC deaminase and EPS has been shown to have an affirmative effect on carotenoid pigments in certain plants. The inoculation of *Capsicum annuum* plants with *Bulkhorderia cepacia* (ACC deaminase activity 12.8 ± 0.44, mM αKB mg^{-1} min^{-1} and

EPS 4.89 ± 0.06 mg/mg protein) resulted in increased chlorophyll a and b content (chlorophyll a 5.7 gm L^{-1} and chlorophyll b 3.4 gm L^{-1}, respectively) as compared to control (chlorophyll a 3.2 gm L^{-1}, chlorophyll b 1.9 gm L^{-1}). Moreover, the plant biomass has also increased (fresh weight of 9 g and dry weight of 3.6 g) as compared to control (fresh weight 6 g, and dry weight 1.6 g) [34].

The co-application of PGPR producing ACC deminase with biochar has been utilized as an effective mechanism against prevalent water stress conditions. Timber-waste biochar was co-applied with *Bacillus amyloliquefaciens* strain to increase the productivity of wheat under simulated water stress conditions. Along with high productivity (59% of 100-grain weight), the grain composition (58% N, 18% P, and 23% K), carotenoid (114% chlorophyll a and 123% chlorophyll b), photosynthetic rate (118%), and transpiration rate (73%) in wheat were also reinforced [10]. To understand the PGPR potential for genetic improvement in plants, the physiology and biochemical characteristics of PGPR isolated from chickpea rhizosphere were implored. The seeds of chickpea were grown in the cultures of *Bacillus subtilis, Bacillus thuringiensis,* and *Bacillus megaterium*. The 25-day old chickpea seedlings were sprayed with plant growth regulators/plant growth retardants (PGRs) i.e., salicylic acid (SA) and putrescine (Put) at the rate of 150 mgL^{-1}. The consortium of PGPR and PGRs significantly improved carotenoid, sugar, and protein contents as compared to irrigated and non-irrigated conditions [14].

The inhibitory effect of water stress in two maize species was mitigated by inoculation with two endophytic bacterial species i.e., *Burkholderia phytofirmans* PsJN and *Enterobacter* sp. FD17 [35]. Maize seedlings were exposed to water stress after the 45th day of vegetative growth. The inoculated bacterial seedlings exhibited 30% relative water content in leaves along with the significant increase in root and shoot biomass, leaf area, chlorophyll content, and photochemical efficiency of PSII (Photosystem II). However, the *Paraburkholderia phytofirmans* PsJN was found to be more effective as compared to *Enterobacter* sp. FD17. The remedial potential of PGPR against water stress conditions can be used as a tool for sustainable agriculture practices (**Table 3**).

Treated Plant/PGPR	Stress Alleviation Effect
Maize/*Azospirillum lipoferum, Bacillus* Spp.	Increase accumulation of soluble sugar, free amino acids, proline and decrease electrolyte leakage; reduced activity of antioxidant enzyme [36, 37]
Soybean/*Pseudomonas putida* H-2–3	Lower the level of abscisic acid and salicylic acid and a higher level of jasmonic acid content; declined superoxide dismutase, flavonoids and radical scavenging activity [38]
Wheat/*Bacillus amyloliquefaciens*, *Azospirillum brasilense*, *Rhizobium leguminosarum*, *Mesorhizobium cicero*, *Rhizobium phaseoli*	Improved homeostasis; catalase, exo-polysaccharides and IAA production [39, 40]
Chickpea/*Pseudomonas putida*	Osmolyte accumulation, ROS scavenging ability and stress-responsive gene expressions [41]
Lettuce/*Azospirillum* sp.	Promote chlorophyll, ascorbic acid content and antioxidant capacity [42]
Rice/*Trichoderma harzianum*	Promote root growth independent of water status and delay drought response [43]

Table 3.
Alleviation of water stress via bioinoculants.

4. Amelioration of salinity via bioinoculants

The soil salinity is naturally occurring phenomena in arid and semi-arid regions. Besides, extensive use of chemical fertilizers and irregular irrigation practices, the continuously changing precipitation pattern is contributing 1–2% salinity to arable lands every year [44]. The plants growing under salinity conditions tend to have high levels of ethylene, stress hormone released in response to both biotic and abiotic stress. As mentioned earlier, PGPR possessing ACC deaminase enzyme can reduce ethylene level by disintegrating ACC (precursor to ethylene) into α-ketobutyrate and ammonia, thereby suppressing the formation of ethylene and ultimately sustaining the plant growth. In doing so, the strains of *Aneurinibacillus aneurinilyticus* and *Paenibacillus sp.* isolated from Garlic (*Allium sativum*) rhizosphere were used for seed bacterization of French bean (*Phaseolus vulgaris*) and evaluated for their ACC deaminase activity [11]. These strains produced more than ~1500 nmol of α-ketobutyrate mg $protein^{-1}$ h^{-1} and more than ~30 μg/ml Indole Acetic Acid (IAA) under saline and nonsalinity stress conditions. Moreover, the seed bacterization in the form of consortia led to a ~ 60% decline in stress stimulated ethylene levels.

Under saline conditions, water uptake is restricted by ionic stress which develops between Na^+ and K^+ due to Na^+ accumulation in aerial parts of plants. Singh and Jha [12] successfully employed inoculation of *Stenostrophomonas maltophilia* SBP-9 on wheat (*Triticum aestivum)* and increased K^+ uptake by 20–28%. The plant-endophytic relationship has been explored by co-inoculation of *Cicer arietinum* (Chickpea) with *Bacillus subtilis* NUU4 and *Mesorhizobium ciceri* IC53 to alleviate salt stress. These bacterial strains increased proline accumulation which in turn sequestered hydrogen peroxide to strengthen the tolerance level against saline conditions. Besides, coinoculation of endophytes also reduced root rot infection caused by *F. Solani* [45]. The PGPR and their relationship with plants proved to be effective in dealing salt stress simultaneous to other biotic and abiotic stress in given global climatic conditions (**Table 4**).

5. Amelioration of oxidative stress via osmoregulation

Global climate change is amplifying the intensity of environmental factors, causing a profound disturbance in the plant's biological processes mainly via intractable generation of ROS. A condition called 'oxidative stress' develops when the available concentration of antioxidants within plant cells become insufficient to neutralize an unrestrained number of reactive species (OH^-, O^- and $H_2O_2^-$) which leads to disruption of cellular components (lipids, protein, nucleic acids, and metabolites). As a defense mechanism, plants and related rhizobacteria synthesize and release osmolytes when required. Osmolytes are low molecular organic compounds (proline: amino acids, sugars, polyols, methylamines, methylsulfonium compunds, and urea) mainly aiming at neutralization of the osmotic pressure of cells under stress conditions. These plants and PGPR synthesized osmolytes synergistically act to capture ROS. The prominent PGPR strains such as *Azospirillum brasilense* SP-7 and *Herbaspirillum seropedicae* Z-152 managed to retain relatively higher water content in vegetal tissues of inoculated/non-irrigated maize plants as compared to non-inoculated/irrigated maize plants. The plants inoculated with *H. seropedicae* had a higher proline accumulation rate (fourfold) as compared to those inoculated with *A. brasilense* (two-fold), thus indicating better osmoregulation under drought conditions. However, the amount of proline in inoculated plants

Treated Plant/PGPR	Stress Alleviation Effect
Groundnut/*Brachybacterium saurashtrense, Brevibacterium casei, Haererohalobacter*	Higher K^+/Na^+ ratio and higher Ca^{2+}, phosphorus, and nitrogen content, higher Shoot and root concentration of auxin [43, 46]
Mung bean/*Rhizobium, Pseudomonas*	ACC-deaminase activity [47]
Wheat/*Azospirillum* Sp., *Pseudomonas* Sp. *Serratia* Sp.	Osmotic adjustments via proline and soluble sugars [48]
	ACC deaminase activity, reduced ethylene level [49]
Maize/*Pseudomonas, Enterobacter*	More N, P, K uptake and high K^+–Na^+ ratios [50]
Rice/*Pseudomonas pseudoalcaligenes, Bacillus pumilus, Bacillus amyloliquefaciens*	Reduced reactive oxygen species, lipid peroxidation and superoxide dismutase activity, modulating differential transcription in at least 14 genes [51, 52]
Lettuce/*Azospirillum*	Promoted ascorbic acid content antioxidant capacity [42]

Table 4.
Alleviation of salinity stress via bioinoculants.

was relatively lower than control plants in both water-stressed and well-watered conditions. This decreased concentration of proline in inoculated plants may be attributed to a balanced amount of osmolytes to the available ROS leading to better osmoregulation in water-stressed conditions. Also, inoculated plants have lower ethylene content than that of control plants. However, the gene expression of ZmVP14 responsible for the biosynthesis of abscisic acid was suppressed in inoculated plants [53], allowing stomata to remain open for better CO_2 assimilation even in drought conditions also supported by other researchers [35].

Inoculation of chickpea varieties with *Pseudomonas putida,* also indicated reduced proline concentration with a reduced level of 114% in BG-362 and 214% in BG-1003 on the seventh day of water stress [41, 53]. However, in *Capsicum annum* inoculated with *Burkholderia cepacia*, the accumulation of proline content was higher (0.143 mmoles gm^{-1}) as compared to *the* control plant (0.065 mmoles gm^{-1}). Co-inoculation of *Azotobacter chroococcum* and *Azospirillum brasilense* significantly improved accumulation of proline and other osmolytes along with physic-chemical characteristics of *Mentha pulegium L.* [54]. Other than proline, some soluble sugars play a vital role in maintaining photosynthesis in leaves under water stress, therefore, also termed 'osmoprotectant'. However, the reduced level of soluble sugar can become deleterious to the cellular membranes of plants. The co-inoculation of PGPR (*Bacillus subtilis*, *Bacillus thuringiensis*, and *Bacillus megaterium*) with PGRs (SA & Put) in chickpea seedlings prompt the accumulation of soluble sugar along with chlorophyll and protein content (majorly proline), thus, regularizing photosynthesis under water stress [14]. The accumulated soluble sugar also acts as a signaling molecule, where it triggers the activation and gene expression of certain genes relevant to photosynthesis in plants. Hence, osmolytes being 'abiotic stress busters' are pivotal in maintaining homeostatic equilibrium at the molecular and cellular level to enhance stress tolerance in plants [55].

6. Amelioration of cold stress via psychrotrophic microbes

Due to global warming, extreme environmental conditions are occurring more frequently including cold stress. In northern parts of the world, the cold weather is unexpectedly intensifying specifically due to the Polar vortex bringing cold Arctic air towards southern parts of the planet, bringing more intense winters in some parts of the world. The biomes of extreme environments serve as excellent habitats for archaea, bacteria, and fungi. Agro-ecosystems at high altitudes are vulnerable to low productivity due to low surface temperature resulting in low soil temperature and in turn low soil fertility [56]. High altitude soils comprised diverse microbes that are capable of surviving cold environments mainly due to synthesis and assimilation of cryoprotective compounds such as metabolites (proline), anthocyanin (secondary metabolite), and carbohydrates (trehalose). The induction of such stress-resistance against abiotic stress is also coined with a term called 'Induced Systematic Resistance' (ISR). An in vitro inoculation of *Vitis vinifera L.* (grapevine) explants with *Burkholderia phytofirmans* strain PsJN successfully developed chilling resistance mainly via accumulation of increased level of sugars, proline, and phenolics as compared to non-inoculated plantlets [57].

Several biotechnological and microbiological attempts have been ongoing for the last two decades to utilize the cold-tolerant (psychrotrophic) or cold-loving (psychrophilic) microbes to improve the agricultural practices in chilling mountainous environments. Psychrophilic microbes inhabiting cold environments majorly belong to phyla *Verrucomicrobia*, *Thaumarchaeota*, *Spirochaetes*, *Proteobacteria*, *Planctomycetes*, *Nitrospirae*, *Mucoromycota*, *Gemmatimonadetes*, *Firmicutes*, *Euryar chaeota*, *Cyanobacteria*, *Chloroflexi*, *Chlamydiae*, *Basidiomycota*, *Bacteroidetes*, *Ascomycota*, and *Actinobacteria.* Like thermotolerant PGPR, the psychrotrophic PGPR explicit various attributes like ACC deaminase activity, solubilization of micronutrients (phosphorus, potassium, and zinc), biological N_2 fixation, and production of ammonia, hydrogen cyanide, indole-3-acetic acid, and Fe-chelating compounds (**Figure 2**). The alleviation of cold stress on the production of cereal crops such as wheat was investigated by inoculation with *Bacillus spp.* CJCL2 and RJGP41 strains isolated from Qinghai-Tibetan plateau [58]. The psychrophilic activity of this strain was also compared to temperate *B. velezensis* FZB42 at a temperature range of 14°C, 10°C, and 4°C after 4 h post-inoculation. The cold-tolerant strain CJCL2 produced the finest biofilm structure as compared to RJGP41 which produced slight biofilms at 96 h post-inoculation at 4°C, whereas FZB42 failed to develop any biofilm at 4°C. The basic mechanism against cold stress was based on the regulation of abscisic acid, lipid peroxidation and proline accumulation, and a lower level of ROS. Moreover, genetic expression for ACC deaminse, glucose dehydrogenase encoding phosphate solubilization, and phytohormones was quite effective and led to the improvement in plant growth under cold stress. The cold-tolerant *Bacillus* spp. maintained osmotic balance in a plant cell by synthesizing glycine betaine, a major osmoprotectant and produced due to OpuAC gene expression. All in all linear four-fivefold increase in the gene expression of cold shock proteins (CspB, CspC, and CspD) observed in CJCL2 and RJGP41 inoculated wheat

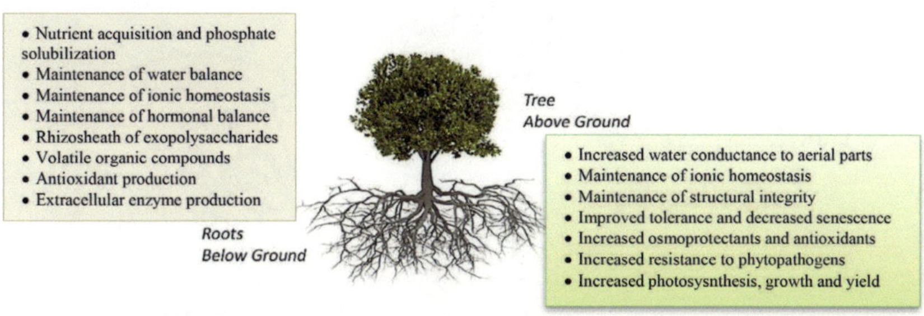

Figure 2.
Key mechanisms targeted by plant growth promoting rhizobacteria.

plants may trigger the modifications in RNA. Also, these proteins may stabilize the secondary structure of nucleotides leading to improved cellular components.

6.1 Plant growth promoting rhizobacteria and root-nodule symbiosis

The PGPR possessing root-nodule symbiosis has also been reported to exhibit psycrotolerant traits at 5°C [59]. These PGPRs were isolated from the root nodule of *Pisum sativum L.* (Pea), cultivated in the Northern Indian plains. Out of nineteen tested strains, four exhibited cold tolerance mainly by producing phytohormone IAA in the range of 62.7–198.1 μg/ml. The inoculation of edible crops growing under adverse climatic conditions with PGPR producing phytohormones showed promising results [60]. The *Serratia nematodiphila* PEJ1011 strain was inoculated to *Capsicum annuum L.* (pepper) at a low temperature of 5°C. The inoculated plants exhibited higher endogenous GA_4 content grown both at normal and low temperatures. Moreover, PGPR increased abscisic acid (phytohormone) level and reduced jasmonic acid and salicylic acid (phytohormones) contents to regulate plant adaptation and growth at low temperature (**Figure 3**).

6.2 Plant growth promoting rhizobacteria and chlorophyll pigments

Chloroplasts containing chlorophyll pigments for photosynthesis are the main organelles affected by the cold. Low temperature also led to stomatal closure in many cold-tolerant plants like *Arabidopsin thliana*. Although the stomatal closure prevents leaf dehydration, CO_2 uptake is inhibited which eventually reduces photosynthesis. The inoculation of *A. thliana* with endophytic *Burkholderia phytofirmans* strain PsJN (Bp PsJN) prevented disruption of the plasma membrane at 0°C, −1°C, and − 3°C and induced cell wall strengthening in leaf cells [62]. Moreover, after night stress, the bacteria led to better photosynthetic pigment content. Similarly, *Vitis vinifera L.* (grapevine) cultivated in temperate and cool climates contains CBF4 gene which is a homolog to *A. thliana* CBF1 accumulates carbohydrates and proline on exposure to low temperature [63]. The *V. vinifera* plantlets were inoculated with endophytic *B. phytofirmans* strain PsJN at 4°C. As expected, the bacterial strain had induced cold resistance principally by enhancing gene expression of CBF (specifically CBF4) genes, along with an accumulation of anti-freeze proteins (PR proteins) and metabolite level. The bacterial strain was found to be highly effective in inhibiting hydrogen peroxide (H_2O_2). However, the inoculated plantlets indicated high H_2O_2 accumulation

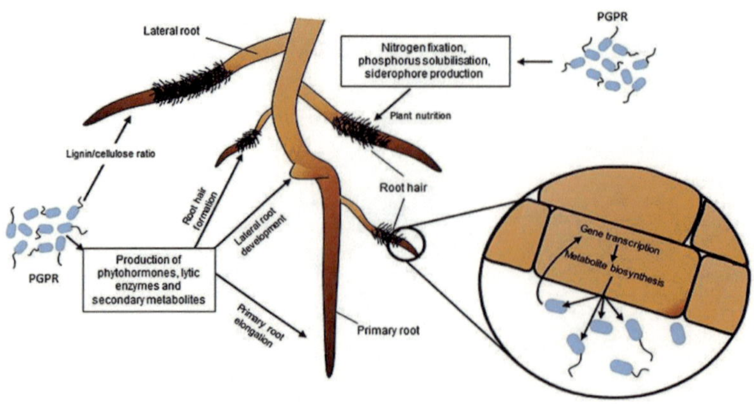

Figure 3.
Impact of plant growth promoting rhizobacteria on nutrient acquisition [61].

DOI: http://dx.doi.org/10.5772/intechopen.97340

in the first three days of cold stress followed by H_2O_2 elimination after a week, further evidenced by a reduction in the level of metabolites after 1 week indicating inhibition of ROS via scavenging process carried out by endophytic PsJN strain.

6.3 Plant growth promoting rhizobacteria and sugar deposition

Certain bacteria, fungi, vascular plants, and invertebrate animals synthesize trehalose or mycose for energy consumption and to tolerate cold and water stress. Trehalose is a disaccharide sugar consisting of two glucose molecules mainly synthesized via phosphorylated intermediate trehalose 6-phosphate (T6P). Being capable of inducing cryoprotective compounds, *B. phytofirmans* strain PsJN was inoculated in grapevine to observe its impact on the accumulation of T6P and trehalose at 4°C. After 120 h of cold exposure, the level of T6P was increased in roots (>0.4 nmol. g^{-1} FW) stems (>0.7 nmol. g^{-1} FW), and leaves (>1.7 nmol. g^{-1} FW) of inoculated plantlets. T6P acts as a signal metabolite in plants responding in particular to changes in sucrose level, thereby a strong correlation had been observed between T6P and sucrose content in all plant organs in both chilled and non-chilled plants. The authors were uncertain about the role of trehalose against chilling because the overall accumulation of trehalose was only found in leaves in low amounts i.e., <15 nmol. g^{-1}. Localized trehalose accumulation might have protected cell membranes against cold-induced dehydration [64, 65]. However, the organ-specific accumulation may indicate trehalose acted as an osmolyte [66]. Previous studies related to drought stress indicated overexpression of genes for the synthesis of trehalose which under water-stress inhibited dehydration [67]. Although psychrotolerant PGPRs are proved to be effective in inducing tolerance against cold stress, the quest of mechanism underlying is still being investigated by the researchers.

7. Conclusions

Despite awareness and efforts being planned or carried out, global temperature, a dynamic entity itself is hard to reverse to its previous levels in 1990. Instead, plant species have to acclimatize themselves to prevailing temperature variations. A natural symbiotic relationship between PGPR and plant growth is known for decades, has been widely considered and accepted as the most convenient and reliable approach to deal with climate-based environmental stresses. This chapter, therefore, quoted and discussed several research studies utilizing PGPR to alter physiological and biochemical processes occurring at inter and intracellular levels to boost the plant's defense mechanism. However, the complexity of the mechanism underlying is still in focus to fully harness the potential of PGPR for sustainable agricultural practices and for sustaining natural plant diversity.

Conflict of interest

The authors declare no conflict of interest.

Abbreviations

PGPR	Plant Growth Promoting Rhizobacteria
1-Aminocyclopropane-1-carboxylate Deaminase	ACC deaminase

ABA	Abscisic Acid
SA	Salicylic Acid
IAA	Indole-3-Acetic Acid
VOCs	Volatile Organic Compounds
ABA	Abscisic Acid (ABA)
HSP	Heat Shock Protein
SOD	Superoxide Dismutase
POD	Peroxidase
CAT	Catalase
ROS	Reactive Oxygen Species
RWC	Relative Water Content
EPS	Exopolysaccharide
PGRs	Plant Growth Retardants (PGRs)
Put	Putrescine
ISR	Induced Systematic Resistance

Author details

Rafia Younas[1], Shiza Gul[1], Rehan Ahmad[1], Ali Raza Khan[1], Mumtaz Khan[2], Tauseef Anwar[3] and Huma Qureshi[4*]

1 Department of Environmental Science, Institute of Soil and Environmental Sciences, Gomal University, Dera Ismail Khan, Pakistan

2 Directorate of Regional Services, Allama Iqbal Open University, Islamabad, Pakistan

3 Department of Botany, The Islamia University of Bahawalpur, Bahawalnagar Campus (New Campus), Bahawalnagar, Pakistan

4 Institute of Biological Sciences, Gomal University, Dera Ismail Khan, Pakistan

*Address all correspondence to: drhuma@gu.edu.pk

References

[1] Estrada F, Botzen WJW, Calderon-Bustamante O. The assessment of impacts and risks of climate change on agriculture (AIRCCA) model: a tool for the rapid global risk assessment for crop yields at a spatially explicit scale. Spatial Economic Analysis. 2020; 15:262-279. DOI: 10.1080/17421772.2020.1754448

[2] Dudal R. Inventory of the major soils of the world with special reference to mineral stress, In: Plant Adaptation to Mineral Stress in Problem Soils, Wright MJ, (Ed.). Comell University, Agricultural Experiment Station, Ithaca, NY. 1976; p. 3-13.

[3] Vinocur B, Altman B. Recent advances in engineering plant tolerance to abiotic stress: achievements and limitations. Current Opinion in Biotechnology. 2005; 16:1-10. DOI: 10.1016/j.copbio.2005.02.001

[4] Bashir T, Naz S, Bano A. Plant Growth Promoting Rhizobacteria in combination with Plant Growth Regulators attenuates the effect drought stress. Pakistan Journal of Botany. 2020; 52:783-792. DOI: 10.30848/PJB2020-3(17)

[5] Khan N, Bano A, Zandi P. Effects of exogenously applied plant growth regulators in combination with PGPR on the physiology and root growth of chickpea (*Cicer arietinum*) and their role in drought tolerance. Journal of Plant Interactions. 2018; 13:239-247. DOI: 10.1080/17429145.2018.1471527

[6] Kumar A, Patel JS, Meena VS, Srivastava R. Recent advances of PGPR based approaches for stress tolerance in plants for sustainable agriculture. Biocatalysis and Agricultural Biotechnology. 2019; 20:101271. DOI: 10.1016/j.bcab.2019.101271

[7] Mega R, Abe F, Kim JS, Tsuboi T, Tanaka K, Kobayashi H, Sakata Y, Hanada K, Tsujimoto H, Kikuchi J, Cutler SR, Okamoto M. Tuning water-use efficiency and drought tolerance in wheat using abscisic acid receptors. Nature Plants. 2019; 5:153-159. DOI: 10.1038/s41477-019-0361-8

[8] Vurukonda SSKP, Vardharajula S, Shrivastava M, Ali SKZ. Enhancement of drought stress tolerance in crops by plant growth promoting rhizobacteria. Microbiological Research. 2015; 184:13-24. DOI: 10.1016/j.micres.2015.12.003

[9] Mukhtar T, Shafiq ur Rehman, Smith D, Sultan T, Seleiman MF, Alsadon AA, Amna, Ali S, Chaudhary HJ, Solieman THI, Ibrahim AA, Saad MAO. Mitigation of heat stress in *Solanum lycopersicum* L. by ACC-deaminase and exopolysaccharide producing *Bacillus cereus*: Effects on biochemical profiling. Sustainability. 2020; 12:2159. DOI: 10.3390/su12062159

[10] Danish S, Zafar-ul-Hye M. Co-application of ACC-deaminase producing PGPR and timber-waste biochar improves pigments formation, growth and yield of wheat under drought stress. Scientific Reports. 2019; 9:5999. DOI: 10.1038/s41598-019-42374-9

[11] Gupta S, Pandey S. ACC Deaminase producing bacteria with multifarious plant growth promoting traits alleviates salinity stress in French bean (*Phaseolus vulgaris*) plants. Frontiers in Microbiology. 2019; 10:1506. DOI: 10.3389/fmicb.2019.01506

[12] Singh RP, Jha PN. The PGPR *Stenotrophomonas maltophilia* SBP-9 augments resistance against biotic and abiotic stress in wheat plants. Frontiers in Microbiology. 2017; 8:1945. DOI: 10.3389/fmicb.2017.01945

[13] Rybakova D, Rack-Wetzlinger U, Cernava T, Schaefer A, Schmuck M,

Berg G. Aerial warfare: A volatile dialogue between the plant pathogen *Verticillium longisporum* and its antagonist *Paenibacillus polymyxa*. Frontiers in Plant Science. 2017; 8:1294. DOI:10.3389/fpls.2017.01294

[14] Khan N, Bano A, Rahman MA, Guo J, Kang Z, Babar MA. Comparative physiological and metabolic analysis reveals a complex mechanism involved in drought tolerance in chickpea (*Cicer arietinum* L.) induced by PGPR and PGRs. Scientific Reports. 2019; 9: 2097. DOI: 10.1038/s41598-019-38702-8

[15] Schmidt R, Köberl M, Mostafa A, Ramadan EM, Monschein M, Jensen KB, Bauer R, Berg G. Effects of bacterial inoculants on the indigenous microbiome and secondary metabolites of chamomile plants. Frontiers in Microbiology. 2014; 5:64. Doi: 10.3389/fmicb.2014.00064

[16] Drigo B, Kowalchuk GA, Van Veen JA (2008) Climate change goes underground: effects of elevated atmospheric CO_2 on microbial community structure and activities in the rhizosphere. Biology and Fertility of Soil. 44:667-679. DOI: 10.1007/s00374-008-0277-3

[17] Rennenberg H, Loreto F, Polle A, Brilli F, Fares S, Beniwal RS, Gessler A. Physiological responses of forest trees to heat and drought. Plant Biology. 2006; 8:556-571. DOI: 10.1055/s-2006-924084

[18] Baniwal SK, Bharti K, Chan KY, Fauth M, Ganguli A, Kotak S, von Koskull-Doring P. Heat stress response in plants: a complex game with chaperones and more than twenty heat stress transcription factors. Journal of Biosciences. 2004; 29:471-487. DOI: 10.1007/BF02712120

[19] Vanaja M, Ramakrishna YS, Rao GGSN, Rao KV, Subbarao VM. Climate change and dryland agriculture. Dryland ecosystems: an Indian perspective. Central Arid Zone Research Institute (CAZRI) and Arid Forest Research Institute (AFRI), Jodhpur. 2007; p. 23-24.

[20] Bensalim S, Nowak J, Asiedu SK. A plant growth promoting rhizobacterium and temperature effects on performance of 18 clones of potato. American Journal of Potato Research. 1998; 75:145-152. DOI: 10.1007/BF02895849

[21] Duc NH, Csintalan Z, Posta K. Arbuscular mycorrhizal fungi mitigate negative effects of combined drought and heat stress on tomato plants. Plant Physiology and Biochemistry. 2018; 132:297-307. DOI: 10.1016/j.plaphy.2018.09.011

[22] El-Daim IA, Bejai S, Meijer J. Improved heat stress tolerance of wheat seedlings by bacterial seed treatment. Plant Soil. 2014; 379:337-350. DOI: 10.1007/s11104-014-2063-3

[23] Mishra PK, Bisht SC, Bisht JK, Bhatt JC. Cold-tolerant PGPRs as bioinoculants for stress management. In Bacteria in Agrobiology: Stress Management; Maheshwari DK. Springer: Berlin/Heidelberg, Germany, 2012; p. 95-118.

[24] Lindow SE. The role of bacterial ice nucleation in frost injury to plants. Annual Review of Phytopathology. 1983; 21:363-384. DOI: 10.1146/annurev.py.21.090183.002051

[25] Theocharis A, Bordiec S, Fernandez O, Paquis S, Dhondt-Cordelier S, Baillieul F, Clément C, Barka EA. *Burkholderia phytofirmans* PsJN primes *Vitis vinifera* L. and confers a better tolerance to low non-freezing temperatures. APS Publications. 2012; 2:241-249. DOI: https://doi.org/10.1094/MPMI-05-11-0124

[26] Park YG, Mun BG, Kang SM, Hussain A, Shahzad R, Seo CW,

Kim AY, Lee SU, Oh KY, Lee DY, Lee IJ, Yun BW. *Bacillus aryabhattai* SRB02 tolerates oxidative and nitrosative stress and promotes the growth of soybean by modulating the production of phytohormones. PLoS ONE, 2017; 12: e0173203. DOI: 10.1371/journal.pone.0173203

[27] Verma P, Yadav AN, Khannam KS, Mishra S, Kumar S, Saxena AK Suman A Appraisal of diversity and functional attributes of thermotolerant wheat associated bacteria from the peninsular zone of India. Saudi Journal of Biological Science. 2019; 26:1882-1895. DOI: 10.1016/j.sjbs.2016.01.042

[28] Ali SKZ, Sandhya V, Grover K, Rao LV, Venkateswarlu B. *Pseudomonas* sp. strain AKM-P6 enhances tolerance of sorghum seedlings to elevated temperatures. Biology and Fertility of Soils. 2009; 46:45-55. DOI: 10.1007/s00374-009-0404-9

[29] Saleem AR, Brunetti C, Khalid A, Rocca GD, Raio A, Emiliani G, De Carlo A, Mahmood T, Centritto M. Drought response of *Mucunapruriens* (L.) DC. inoculated with ACC deaminase and IAA producing rhizobacteria. PLoS ONE. 2018; 13:e0191218. DOI: 10.1371/journal.pone.0191218

[30] Babalola OO. Beneficial bacteria of agricultural importance. Biotechnology Letters. 2010; 32:1559-1570. DOI: 10.1007/s10529-010-0347-0

[31] Mayak S, Tirosh T, Glick BR. Plant growth-promoting bacteria that confer resistance to water stress in tomatoes and peppers. Plant Science. 2004; 166:525-530. DOI: 10.1016/j.plantsci.2003.10.025

[32] Chandra D, Srivastava R, Gupta VV, Franco CM, Sharma AK. Evaluation of ACC deaminase producing rhizobacteria to alleviate water stress impacts in wheat (*Triticum aestivum* L.) plants. Canadian Journal of Microbiology. 2019; 65:387-403. DOI: 10.1139/cjm-2018-0636

[33] Naseem H, Ahsan M, Shahid MA, Khan N Exopolysaccharides producing rhizobacteria and their role in plant growth and drought tolerance. Journal of Basic Microbiology (2018); 58:1009-1022. DOI: 10.1002/jobm.201800309

[34] Maxton A, Singh P, Masih SA. ACC deaminase-producing bacteria mediated drought and salt tolerance in *Capsicum annuum*. Journal of Plant Nutrition. 2018; 41:574-583. DOI: 10.1080/01904167.2017.1392574

[35] Naveed M, Mitter B, Reichenauer TG, Wiezorek K, Sessitsch A Increased drought stress resilience of maize through endophytic colonization by *Burkholderia phytofirmans* PsJN and *Enterobacter sp*. FD17. Journal of Environmental and Experimental Botany. 2014; 97:30-39. DOI: 10.1016/j.envexpbot.2013.09.014

[36] Bano Q, Ilyas N, Bano A, Zafar N, Akram A, Hassan F. Effect of Azospirillum inoculation on maize (*Zea mays* L.) under drought stress. Pakistan Journal of Botany. 2013; 45: 13-20. DOI: 10.3390/microorganisms5030041

[37] Vardharajula S, Ali SZ, Grover M, Reddy G, Bandi V. Drought-tolerant plant growth promoting *Bacillus* spp.: effect on growth, osmolytes, and antioxidant status of maize under drought stress. Journal of Plant Interactions. 2011; 6:1-14. DOI: 10.1080/17429145.2010.535178

[38] Kang S, Radhakrishnan R, Khan AL, Kim M, Park J, Kim B, Shin D, Lee I. Gibberellin secreting rhizobacterium, *Pseudomonas putida* H-2-3 modulates the hormonal and stress physiology of soybean to improve the plant growth under saline and drought conditions. Plant Physiology and Biochemistry. 2014; 84:115-124. DOI: 10.1016/j.plaphy.2014.09.001

[39] Kasim WA, Osman ME, Omar MN, El-Daim IA, Bejai S, Meijer J. Control of drought stress in wheat using plant-growth-promoting bacteria. Journal of Plant Growth Regulation. 2013; 32:122-130. DOI: 10.1007/s00344-012-9283-7

[40] Hussain MB, Zahir ZA, Asghar HN, Asghar M. Can catalase and exopolysaccharides producing rhizobia ameliorate drought stress in wheat? International Journal of Agriculture and Biology. 2014; 16: 3-13.

[41] Tiwari S, Lata C, Chauhan PS, Nautiyal CS. *Pseudomonas putida* attunes morphophysiological, biochemical and molecular responses in *Cicer arietinum* L. during drought stress and recovery. Plant Physiology and Biochemistry. 2016; 99:108-117. DOI: 10.1016/j.plaphy.2015.11.001

[42] Gabriela F, Casanovas EM, Quillehauquy V, Yommi AK, Goñi MG, Roura SI, Barassi CA. *Azospirillum* inoculation effects on growth, product quality and storage life of lettuce plants grown under salt stress. Scientia Horticulturae. 2015; 195:154-162. DOI: 10.1016/j.scienta.2015.09.015

[43] Shukla PS, Agarwal PK, Jha B. Improved salinity tolerance of *Arachis hypogaea* (L.) by the interaction of halotolerant plant-growth-promoting rhizobacteria. Journal of Plant Growth Regulation. 2012; 31, 195-206. DOI: 10.1007/s00344-011-9231-y

[44] FAO. Salt-Affected soils from sea water intrusion: strategies for rehabilitation and management: Report of the regional workshop. Bangkok. 2005.

[45] Egamberdieva D, Wirth SJ, Shurigin VV, Hashem A, Abd-Allah EF. Endophytic bacteria improve plant growth, symbiotic performance of chickpea (*Cicer arietinum* L.) and induce suppression of root rot caused by *Fusarium solani* under salt stress. Frontiers in Microbiology. 2017; 8:1887. DOI: 10.3389/fmicb.2017.01887

[46] Shukla N, Awasthi RP, Rawat L, Kumar J. Biochemical and physiological responses of rice (*Oryza sativa* L.) as influenced by *Trichoderma harzianum* under drought stress. Plant Physiology and Biochemistry. 2012; 54, 78-88. DOI: 10.1016/j.plaphy.2012.02.001

[47] Ahmad M, Zahir ZA, Asghar HN, Asghar M. Inducing salt tolerance in mung bean through coinoculation with rhizobia and plant-growth-promoting rhizobacteria containing 1-aminocyclopropane-1-carboxylate deaminase. Canadian Journal of Microbiology. 2011; 57:578-589. DOI: 10.1139/w11-044

[48] Chang P, Gerhardt KE, Huang XD, Yu XM, Glick BR, Gerwing PD, Greenberg BM. Plant growth-promoting bacteria facilitate the growth of barley and oats in salt-impacted soil: implications for phytoremediation of saline soils. International Journal of Phytoremediation. 2014; 16:1133-1147. DOI: 10.1080/15226514.2013.821447

[49] Zahir ZA, Ghani U, Naveed M et al. Comparative effectiveness of *Pseudomonas* and *Serratia* sp. containing ACC-deaminase for improving growth and yield of wheat (*Triticum aestivum* L.) under salt-stressed conditions. Archives of Microbiology. 2009;191: 415-424. DOI: 10.1007/s00203-009-0466-y.

[50] Nadeem SM, Zahir ZA, Naveed M, Arshad M. Rhizobacteria containing ACC-deaminase confer salt tolerance in maize grown on salt-affected fields. Canadian Journal of Microbiology. 2009; 55:1302-1309. DOI: 10.1139/w09-092

[51] Jha Y, Subramanian RB. PGPR regulate caspase-like activity, programmed cell death, and antioxidant enzyme activity in paddy under salinity.

Physiology and Molecular Biology of Plants. 2014; 20:201-207. DOI: 10.1007/s12298-014-0224-8

[52] Nautiyal CS, Srivastava S, Chauhan PS, Seem K, Mishra A, Sopory SK. Plant growth-promoting bacteria *Bacillus amyloliquefaciens* NBRISN13 modulates gene expression profile of leaf and rhizosphere community in rice during salt stress. Plant Physiology and Biochemistry. 2013; 66:1-9. DOI: 10.1016/j.plaphy.2013.01.020

[53] Curá JA, Franz DR, Filosofía JE, Balestrasse KB, Burgueño LE. Inoculation with *Azospirillum* sp. and *Herbaspirillum* sp. bacteria increases the tolerance of maize to drought stress. Microorganisms. 2017; 5:41. DOI: 10.3390/microorganisms5030041

[54] Asghari B, Khademain R, Sedaghati B. Plant growth promoting rhizobacteria (PGPR) confer drought resistance and stimulate biosynthesis of secondary metabolites in pennyroyal (*Mentha pulegium L.*) under water shortage condition. Scientia Horticulturae. 2020; 263:109132. DOI: 10.1016/j.scienta.2019.109132

[55] Iqbal N, Nazar R, Khan NA. Osmolytes and plants acclimation to changing environment: Emerging omics technologies. Springer, India. 2016.

[56] Bhandari G. Microbes adapted to cold and their use as biofertilizers for mountainous regions. Recent Advancements in Microbial Diversity. 2020; 197-215.

[57] Barka EA, Nowak J, Clement C. Enhancement of chilling resistance of inoculated grapevine plantlets with a plant growth promoting rhizobacterium *Burkholderia phytofirmans* strain PsJN. Applied and Environmental Microbiology. 2006; 72:7246-7252. DOI: 10.1128/AEM.01047-06

[58] Zubair M, Hanif A, Farzand A, Sheikh TMM, Khan AR, Suleman M, Ayaz M, Gao X. Genetic screening and expression analysis of Psychrophilic *Bacillus spp.* reveal their potential to alleviate cold stress and modulate phytohormones in wheat. Microorganisms. 2019; 7:337. DOI: 10.3390/microorganisms7090337

[59] Meena RK, Singh RK, Singh NP, Meena SK, Meena VS. Isolation of low temperature surviving plant growth–promoting rhizobacteria (PGPR) from pea (*Pisum sativum* L.) and documentation of their plant growth promoting traits. Biocatalysis and Agricultural Biotechnology. 2015; 4:806-811. DOI: 10.1016/j.bcab.2015.08.006

[60] Kang SM, Khan AL, Waqas M, You YH, Hamayun M, Joo GJ, Shahzad R, Choi KS, Lee IJ. Gibberellin-producing *Serratia nematodiphila* PEJ1011 ameliorates low temperature stress in *Capsicum annuum L.* European Journal of Soil Biology. 2015; 68:85-93. DOI: 10.1016/j.ejsobi.2015.02.005

[61] Vacheron J, Desbrosses G, Bouffaud M, Touraine B, Moenne-Loccoz Y, Muller D, Legendre L, Wisniewski-Dyé F, Prigent-Combaret C. Plant growth-promoting rhizobacteria and root system functioning. Frontiers in Plant Science. 2013; 4:1-19. DOI: 10.3389/fpls.2013.00356

[62] Su F, Jacquard C, Villaume S, Michel J, Rabenoelina F, Clément C, Barka EA, Dhondt-Cordelier S, Vaillant-Gaveau N. *Burkholderia phytofirmans PsJN* reduces impact of freezing temperatures on photosynthesis in *Arabidopsis thaliana*. Frontiers in Plant Science. 2015; 6:810. DOI: 10.3389/fpls.2015.00810

[63] Theocharis A, Bordiec S, Fernandez O, Paquis S, Dhondt-Cordelier S, Baillieul F,

Barka EA. *Burkholderia phytofirmans* PsJN primes *Vitis vinifera* L. and confers a better tolerance to low nonfreezing temperatures. Molecular Plant-Microbe Interactions. 2012; 25:241-249. DOI: 10.1094/MPMI-05-11-0124

[64] Crowe JH. Trehalose as a 'chemical chaperone' fact and fantasy. In: Back N, Cohen IR, Lajtha A, Lambris JD, Paoletti R (eds). Molecular aspects of the stress response: chaperones, membranes and networks. Springer, New York. 2007; 143-158.

[65] Ruelland E, Vaultier MN, Zachowski A, Hurry V, Kader JC, Delseny M. Cold signaling and cold acclimation in plants. Advances in Botanical Research Advances in Botanical Research. 2009; 49:35-150.

[66] Fernandez O, Béthencourt L, Quero A, Sangwan RS, Clément C. Trehalose and plant stress responses: friend or foe? Trends in Plant Science. 2010; 15:409-417.

[67] Karim S, Aronsson H, Ericson H, Pirhonen M, Leyman B, Welin B, Mantyla E, Palva E, van Dijck P, Holmstrom KO. Improved drought tolerance without undesired side effects in transgenic plants producing trehalose. Plant Molecular Biology. 2007; 64:371-386. DOI: 10.1007/s11103-007-9159-6

Chapter 15

Strategies and Programs for Improved Nutrient Use Efficiency, Doubling Farmer's Income, and Sustainable Agriculture: Indian Context

Ravinder Juttu, Kamalakar Jogula, Subhashree Priyadarshini, Sharan Bhoopal Reddy, Prasanta Kumar Patra, Bairi Raju, J.D. Saritha, Princy Thakur, S. Rakesh, Jogarao Poiba, A. Krishna Chaitanya and Kiran Pilli

Abstract

Since the Green Revolution era, the farming sector exploited the soils for food, fiber, fodder, etc., with high input responsive varieties that excavated vast amounts of chemical fertilizers. The burgeoning population of the country calls for a commensurate increase in food production to satisfy the demands of its inhabitants. Further, due to innovative mechanization in agriculture, specialization, and government policy programs, the productivity of food has soared. Subsequently, it ensued greater productions and minimized food prizes. Regrettably, intensive agricultural operations degraded the soil quality and now reached such a stage where without external inputs, growers unable to achieve their targeted yields. India has lost 68% innate productive capacity of agricultural soils. This plunder of land's quality continues unabated, further resulting in low nutrient use efficiency and insufficient yields of agroecosystems. Therefore, this is high time to realize the dreadful impacts of intensive crop production on the natural ecosystem. Irrefutably, both soil and its nutrients are the wondrous gifts of nature to humankind; utilizing them sustainably is imperative. The present chapter highlights the impacts of non-judicious nutrient management on soil productivity, nutrient use efficiency, and novel technologies required to promote sustainable agriculture and achieve the target of doubling farmer's income in India.

Keywords: Food security, sustainability, nutrient use efficiency, technologies

1. Introduction

The agriculture sector is the primary source of livelihood for over 58% of the Indian population and a key contributor to the health of the country's economy as it contributes approximately 17.1% to India's gross value added (GVA). It generates an

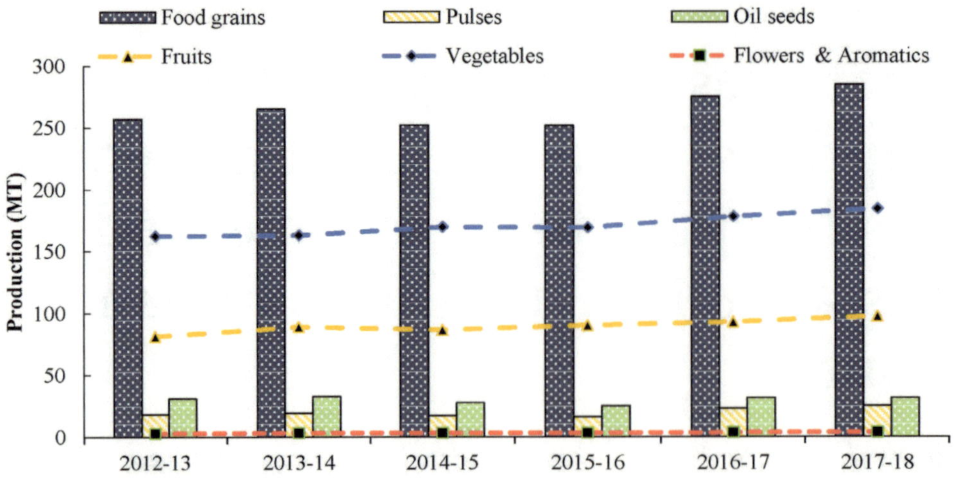

Figure 1.
Production rates of different agricultural produce in India [1].

employment opportunity for 44% of its workforce. Agriculture in India, continues to make impressive progress, while food grain production at record 296.7 million tons (Mt) and total oilseeds production at record 33.4 Mt., during 2019–2020. Such increments were also observed in other major crops. Current production estimates show that food grains, pulses, oilseeds, fruits, vegetables, and flower and aromatics are about 284.83, 25.23, 31.31, 97.35, 184.39, and 3.65 Mt., respectively. The production rates of different agricultural produce in India for the period from 2012 to 2013 to 2017–2018 are presented in **Figure 1.** Despite this remarkable growth in production in India, about 14.8% of the population and 38.4% of children remain malnourished. As per evaluation of the Global Food Security Index (GFSI), India ranked 76th out of 113 countries.

Only agriculture can help us achieving food security in the country. Since the green revolution (GR) in India, growers are cultivating high-yielding varieties under the irrigated condition with high amount of fertilizer nutrients and pesticides, saved millions of lives from starvation, and transformed India into a self-sufficient country in food production; also from the status of a "hungry nation" to that of a "food exporting nation" [2]. But, because of burgeoning population pressure and restricted land usage, it has become a serious challenge. Intensive agricultural operations involving continuous tillage and chemical fertilizers and pesticides are a concern in terms of environmental issues and soil degradation in the post-green revolution period. Increasing food demand has laid more stress on the agricultural soils for increased productivity. The total NPK fertilizer consumption during 2018–2019 was 27.23 Mt. and it is likely to increase to around 48.0 Mt. by 2050.

The persistent decline in nutrient use efficiency, soil fertility status, and environmental quality are the key constraints coming in the way of achieving sustainability in Indian agriculture. Specifically, in the Indian soils, nutrient use efficiency (NUE) is very low. It varies from 30 to 50% for nitrogen (N), 15 to 20% for phosphorus (P), 60 to 70% for potassium (K), 8 to 10% for sulphur (S), and 1 to 2% for micronutrients [3]. The unutilized N is lost through several mechanisms such as leaching, denitrification, volatilization, etc., pollute the groundwater and atmosphere. Considerable amounts of P and K are also lost through soil erosion. As per the estimation, annually, over 5.3 billion tonnes (Bt) of soil is lost through water erosion that ultimately results in the loss of about 8 million tonnes (Mt) of plant nutrients (NPK). Finally, there exists a huge yield gap (difference between the achievable and the actual yield) in most of the crops of India.

Yet the production of food, fiber, and raw materials must be enhanced in a sustainable manner to the demands of the ever-growing and progressively affluent population of India. Whereas the inorganic inputs supplying for the food production may never be completely replaced by the organic amendments but with appropriate management strategies/technologies, we could be able to achieve higher nutrient use efficiency, minimize the adverse impacts on the environment, and double the farmer's income that brings sustainability in the agricultural production system.

2. Technological options suitable for Indian agriculture

2.1 Enhancement of soil organic carbon

Improving SOC in agricultural soils is now a global challenge for environmental safety [4, 5] as it is the most realistic approach to regulate soil degradation [6] and improve soil productivity to achieve higher crop yields [7–10]. It has several benefits on soil quality maintenance as showed in **Figure 2**. Mahmood et al. [11] revealed in his experiment that incorporation of OC into the soil system through sheep manure (SM), farmyard manure (FYM), and poultry manure (PM) alone or amalgamation with chemical fertilizers significantly improved the soil organic carbon status as well as the available total nitrogen (TN), total phosphorus (TP) and total potassium (TK) (**Figure 3**) over the control and complete inorganic treatment. Improvement in soil fertility always has a direct benefit on crop yields as the availability of all essential soil nutrients increases through the addition of C input. There was an increase in crop yields (kg ha^{-1}) for every Mg ha^{-1} increase in SOC stock in the root zone. A field experiment conducted to evaluate the effect of C input under various crop production systems showed that an improvement in crop yields (kg ha^{-1}) such as 170 for pearl millet, 145 for soybean, 150 for castor, 160 for upland rice, 18 for

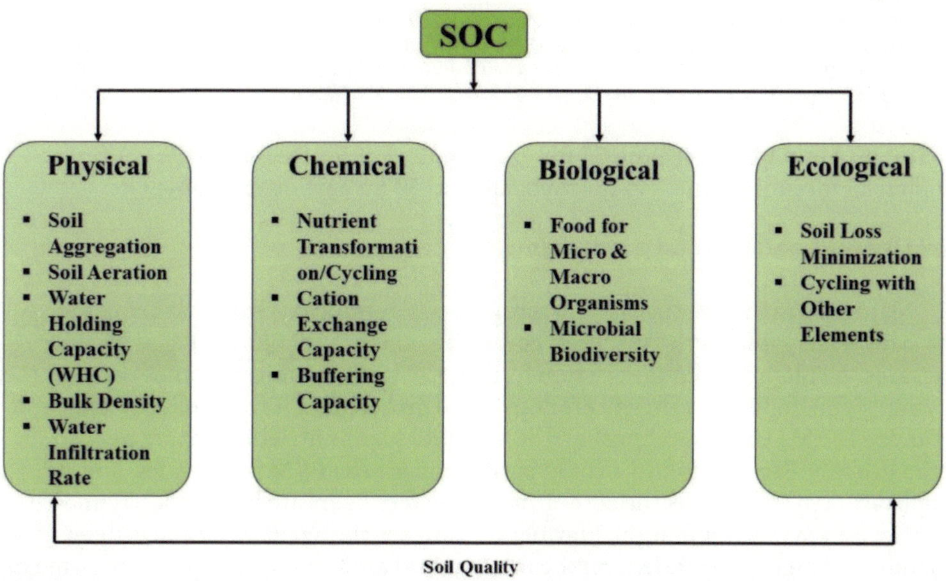

Figure 2.
Important SOC functions in soil quality maintenance.

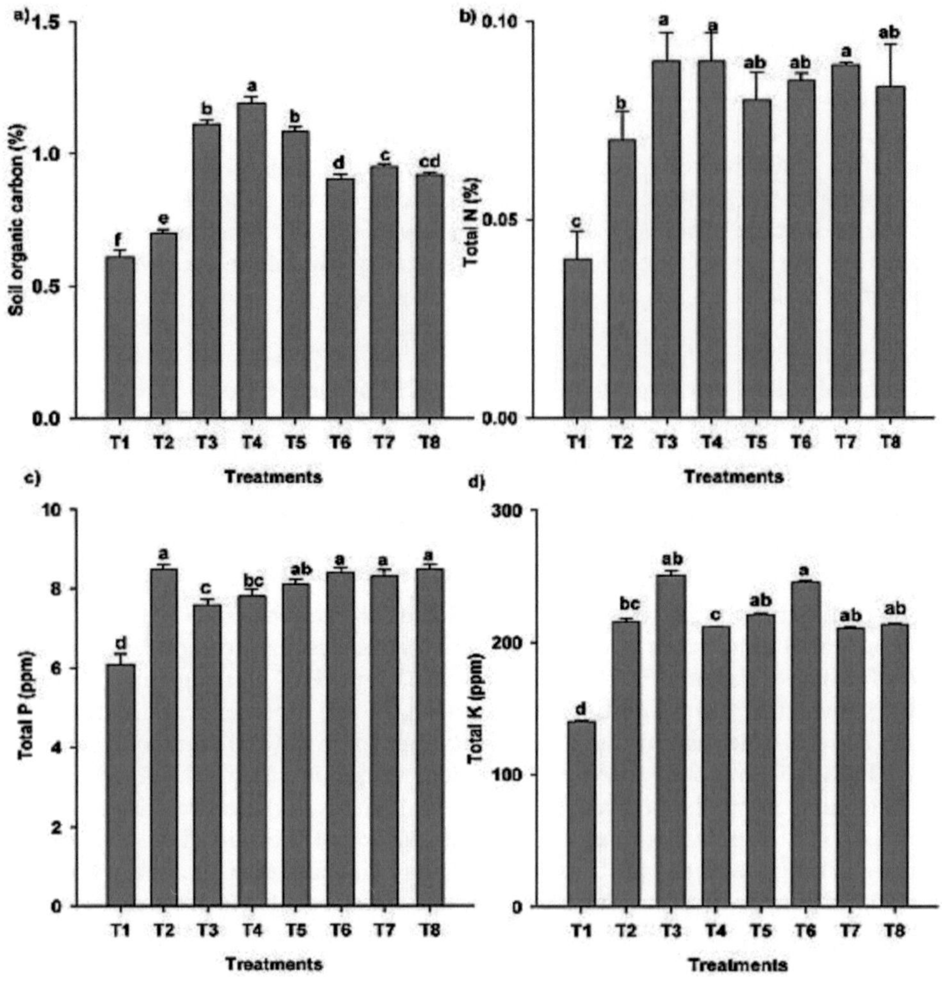

Figure 3.
Residual Impact of sheep manure (SM), farmyard manure (FYM), and poultry manure (PM) on (a) SOC (%), (b) TN (%), (c) TP (ppm), (d) TK (ppm) after crop harvest (Source: [11]). ***T1:*** *Unfertilized control;* ***T2:*** *NPK at 250–150-125 kg ha*$^{-1}$*;* ***T3:*** *SM at 15 t ha*$^{-1}$*;* ***T4:*** *FYM at 16 t ha*$^{-1}$*;* ***T5:*** *PM at 13 t ha*$^{-1}$*;* ***T6:*** *NPK at 150–85-50 + 8 t ha*$^{-1}$ *SM;* ***T7:*** *NPK at 150–85-50 + 8.5 t ha*$^{-1}$ *FYM;* ***T8:*** *NPK at 150–85-50 + 7 t ha*$^{-1}$ *PM 2.2.*

lentil, 90 for winter sorghum, 33 for groundnut, 124 for finger millet, 101 finger millet, 13 for groundnut, 145 for soybean, and 59 for safflower (**Table 1**).

2.2 Organic manures and green manures

Manures are the decomposed heterogeneous organic mixture that are made up of farm wastages like crop residues, cow dung, and household wastages. Manure releases the plant nutrients very slowly, thus the initial requirements of the crop met by supplying fertilizer nutrients for optimum growth and development. Farmyard manure (FYM) contains almost all the essential plant nutrients that are needed for crop growth. Farmers in India could easily manage the FYM preparation and its application in the field as the cost of inorganic fertilizers are high which is unable to afford by small and marginal farmers. However, the availability and efficiency of manure are highly dependent on the method and amount of its application, time to incorporate, and decomposition rate by soil microorganisms.

Green manures (GM) where leguminous crops are incorporated into the soil during the flowering stage that makes the soil fertile and increases crop productivity.

Cropping system	Mean annual C input ($Mg\ C\ ha^{-1}\ year^{-1}$)	Yield increased ($kg\ ha^{-1}$)	Reference
Pearl millet- Cluster Bean- Castor	0.2–1.9	Pearl millet- 170 Cluster Bean- 140 Castor- 150	[12]
Groundnut	0.5–3.5	Groundnut- 13	[13]
Finger millet	0.3–3.1	Finger millet- 101	[14]
Groundnut-Finger millet	0.3–3.0	Groundnut- 33 Finger millet- 124	[15]
Soybean-Safflower	1.9–7.0	Soybean- 145 Safflower- 59	[16]
Winter Sorghum	0.6–3.4	Winter Sorghum- 90	[17]
Rice-Lentil	1.1–5.6	Upland rice- 160 Lentil- 18	[18]

Table 1.
Effect of mean annual C input on Crop yields under various cropping systems.

These manures are a great source of biologically fixed N and organic carbon. The area for green manuring crops is limited to 7 Mha in India [19] and has also not expanded over the last few decades. Probably, the low price of urea N, intensification in crop production, scarcity of land, and irrigation water are the main factors for the long-term reduction in GM use. Legume crops like mungbean/cowpea or typical GM crops like *dhaincha*/sunn hemp can be grown and incorporated into the soil. GMs have the capacity to meet the N demands of the crops. A 40–45 days old GM crop can supply 100–125 kg N which is equal to the N requirement for cereal crops [20] Even after harvesting the pods, residues can be incorporated into the soil, which saves around 60 kg N ha^{-1} in succeeding crops like rice or maize [20, 21].

2.3 Integrated nutrient management

Integrated nutrient management (INM) is a technique of combined usage of chemical fertilizers, organic amendments, and bio-fertilizers in farming which is an economically feasible and environmentally benign way of managing plant-available nutrients. This concept originated at the beginning of the 1990s because of the widespread emergence of multi-nutrient deficiencies and soil degradation. Thus, INM comprises major objectives such as soil fertility maintenance, sustenance of crop productivity, and improvement of the farmers' profitability. The amalgamation of inorganic fertilizers with organic amendments aids in the provision of improving soil productivity by improving soil C storage [10, 22]. The INM-induced SOM build-up aids in the provision of improved soil structure and water holding capacity that directly enhances crop yields.

A long-term field experiment showed higher grain yield and sustainability yield index (SYI) of maize and black gram under the INM treatment (conjunctive use of fertilizers and organic amendments) as compared to sole inorganic treatment and control (**Figure 4**) [10]. Such improvements in yields may be due to the continuous addition of C inputs that creates a congenial environment for plant growth by modifying the soil's physical properties [23, 24]. The importance of INM practice for increasing soil health, nutrient use efficiency, crop yields, and decreasing environmental pollution has been recorded by several researchers in the Indian subcontinent [25, 26]. The location-specific INM strategies working better under various cropping systems are summarized in **Table 2**.

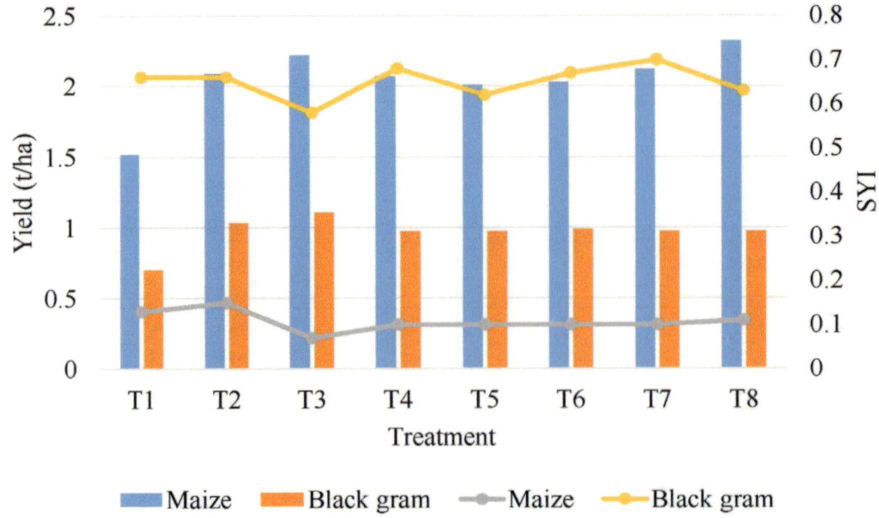

Figure 4.
Mean crop yields and sustainability yield index of maize-black gram system as influenced by the INM approach (Source: [10]). T1: Control; T2: 100% RDF of NP; T3: 25 kg ha^{-1} N (FYM) + 25 kg N (Urea) + 30 kg P ha^{-1}; T4: 25 kg ha^{-1} N (Compost) + 25 kg N (Urea) + 30 kg P ha^{-1}; T5: 25 kg ha^{-1} N (Crop residue) + 25 kg N (Urea) + 30 kg P ha^{-1}; T6: 15 kg ha^{-1} N (FYM) + 10 kg N (Crop Residue) + 25 kg N (Urea) + 30 kg P ha^{-1}; T7: 15 kg ha^{-1} N (FYM) + 10 kg N (Compost) + 25 kg N (Urea) + 30 kg P ha^{-1}; T8: 15 kg ha^{-1} N (FYM) + 10 kg N (Green Leaf) + 25 kg N (Urea) + 30 kg P ha^{-1}.

2.4 Conservation agriculture

Traditional agriculture, based tillage, and other management operations lead to soil erosion problems, surface and groundwater pollution [27]. Conservation agriculture (CA) technology involving three basic principles such as minimum soil disturbance, efficient and diversified crop rotations, and surface crop residue retention aids in the provision of enhancing/improving soil organic carbon storage. Tillage and residue management greatly influence the soil's physicochemical and biological properties [28]. Zero tillage (ZT) for crop production has been identified as an important practice to increase soil aggregation and C sequestration [28] as compared with traditional systems i.e., conventional tillage (CT).

Adoption of ZT in wheat production in India, reduced the cost of production by Rs 2,000 to 3,000 ha^{-1} ($ 33 to 50) [29]; enhanced the soil quality, [30]; improved C sequestration and mitigation of Green House Gas emissions [31]; reduction in weed population *Phalaris minor* in wheat (Malik et al., 2005), enhanced water and nutrient use efficiency [31, 32] and overall increments in production and productivity (4–10%) [33]. The practice of ZT significantly increased the soil total nitrogen content (**Figure 5**) in both rice-wheat and rice-maize cropping systems in both districts studied in West Bengal [34]. CA adoption in India is still in the initial phases. Over the past few years, the adoption of ZT expanded to cover about 1.5 Mha [32].

2.5 Pulses in crop rotation

Meeting the N demands of the crop efficiently with fewer N losses and more use efficiency is a critical challenge for the food production community [35]. Many researchers reported that pulse crops can significantly improve soil nitrogen availability, soil water conservation, and increase total system productivity. Gan et al. [36] reported that the practice of growing pulses helps in biological fixation of atmospheric N_2, increases total grain production by 35.5%, protein yield by 50.9%,

Location/system	Kharif	Rabi
Rice-wheat		
Jammu, Jammu and Kashmir Palampur, Himachal Pradesh	50% NPK through fertilizers +50% NPK through FYM 50% NPK through fertilizers +50% N through organic sources such as FYM green manure and wheat straw	100% NPK through fertilizers
Kalyani, West Bengal	50% NPK through fertilizers +50% N through FYM/green manure or rice straw	100% NPK through fertilizers
Navsari, Gujarat	75% NPK through fertilizers +25% N through FYM or green manure	
Faizabad, Uttar Pradesh	50% NPK through fertilizers +50% N through FYM or green manure	100% NPK through fertilizers
Rice-rice		
Jorhat, Assam Karamana, Kerala	75% NPK through fertilizers +25% N through rice straw 50–75% NPK through fertilizers +25–50% through FYM or crop residue or green manure	75% NPK through fertilizers
Siruguppa, Karnataka	50–75% NPK through fertilizers +25–50% N through rice straw or *glyricidia* green leaf manure	100% NPK through fertilizers
Rajendranagar, Telangana	50% NPK through fertilizers +50% N through *glyricidia* green leaf manure	100% NPK through fertilizers
Rice – Maize		
Kathalgere, Karnataka	75% NPK through fertilizers +25% N through FYM or paddy straw or*glyricidia* green leaf manure	100% NPK through fertilizers
Pearl Millet – Wheat		
S.K. Nagar, Gujarat	50–75% NPK through fertilizers +25–50% N through FYM/wheat straw/sunhemp	75% NPK through fertilizers
Bichpuri, Uttar Pradesh	50–75% NPK through fertilizers +25–50% through FYM or green manuring	75% NPK through fertilizers
Sorghum – Wheat		
Akola, Maharashtra	50% NPK through fertilizers +50% N through FYM or wheat straw or *Leucaena* loppings	100% NPK through fertilizers
Rahuri, Maharashtra	50% NPK through fertilizers +50% N through FYM	100% NPK through fertilizers

Table 2.
Best INM treatments practicing under various cropping systems of India.

and fertilizer-N use efficiency (FUE) by 33.0% over the summer fallow system. Diversifying cropping systems with pulses such as dry pea (*Pisum sativum* L.), lentil (*Lens culinaris* Medikus), and chickpea (*Cicer arietinum*), etc. can serve as an effective alternative to summer-fallowing in rainfed dry areas. These pulses could increase the systems' productivity and decreases the negative impacts on the environment.

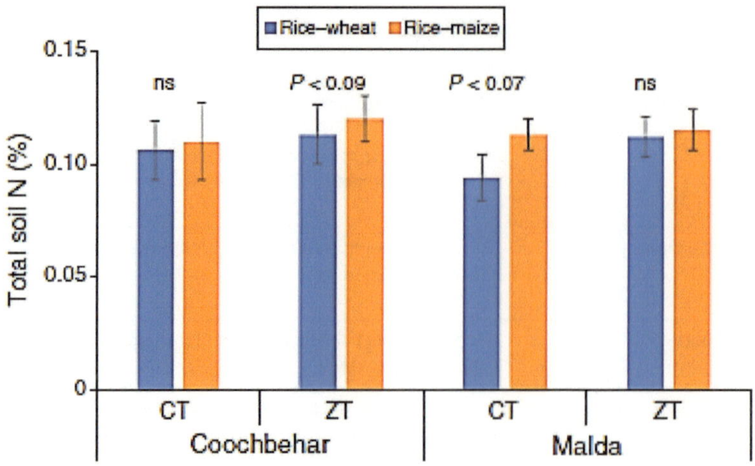

Figure 5.
Effect of ZT and CT on total soil N (0–10 cm) under rice–wheat and rice–maize systems at different districts of West Bengal (Source: [34]).

The inclusion of pulse crops in the crop rotation aids in the provision of improved crop yields and decreased N fertilizer requirements [37] and enhanced nitrogen use efficiency [38]. Pulse crops provide a large portion of N requirements to subsequent cereal crops and maintain economic returns [39]. Reduction in N fertilizers requirements naturally reduced the cost of cultivation thereby benefiting the farmers with less expenditure. Long-term experiments revealed that crop diversification with pulses and oilseed can help farmers to improve overall agricultural sustainability [40].

2.6 Fertigation and foliar spraying

Fertigation is a technique of supplying plant nutrients along with irrigation which helps in increasing crop yields or N fertilizer efficiency in many conditions with different crops [41]. Supplement of N and P fertilizer through fertigation technique significantly enhanced the wheat grain yield by 16% as compared to top-dressed N [42]. The fertigation process allows the soil to absorb up to 90% of supplied nutrients, while it is only about 10 to 40% under dry fertilizer or granular application. It ensures saving in fertilizer quantity of about 40–60%, because of better fertilizer use efficiency and reduced leaching losses [43]. Application of liquid biofertilizers and mineral fertilizers along with drip fertigation in green gram cultivation significantly increased the number of pods plant, number of seeds pod^{-1}, test weight, seed yield, and haulm yield [44]. Manikandan and Sivasubramaniam [45] reported that drip fertigation with 100% recommended dose of fertilizer through water-soluble fertilizers + foliar feeding with 0.5% $ZnSO_4$ resulted in the highest onion bulb yield and quality. Various advantages of fertigation in agriculture production are illustrated in **Figure 6**.

The positive effects of foliar spraying of zinc nano-fertilizer on vegetative growth parameters of pearl millet [46] on and snap bean plants [47] are reported. Foliar application of micronutrients increases the vegetative growth, consequently higher production capacity which reflected in the quality of barley [48] and faba bean [49].

2.7 Water-soluble fertilizers

Water-soluble fertilizers are 100% soluble in water which is suitable for foliar application due to their low salt index to reduce the potential for the burning of

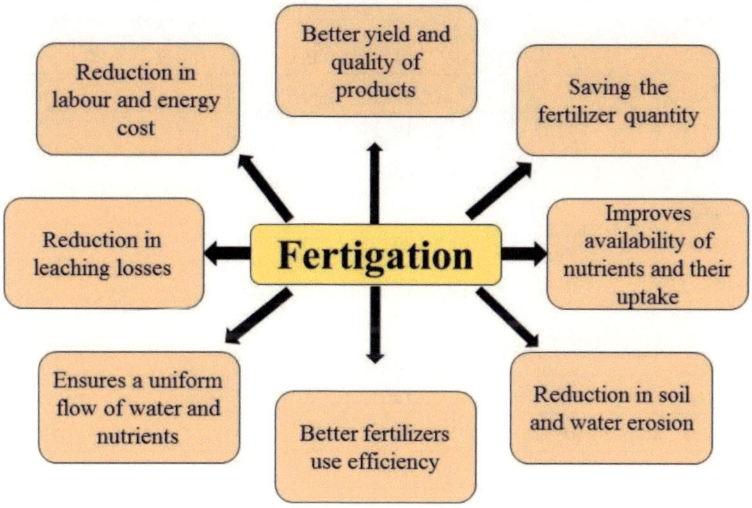

Figure 6.
Advantages of fertigation technique in agriculture.

plant tissue. It is also used in fertigation, sprinkler, or drip irrigation systems to increase yield and to improve the quality of fruits and vegetable crops. These fertilizers should meet certain criteria such as 100% solubility, high purity, low salt index, (EC = 0.9–1.2), pH acidic (5.5 to 6.5), and no inert matter, free from sodium and chloride, driven by R&D, suitable for fertigation and foliar application, higher nutrient use efficiency, etc. These fertilizers are mostly the combination of N, P, K, Ca, Mg, S, and micronutrients with different ratios developed to suit the type of crop, quality of water, soil fertility, and climatic conditions [50]. The Fertilizer Control Order (FCO) approved water-soluble fertilizers and their nutrient composition is presented in **Table 3**.

2.8 Biofertilizers

Biofertilizers are the source of microbial inoculants prepared in a controlled laboratory condition that acts as a substituent for chemical fertilizer and helps to achieve sustainable agriculture [51] boosting farm productivity [52]. Several studies indicated the use of biofertilizers in agriculture enhanced crop yields at greater levels. Usage of biofertilizers such as *Azotobacter*, *Azospirillum, Rhizobium* for N, and phosphate solubilizing bacteria (PSB) for P, vesicular-arbuscular micorhizae (VAM) for other nutrients availability in crop cultivation helps in improving crop yields and quality. Soil inoculation with *Azotobacter*, *Azospirillum,* and PSB produced maximum crop yields by 5–10% over farmers' practice [53]. In another study under jute, the yield was increased by 19% due to biofertilization over RDF, rice by 8% and green gram by 12%. Rao [53] studied the effect of BF on nutrient recovery. The study revealed that NPK recovery increased from 62.0% to 74.0% in recommended fertilizers + BF treatment. Combining soil test based fertilizer recommendation with organics and biofertilizers under maize cultivation considerably enhanced the recovery of N from 18–66%, P from 9–36%, K from 33–88%, and S from 17–34% [54] (**Table 4**). ICAR is also promoting the development of biofertilizers consisting of *Azospirillum lipoferum*, *Azotobacter chroococcum* and plant growth promoting *Rhizobacteria* (PGPR Mix I). But in India, the current supply position is very low (<100, 000 t), as the total anticipated biofertilizers demand is 1 Mt. [53].

Product (grade)	Nutrient composition (%)						
	N	P	K	S	Ca	Mg	Zn
NPK (13–40-13)	13	40	13				
NPK (18–18-18)	18	18	18				
NPK (13–5-26)	13	5	26				
NPK (6–12-36)	6	12	36				
NPK (20–20-20)	20	20	20				
NPK (19–19-19)	19	19	19				
NPK (12–30-15)	12	30	15				
NPK (12–32-14)	12	32	14				
Potassium nitrate (13–0-45)	13	0	45				
Mono potassium phosphate (0–52-34)	0	52	34				
Calcium nitrate	15.5				18.8		
Potassium magnesium sulphate			22	20		18	
Mono ammonium phosphate (12–61-0)	12	61	0				
Urea phosphate (17–44-0)	17	44	0				
Urea phosphate with SOP (18–18-18)	18	18	18	6.1			
NPK Zn (7.6–23.5-7.6-3.5)	17.6	23.5	7.6				3.5

Table 3.
FCO approved 100% water-soluble fertilizers (Source: [50]).

Treatment	Nutrient (%)			
	N	P	K	S
Soil test based fertilizer	18	9	33	17
Soil test based fertilizer + Organics	59	30	80	26
Soil test based fertilizer + Organics + Bio fertilizers	66	36	88	34

Table 4.
Apparent recovery of nutrients as affected by different treatments under maize cultivation (Source: [54]).

2.9 Nano fertilizers

Regular synthetic fertilizers are highly vulnerable to the losses such as leaching, volatilization, percolation, etc. which ultimately results in low NUE which is below 30%. "Nano fertilizers" are prepared by extracting the nutrients from different parts of the plant through chemical, physical, mechanical, or biological methods using nanotechnology. Nanotechnology has a long-term impact on agriculture and food production as the usage of these fertilizers in farming improves crop growth, yield, and quality parameters while increasing the nutrient use efficiency (NUE), and reducing the wastage and cost of cultivation. A significant higher selenium uptake was observed in the plots where nano-sized particles were applied [55]. Nano-fertilizers in agriculture enhanced nutrient uptake and crop productivity [56, 57]. The percent yield increased in different crops due to the addition of Nano-fertilizers illustrated in the table below (**Table 5**).

2.10 Customized fertilizers

Customized fertilizers (CF) are multi-nutrient (macro-N, P, K, secondary-Ca, Mg, S, and micro nutrient-Zn, Cu, B, Fe, Mn, etc.) produced from both inorganic

Nano fertilizer	Crops	% Yield increased
Aqueous solution on nano iron	Cereals	8–17
Nano silver + allicin	Cereals	4–8.5
Nano fertilizer + urea	Rice	10.2
Nano-encapsulated phosphorous	Vegetables	12.0–19.7
Rare earth oxides nanoparticles	Vegetables	7–45
Nano chitosan-NPK fertilizers	Wheat	14.6
Nano chitosan	Tomato	20.0
Nano powder of cotton seed and ammonium fertilizer	Sweet potato	16
Nanoparticles of ZnO	Cucumber	6.3

Table 5.
% yield increased by the application of nano fertilizer in different crop production (Source: [56]).

and organic sources, manufactured through a systematic process of granulation designed to facilitate the availability of a complete range of nutrients to the plant growth during its growth stages [58]. It has various advantages besides soil health enhancement and maximum crop yields (**Figure 7**). On the basis of nutrient uptake, total soil fertility status, crop nutrient requirement, and fertilizer nutrient to be applied and its use efficiency, grades of the CF are prepared [59]. Different forms of CF available across various geographical areas of India presented in **Table 6**.

2.11 Sensors based technologies for irrigation and fertilization

A new approach of collecting real soil moisture using sensors offers real potential for reliably monitoring the status of soil water in croplands [60]. The use of sensor technology for an automatic irrigation system is highly economical

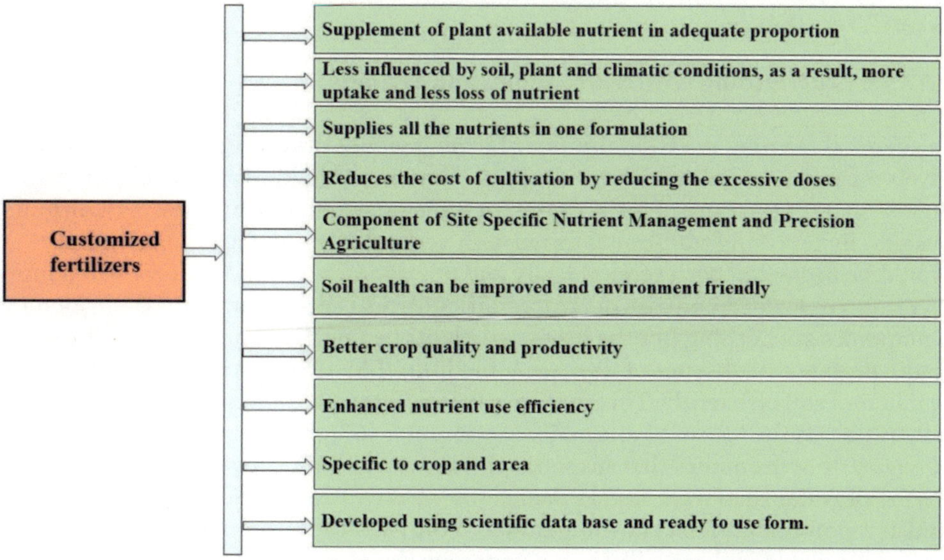

Figure 7.
Advantages of using customized fertilizers in agriculture.

Crops	Formulations (N:P:K:S:Zn:B)/ N:P:K: Zn/ N:P:K:S:Mg:Zn:B:Fe/ N:P:K:S:Zn:B)	Geography
Rice	8:15:15:0.5:0.15:0	GB Nagar, Ghaziabad, Rampur, Shahjahanpur, Mainpuri and US Bagar
Wheat	10:18:25:3:0.5:0	Muzaffarnagar, Barielly, Bijnore, Hathras, Pilibhit, Mathura, Meerut and Etah
Maize	14:27:10:4:0.5	Karimnagar, Warangal and Ranga Reddy
Potato	8:16:24:6:0.5:0.15	Agra, Aligarh, Budaun, Bulandshahar and Baghpath
Sugarcane	7:20:18:6:0.5:0	Moradabad, KR Nagar, Farukhabad and Ferozabad
Groundnut	15:15:15:9:0.5:0.2	Andhra Pradesh
Grape, Cotton, Onion, Banana, Tomato, Gourds & Leafy and Vegetable	15:15:15:5:2:0.5:0:0.2	Nasik, Dhule, Jalgaon, Pune, Ahmednagar and Aurangabad

Table 6.
Customized Fertilizer Formulations available for different crops in India (Source: [57]).

as it aids in the provision of optimum use of water, saving money, electricity, and time of the farm. Many researchers reported significant water saving through this technology [61]. The use of sensors with drip and sprinkler irrigation systems can effectively improve the water application efficiency up to 80–90% as compared to the surface irrigation method (40–45%) [62]. In Egypt, potato yields were increased and a loss of 2 billion pounds was recovered in a year through the wireless sensor network technology [63]. Sensors in the prediction of crop nitrogen requirements are also practically significant in agricultural production and environmental safety. Usage of an optical sensor-based algorithm that employs yield prediction and N responsiveness by location can enhance the crop yields and minimize the environmental contamination caused by the application of excessive N fertilizers [64].

2.12 Vertical farming with hydroponics and aeroponics

Vertical farming is a type of indoor farming where the crop grows in multiple levels on a vertical axis which results in maximum production and efficiency per square foot [65]. It is a potential option to achieve sustainability in the agricultural system. By replacing traditional farms with vertical farming techniques, society would be protected both economically and environmentally. It reduces the amount of resources needed and also decreases agriculture's carbon footprint. Hydroponics/aquaponics are nothing but the produced plants in a nutrient-enriched solution in the presence or absence of a growing medium [66]. This system reduces labor, water, and soil efficiently. This system is more sustainable and profitable in food generation; is the future of alternative agriculture [67]. Aeroponics is a subcategory of hydroponics that suspends the roots in the air, thus there is around 95% saving of water than traditional systems [66]. Cultivation of these hydroponics and aeroponics in vertical farming under a controlled environment makes more profitable yields as there will be no damages to the plant by the external factors. By implementing vertical farms in communities, the cost of food would decrease, the economy could thrive, transportation costs are cut dramatically, and therefore so are food prices, create employment and increase educational opportunities. Further, people would be able to become economically and nutritionally stable as it makes a huge impact on both food insecurity and poverty.

3. Programs and policies executed by GOI for sustainable agriculture

The convergence of various policy programs has been initiated by the GOI to ensure the effective utilization of existing resources are briefly discussed here. The National Mission of Sustainable Agriculture (NMSA) under the National Action Plan on Climate Change (NAPCC) was launched in 2010 in order to encourage the judicious management of existing resources. The Paramparagat Krishi Vikas Yojana (PKVY) mission was executed in conjunction with the Indian Council of Agricultural Research (ICAR) and state governments of India to extensively leverage adaptation of climate-smart practices and technologies. In 2015, GOI has launched the Soil Health Card (SHC) scheme to protect the soil health for future agriculture with the main objective of analyzing soil samples of farmers' fields and recommending fertilizers accordingly. Additionally, Neem-Coated Urea (NCU) was introduced to the farmers of India for a slow supplement of nitrogen (N0 by reducing the N losses and excess addition of urea fertilizers. Programs such as the National Project on Organic Farming (NPOF) and National Agroforestry Policy (NAP) was introduced in 2004 and 2014 respectively to encourage the farmers with more profit and ecosystem service through supplements of plant nutrients in the form of organic amendments, improvement of soil carbon storage, and soil protection from erosion loss. States like Andhra Pradesh, Himachal Pradesh, Sikkim, etc. have already adopted and promoted organic farming practices on a wider scale. "Sikkim" state recognized as an "*Organic State*" of India. A "4 per 1000/4 per mille" initiative launched by France in 2015 as a part of the Global Climate Action Plan (GCAA) adopted by the United Nations Framework Convention on Climate Change (UNFCCC) at a conference of the parties (COP) 22 also recognized the importance of SOC in achieving sustainability in agriculture system. It considered the technologies such as agroforestry, conservation agriculture system intensification (CASI), and landscape management to improve SOC.

4. Conclusion and way forward

The post-Green Revolution period witnessed a drastic change in environmental conditions and the status of existing natural resources. A gradual decline in soil fertility has occurred by the non-judicious management of chemical fertilizers that further exacerbated by the progressively decreasing usage of organic amendments. Added to this, abysmally low NUE of applied fertilizers impacted agricultural productivity and sustainability to a great extent. Supplement of plant nutrients in balanced proportion is important; at least in such a way that the critical growth stages of the crop meet the required amount of nutrients results in achieving maximum crop yields that satisfy the growing population. Hence, these constitute a vital component of sustainable food production. Further, agricultural intensification is in critical need of improvements in the flow of plant nutrients to the crops from the soil through efficient nutrient uptake. Improved technologies involved effective nutrient management strategies are the need of the hour to accomplish the targeted food grain production while balancing the stability of the agriculture system, farmers' income, and feed the over-exploiting population of the country.

- Creating awareness among the farmers regarding fertilizer management, nutrient flows, and use efficiencies are essential.
- Strengthening the database on nutrient recommendations specific to soil type, cropping system, and climatic regions is crucial to manage soil productivity.
- Efficient modern technologies are critical in order to ensure the food security of the country.
- Evaluation of technology should be based on locally available resources.
- Identifying policy interventions needed to promote soil management practices that help in achieving maximum crop yields and nutrient use efficiency.
- Developing site-specific holistic land management practices to sustain production rates.
- Implementing a protocol for payment to farmers for strengthening of ecosystem services generated through adoption of sustainable agriculture.
- Bringing sustainability into the agricultural ecosystem needs to be meaningful and result-oriented.

Conflict of interest

The authors declare no conflict of interest.

Abbreviations

%	Percentage
B	Boron
C	Carbon
Ca	Calcium
CASI	Conservation agriculture system intensification
CF	Customized fertilizers
COP	conference of the parties
Cu	Copper
FCO	Fertilizer Control Order
Fe	Iron
GCAP	Global Climate Action Plan
GFSI	Global Food Security Index
GM	Green manure
GOI	Government of India
GR	Green revolution
GVA	Gross value added
ha	Hectare
ICAR	Indian Council of Agricultural Research
INM	Integrated Nutrient Management
K	Potassium
Kg	Kilo gram
Mg	Magnesium

Mn	Manganese
Mt.	Million tons
N	Nitrogen
N_2	Atmospheric Nitrogen
NAP	National Agroforestry Policy
NAPCC	National Action Plan on Climate Change
NCU	Neem-Coated Urea
NMSA	National Mission of Sustainable Agriculture
NPOF	National Project on Organic Farming
NUE	Nutrient Use Efficiency
P	Phosphorus
PKVY	Paramparagat Krishi Vikas Yojana
PSB	Phosphate solubilizing bacteria
S	Sulphur
SHC	Soil Health Card
SOC	Soil organic carbon
SYI	Sustainability yield index
t	Tons
UNFCCC	United Nations Framework Convention on Climate Change
yr.	Year
Zn	Zinc

Author details

Ravinder Juttu[1], Kamalakar Jogula[2], Subhashree Priyadarshini[3], Sharan Bhoopal Reddy[4], Prasanta Kumar Patra[5], Bairi Raju[2], J.D. Saritha[6], Princy Thakur[7], S. Rakesh[7]*, Jogarao Poiba[8], A. Krishna Chaitanya[9] and Kiran Pilli[5]

1 Department of Soil Science and Agricultural Chemistry, Regional Sugarcane and Rice Research Station, Professor Jayashankar Telangana State Agricultural University, Nizamabad, India

2 Department of Soil Science and Agricultural Chemistry, Agricultural College, Professor Jayashankar Telangana State Agricultural University, Warangal, Telangana, India

3 Department of Agricultural Entomology, College of Agriculture, Professor Jayashankar Telangana State Agricultural University, Rajendranagar, Telangana, India

4 Department of Soil Science and Agricultural Chemistry, University of Agriculture Sciences, Raichur, Karnataka, India

5 Department of Agricultural Chemistry and Soil Science, Bidhan Chandra Krishi Viswavidyalaya, Nadia, West Bengal, India

6 Department of Soil Science and Agricultural Chemistry, Agricultural College, Professor Jayashankar Telangana State Agricultural University, Palem, Telangana, India

7 Department of Soil Science and Agricultural Chemistry, Uttar Banga Krishi Viswavidyalaya, Coochbehar, West Bengal, India

8 Regional Agricultural Research Station, Visakhapatnam, Andhra Pradesh, India

9 Department of Soil Science and Agricultural Chemistry, Regional Agricultural Research Station, Professor Jayashankar Telangana State Agricultural University, Jagtial, Telangana, India

*Address all correspondence to: rakisavan.940@gmail.com

References

[1] ICAR-Data Book 2019.

[2] Siegel BE. Hungry Nation: Food, Famine and Making of Modern India. Cambridge University Press. 2018

[3] Sarkar D, Rakshit A, Al-Turki AI, Sayyed RZ, Datta, R. Connecting bio- priming approach with integrated nutrient management for improved nutrient use efficiency in crop species. Agriculture. 2021;11(4): 372. doi:10.3390/agriculture11040372.

[4] Milne E, Banwart SA, Noellemeyer E, Abson D, Ballabio C, Bampa F, Bationo A, Batjes N, Bernoux M, Bhattacharyya T, Black H, Buschiazzo D, Cai Z, Cerri C, Cheng K, Compagnone C, Conant R, Coutinho H, Zheng J. Soil carbon, multiple benefts. Environmental Development. 2015;13, 33-38.

[5] Ramesh T, Bolan NS, Kirkham MB, Wijesekara H, Manjaiah KM, Srinivasarao, Ch, Sandeep S, Rinklebe J, Ok YS, Choudhury BU, Want H, Tang C, Song Z, Freeman OW. Soil organic carbon dynamics: impact of land use changes and management practices: a review. Advances in Agronomy. 2019;156, 1-125.

[6] Bationo A, Kihara J, Vanlauwe B, Waswa B, Kimetu J. Soil organic carbon dynamics, functions and management in West African agro-ecosystems. Agricultural Systems. 2007;94(1), 13-25.

[7] Syers JK. Managing soils for long-term productivity. Philosophical Transactions of the Royal Society B. 1997;352,1011-1021

[8] Lal R. Enhancing crop yields in the developing countries through restoration of the soil organic carbon pool in agricultural lands. Land Degradation and Development. 2006; 17,197-209.

[9] Cowie AL, Penman TD, Gorissen L, Winslow MD, Lehmann J, Tyrrell TD, Twomlow S, Wilkes A, Lal R, Jones JW, Paulsch A, Kellner K, Akhtar-Schuster M. Towards sustainable land management in the drylands: scientifc connections inmonitoring and assessing dryland degradation, climate change and biodiversity. Land Degradation and Development 2011;22, 248-260.

[10] Srinivasarao Ch, Sumanta K, Kumpawat BS, Kothari AK, Sodani SN, Sharma SK., Vikas Abrol, Ravindra CG, Pravin BT, Yashavanth, B.S. Soil organic carbon dynamics and crop yields of maize (*Zea mays*)–black gram (*Vigna mungo*) rotation-based long term manorial experimental system in semi-arid Vertisols of western India. Tropical Ecology 2019. https://doi.org/10.1007/s42965-019-00044-x

[11] Mahmood F, Khan I, Ashraf U, et al. Effects of organic and inorganic manures on maize and their residual impact on soil physico-chemical properties. Journal of Soil Science and Plant Nutrition. 2017;17(1): 22-32.

[12] Srinivasarao Ch. et al., Long-term manuring and fertilizer effects on depletion of soil organic carbon stocks under pearl millet–cluster bean–castor rotation in western India. Land Degradation and Development. 2011;25, 173-183.

[13] Srinivasarao Ch. et al. Soil carbon sequestration and agronomic productivity of an Alfisol for a groundnut based system in a semi-arid environment in South India. European Journal of Agronomy. 2012a;43, 40-48.

[14] Srinivasarao Ch. et al., Critical carbon inputs to maintain soil organic carbon stocks under longterm finger-millet (Eleusine coracana [L.] Gaertn.) cropping on Alfisols in semiarid tropical

India. Journal of Plant Nutrition and Soil Science. 2012b;175, 681-688.

[15] Srinivasarao Ch. et al., Long-term effects of crop residues and fertility management on carbon sequestration and agronomic productivity of groundnut–finger millet rotation on an Alfisol in southern India. International Journal of Agricultural Sustainability. 2012c; 10, 1-15.

[16] Srinivasarao Ch. et al., Sustaining agronomic productivity and quality of a Vertisolic soil (Vertisol) under soybean–safflower cropping system in semi-arid central India. Canadian Journal of Soil Science. 2012d;92, 771-785.

[17] Srinivasarao Ch. et al, Grain yield and carbon sequestration potential of post monsoon sorghum cultivation in Vertisols in the semi-arid tropics of central India. Geoderma, 2012e;175-176, 90-97.

[18] Srinivasarao Ch, Venkateswarlu B, Lal R, Singh AK, Vittal KPR, Kundu S, Singh SR, Singh SP Long-term effects of soil fertility management on carbon sequestration in a rice–lentil cropping system of the Indo-Gangetic Plains. Soil Science Society of American Journal. 2012f;76:168-178

[19] Indoria AK, Sharma KL, Sammi RK, Srinivasarao Ch, Srinivas K, Balloli SS, Osman M, Pratibha G, Raju NS. Alternative sources of soil organic amendments for sustaining soil health and crop productivity in India—impacts, potential availability, constraints and future strategies. Current Science. 2018; 115:2052-2062

[20] Yadvinder-Singh, Khind CS, Bijay-Singh. Efficient management of leguminous green manures in wetland rice. Advances in Agronomy 1991;45: 135-189.

[21] Yadvinder-Singh, Bijay-Singh, Thind, H.S. (Editors). Green Manure Approaches to Crop Production and Sustainable Agriculture. Angkor Publishers Pvt. Ltd, Noida. 2010;1-130.

[22] Srinivasarao Ch, Gopinath KA, Prasad JVNS, Prasanna K, Singh AK. Climate resilient villages for sustainable food security in tropical India: concept, process, technologies, institutions, and impacts. Advances in Agronomy. 2016;140:101-214.

[23] Ferreras L, Gomez E, Toresani S, Firpo I, Rotondo R. Effect of organic amendments on some physical, chemical and biological properties in a horticultural soil. Bioresource Technology. 2006; 97:635-640

[24] Hati MK, Swarup A, Dwivedi AK, Mira AK, Bandyopadhyay KK. Changes in soil physical properties and organic carbon status at the topsoil horizon of a Vertisol of central India after 28 years of continuous cropping, fertilization and manuring. Agricultural Ecosystem and Environment. 2007; 119:127-134

[25] Sharma S, Padbhushan R, Kumar U. Integrated nutrient management in rice–wheat cropping system: an evidence on sustainability in the Indian subcontinent through meta-analysis. Agronomy. doi:10.3390/agronomy 9020071. 2019;9.

[26] Bijay-Singh, Ali M. Advances in integrated plant nutrient management. In Achieving Sustainable Crop Nutrition (Z. Rengel, Ed.), Burleigh Dodds Science Publishing, Cambridge, UKDOI: 10.19103/ AS.2019.0062.25. 2020.

[27] Wolff P, Stein TM. Water efficiency and conservation in agriculture – opportunities and limitations. Agriculture and Rural Development. 1998; 5(2): 17-20

[28] Jat H.S, Datta A, Choudhary M, Yadav AK, Choudhary V, Sharma PC, Gathala MK, Jat ML, McDonald A,

Effects of tillage, crop establishment and diversification on soil organic carbon, aggregation, aggregate associated carbon and productivity in cereal systems of semi-arid northwest India. Soil & Tillage Research. 2019; 190:128-138.

[29] Malik et al. Accelerating the Adoption of Resource Conservation Technologies in Rice Wheat System of the Indo-Gangetic Plains, Proceedings of Project Workshop, Directorate of Extension Education, Chaudhary Charan Singh Haryana Agricultural University (CCSHAU), June 1-2, 2005. Hisar, India: CCSHAU (2005) RWC-CIMMYT Agenda Notes. 13th Regional Technical Coordination Committee Meeting. RWC-CIMMYT, Dhaka, Bangladesh (2005)

[30] Gathala MK, Ladha JK, Saharawat YS, Kumar V, Kumar V, Sharma PK. Effect of Tillage and Crop Establishment Methods on Physical Properties of a Medium-Textured Soil under a Seven-Year Rice – Wheat Rotation. Soil Science Society of America Journal. 2011a; (75):1851-1862

[31] Saharawat YS, Ladha JK, Pathak H, Gathala M, Chaudhary N, Jat ML. Simulation of resource-conserving technologies on productivity, income and greenhouse gas emission in rice-wheat system. Journal of Soil Science and Environmental Management. 2012; 3 (1):9-22

[32] Jat ML, Malik RK, Saharawat YS, Gupta R, Bhag M, Raj Paroda. Proceedings of Regional Dialogue on Conservation Agricultural in South Asia. Proceedings of Regional Dialogue on Conservation Agricultural in South Asia, New Delhi, India, APAARI, CIMMYT, ICAR. 2012; p. 31.

[33] Gathala MK, Ladha JK, Kumar V, Saharawat YS, Kumar V, Sharma PK, Sharma S, Pathak H. Tillage and Crop Establishment Affects Sustainability of South Asian Rice-Wheat System. Agronomy Journal. 2011b; 103 p. 961672.

[34] Sinha AK, Ghosh A, Dhar T et al. 2019. Trends in key soil parameters under conservation agriculture- ased sustainable intensification farming practices in the Eastern Ganga Alluvial Plains. Soil Research. https://doi.org/10.1071/SR19162

[35] Conijn JG, Bindraban PS, Schröder JJ, Jongschaap REE. Can our global food system meet food demand within planetary boundaries?, Agriculture, Ecosystems & Environment. 2018; 251: 244-256. ISSN 0167-8809, https://doi.org/10.1016/j.agee.2017.06.001.

[36] Gan Y, Chantal Hamel1, John T, O'Donovan, Herb Cutforth, Robert P, Zentner, Con A. Campbell, Yining Niu, Lee Poppy. Diversifying crop rotations with pulses enhances system productivity Scientific Reports. 2015; 5:14625 | DOI: 10.1038/srep14625

[37] Kirkegaard JA, Ryan MH. Magnitude and mechanisms of persistent crop sequence effects on wheat. Field Crops Research. 2014; 164:54-165.

[38] Kröbel R. et al. Water use efficiency of spring wheat in the semi-arid Canadian prairies: Effect of legume green manure, type of spring wheat, and cropping frequency. Canadian Journal of Soil Science. 2014; 94:223-235.

[39] Cox HW, Kelly RM, Strong WM. Pulse crops in rotation with cereals can be a profitable alternative to nitrogen fertilizer in central Queensland. Crop and Pasture Science. 2010; 61:752– 762. http://doi.org/10.1071/CP09352

[40] Zentner RP. et al. Economics of spring wheat production systems using conventional tillage management in the

Brown soil zone - Revisited. Canadian Journal of Plant Science. 2007; 87, 27-40.

[41] Salo T, Suojala T, Kallela M. The effect of fertigation on yield and nutrient uptake of cabbage, carrot and onion. Acta Horticulturae. 2002; 571:235-241. DOI: 10.17660/ActaHortic.2002.571.28

[42] Vishandas, Hassan ZU, Rashad M, Shah AN. Phosphorus fertigation at first irrigation due to its unavailability at sowing time prevents yield losses in (*Triticum aestivum* L.). Pak. J Bot 2006;38(5):1439-1447.

[43] Kumar A, Singh AK. Improving nutrient and water use efficiency through fertigation. J Water Mangnt. 2002; 10:42-48.

[44] Shravani K, Triveni S, Latha PC, Ramulu V, Tejashree M, Damodharachari K. Influence of Biofertilizers under Drip Fertigation System on NPK Uptake and Yield Performance of Greengram (Vigna radiata L.). International Journal of Current Microbiology and Applied Sciences 2019; 8(5):2409-2417.

[45] Manikandan S, Sivasubramaniam K. Influence of drip fertigation and sowing season on plant growth, physiological characters and yield of pigeon pea (Cajanus cajan L.). African Journal of Agricultural Research. 2015;10(27):626-2632.

[46] Tarafdar J, Ramesh R, Himanshu M, Indira R. Development of zinc nanofertilizer to enhance crop production in pearl millet. Journal of Agricultural Reseacrh. 2014; https://doi.org/10.1007/s40003-014-0113–/

[47] Nahla M, Abdelhakeem MS, Mohamed A. Zinc foliar spray on snap beans using nano-Zn with N-soil application using mineral,organic and biofertilizer. Middle East Journal of Agricultural Research. 2017; 6(4):1301-1312

[48] Mohsen J, Tahereh A, Naserand S, Shahryar D. Impact of foliar application of nano micronutrient fertilizers and titanium dioxide nanoparticles on the growth and yield components of barley under supplemental irrigation. Acta Agriculturae Slovenica. 2016; 107(2):265-276

[49] Gomaa MA, Kandil EE, AbuoZeid AA, Bilkees MAS. Response of some faba bean varieties to fertilizers manufactured by nanotechnology. Journal of Advanced Agricultural Research. 2016; 21(3):384-399.

[50] Malhotra SK, Water soluble fertilizers in horticultural crops— An appraisal. Indian Journal of Agricultural Sciences. 2016;86 (10): 1245-56, October 2016/Review Article.

[51] Ebrahimpour F, Eidizadeh KH, Damghani AM. Sustainable nutrient management in maize with integrated application of biological and chemical fertilizers. International Journal of Agricultural Science. 2011; 1: 423-426

[52] Thavaprakaash N, Velayudham K, Muthukumar VB. Effect of crop geometry, intercropping systems and integrated nutrient management practices on productivity of baby corn (*Zea mays* L.) based intercropping systems. Journal of Agriculture and Biological Science. 2005;1: 295-302

[53] Rao D.L.N. Promoting biofertilizers in IPNS with improved technology and extension in India. In SAARC Training Manual on Integrated Nutrient Management for Improving Soil Health and Crop Productivity (P. Dey, S. Srivastava, N.K. Lenka, K.C. Shinogi, A.K. Vishwakarma and A. K. Patra, Eds.). ICAR- Indian Institute of Soil Science, Bhopal. 2018;71-76.

[54] Sarkar AK, Pattanayak SK, Surendra S, Mahapatra P, Arvind K,

Ghosh GK. Integrated Nutrient Management Strategies for Acidic Soils. Indian Journal of Fertilisers.2020;16 (5): 476-491.

[55] Yoo HS, Kim TG, Park TG. Surface-functionalized electrospun nanofibers for tissue engineering and drug delivery. Advanced Drug Delivery Reviews. 2009; 61:1033-1042

[56] Iqbal MA. Nano-Fertilizers for Sustainable Crop Production under Changing Climate: A Global Perspective, Sustainable Crop Production, Mirza Hasanuzzaman, Marcelo Carvalho Minhoto Teixeira Filho, Masayuki Fujita and Thiago Assis Rodrigues Nogueira, IntechOpen, 2019. DOI: 10.5772/intechopen.89089.

[57] Singh NB, Amist N, Yadav K, Singh D, Pandey JK, Singh SC. Zinc oxide nano particles as fertilizer for the germination, growth and metabolism of vegetable crops. Journal of Nano-engineering and Nanomanufacturing. 2013; 3:353-364.

[58] Rakshit R, Amitava R, Anupam Das. Customized Fertilizers: Marker in Fertilizer Revolution. International Journal of Agriculture, Environment and Biotechnology.2012;5(1): 67-75.

[59] Susan John K, Anju PS, Chithra S, Shanida Beegum SU, Suja G, Anjana Devi IP, Ravindran CS, James George, Sheela, MN, Ravi V, Manikantan Nair M, Pallavi Nair K, Remya D. Recent advances in the integrated nutrient management (INM) practices of tropical tuber crops. Technical Bulletin Series No. 75, ICAR - Central Tuber Crops Research Institute, Thiruvananthapuram, Kerala, 2019.68p.

[60] Xiao D, Feng J, Wang N, Luo X, Hu Y. Integrated soil moisture and water depth sensor for paddy fields. Computers and Electronics in Agriculture. 2013;214-221.

[61] Shah NG, Das I. Precision Irrigation Sensor Network Based Irrigation, Problems, Perspectives and Challenges of Agricultural Water Management, IIT Bombay, India. 2012;217-232.

[62] Carpena RM, Dukes MD. Automatic Irrigation Based on Soil Moisture for Vegetable Crops. 2005. AE354 http://edis.ifas.ufl.edu/ae354

[63] Abd El-kader SM, Mohammad El-Basioni. BM. Precision farming solution in Egypt using the wireless sensor network technology. Journal of Egyptian Informatics. 2013. http://dx.doi.org/10.1016/j.eij.2013.06.004.

[64] Raun WR, Solie JB, Stone ML, Martin KL, Freeman KL, Mullen RW, Zhang H, Schepers JS, Johnson GV. Optical Sensor-Based Algorithm for Crop Nitrogen Fertilization. Communications in Soil Science and Plant Analysis. 2005;36: 2759-2781.

[65] Bromfield E. "Why Vertical Farms?" Foundation Earth. 2018. http://www.fdnearth.org/files/2012/09/Why-vertical-farms_Final.pdf

[66] Al-Kodmay K. The Vertical Farm: A Review of Developments and Implications for the Vertical City." Buildings.2018. http://www.mdpi.com/journal/buildings.

[67] Patterson S. Hydroponics vs. Aquaponics: The Pros and Cons of Two Soilless Farming Methods. Off the Grid News. 2013. http://www.offthegridnews.com/survival-gardening-2/hydroponics-vs-aquaponics-the-pros-and-cons-of-two-soilless-farming-methods/print/

Chapter 16

Biostimulants as Plant Growth Stimulators in Modernized Agriculture and Environmental Sustainability

Muhammad Amjad Bashir, Abdur Rehim, Qurat-Ul-Ain Raza, Hafiz Muhammad Ali Raza, Limei Zhai, Hongbin Liu and Hongyuan Wang

Abstract

Plant growth stimulators (growth regulators + biostimulants; PGS) are chemical substances (organic/inorganic), helpful in plant growth and development. These are not considered as the replacement of fertilizers but can help in improved crop and soil quality. Both compounds can amplify the root biomass, nutrients translocation, enzymatic activities, crop yield, physiology, and nutrient uptake. Biostimulants are rich in minerals, vitamins, plant hormones, oligosaccharides, and amino acids. These compounds have a serious role to improve soil health, fertility, sorption, and desorption of nutrients. Hence, have a vital character in nutrients cycling, abiotic stress control, heavy metals bioavailability, and greenhouse gaseous emission. This chapter focuses on the discussions about the influence of plant growth regulators and biostimulants in crop production, soil health, heavy metal cycling, greenhouse gases emission with environmental sustainability. Whereas, the impact of biostimulants on greenhouse gases is a research gap.

Keywords: PGR, Biostimulants, Sustainable agriculture, sustainability

1. Introduction

Modernized agricultural practices are focusing on sustainable environmental systems. The major challenges for agriculture scientists and experts are to improve the crop quality and yield with minimum inputs focusing on environmental sustainability. To fulfill this aim, various breeding programs are introduced but it is a time-consuming and species-specific method. Apart from this, a less time-consuming and cheaper method identification is a vital need, as the use of an organic substance can stimulate healthy plant metabolism and improve their growth and development functions [1].

In addition to increasing population, food security, and environmental pressure, modern agriculture is also facing the challenges of soil degradation and reduction of arable lands. About 24 billion tons of fertile soil has been depleted worldwide due to inadequate agriculture practices and erosion. In addition, drought stress, salinization of cultivated lands, and natural disasters have poorly impacted agriculture. Approximately, 60–70% yield gap is due to abiotic stresses specifically salinity, heat stress, drought, nutrient deficiency, and hypoxia [2].

Biostimulant is a material, which when applied in minute quantity promotes plant growth. Here, the word "minute" describes how the biostimulants are different from soil amendments and nutrients that perform a similar task but are applied in higher amounts [3]. Do not confuse biostimulants with fertilizers because they do not supply nutrients directly to the plants. But facilitate the plant and soil metabolic processes to improve nutrient availability [4]. According to the new regulation (EU; 2019/1009), biostimulant is a fertilizing product that improves the plant's nutritional processes independent of its own nutrient content. Moreover, its goal is to achieve one or more following [2] characteristics of the plant and/or the plant rhizosphere.

i. nutrient use efficiency

ii. tolerance resistance to (a) biotic stress

iii. quality characteristics

iv. availability of confined nutrients in the soil or rhizosphere

Initially, plant biostimulants were used for organic production but as its benefits explored, it is now being adopted in sustainable agricultural practices and integrated cropping systems [2]. Biostimulants are the extracts derived from organic raw substances containing bioactive compounds. The common components of biostimulants are humic substances, mineral elements, amino acids, chitin, chitosan, vitamins, poly-, and oligosaccharides [1]. Biostimulants vary in their formulation and ingredients, but the major classification is on the basis of source and content (includes: hormone-containing products, amino acid-containing products), and humic substances [3].

Various substances such as enzymes, micronutrients, proteins, amino acids, phenols, humic and fulvic acid, salicylic acid, protein hydrolases, and other compounds are sources as biostimulants. Moreover, living organisms i.e., bacteria and fungi that can induce changes among the organism present in plant or soil system are included in the group of biostimulants. Biostimulants are applied via soil application in the forms of granules, powders, capsules, or solutions, or as foliar sprays. Moreover, it can be also applied via fertigation through irrigation systems and foliar application [4]. Different substances are included under the term PGS. **Table 1** summarizes the definition and examples of PGS discussed.

Terminology	Definition	Examples
Biostimulant	It is a fertilizing substance when applied in minute amount improves the plant's growth and nutrition	Seaweed extract, Protein hydrolysates.
Plant growth regulators	It is a synthetic substance that has the potential to improve plant growth and alter biological processes in plants.	Abscisic acid, Cytokinins.
Plant growth hormones	Plant growth regulators when naturally produced by and inside the plant are termed as plant hormones.	-do-

Table 1.
Concept of different plant growth stimulators.

In this chapter, we have focused on the plant growth stimulators (PGS) which comprise both plant biostimulants and growth regulators. The chapter consists of 8 sections, which describes the introduction and significance of the plant growth stimulators (Section 1), the influence of PGS on crop productivity (Section 2), the role of PGS in abiotic stress conditions (Section 3), the efficacy of PGS in soil health (Section 4), the role of PGS in heavy metals cycling (Section 5), its impact on greenhouse gas emission (Section 6), the constraints, challenges and future aspects (Section 7), and the conclusion of the chapter (Section 8), respectively.

2. Influencing the crop productivity with PGS

Plant-derived biostimulants have been reported as an innovative tool to cope with agriculture challenges and environmental sustainability. Moreover, it has been reported that plant biostimulants impact plant growth hormones that improve the plant metabolic activities and ultimately enhance crop productivity [5]. Moreover, it is reported that plant biostimulants improve chlorophyll synthesis, the mineral status, and also synthesis and accumulate antioxidant metabolites. These antioxidants reactivate photosynthetic activity and improve plant growth [2].

Biostimulants are also responsible to increase the leaf chlorophyll content. Its application to vegetables and floriculture crops has been reported to build tolerance against biotic and abiotic stresses by improving the internal and external quality. Moreover, it also reduces the fertilizer requirement thereby is recognized as a step towards environmental sustainability [1, 6]. In addition, biostimulants also enhance the thiamine levels of green beans in proportion to the thiamine content it contains [7]. Furthermore, biostimulants also influence the mechanical properties of fruits and vegetables such as firmness. It might be due to the stiffness of the cell wall that results in extensibility reduction. Biostimulants also significantly improve the cell wall flexibility that helps increase the shelf life of fruits and vegetables, thus facilitate transportation and storage processes [4].

Non-microbial and microbial plant biostimulants also have a positive impact on crop productivity. It increases plant growth and development, nutrient uptake, and translocation consequently increases the yield and biomass production in horticulture and agronomic crops. Moreover, it improves nutrient soil solubilization (both macro and micronutrients), the plant root system architecture, and enhances soil exploration. Thus, it has also been shown to influence nutrient use efficiency specifically nitrogen in plants [2]. Plant biostimulants are also responsible to improve nutrient assimilation by improving the gene expression of functioning in the plant metabolism or due to the improved nutrient uptake and transport [8].

Biostimulants based on chitosan positively impact strawberry pulp firmness and improves the shelf-life by increasing the concentration of the phenolic compound in plants [9]. Moreover, it has been demonstrated that the application of biostimulants in the absence of fertilizers improves the radish shoot and root biomass [6]. Whereas the foliar application of seaweed extract influenced the growth of soybean possibly due to the identified minerals and plant growth regulators present in the biostimulant [10]. Moreover, seaweed is also reported to improve plant water and nutrient use efficiency due to the phytohormones involved. Furthermore, the potential outcomes of seaweed to minimized abiotic stresses and nutrient deficiencies can be expected [11]. Biostimulants have a promising role in improving plant growth, metabolic activities, better stress resistance, and reduction in fertilizer use [12]. Furthermore, **Table 2** summarizes the role of PGS and plants responses respectively.

Crop	PGS	Group	Plant Response	Reference
Zea mays L. (Hydroponics)	Biostimulant	Protein Hydrolysates	Mitigate abiotic stress and regulates the transcription of the gene involved in nitrate transport	[13]
Solanum lycopersicum L.	Biostimulant	Microbe-based	Improve uptake of macro-and microelements (potassium, sodium, and manganese)	[14]
Ocimum basilicum L., *Solanum lycopersicum* L., and *Chrysanthemum indicum* L.	Biostimulant	Vegetable derived protein	Enhances Adventitious Rooting	[5]
Glycine max L.	Biostimulant	Seaweed Extract	Altered the nutraceutical and antioxidative potential and improved the growth and yield	[10]
Crocus sativus L. (Soil less condition)	Biostimulant	Arbuscular mycorrhizal fungi	Increased the polyphenol content	[15]
Lactuca sativa L.	Biostimulant	Amino Acids	Improves the Growth and Yield, Enhance Photosynthetic Assimilation and Nutrient Availability	[16]
Zea mays L.	Biostimulant	Humates and lignosulfonates	Increase root growth, enhance photosynthesis and stimulate N metabolism	[17]
Strawberry tray plants (*Fragaria × ananassa* Duch.)	Biostimulant	10 different biostimulants	Increased pulp firmness, high nutritional value, and yield and fruit quality	[9]
Trifolium pratense L. and *Lolium perenne* L.	Biostimulant (Seed coating)	Soy flour, diatomaceous earth, micronized and concentrated vermicompost	Enhance seedling growth, increased the integrity and compressive strength of seeds.	[18]
Lactuca sativa L.	Biostimulant	Seaweed extract, legume-derived protein hydrolysate and tropical plant extract	Increase leafy vegetable productivity in low fertility soils, better physiological and biochemical status	[19]
Raphanus sativus L.	Biostimulant	Biostimulant, Vitamin B12, and CoQ10	Improved root and shoot biomass	[6]

Table 2.
Plant growth stimulators and plant responses in different crops.

3. Limiting the abiotic stress by PGS

Plants face multiple stressful events throughout their lifecycle. Stresses are classified based on the nature of the trigger factor as biotic and abiotic stresses. Biotic stresses are caused by a living organism such as insects microorganisms, weeds, etc. that impact plant growth, and productivity. However, the latter one is general associated with different environmental components that negatively impact plant development and survival. Drought, non-optimal temperatures, low soil fertility, and salinity are the most common abiotic stresses that limit agriculture production globally. Whereas, in the developing countries where the life of the majority depends on the agriculture sector, drought and nutrient deficiencies major issues [20].

Besides this, the rapid increase in population leading to urbanization and increases soil erosion has poorly impacted the prime cropland. Therefore, it is the need for time to utilize the less productive soils and enhance the crop yield and productivity [16]. Biostimulants ensure a promising role to improve the productivity of vegetables and also develop tolerance against stresses. Moreover, it positively impacts the plant metabolic activities in optimal and sub-optimal environmental conditions. In addition, it is very crucial to identify the proper timing for the application of biostimulants. It depends on the critical stages of development and crop species. To avoid unexpected results, minimize production cost and wastage of products, it is good to determine the exact dose and application time of biostimulants [20].

Diverse changes in temperature and precipitation have been reported due to climate change, this has resulted in severe drought conditions [16]. Drought has adverse effects on the plant gas exchange and causes changes in transpiration and photosynthetic rates, which ultimately results in yield losses. Biostimulants can be used to overcome water stress as it is effective to improve water use efficiency in plants [20]. It is reported that the micro-algal-biostimulants minimized the drought damaging effects on tomatoes and improve plant growth. It might be due to the presence of plant growth hormones, like abscisic acid that regulates transpiration and reduce water losses, present in biostimulant [21]. Microalgae improve the total flavonoid and phenolic content in plants which also increase the enzymatic activities of antioxidants such as catalase, superoxide dismutase, ascorbate peroxidase that consequently mitigate drought-induced oxidative damage [22].

Furthermore, biostimulants are also useful against nutrient deficiency. The results have shown that the application of biostimulants cannot replace fertilizers but it can contribute to overcoming nutrient deficiency and imbalanced conditions. It is responsible for improving plant root morphology which ultimately improves nutrient uptake, translocation, and assimilation [23]. Cold stress or low temperatures adversely affect plant metabolism and delay physiological processes. It also damages the cell membrane by destabilizing phospholipid layers. Biostimulants also help to stimulate biosynthetic pathways that increase the osmotic molecule accumulation, membrane thermostability, and overcome chilling injury. Moreover, the seed priming with chitosan also improved germination and plant growth under temperature stress [24].

The most abundant environmental stress is a salinity that adversely impacts plant metabolism and growth. With the application of biostimulants to salt stress environmental conditions, the damage can be minimized as it induces the accumulation of osmolytes to increase the osmotic potential of plant cells and enhance the level of protective molecules against oxidative stress [25]. Moreover, it has been reported that the protein hydrolysates-based biostimulant significantly mitigates single as well as multiple stresses (nutrient stress + hypoxia or nutrient stress + salinity) in maize under hydroponic conditions [2].

Plant growth regulators also play a constructive role to mitigate the abiotic stress damages and improve plant development. Such as, salicylic acid is effective against drought as it increases the restoration process in plants. Moreover, putrescine fights against oxidative, drought, and salinity stress probably due to its acid-neutralizing abilities. Thus, plant growth regulators, plant growth-promoting rhizobacteria, and biostimulants could play a significant performance against environmental stresses [16].

4. Efficacy of PGS in soil health

Plant growth regulators also effectively influence the soil properties and ultimately give a huge benefit to agriculture management. As describes earlier, plant biostimulants improve nutrient availability, uptake, translocation, and assimilation, which is beneficial to organic farming. In addition, incredible results can be obtained with the integrated application of biostimulants with chemical fertilizers in soil [26]. It contributes to improving the cation exchange capacity of the soil and enhances the solubility of the nutrients in soil solution which subsequently increases the nutrient availability for plant uptake [8]. Whereas, biostimulants are not nutrients and has no direct impact on nutrients bioavailability but they have a potential to reduce the application of mineral nutrients [6].

Protein hydrolysates, an important plant biostimulant is prepared using protein sources by the process of partial hydrolysis. It comprises amino acids, mixtures of poly- peptides, and oligopeptides. When applied as foliar spray or in the soil/root system, protein hydrolysates improve the microbial biomass and activity, soil respiration by providing a rich source of C and N to microbes. Moreover, it forms complexes and chelates with soil micronutrients (i.e., Zn, Mn, Fe, and Cu), improves nutrient availability, and ultimately improves plant nutritional status [8].

Another well-known plant biostimulant is "seaweed extracts". Due to its complex biochemical composition (minerals, antioxidants, polysaccharides, hormones, vitamins, pigments, fats, oils, acids), it is highly difficult to understand its mechanism. Likewise, protein hydrolysates, seaweed extract can be applied to soil as well as plants (foliar spray). It is responsible for improving soil retention, soil microflora as well as soil remediation. Moreover, it can be a rich source of nutrients and probably have hormonal effects. Whereas, multidisciplinary approaches are required to understand the complex interaction between bioactive compounds present in the extract [11].

5. Role of PGS in heavy metals cycling

The increase in anthropogenic activities including industrial and mining activities, urbanization, use of chemicals in agriculture has potentially increased the concentration of toxic elements in soil throughout the globe [27]. Moreover, wastewater mishandling also contributes to heavy metal contamination and induces toxic effects on plant metabolic activities, soil environment, groundwater, and ultimately human health [28]. One of the most toxic trace elements is chromium (Cr). It is harmful to the soil microbes as it depresses their microbial and enzymatic activities, as well as humans if enters the food chain [29]. Therefore, it is important to minimize or treat the negative impacts of Cr. Plant growth stimulators can also play a significant role in this regard.

The application of micronutrient-amino acid is responsible to reduce the Cr stress in plants. It is reported that iron-lysine (Felys) application significantly improved plant growth and biomass. Additionally, with the increased nutrient uptake, gaseous exchange parameters also increased whereas it contributes to form complexes that reduce oxidative stress in plants. Therefore, the Cr toxicity caused due to contaminated water can be treated using Felys. Whereas, the mechanism behind the Felys attributes needs to be explored [30].

The foliar application of Zinc-lysine (Znlys) also improved rice growth and contributed to stimulating the anti-oxidant defense system, decrease oxidative stress, increase Zinc (Zn) uptake, and decreases Cr concentration in plants. Whereas, further studies are required to explore the mechanism of Znlys in mitigating Cr level in plants [31]. Moreover, Znlys is also effective to reduce cadmium (Cd) concentration in wheat along with improving Zn contents and plant growth in Zn deficient and Cd contaminated soil. It is useful to protect people from Cd risk and Zn deficiency [32].

Apart from biostimulants, plant growth regulators also contribute to phytoremediation. Plant growth regulators can be defined as the substance that can impact plant growth and is capable of altering biological processes in plants. Such compounds when produced inside the plant are known as plant hormones but when used as synthetic compounds to play a similar role are termed as plant growth regulators [33]. Auxins perform various biochemical changes in plant cells and their membranes which reduces the toxicity of metal ions [34]. Different hypotheses report that auxins cause changes in the cell membrane and alter its properties against toxic elements [33]. It also improves the rate of transpiration and ultimately heavy metal adsorption.

The use of cytokinins on *Alyssum murale* also improves the plant aerial biomass and transpiration due to the adsorption of heavy metals in soil solution. Likewise, the application of cytokinin on *Helianthus annuus* is helpful in the adsorption of Zn and lead (Pb). Studies reported that the usage of salicylic acid helps reduce the negative impacts of copper (Cu) and Pb in plants. Thus the use of plant growth stimulators is significant in heavy metal recycling [33].

6. Greenhouse gases emission and PGS

Greenhouse gases (GHGs) are defined as the gases that obtain heat energy from the sunlight and limit the backward movement of radiations from the earth's crust thus resulting in global warming. The agriculture sector contributes a lot in this aspect by the emission of methane (CH4), nitrous oxide (N_2O), and carbon dioxide (CO_2). One of the major sources of GHGs emissions is paddy systems [35]. Moreover, the long-term application of manures also increases GHGs emissions [36].

With the increase in population, the demand for food has also increased. To fulfill this challenge, the farmers tend to apply more fertilizer to attain more yield. However, field fertilization is supposed to be a major source of GHGs emissions. It is reported that agriculture produces about 90% N_2O and 20% CO_2 worldwide. Depending upon the global warming potential, carbon dioxide and nitrous oxide are two important GHGs. Moreover, after CH_4, and CO_2, N_2O is an important GHG that contributes about 6–8% of global warming [36]. Therefore, it is the need of today to introduce such agriculture practices that ensure high yield with minimum negative impacts on global warming and climate change.

Microbes-biostimulants specifically N-fixing microorganisms are responsible to increase N_2O emission. Whereas few pieces of research have been reported in this regard. Further studies need to be conducted to explore the impacts of different types of biostimulants on agro-ecosystems [37]. Additionally, the foliar application of plant growth regulators (abscisic acid and kinetin) has been shown to mitigate the N_2O emission and also manipulate plant growth and development processes in the wheat cropping system. It is possibly due to the manipulation of anatomical and physiological processes. Thus, the application of plant growth regulators might be an effective tool for modernized agriculture and environmental sustainability [38]. Apart from this, there is a huge gap to study PGS for GHG emission and reduced agro-environmental pollution, especially from non-point sources.

7. Constraints, challenges, and future aspects

Agriculture production is facing multiple challenges including food security, climate change, soil restoration, and environmental sustainability. Plant growth stimulators have the potential to overcome the biotic and abiotic stresses, improve crop productivity, better soil nutrient cycling, mitigation of heavy metal uptake, and GHGs emission. Due to the complex composition of biostimulants, their potential role and mechanism are still not clear. Therefore, further studies about the possible use of biostimulants to minimize GHGs emission needs to be studied.

Biostimulants are also expected to increase NUE, therefore, careful monitoring, climatic aspects, and related modeling need to be studied. Limited literature related to the impacts of biostimulants under sub-optimal nitrogen regimens is available. Further studies can bring huge benefits to modern agriculture. Moreover, open field studies should also be carried out in the future, which then moves back to the lab for further elucidation. Few field-based research using biostimulants has been reported which is a major constraint to adopt its application in farming. Furthermore, the farming community should be introduced with the product along with the cost–benefit analysis, thus moving towards environment sustainability.

Currently, biostimulants are gaining popularity in the market. Variety of biostimulants with different active ingredients (humic substances, seaweed extracts, microbial amendments, and amino acids) are estimated to account for $2.6 billion and will reach about $5 billion by 2025 [2]. Therefore, it is time to introduce new and beneficial biostimulants that have practical application and yield benefits for sustainable agriculture. Integration of plant biostimulants and growth regulators with nutrients and waste organics for different regions can be introduced with also helps in minimizing pollution and waste recycling.

8. Conclusion

A sustainable agriculture management system needs to be adopted for dealing the adverse climatic conditions. Plant growth stimulators have positively contributed towards modern agriculture and have the potential to improve crop yield, plant growth, and development, handle biotic and abiotic stresses, minimize heavy metal translocation, and contribute to mitigating GHGs emissions. Biostimulants are rich in multi-nutrients but cannot replace fertilizers. Although it has a potential to improve soil quality and plant productivity even under stress conditions. Whereas, few researches regarding GHGs emission using biostimulants has been reported. It is a huge research gap. Whereas, different mechanisms need to be studied further to develop a better understanding and introducing useful products.

Acknowledgements

This study was financially supported by the National Natural Science Foundation of China (31972519).

Author details

Muhammad Amjad Bashir[1,2], Abdur Rehim[1*], Qurat-Ul-Ain Raza[1], Hafiz Muhammad Ali Raza[1], Limei Zhai[2], Hongbin Liu[2] and Hongyuan Wang[2*]

1 Department of Soil Science, Bahauddin Zakariya University, Multan, Pakistan

2 Key Laboratory of Nonpoint Source Pollution Control, Ministry of Agriculture, Institute of Agricultural Resources and Regional Planning, Chinese Academy of Agricultural Sciences, Beijing, P.R. China

*Address all correspondence to: abdur.rehim@bzu.edu.pk
and wanghongyuan@caas.cn

References

[1] Bulgari, R.; Cocetta, G.; Trivellini, A.; Vernieri, P.; Ferrante, A. Biostimulants and crop responses: A review. Biol. Agric. Hortic. 2015, *31*, 1-17.

[2] Rouphael, Y.; Colla, G. Toward a sustainable agriculture through plant biostimulants: From experimental data to practical applications. Agronomy **2020**, *10*, doi:10.3390/agronomy10101461.

[3] du Jardin, P. Plant biostimulants: Definition, concept, main categories and regulation. Sci. Hortic. (Amsterdam). **2015**, *196*, 3-14, doi:10.1016/j.scienta.2015.09.021.

[4] Drobek, M.; Frąc, M.; Cybulska, J. Plant biostimulants: Importance of the quality and yield of horticultural crops and the improvement of plant tolerance to abiotic stress-a review. Agronomy **2019**, *9*, 335, doi:10.3390/agronomy9060335.

[5] Kim, H.J.; Ku, K.M.; Choi, S.; Cardarelli, M. Vegetal-derived biostimulant enhances adventitious rooting in cuttings of Basil, tomato, and chrysanthemum via brassinosteroid-mediated processes. Agronomy **2019**, *9*, doi:10.3390/agronomy9020074.

[6] Rehim, A.; Amjad Bashir, M.; Raza, Q.-U.-A.; Gallagher, K.; Berlyn, G.P. Yield Enhancement of Biostimulants, Vitamin B12, and CoQ10 Compared to Inorganic Fertilizer in Radish. Agronomy **2021**, *11*, 697, doi:10.3390/agronomy11040697.

[7] Russo, R.O.; Berlyn, G.P. Vitamin-Humic-Algal Root Biostimulant Increases Yield of Green Bean. *HortScience* **1992**, *27*, 847, doi:10.21273/hortsci.27.7.847.

[8] De Pascale, S.; Rouphael, Y.; Colla, G. Plant biostimulants: Innovative tool for enhancing plant nutrition in organic farming. *Eur. J. Hortic. Sci.* **2017**, *82*, 277-285, doi:10.17660/eJHS.2017/82.6.2.

[9] Soppelsa, S.; Kelderer, M.; Casera, C.; Bassi, M.; Robatscher, P.; Matteazzi, A.; Andreotti, C. Foliar applications of biostimulants promote growth, yield and fruit quality of strawberry plants grown under nutrient limitation. Agronomy **2019**, *9*, 483, doi:10.3390/agronomy9090483.

[10] Kocira, S.; Szparaga, A.; Kuboń, M.; Czerwińska, E.; Piskier, T. Morphological and biochemical responses of Glycine max (L.) Merr. To the use of seaweed extract. Agronomy **2019**, *9*, 1-23, doi:10.3390/agronomy9020093.

[11] El Boukhari, M.E.M.; Barakate, M.; Bouhia, Y.; Lyamlouli, K. Trends in seaweed extract based biostimulants: Manufacturing process and beneficial effect on soil-plant systems. Plants **2020**, *9*, doi:10.3390/plants9030359.

[12] Russo, R.O.; Berlyn, G.P. The use of organic biostimulants to help low input sustainable agriculture. J. Sustain. Agric. **1990**, *1*, 19-42.

[13] Trevisan, S.; Manoli, A.; Quaggiotti, S. A novel biostimulant, belonging to protein hydrolysates, mitigates abiotic stress effects on maize seedlings grown in hydroponics. Agronomy **2019**, *9*, 1-15, doi:10.3390/agronomy9010028.

[14] Allaga, H.; Bóka, B.; Poór, P.; Nagy, V.D.; Szucs, A.; Stankovics, I.; Takó, M.; Manczinger, L.; Vágvölgyi, C.; Kredics, L.; et al. A composite bioinoculant based on the combined application of beneficial bacteria and fungi. Agronomy **2020**, *10*, doi:10.3390/agronomy10020220.

[15] Caser, M.; Victorino, Í.M.M.; Demasi, S.; Berruti, A.; Donno, D.;

Lumini, E.; Bianciotto, V.; Scariot, V. Saffron cultivation in marginal alpine environments: How AMF inoculation modulates yield and bioactive compounds. Agronomy **2019**, *9*, 1-14, doi:10.3390/agronomy9010012.

[16] Khan, N.; Bano, A.M.D.; Babar, A. *Impacts of plant growth promoters and plant growth regulators on rainfed agriculture*; 2020; Vol. 15; ISBN 1111111111.

[17] Ertani, A.; Nardi, S.; Francioso, O.; Pizzeghello, D.; Tinti, A.; Schiavon, M. Metabolite targeted analysis and physiological traits of zea mays L. In response to application of a leonardite-humate and lignosulfonate-based products for their evaluation as potential biostimulants. Agronomy **2019**, *9*, 1-19, doi:10.3390/agronomy9080445.

[18] Qiu, Y.; Amirkhani, M.; Mayton, H.; Chen, Z.; Taylor, A.G. Biostimulant seed coating treatments to improve cover crop germination and seedling growth. Agronomy **2020**, *10*, 1-14, doi:10.3390/agronomy10020154.

[19] Mola, I. Di; Cozzolino, E.; Ottaiano, L.; Giordano, M.; Rouphael, Y.; Colla, G.; Mori, M. Effect of vegetal- And seaweed extract-based biostimulants on agronomical and leaf quality traits of plastic tunnel-grown baby lettuce under four regimes of nitrogen fertilization. Agronomy **2019**, *9*, doi:10.3390/agronomy9100571.

[20] Bulgari, R.; Franzoni, G.; Ferrante, A. Biostimulants application in horticultural crops under abiotic stress conditions. Agronomy 2019, *9*.

[21] Oancea, F.; Velea, S.; Fatu, V. Micro-algae based plant biostimulant and its effect on water stressed tomato plants. Rom. J. Plant Prot. **2013**, *6*, 104-117.

[22] Kusvuran, S. Microalgae (Chlorella vulgaris Beijerinck) Alleviates Drought Stress of Broccoli Plants by Improving Nutrient Uptake, Secondary Metabolites, and Antioxidative Defense System. Hortic. Plant J. **2021**, doi:10.1016/j.hpj.2021.03.007.

[23] Toscano, S.; Romano, D.; Massa, D.; Bulgari, R.; Franzoni, G.; Ferrante, A. Biostimulant applications in low input horticultural cultivation systems. *Italus Hortus* **2018**, *25*, 27-36, doi:10.26353/J.ITAHORT/2018.1.2736.

[24] Ashraf, M.A.; Akbar, A.; Askari, S.H.; Iqbal, M.; Rasheed, R.; Hussain, I. Recent advances in abiotic stress tolerance of plants through chemical priming: An overview. Adv. Seed Priming **2018**, 51-79, doi:10.1007/978-981-13-0032-5_4.

[25] Semida, W.M.; Abd El-Mageed, T.A.; Hemida, K.; Rady, M.M. Natural bee-honey based biostimulants confer salt tolerance in onion via modulation of the antioxidant defence system. J. Hortic. Sci. Biotechnol. **2019**, *94*, 632-642, doi:10.1080/14620316.2019.1592711.

[26] Nephali, L.; Piater, L.A.; Dubery, I.A.; Patterson, V.; Huyser, J.; Burgess, K.; Tugizimana, F. Biostimulants for plant growth and mitigation of abiotic stresses: A metabolomics perspective. Metabolites **2020**, *10*, 1-26, doi:10.3390/metabo10120505.

[27] Visconti, D.; Caporale, A.G.; Pontoni, L.; Ventorino, V.; Fagnano, M.; Adamo, P.; Pepe, O.; Woo, S.L.; Fiorentino, N. Securing of an industrial soil using turfgrass assisted by biostimulants and compost amendment. Agronomy **2020**, *10*, 1-25, doi:10.3390/agronomy10091310.

[28] Bashir, M.A.; Rehim, A.; Liu, J.; Imran, M.; Liu, H.; Suleman, M.; Naveed, S. Soil survey techniques determine nutrient status in soil profile and metal retention by calcium carbonate. Catena **2019**, *173*, 141-149, doi:10.1016/j.catena.2018.10.015.

[29] Hassan, Z.; Ali, S.; Farid, M. Effect of Chromium (Cr) on the Microbial and Enzymatic Activities in the Soil : A Review. In *Biodiversity conservation in changing climate*; 2017; pp. 1-20 ISBN 9789385995033.

[30] Zaheer, I.E.; Ali, S.; Saleem, M.H.; Imran, M.; Alnusairi, G.S.H.; Alharbi, B.M.; Riaz, M.; Abbas, Z.; Rizwan, M.; Soliman, M.H. Role of iron–lysine on morpho-physiological traits and combating chromium toxicity in rapeseed (*Brassica napus* L.) plants irrigated with different levels of tannery wastewater. *Plant Physiol. Biochem.* **2020**, *155*, 70-84, doi:10.1016/j.plaphy.2020.07.034.

[31] Hussain, A.; Ali, S.; Rizwan, M.; Zia ur Rehman, M.; Hameed, A.; Hafeez, F.; Alamri, S.A.; Alyemeni, M.N.; Wijaya, L. Role of Zinc–Lysine on Growth and Chromium Uptake in Rice Plants under Cr Stress. *J. Plant Growth Regul.* **2018**, *37*, 1413-1422, doi:10.1007/s00344-018-9831-x.

[32] Rizwan, M.; Ali, S.; Hussain, A.; Ali, Q.; Shakoor, M.B.; Zia-ur-Rehman, M.; Farid, M.; Asma, M. Effect of zinc-lysine on growth, yield and cadmium uptake in wheat (Triticum aestivum L.) and health risk assessment. Chemosphere **2017**, *187*, 35-42, doi:10.1016/j.chemosphere.2017.08.071.

[33] Rostami, S.; Azhdarpoor, A. The application of plant growth regulators to improve phytoremediation of contaminated soils: A review. Chemosphere **2019**, *220*, 818-827, doi:10.1016/j.chemosphere.2018.12.203.

[34] Hac-Wydro, K.; Sroka, A.; Jabłońska, K. The impact of auxins used in assisted phytoextraction of metals from the contaminated environment on the alterations caused by lead(II) ions in the organization of model lipid membranes. Colloids Surfaces B Biointerfaces **2016**, *143*, 124-130, doi:10.1016/j.colsurfb.2016.03.018.

[35] Bashir, M.A.; Liu, J.; Geng, Y.; Wang, H.; Pan, J.; Zhang, D.; Rehim, A.; Aon, M.; Liu, H. Co-culture of rice and aquatic animals: An integrated system to achieve production and environmental sustainability. J. Clean. Prod. 2020, *249*, 119310.

[36] Zhang, T.; Liu, H.; Luo, J.; Wang, H.; Zhai, L.; Geng, Y.; Zhang, Y.; Li, J.; Lei, Q.; Bashir, M.A.; et al. Long-term manure application increased greenhouse gas emissions but had no effect on ammonia volatilization in a Northern China upland field. Sci. Total Environ. **2018**, *633*, 230-239, doi:10.1016/j.scitotenv.2018.03.069.

[37] Souza, E.F.C.; Rosen, C.J.; Venterea, R.T. Contrasting effects of inhibitors and biostimulants on agronomic performance and reactive nitrogen losses during irrigated potato production. F. Crop. Res. **2019**, *240*, 143-153, doi:10.1016/j.fcr.2019.05.001.

[38] Bordoloi, N.; Baruah, K.K.; Thakur, A.J. Effectiveness of plant growth regulators on emission reduction of greenhouse gas (Nitrous oxide): An approach for cleaner environment. J. Clean. Prod. **2018**, *171*, 333-344, doi:10.1016/j.jclepro.2017.09.284.

Chapter 17

Role of Nanoparticles in Abiotic Stress

Mohd. Tariq, Shipra Choudhary, Harjeet Singh, Mohd. Asif Siddiqui, Hirdesh Kumar, Asad Amir and Neelesh Kapoor

Abstract

Nanotechnology is currently seeking much attention of researchers because of their wide applications in diverse sectors including agriculture. The influence of nanoparticles on physiological state of plants at the different levels of their organization, beginning from molecular, has been studied at various plants. It is known that nanoparticles in different concentrations can impact both positive and negative biological effects. Nanomaterials confer profound uses for sustainable crop production, reducing loss of nutrients, suppression of diseases and thereby enhancing the yields. Concerning the role of nanomaterials in alleviating the damage of plant abiotic stresses or in inhibiting plant growth and its toxicity, further studies are essential under different levels including plant molecular and cellular levels. A wide variety of research has been conducted to study plant responses to waterlogging stress that include various disciplines like molecular, biochemical, and physiological, anatomical and morphological examinations. Nano technological implications for curbing water-logged conditions recently came into limelight and have drawn much attention in the last few years. Nanotechnology is defined as the systems and processes which operate at a scale of 100 nm or less. Nanotechnology has many applications in the field of agriculture. There are majority of nano-materials which are known for its plant growth promoting effects. Nanoparticles have unique physiochemical properties such as high reactivity, particle morphology, and large surface area. They also boost the plant metabolism.

Keywords: abiotic stress, crop plants, heat stress, heavy metals, nanoparticles, salinity

1. Introduction

Population explosion during the last few decades has led to increased pressure on the agriculture sector by an upsurge of continuously increasing food demand. Natural resources of the world are continuously diminishing at a much faster pace than their renewal and the agriculture sector is no exception to this presently prevailing scenario. Sustainability issues due to population explosion, climate change, urbanization, habitat loss assisted by environmental issues are some of the global challenges faced by the green plants including the agriculture sector [1]. Plants, the vital component of our planet remain always exposed to different environmental variations and numerous stress factors throughout their life. Unlike animals, plants are deprived of motility

IntechOpen

to a better place on the arrival of any kind of stress either *biotic* or *abiotic*. To combat such stresses, nature has provided these living entities with certain defensive mechanisms that help these sessile organisms to endure these unpleasant situations. Though plants develop several mechanisms which involve avoidance, escapism, and tolerance, to deal against adverse conditions their responses could vary appreciably even in the same plant species. For this reason, the identification of tolerant plant species is always the major concern towards sustainable agriculture and crop production [2]. Major abiotic stresses which affect plants include *heat, salinity, cold drought, flooding/submergence (anoxia), chemical toxicities*, and *excess light* [3].

Technological advancements in the last few decades have led to profound structural changes in the agriculture sector and improvisation of plant health dealing with different *abiotic* stresses, improvements required to increase the production rate in ways that promote food security and public health improvement remains the matter of concern. So, there is a major concern among scientific communities to raise world food crop production by 70% [4]. In such varying environmental scenarios, it is needful to recognize an area of research to conquer the technical challenges in addressing the yield barrier, resource use efficiency, and development of environmentally accepted technology [5].

Nanotechnology and nano-sciences have come out as powerful and promising tool dealing with nearly all the aspects of the masses and people's life in 21st century that include medicine, agriculture, industrial, environment, electronics with application in numerous preparations [6, 7]. Precise potential to control and fabricate matters at nano-scale remain the beauty of this newly emerging scientific discipline. Nanotechnology has emerged out broadly into the '*agri-food sector*' which include the nanosensors, tracking devices, targeted delivery of required components, food safety, new product developments, precision processing, smart packaging, and others [8–11]. Nanotechnology offers a wide research area and provides possibilities for a large scope of diverse applications and advantages in fields of biotechnology and agriculture-based research such as disease prevention [12], nutrient management by nano-fertilizers [13], nano-pesticides or nano-herbicides [14, 15], mitigating abiotic stress [2]. Also, nanotechnology holds good promises for solving the problem associated with abiotic stresses to obtain sustainability in the field of agriculture [2].

Improving plant traits against different diseases and abiotic and biotic stresses such as drought, salinity, plant diseases, and others is one of the primary objectives of biotechnological research. Nanotechnology-enabled gene sequencing is expected to introduce rapid and cost-effective capability within a decade [16], thereby leading to more effective identification and usage of plant gene trait resources that could help plants in overcoming adversities due to different abiotic stresses. Considering these issues in this article, we are dealing with how nanotechnology can be made useful for mitigating various abiotic stresses of crops and various mechanisms associated with them [1].

2. Abiotic stress in crops and current scenario

Plants are constantly exposed to various stress factors throughout their life span. As per the data available, the relative decreases in potential maximum yields associated with abiotic stress factors vary between 54 and 82% [17]. Crops confront various types of abiotic stress and it has been well documented as well that among stresses, extreme temperatures (freezing, cold, heat), water availability (drought, flooding), and ion toxicity (salinity, heavy metals) are the major causes which adversely affect the plant growth and productivity worldwide [18–21]. These abiotic stresses are interconnected to osmotic stress that results in the disruption of ion

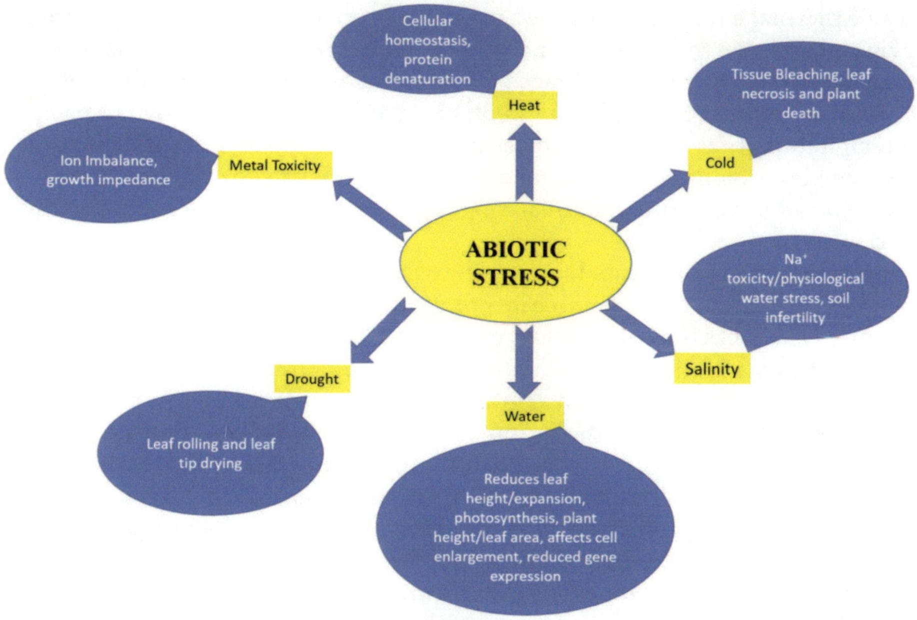

Figure 1.
Types of abiotic stress with their effects on growth of plants.

distribution and cell homeostasis. Crop plants are adversely affected by abiotic stress conditions. On account of the current scenario, more than about 50% loss of yield/year is sole because of abiotic stresses such as drought, salinity, heat, and cold. In developing countries, drought and low soil fertility has been proved to be a major cause for affecting crop production [22]. Recently, several transcription factors (TFs) due to their efficacy as a master regulator, have been proving as a potential candidate for genetic engineering to bread stress-tolerant crops and improve stress tolerance [23]. Six Asian region countries namely Bangladesh, China, India, Indonesia, Pakistan, and Thailand are actively involved in the Research and Development related activities for the development of abiotic stress-tolerant crops [24]. Various abiotic factors along with their probable significance are depicted in **Figure 1**.

3. Role of nanotechnology in abiotic stress

Nanotechnology is a platform for developing tools and technology for the improvement of the bio system [25]. Nanoparticles (NPs) are small molecular aggregates with dimensions of 1-100 nm [26]. NPs have been investigated to improve plant growth, development, and productivity and thus, proving their use to overcome various abiotic and biotic stress of crops [27]. In the last decade, the science of nanotechnology has attained, a promising position to mitigate the constraints associated with the aforementioned stresses to achieve a secure future of agriculture worldwide and nano technological findings possess immense potential to open up numerous ways in the field of biotechnology and agriculture [28, 29]. Some of them are discussed here in detail:

3.1 Heat stress

The negative impacts of heat stress (also Thermal stress) on plants are substantial, detrimental, and often account for reduced crop yield and productivity as well.

Adverse thermal environments pose a great challenge for crop plants to sustain and survive. In addition to it, another major concern remains the global climatic change i.e., an overall increase in the average global temperature of the earth that had led to increased thermal stress on plants and other organisms along with altered patterns of precipitation. Leaving aside all these problems, defining and quantifying heat stress remains a daunting task. In general, heat stress is categorized relative to some estimate of an optimal thermal range that is characteristic of each species in question.

Heat stress involves elevated temperature at such a harsh level for a long enough time that could result in irreparable loss to the development of plants [30, 31]. Heat stress enhances the Reactive Oxygen Species (ROS) generation and causes oxidative stress, as a result of which membrane lipid degeneration and leakage of membrane ion occur which led to degradation of the protein [32–35], in addition to decreased rate of photosynthesis and chlorophyll content [36]. Several studies have been conducted by many workers from time to time to access the applicability of nanotechnology to minimize heat stress. Selenium nanoparticle application in the low concentration found reducing the effect of heat stress by increasing hydration ability, chlorophyll content, and development of plant [37]. Also, Selenium nanoparticles at low concentrations exhibit antioxidative properties to plants, while oxidative stress had been induced by the high concentration of Se nanoparticles [38, 39]. Plants synthesize several heat shock proteins and molecular chaperones during the period of heat stress [40]. Other proteins are assisted by heat shock proteins in sustaining their fidelity in stress conditions [30] and are involved in heat stress resistance. It was already in reports that multiwall carbon nanotubes could upregulate gene expression of heat shock proteins *viz.* HSP90 [41]. Also, maize plants exposed to CeO2 nanoparticles depicted excessive generation of H2O2 and upregulation of HSP70 [42]. Furthermore, TiO2 nanoparticles treatment reduced the effect of heat stress by stomatal opening regulation [43].

3.2 Salinity

Salinity, a major type of abiotic stress factor, limits the production of food and deteriorates the quality of ever-increasing growth in food crops. For scientific communities, increased salinity remains a major constraint to attain sustainable crop production. Worldwide, 20% of cultivated land is facing salinity stress and the amount is increasing day by day. The majority of crop plants species belong to the category of glycophytes, which are highly susceptible to salt stress hence are the most critical environmental abiotic stress that can ruin crop production [44, 45]. Most salinity problems arise due to excess sodium chloride (NaCl) which is widely distributed along with coastal and arid region soils and water supplies. Higher levels of NaCl impose at least three types of problems for higher plants. These include: (i) the osmotic pressure in the external solution can exceed the osmotic pressure in the plant cells and therefore require an osmotic adjustment by the cells to avoid desiccation; (ii) sodium, in excess, can disrupt the uptake and transport of nutritional ions such as K and Ca; and (iii) both Na and Cl can exert direct toxic effects on membranes and enzyme systems [46]. Besides the aforementioned problems, lowering of soil osmotic potential, creation of nutritional imbalance, enhancing specific ionic toxicity (salt stress), or one or more combination of these factors, are some of the common implications that salinity stress exerts on crop plants. Most vital processes of plants like photosynthesis, protein synthesis, and lipid metabolisms, etc. are badly affected by salinity stress [47]. Salt stress is associated with oxidative stress too. However, to confront salt stress-induced oxidative stress, plants are very well equipped with a defense system of various antioxidant enzymes that include superoxide dismutase (SOD) and peroxidase (POD). The SOD constitutes the first line of defense against ROS [48] and dismutase superoxide radicals to H2O2, whereas POD reorganizes H2O2 into water and oxygen [49]. Besides oxidative stress, salt stress also creates osmotic stress, which reduces the ability of plants to take up water and minerals [50]. Also, plants have been found to abide by osmotic stress by the provision of accumulation of osmolytes, such as proline (Pro) and Glycine Betaine (GB) [51]. Application of nano-fertilizers is a quite hopeful method that can potentially increase plant resource use efficiency and help in reducing environmental toxicity due to the accumulation of unused chemical fertilizers and pesticides in the soil. Therefore, the application of nano-fertilizers could serve as an alternative approach to overcome soil toxicity issues and other associated stresses.

Adverse effects of salinity stress on crop plants have been extensively studied by many workers from time to time. Hussein and Abou-Baker [52] conducted experiments to study the foliar application of nano zinc to mitigate the adverse effect of salinity and confirmed that diluted seawater could be used in the irrigation of the cotton plant. They reported that increasing the application rate of nano-Zn may reduce phosphorous (P) absorption and translocation to leaves and consequently reduce the P/Zn ratio. They suggested that an additional dose of P-fertilizer with nano-Zn could be used to avoid the P/Zn imbalance. Avestan et al., [53] in their investigations proposed that salinity stress treatments were detrimental to morphological and physiological parameters of strawberry plants. They found that nSiO2 treatments suppressed the negative effects of salinity, possibly by improving the Epicuticular Wax Layer (EWL); and nSiO2 treatments enabled salt-stressed plants to better maintain their chlorophyll content and leaf relative water content (RWC) and relative water protection (RWP) relative to controls (no SiO2). They concluded their findings by suggesting three possible directions for future research: (1) Further exploring how variation in the timing of silicon treatments influences EWL deposition by testing EWL at multiple plant developmental stages; (2) investigation of whether there is genetic variation for EWL deposition in strawberry; and (3) testing to distinguish the benefit of greater EWL deposition in saline conditions relative to the benefit of the other signaling and physiological changes that are linked to increased silicon uptake.

Khan [54] in his studies investigated the effect of nano TiO2 in several plant developmental processes including defense against environmental stresses. They concluded that the cumulative effect of the parameters under consideration contributed to improved growth and yield of tomato plants. Therefore, based on the assessment of results it was propounded that nano-TiO2 at the rate of 20 mg/l proved best in enhancing the growth, yield, and quality of tomatoes. In one more study, conducted by Yassen et al., [55] on the cucumber (*Cucumis sativa*) effect of silicon dioxide nanoparticles was assessed where the results indicated an increase in nitrogen and phosphorus, content and uptake and decrease in Na content and uptake when adding SiO2 nano fertilizer. The findings of the study suggested that silicon dioxide nano fertilizer can exert a positive effect on the growth and yield of cucumber.

3.3 Heavy metal stress

Nano biotechnology growing as a technology that could make the environment cleans. Nanoparticles, often regarded as particles having a significant amount of surface area with unique physical and chemical properties and having applications in reducing the negative effects of heavy metals on the natural wealth [56, 57]. Some workers have exploited nanotechnology to explore plant phytotoxicity caused by heavy metals in various environments. Although nanoparticles are cost-effective in reducing heavy metal toxicity in plants [58], mitigation of heavy metal-induced root growth inhibition and oxidative stress in the plant has been barely studied [58, 59].

Heavy metal ions were productively adsorbed by magnetic nanoparticles (Fe_3O_4) [57]. In addition, Nanoscale zero-valent iron (nZVI) nanomaterials are core-shell structures that are in use for decreasing metal toxicity. Ronavari et al. [60] reported that nZVI nanoparticles are for immobilizing heavy metal ions due to their distinct structure. Also, Fajardo et al. [61] found that lead and zinc mobility and availability decreased when soils were treated with nZVI. The addition of nZVI and active carbon efficiently immobilized copper, lead, cadmium, and chromium in sediments, thus, decreasing the bioavailability and toxicity of heavy metals [62].

Nano hydroxyapatite (nHAp) particles are also in use to remediate metal toxicity. nHAp have been successfully applied to remediate soils contaminated by metals and to purify wastewater due to their outstanding ability to absorb heavy metals like copper (II), zinc (II), lead (II), and cadmium (II) [63]. Zhang et al. [64] found that nHAp effectively decreased the exchangeable fractions of Pb and Cd in contaminated sediments, especially for Pb, and dramatically decreased the metal(loid) ion concentration in pore water.

Carbon nanotubes (CNTs) were discovered by [65] and can be used as absorbents. They can be (i) single-walled carbon nanotubes (SWCNTs) and (ii) multi-walled carbon nanotubes (MWCNTs) [66, 67] and are promising nanomaterial to remove organic and inorganic toxic compounds [68, 69].

3.4 Drought stress

Plants have been always combating water stress for millions of years, ever since they first left the water bodies and conquered and colonized dry land. When drought strikes, higher plants are the first victims that have always been obliged to endure it or to adjust their life cycles to avoid it. Thus, a major means of propulsion behind the evolution and emergence of land plants has been their need to search for water, to absorb it, to transport it, and retain it. Even so, drought is still the major constraint to crop production [46, 70, 71]. The term '*Drought*' does not merely represent lack of rainfall instead for plant physiologists, it is a concurrence of various environmental stresses that includes: (i) low soil moisture availability; (ii) high evaporative load, (iii) high temperature, (iv) high solar irradiance, (v) increased soil hardness, (vi) unavailability of nutrients and (vii) accumulation of salts in the topsoil region.

Taran et al., [72], in their studies have shown that Cu-Zn-nanoparticles reduced the negative effect of drought action upon plants of steppe ecotype Acveduc. In particular, increased activity of antioxidative enzymes reduced the level of accumulation of Thiobarbituric Acid Reactive Substances (TBARS) and stabilized the content of photosynthetic pigments, and increased relative water content in leaves. Colloidal solution of Cu-Zn-nanoparticles had a less significant influence on these indexes in seedlings of the *Stolichna* variety under drought. They studied the use of binary compositions of nanoparticles in agro-technologies to enhance the biological productivity of agriculture systems. Ashkavand et al., [73] studied the effect of SiO_2 nanoparticles on drought resistance in hawthorn seedlings and concluded that silicon nanoparticles (SNPs) can increase plant resistance to drought stress. It could be explained by the improvement of photosynthesis rate and stomatal conductance by SNPs pretreatments. Application of silicon on two sorghums (*Sorghum bicolor* (L.) Moench) cultivars possessing different drought susceptibility exhibited improved drought tolerance irrespective of their drought susceptibility by lowering shoot to root (S/R) ratio, which perhaps could be an indicator of improved root growth and the maintenance of the photosynthetic rate. These findings could be attributed to improving the drought tolerance of sorghum via the augmenting water uptake efficiency of plants [2, 74]. Applications of silver nanoparticles

(AgNPs) has also been appreciated in diminishing negative effects of drought stress on lentil (*Lens culinaris* Medic). Significant effects of different concentrations of Polyethylene glycol (PEG) and silver nanoparticles on germination rate and germination percentage, root length, root fresh, and dry weight in lentil seeds were reported [75]. In a study, conducted by Sedghi et.al [76], it was observed that nano zinc oxide has the potential to enhance seed germination percentage thereby, overcoming water stress.

3.5 Water logging

Over irrigation, prolonged periods of precipitation coupled with poor soil drainage system gives rise to a condition called *'water logging'*. Both, natural vegetation and agriculture crops are equally affected by this worldwide occurring condition of waterlogging. The waterlogged soils offers/presents an unpleasant and uneasy environment for normal growth and development of plants because: (i) air spaces occupied by water delays the exchange and diffusion of gases between the roots (rhizosphere) and atmosphere [77]; (ii) levels of dissolved oxygen are depleted from soil solution by respiration of soil inhabitants and roots [78] and (iii) flooding of fields is often associated with the release of toxic compounds and obnoxious gases. Depending upon the height of the water column produced, flooding can be classified as (i) *waterlogging*, when it is superficial and encase only the roots, and (ii*) submergence*, when water completely covers the aerial plant tissues [79]. In both types of flooding, the movement of oxygen from the air to plant tissues is highly disrupted [80], producing a natural condition known as hypoxia (<21% O2) [79].

Depending upon certain parameters like temperature, microbial respiration activity, frequency, and duration of soil saturation, the depletion of dissolved oxygen in waterlogged soils leads to conditions called *'hypoxia* 'and *'anoxia'* within few hours to days. In recent years, flooding stress and its subordinates like submergence, waterlogging, hypoxia, and anoxia, were investigated extensively in plants, especially in *Arabidopsis* and rice, to pinpoint molecular elements that may play a vital role in flood tolerance. Roots of the plants remain the first victims that are worst hit by flooding. Plant roots facing waterlogging stress follow glucose metabolism according to the classical scheme of alcoholic fermentation in an oxygen deficit medium *(anaerobiosis)*, where self-poisoning of tissues takes place as a result of the formation of end products of fermentation mainly ethanol. Maintenance of an appropriate oxygen supply and energy balance is paramount for the survival of the root system to waterlogging stress.

Nanotechnology has provided new discernment to the problems arising in plants and food science (post-harvest products) and offers novel approaches to the rational selection of raw materials. Silver Nano Particles (SNPs) are the most commonly used nanomaterials in the field of nanotechnology after carbon nano-tubes that every day is added in its application to the nano-world. In this sense, nanoparticles are useful tools as an excessive water supply induces hypoxia in plants [80], increases the vulnerability to pathogen attack [81], and limits the flow of light to the plant [82].

During recovery after a flooding event, plants experience oxidative stress [83] and must remobilize nutrients to achieve a normal homeostatic state [84]. Concerning the protection of plants against oxidative stress, nanomaterials are found to mimic the role of first-line defense antioxidative enzymes like peroxidase, superoxide dismutase, and catalase, which are supposed to form the antioxidant defense grid [85]. Also, plants respond to flooding and the associated stress by changes in gene expression that are finely regulated at a multilevel scale from epigenetics [86] to transcriptional [80, 87] and translational regulation [88].

Rezvani et al., [89] conducted experiments to study the effect of Nano silver ions (as an ethylene inhibitor on the growth of Saffron (*Crocus sativus*) under flooding conditions. Corms of saffron were soaked with different concentrations of nano-silver ranging from 0 to 120 ppm (0, 40, 80, and 120) and planted under flooding stress or non-flooding stress conditions and the results of the investigations showed that the number of roots, root length, fresh and dry weight of roots and leaves were reduced by 10-day flooding stress. Soaking the saffron corms with 40 or 80 ppm concentration of Nano silver rewarded the effect of flooding stress on the root number by increasing it. Also, it was found that 40 ppm of nano-silver increased the root length in stress. 80 ppm concentration of nano-silver was found to increase leaves dry weight. In another study conducted on the same plant (*C. sativus*) under flooding stress, foliar application of Nano silver was accessed by Sorooshzadeh *et al.,* [90]. Results of the investigations showed that flooding stress led to a significant reduction in weight and height of the plant and the number of corms per plant was increased by increasing the concentration of nano-silver. In all, they concluded that flooding stress and Nano silver had a significant interaction effect on all parameters under consideration of the study.

4. Mechanism of abiotic stress control by nanoparticles (NPs)

Developing technology for improving food production, minimizing crop productivity loss is the prerequisite for obtaining sustainability in the field of agriculture. Abiotic stress of plants is considered a major emerging problem in the field of agriculture, its diverse types include salinity drought, waterlogging, submergence, heavy metal stresses, and mineral and metal toxicity/deficiencies that minimize crop growth and productivity [91–93]. A decrease in productivity is mainly attributed to these factors. Plant throughout their lifespan has to face various types of abiotic stress and has to come up with strong defense mechanisms to cope up with them. Investigation on NPs has reported that they help plants to overcome abiotic stress by their concentration-dependent impact on plant growth and development [73, 94–96]. It is also reported that various antioxidant enzymes like catalase (CAT), peroxidase (POD), superoxide dismutase (SOD) were found to enhance their activity using NPs [97]. Depending upon their chemical composition, size,

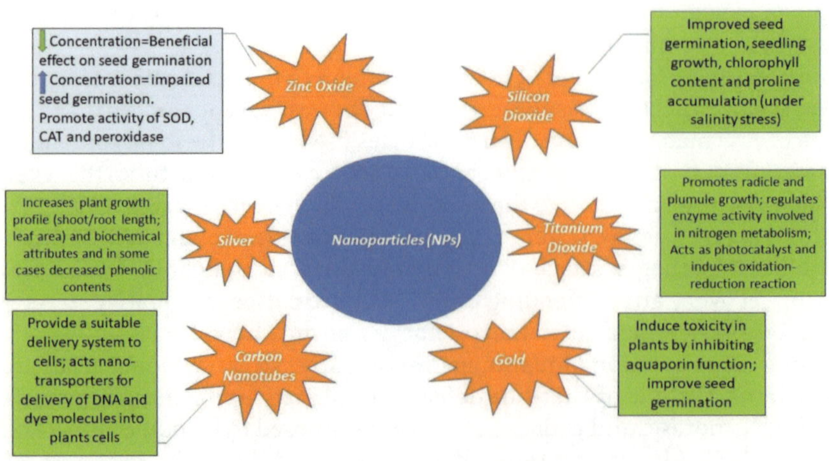

Figure 2.
Various nanoparticles with their effect on plant growth.

surface covering, reactivity, NPs interact with plants in various ways causing many morphological and physiological changes and play a very vital role in improving crop plants. NPs has both positive and negative effect on plant growth and development [98]. Some NPs along with their possible effect (negative and positive) on a plant are depicted in **Figure 2.**

5. Conclusion

Nanotechnology, a multi-disciplinary approach, has emerged out as a powerful discipline in the last few years and is revolutionizing various fields like medicine, agriculture, industrial, environment, electronics, etc. Nanotechnology is emerging as a tool for agriculture by empowering it with tools to conquer nutritional poverty and food scarcity. Nanoparticles are proven beneficial to boost plant growth, development, and increase yield capacity and help to overcome biotic and abiotic stress. The use of nanotechnology will lay a strong platform and will permit a secure future towards sustainability, crop productivity, and overcome abiotic stresses, where loss can be minimized and yield could be enhanced. The most effective way for understanding the action of the mechanisms of NPs applications is to apply the present knowledge by collaborating with various disciplines that may include molecular biology, plant physiology, plant breeding, cytology, soil physics along nanotechnology. Such associations could be helpful for the encouragement of multi-disciplinary projects that may be carried worldwide. Nanotechnology promises new insights into the mechanism of various abiotic stress tolerance in plants to complement physiological studies. Also, there is a need to detangle various factors responsible for abiotic stress. The implementation of action mechanisms of NPs will require information and expertise from the aforementioned disciplines to combat various stress effects. The applicability of nanotechnology needs to be commercialized from laboratory to agricultural fields.

Author details

Mohd. Tariq[1,2†*], Shipra Choudhary[2†], Harjeet Singh[3†], Mohd. Asif Siddiqui[2], Hirdesh Kumar[2], Asad Amir[2] and Neelesh Kapoor[4]

1 G. B. Pant National Institute of Himalayan Environment, Almora, Uttarakhand, India

2 Department of Biotechnology and Microbiology, Meerut Institute of Engineering and Technology, Meerut, Uttar Pradesh, India

3 Cattle Genetics and Breeding Laboratory, Central Institute for Research on Cattle, Meerut, Uttar Pradesh, India

4 Department of Fingerprinting, College of Biotechnology, Sardar Vallabhbhai Patel University of Agriculture and Technology, Meerut, Uttar Pradesh, India

*Address all correspondence to: tariq.du14@gmail.com

† These authors have contributed equally.

References

[1] Chen, H. and R. Yada (2011). Nanotechnologies in agriculture: New tools for sustainable development. Tends in Food Science and Technology. **22**: 585-594.

[2] Saxena, R., Tomar, R. S., and Kumar, M. (2016). Exploring nanobiotechnology to mitigate abiotic stress in crop plants. Journal of Pharmaceutical Sciences and Research, *8*(9), 974.

[3] Lavania, D., Singh, A.K., Siddiqui, M.H., Al-Whaibi, M.H. and A. Grover (2015). Abiotic stress tolerant transgenic plants and nanotechnology. In: *Nanotechnology and Plant Sciences* (Eds. M.H. Siddiqui *et al.*) Springer International Publishing Switzerland, pp. 165-181.

[4] FAO (2009). High Level Expert Forum-How to Feed the World in 2050. Economic and Social Development, Food and Agricultural Organization of the United Nations, Rome, Italy.

[5] Jalil, S.U. and M.I. Ansari (2019). Nanoparticles and abiotic stress tolerance in plants: synthesis, action, and signaling mechanisms. In: *Plant Signaling Molecules.* Elsevier Publishing, pp.549-561.

[6] Gruere, G., Narrod, C., and Abbott, L. (2011). Agriculture, food and water nanotechnologies for the poor: Opportunities and constraints. International Food Policy Research Institute (IFPRI). Policy Brief 19 June, 2011.

[7] Scott, N.R., and Chen, H. (2003). Nanoscale science and engineering on agriculture and food systems. In: Roadmap Report of National Planning Workshop. Washington D.C. November 18-19, 2002.

[8] Dasgupta N, Ranjan S, Mundekkad D et al (2015) Nanotechnology in agro-food: from field to plate. Food Res Int 69:381-400

[9] Huang Q, Yu H, Ru Q (2010) Bioavailability and delivery of nutraceuticals using nanotechnology. J Food Sci 75: R50–R57.

[10] McClements DJ, Decker EA, Park Y, Weiss J (2009) Structural design principles for delivery of bioactive components in nutraceuticals and functional foods. Crit Rev Food Sci Nutr 49:577-606. doi:10.1080/10408390 902841529

[11] Ranjan S, Dasgupta N, Chakraborty AR et al (2014) Nanoscience and nanotechnologies in food industries: opportunities and research trends. J Nanoparticle Res 16:2464. doi:10.1007/ s11051-014- 2464-5

[12] Carmen IU, Chithra P, Huang Q et al (2003) Nanotechnology: a new frontier in food science. Food Technol 57:24-29

[13] Priester JH, Ge Y, Mielke RE et al (2012) Soybean susceptibility to manufactured nanomaterials with evidence for food quality and soil fertility interruption. PNAS 109:E2451–E2456

[14] Marchiol L (2012) Synthesis of metal nanoparticles in living plants. Italian J Agro 7: 37.

[15] Prasad R (2014) Synthesis of silver nanoparticles in photosynthetic plants. Journal of Nanoparticles Article ID 963961, pp: 1-8.

[16] Branton, D., Deamer, D.W., Marziali, A., Bayley, H., Benner, S.A., Butler, T., et al. (2008). the potential and challenges of nano pores sequencing. Nature Biotechnology, **26** (10): 1146-1153.

[17] Bray EA, Bailey-Serres J, Weretilnyk E. In: Gruissem W, Buchannan B, Jones R (Eds.)

Biochemistry and Molecular Biology of Plants, Responses to abiotic stresses, American Society of Plant Biologists, Rockville, MD, 2000; pp. 1158-1249.

[18] Gill SS, Khan NA, Anjum NA, Tuteja N. Amelioration of Cadmium Stress in Crop Plants by Nutrient Management: Morphological, Physiological and Biochemical Aspects. Plant Stress (Spl issue) 2010; (In press)

[19] Khan NA, Singh S. (Eds.) Abiotic Stress and Plant Responses. IK International Publishing House Pvt. Ltd., New Delhi, 2008; pp. 1-299.

[20] Mahajan S, Tuteja N. Cold, salinity and drought stress: an overview. Arch Biochem Biophys 2005; 444: 139-158.

[21] Tuteja N. Mechanisms of high salinity tolerance in plants. Meth Enzymol 2007; 428: 419-438.

[22] Clair, S. B. S., and Lynch, J. P. (2010). The opening of Pandora's Box: climate change impacts on soil fertility and crop nutrition in developing countries. Plant and Soil, *335*(1-2), 101-115.

[23] Wang H, Wang H, Shao H, Tang X (2016) Recent Advances in Utilizing Transcription Factors to Improve Plant Abiotic Stress Tolerance by Transgenic Technology. Front Plant Sci 7: 67.

[24] Verma, A. K., and Deepti, S. (2016). Abiotic stress and crop improvement: current scenario. Adv Plants Agric Res, *4*(4), 00149.

[25] Fortina, P., Kricka, L.J., Surrey, S., Grodzinski, P., 2005. Nanobiotechnology: the promise and reality of new approaches to molecular recognition. Trends Biotechnol. 23, 168.

[26] Roco, M.C., 2003. Broader societal issue of nanotechnology. J. Nanopart. Res. 5, 181-189.

[27] Giraldo, J. P., Landry, M. P., Faltermeier, S. M., McNicholas, T. P., Iverson, N. M., Boghossian, A. A., ... and Strano, M. S. (2014). Plant nanobionics approach to augment photosynthesis and biochemical sensing. Nature materials, *13*(4), 400-408

[28] Perez-de-Luque, A. and R. Diego (2009). Nanotechnology for parasitic plant control. Pest Manag. Sci. **65**: 540-545.

[29] Torney, F., Trewyn, B.G., Lin, V.S.Y. and K. Wang (2007). Mesoporous silica nanoparticles deliver DNA and chemicals into plants. Nat. Nanotechnol. **2**: 295-300.

[30] Wahid, A. (2007). Physiological implications of metabolites biosynthesis in net assimilation and heat stress tolerance of sugarcane (*Saccharum officinarum*) sprouts. J. Plant Res. **120**: 219-228.

[31] Zandalinas, S. I., Mittler, R., Balfagón, D., Arbona, V., and Gómez-Cadenas, A. (2018). Plant adaptations to the combination of drought and high temperatures. Physiologia plantarum, *162*(1), 2-12.

[32] Banerjee, A., and Roychoudhury, A. (2018). Abiotic stress, generation of reactive oxygen species, and their consequences: An overview. *Revisiting the role of reactive oxygen species (ROS) in plants: ROS Boon or bane for plants*, 23-50.

[33] Karuppanapandian, T., Wang, H.W., Prabakaran, N., Jeyalakshmi, K., Kwon, M. and K. Manoharan (2011). 2,4-dichlorophenoxyacetic acid-induced leaf senescence in mung bean (*Vigna radiate* L. Wilczek) and senescence inhibition by co-treatment with silver nanoparticles. *Plant Physiol.* Biochem. **49**: 168-217.

[34] Moller, I.M., Jensen, P.E. and A. Hansson (2007). Oxidative modifications to cellular components in plants. Ann. Rev. Plant Biol. **58**: 459-481.

[35] Savicka, M. and N. Skute (2010). Effects of high temperature on

malondialdehyde content, superoxide production and growth changes in wheat seedlings (*Triticum aestivum* L.). Ekologija **56**: 26-33.

[36] Prasad, P.V.V., Pisipati, S.R., Momcilovic, I. and Z. Ristic (2011). Independent and combined effects of high temperature and drought stress during grain filling on plant yield and chloroplast EF-Tu Expression in spring wheat. J. Agron. Crop Sci. **197**: 430-441.

[37] Haghighi, M., Abolghasemi, R. J.A. Teixeira da Silva (2014). Low and high temperature stress affect the growth characteristics of tomato in hydroponic culture with Se and nano-Se amendment. Sci. Hortic. **178**: 231-240.

[38] Hartikainen, H., Xue, T. and V. Piironen (2000). Selenium as an antioxidant and prooxidant in ryegrass. Plant Soil **225**: 193-200.

[39] Hasanuzzaman, M., Nahar, K. and M. Fujita (2014). Silicon and selenium: two vital trace elements that confer abiotic stress tolerance to plants. Emerging Technologies and Management of Crop Stress Tolerance. Elsevier, The Netherlands, pp. 377-422

[40] Schulze, E.-D., Beck, E. and K. Muller-Hohenstein (2005). Plant Ecology. Springer, Berlin.

[41] Khodakovskaya, M.V., de Silva, K., Nedosekin, D.A., Dervishi, E., Biris, A.S. E.V. Shashkov (2011). Complex genetic, photothermal, and photoacoustic analysis of nanoparticle-plant interactions. Proc. Natl. Acad. Sci. U. S. A. **108**: 1028-1033.

[42] Zhao, L., Peng, B., Hernandez-Viezcas, J.A., Rico, C., Sun, Y. and J.R. Peralta- Videa (2012). Stress response and tolerance of Zea mays to CeO_2 nanoparticles: cross talk among H2O2, heat shock protein and lipid peroxidation. ACS Nano **6**: 9615-9622.

[43] Qi, M., Liu, Y. and T. Li (2013). Nano-TiO2 improves the photosynthesis of tomato leaves under mild heat stress. Biol. Trace Elem. Res. **156**: 323-328.

[44] Flowers, T. J. (2004). Improving crop salt tolerance. Journal of Experimental botany, *55*(396), 307-319.

[45] Munns, R., and Tester, M. (2008). Mechanisms of salinity tolerance. Annu. Rev. Plant Biol., *59*, 651-681.

[46] Basra, A.S. and R. K Basra. (1997). In: "Mechanisms of environmental stress resistance in plants". *Drought resistance in plants* (eds). Basra, A.S. and R.K Basra, Harwood Academic Publishers, Netherlands, pp 1-42

[47] Parida, A. K., and Das, A. B. (2005). Salt tolerance and salinity effects on plants: a review. Ecotoxicology and environmental safety, *60*(3), 324-349.

[48] Alscher, R. G., Erturk, N., and Heath, L. S. (2002). Role of superoxide dismutases (SODs) in controlling oxidative stress in plants. Journal of experimental botany, 53(372), 1331-1341.

[49] Weydert, C. J., and Cullen, J. J. (2010). Measurement of superoxide dismutase, catalase and glutathione peroxidase in cultured cells and tissue. Nature protocols, *5*(1), 51.

[50] Munns, R., James, R. A., and Läuchli, A. (2006). Approaches to increasing the salt tolerance of wheat and other cereals. Journal of experimental botany, *57*(5), 1025-1043.

[51] Wutipraditkul, N., Wongwean, P., and Buaboocha, T. (2015). Alleviation of salt-induced oxidative stress in rice seedlings by proline and/or glycinebetaine. Biologia Plantarum, *59*(3), 547-553.

[52] Hussein, M. M., and Abou-Baker, N. H. (2018). The contribution of nano-zinc to alleviate salinity stress on cotton

plants. Royal Society open science, 5(8), 171809.

[53] Avestan, S., Ghasemnezhad, M., Esfahani, M., and Byrt, C. S. (2019). Application of Nano-Silicon Dioxide Improves Salt Stress Tolerance in Strawberry Plants. Agronomy, 9(5), 246.

[54] Khan, M. N. (2016). Nano-titanium Dioxide (Nano-TiO2) mitigates NaCl stress by enhancing antioxidative enzymes and accumulation of compatible solutes in tomato (*Lycopersicon esculentum* Mill.). J. Plant Sci, *11*, 1-11.

[55] Yassen, A., Abdallah, E., Gaballah, M., and Zaghloul, S. (2017). Role of Silicon dioxide nano fertilizer in mitigating salt stress on growth, yield and chemical composition of Cucumber (*Cucumis sativus* L.). *Int.* J. Agric. Res, *22*, 130-135.

[56] Dickinson, M. and T.B. Scott (2010). The application of zero-valent iron nanoparticles for the remediation of a uranium-contaminated waste effluent. J. Hazard. Mater. **178**: 171-179.

[57] Shen, Y., Tang, J., Nie, Z., Wang, Y., Ren, Y. and L. Zuo (2009). Preparation and application of magnetic Fe_3O_4 nanoparticles for wastewater purification. Sep. Purif. Technol. **68**: 312-319.

[58] Wang, M., Chen, L., Chen, S. and Y. Ma (2012). Alleviation of cadmium-induced root growth inhibition in crop seedlings by nanoparticles. Ecotoxicol. Environ. Saf. **79**: 48-54.

[59] Lin, L., Zhou, W., Dai, H., Cao, F., Zhang, G. and F. Wu (2012). Selenium reduces cadmium uptake and mitigates cadmium toxicity in rice. J. Hazard. Mater. **235**: 343-351.

[60] Ronavari, A., Balazs, M., Tolmacsov, P.,Molnar, C., Kiss, I., Kukovecz, A. and Z. Konya (2016). Impact of the morphology and reactivity of nanoscale zero-valent iron (NZVI) on dechlorinating bacteria. Water Res. **95**: 165-173.

[61] Fajardo, C., Ortiz, L.T., Rodriguez-Membibre, M.L., Nande, M., Lobo, M.C. and M. Martin (2012). Assessing the impact of zero-valent iron (ZVI) nanotechnology on soil microbial structure and functionality: a molecular approach. Chemosphere **86**: 802-808.

[62] Chen, W.F., Zhang, J., Zhang, X., Wang, W. and Y. Li (2016). Investigation of heavy metal (Cu, Pb, Cd, and Cr) stabilization in river sediment by nano-zero-valent iron/activated carbon composite. Environ. Sci. Pollut. Res. Int. **23**: 1460-1470.

[63] Silva, M.M., Pérez, D.V., Wasserman, J.C., Santos-Oliveira, R. and M.A.V. Wasserman (2017). The effect of nanohydroxyapatite on the behavior of metals in a microcosm simulating a lentic environment. Environ. Nanotechnol. Monit. Manage. **8**: 219-227.

[64] Zhang, Z., Li, M., Chen, W., Zhu, S., Liu, N. and L. Zhu (2010). Immobilization of lead and cadmium from aqueous solution and contaminated sediment using nanohydroxyapatite. Environ. Pollut. **158**: 514-519.

[65] Iijima, S. (1991). Helical microtubules of graphitic carbon. Nature **354**: 56-58.

[66] Ihsanullah, A.A., Al-Amer, A.M., Laoui, T., Al-Marri, M.J., Nasser, M.S., Khraisheh, M. and M.A. Atieh (2016). Heavy metal removal from aqueous solution by advanced carbon nanotubes: critical review of adsorption applications. Sep. Purif. Technol. **157**:141-161.

[67] Trojanowicz, M. (2006). Analytical applications of carbon nanotubes: a review. Trac-Trends in Analytical Chemistry **25**: 480-489.

[68] Gong, J.L., Wang, B., Zeng, G.M., Yang, C.P., Niu, C.G., Niu, Q.Y., Zhou,

W.J. and Y. Liang (2009). Removal of cationic dyes from aqueous solution using magnetic multi-wall carbon nanotube nanocomposite as adsorbent. J. Hazard. Mater. **164**: 1517-1522.

[69] Tang, W.W., Zeng, G.M., Gong, J.L., Liang, J., Xu, P., Zhang, C. and B.B. Huang (2014). Impact of humic/fulvic acid on the removal of heavy metals from aqueous solutions using nanomaterials: a review. Sci. Total Environ. **468-469**: 1014-1027.

[70] Boyer, J.S. (1982). Plant productivity and environment. Science **218:** 443-448.

[71] McWilliam, J.R. (1986). The national and international importance of drought and salinity effects on agricultural production. Australian Journal of Plant Physiology **13:** 1-14.

[72] Taran, N., Storozhenko, V., Svietlova, N., Batsmanova, L., Shvartau, V., and Kovalenko, M. (2017). Effect of zinc and copper nanoparticles on drought resistance of wheat seedlings. Nanoscale research letters, *12*(1), 60.

[73] Ashkavand, P., Tabari, M., Zarafshar, M., Tomášková, I., and Struve, D. (2015). Effect of SiO_2 nanoparticles on drought resistance in hawthorn seedlings. Forest Research Papers, 76(4), 350-359.

[74] Hattori T, Inanaga S, Araki H, An P., Morita S, Luxová M, Lux A., "Application of silicon enhanced drought tolerance in Sorghum bicolour,". Physiologia Plantarum, 2005,123, 459-466.

[75] Hojjat, S. S., and Ganjali, A. (2016). The effect of silver nanoparticle on lentil seed germination under drought stress. Int J Farm Allied Sci, *5*(3), 208-212.

[76] Sedghi, M., Hadi, M., and Toluie, S. G. (2013). Effect of nano zinc oxide on the germination parameters of soybean seeds under drought stress. *Annales of West University of Timisoara. Series of Biology*, *16*(2), 73.

[77] Jackson, M. B., Fenning, T. M., and Jenkins, W. (1985). Aerenchyma (gas-space) formation in adventitious roots of rice (Oryza sativa L.) is not controlled by ethylene or small partial pressures of oxygen. Journal of Experimental Botany, *36*(10), 1566-1572.

[78] Drew, M. C., and Lynch, J. (1980). Soil anaerobiosis, microorganisms, and root function. Annual Review of Phytopathology, *18*(1), 37-66.

[79] Sasidharan, R., Bailey-Serres, J., Ashikari, M., Atwell, B. J., Colmer, T. D., Fagerstedt, K., ... and Holdsworth, M. J. (2017). Community recommendations on terminology and procedures used in flooding and low oxygen stress research. New Phytologist, *214*(4), 1403-1407.

[80] Lee, S. C., Mustroph, A., Sasidharan, R., Vashisht, D., Pedersen, O., Oosumi, T., ... and Bailey-Serres, J. (2011). Molecular characterization of the submergence response of the Arabidopsis thaliana ecotype Columbia. New Phytologist, *190*(2), 457-471.

[81] Hsu, F. C., Chou, M. Y., Chou, S. J., Li, Y. R., Peng, H. P., and Shih, M. C. (2013). Submergence confers immunity mediated by the WRKY22 transcription factor in Arabidopsis. The Plant Cell, *25*(7), 2699-2713.

[82] Jackson, M. B., and Ram, P. C. (2003). Physiological and molecular basis of susceptibility and tolerance of rice plants to complete submergence. Annals of botany, *91*(2), 227-241.

[83] Yeung, E., van Veen, H., Vashisht, D., Paiva, A. L. S., Hummel, M., Rankenberg, T., ... and Schuurink, R. C. (2018). A stress recovery signaling network for enhanced flooding tolerance in Arabidopsis thaliana. Proceedings of the National Academy of Sciences, *115*(26), E6085-E6094.

[84] Tsai, K. J., Lin, C. Y., Ting, C. Y., and Shih, M. C. (2016). Ethylene-regulated

glutamate dehydrogenase fine-tunes metabolism during anoxia-reoxygenation. Plant physiology, *172*(3), 1548-1562.

[85] Zaytseva, O., and Neumann, G. (2016). Carbon nanomaterials: production, impact on plant development, agricultural and environmental applications. Chemical and Biological Technologies in Agriculture, *3*(1), 17.

[86] Tsuji, H., Saika, H., Tsutsumi, N., Hirai, A., and Nakazono, M. (2006). Dynamic and reversible changes in histone H3-Lys4 methylation and H3 acetylation occurring at submergence-inducible genes in rice. Plant and cell physiology, *47*(7), 995-1003.

[87] Mustroph, A., Lee, S. C., Oosumi, T., Zanetti, M. E., Yang, H., Ma, K., ... and Bailey-Serres, J. (2010). Cross-kingdom comparison of transcriptomic adjustments to low-oxygen stress highlights conserved and plant-specific responses. Plant Physiology, *152*(3), 1484-1500.

[88] Juntawong, P., Girke, T., Bazin, J., and Bailey-Serres, J. (2014). Translational dynamics revealed by genome-wide profiling of ribosome footprints in Arabidopsis. Proceedings of the National Academy of Sciences, *111*(1), E203-E212.

[89] Rezvani, N., Sorooshzadeh, A., and Farhadi, N. (2012). Effect of nano-silver on growth of saffron in flooding stress. World Acad. Sci. Eng. Technol, *6*(1), 517-522.

[90] Sorooshzadeh, A., Hazrati, S., Oraki, H., Govahi, M., and Ramazani, A. (2012). Foliar application of nano-silver influence growth of saffron under flooding stress. In *Proceeding of the 4th international conference "Nanocon-2012*(pp. 510-512).

[91] Tripathi, D.K., Singh, S., Singh, S., Mishra, S., Chauhan, D.K., Dubey, N.K., 2015. Micronutrients and their diverse role in agricultural crops: advances and future prospective. Acta Physiol. Plant. 37 (7), 1-14.

[92] Tripathi, D.K., Singh, V.P., Kumar, D., Chauhan, D.K., 2012. Impact of exogenous silicon addition on chromium uptake, growth, mineral elements, oxidative stress, antioxidant capacity, and leaf and root structures in rice seedlings exposed to hexavalent chromium. Acta Physiol. Plant 34, 279-289. https://doi.org/10.1007/s11738- 011-0826-5.

[93] Upadhyaya, H., Panda, S.K., Dutta, B.K., 2008. Variation of physiological and antioxidative responses in tea cultivars subjected to elevated water stress followed by rehydration recovery. Acta Physiol. Plant. 30 (4), 457-468.

[94] Dimkpa, C.O., Bindraban, P.S., Fugice, J., et al., 2017. Composite micronutrient nanoparticles and salts decrease drought stress in soybean. Agron. Sustain. Dev. 37, 5. https://doi.org/10.1007/s13593-016-0412-8. doi:10.1111/j.1750-3841.2009.01457.x

[95] Mishra, S., Keswani, C., Abhilash, P.C., Fraceto, L.F., Singh, H.B., 2017. Integrated Approach of Agrinanotechnology: Challenges and Future Trends. Front. Plant Sci.8 471. https://doi.org/10.3389/fpls.2017.00471.

[96] Mishra, S., Singh, H.B., 2016. Preparation of biomediated metal nanoparticles. Indian Patent Filed, 201611003248

[97] Laware, S.L., Raskar, S., 2014. Effect of titanium dioxide nanoparticles on hydrolytic and antioxidant enzymes during seed germination in onion. Int. J. Curr. Microbiol. Appl. Sci. 3 (7), 749-760.

[98] Ma X, Geiser-Lee J, Deng Y, Kolmakov A (2010) Interactions between engineered nanoparticles (ENPs) and plants: phytotoxicity, uptake and accumulation. Sci Total Environ 408(16):3053-3061

Chapter 18

Evolution of the Soil-Based Agriculture and Food System to Biologically-Based Indoor Systems

Norman R. Scott

Abstract

There is no area of human activity more basic to society than a sustainable agricultural, food and natural resource system. The 'major' question is, how will food be produced sustainably for the expected global population growth to 9.5–10 billion people by 2050? The agriculture and food system is a highly complex adaptive system, operating across the spectrum of economics, biophysics and sociopolitics. There is a need to move beyond contentious debates between many constituencies, rooted in ideological solutions, to acceptance of a broad array of different approaches. This chapter focuses on the evolution from long and traditionally soil-based systems to biologically-based indoor systems, largely independent of soil with unique characteristics. Science and technology advancements have been critical to achievements of the existing land/soil-based systems and are equally critical in development of the emerging biologically-based indoor systems of controlled environment agriculture (greenhouses and vertical farms) and plant-based food alternatives, cell-cultured foods and 3D printed foods. Thus, there is no system more in need of and more likely to benefit from a comprehensive application of convergence thinking across disciplines and stakeholders.

Keywords: Agriculture, Food, Sustainability, Systems, Digital Ag, Computing and information science, Renewable energy, Sensors, Robots, Drones, Regenerative agriculture, Circularity, Nanotechnology, Biotechnology, Plant-based food alternatives, Cell-cultured foods, 3D printed foods

1. Introduction

With the projections that global population will grow to as much as 10 billion by 2050, there has developed an increasing concern in how this population will be fed, how will food be produced and can it be done sustainably, what will constitute a healthy diet, will the environment be destroyed in the process, will natural resources and ecosystems be compromised, will the food system reduce or increase hunger and poverty, and will the system enhance or decrease equity and access to food for a healthy and productive global population? These and many more critical questions challenge all of us who are a participant in the food and agriculture system (FAS) and every one of us is involved at some level ranging from our daily consumption to innovative scientific research. Thus, one might first ask, what is meant by a food and agriculture system? A report of the National Academy of

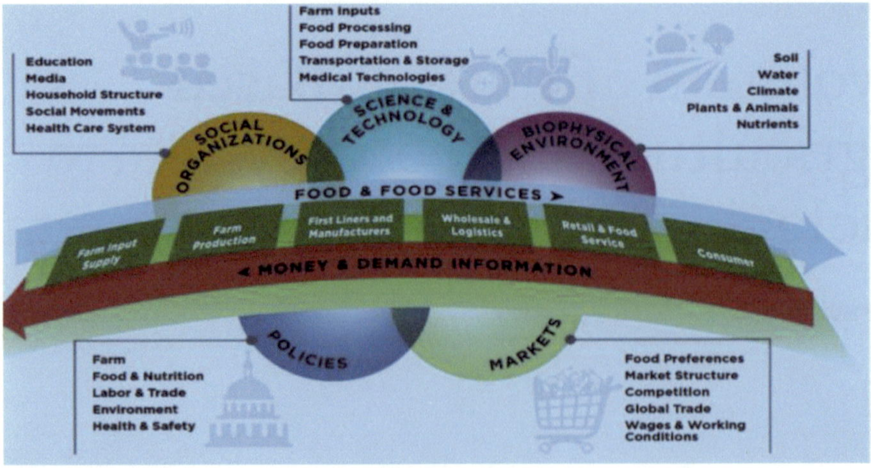

Figure 1.
Links between the food supply chain and the larger biophysical and social/institutional context [1].

Sciences, Engineering and Medicine (NASEM), characterizes the food system as a complex adaptive system that operates across a broad spectrum of economics, biophysical, and sociopolitical contexts [1]. This is captured in **Figure 1**.

The area of Food and Agriculture System (FAS) has been addressed by a number of excellent reports addressing the system from numerous perspectives. Specifically, the World Resources Foundation Report, *Creating a Sustainable Food Future* [2] presents a menu of solutions to feed nearly 10 Billion people by 2050. The report explores 22 items in broad terms that are suggested to stabilize the climate, promote economic development and reduce poverty. An Expert Panel presents a comprehensive report, *Socio-Technical Innovation Bundles for Agri-Food Systems Transformation* [3]. The Panel presents a vision within four core objectives: healthy and nutritious diets (H), equitable and inclusive value chains (E), resilience to shocks and stressors (R) and climate and environmental sustainability (S) which they characterize as HERS.

A report from EAT-Lancet Commission, *Food Planet Health: Healthy Diets from Sustainable Food Systems* [4] emphasizes a goal of transformation from current diets to healthy diets, sustainable food production and reductions in food loss and waste. Walker and Buhler [5] emphasize the role of biotechnology in catalyzing holistic agriculture innovation across a biological scale with a focus on smart machines, advanced sensors, big data, digital science, artificial intelligence, and controlled environment agriculture. A workshop sponsored by the National Academy of Science, Engineering and Medicine (NASEM) explored the future of food in a review of current and emerging knowledge about innovations for food systems [6].

2. "Toward" a sustainable food and agriculture system

Agricultural productivity has been a consistent and important focus during the 20th century and the 21st century, with good reason, to feed a growing world population. However, while providing safe and affordable food remains a driving force for the FAS, there are emerging and numerous factors that challenge our present and future FAS. Some of these are: impacts of the FAS on the environment; trust in science and technology; increasing urbanization; climate change; changing food preferences; globalization; integrated value chains; international regulations;

economic viability of rural communities; and more recently a recognition of the disruption that major events such as a pandemic can create for the FAS. Friedman [7] captures these as a time when the three largest forces on the planet (technology, globalization and climate change) are all accelerating at once, creating the greatest inflection point in history.

Since the 1990's to present, the sustainability of the FAS has become an accepted concept to capture the intersection of environment, economics and equity (or social responsibility). The concept of sustainability arising from the report, *Our Common Future,* [8] has become widely applied to many systems. In the opinion of many people, it has become over utilized and as such does not have significant meaning. Relative to the FAS, the NRC report [9], *Toward Sustainable Agricultural Systems in the 21st Century,* states, "The transformative approach to improving agricultural sustainability.... would facilitate the adoption of production approaches that capitalize on synergies, efficiencies, resilience characteristics associated with complex natural systems and their linked social, economic, and biophysical systems".

The preference of this author is to describe sustainable development as a "process of change in which the direction of investment, the orientation of technology, the allocation of resources, the development and functioning of institutions, and the advancement of human and community well-being meets present needs and aspirations without compromising the ability of future generations to meet their own needs and aspirations" (adapted and modified from [8, 10]). It suggests an imperative for action by which goals for development of the FAS can be measured. Moreover, in addition to implications for resources, it embodies attributes of the environment, economic and social responsibility now and into the future.

At an international level, the United Nation's Sustainability Development Goals [11] have been used widely to encourage development that meets sustainability objectives (**Figure 2**). At first glance it would appear that SDG 2 (zero hunger) would be the primary SDG for the FAS. But on further reflection, it is clear that because the FAS consists of a web of interactions across many complex interlinkages, todays FAS is strongly connected to the other 16 SDGs as well [3]. All of the elements of growing, harvesting, storing, processing, distributing, consuming and managing food losses and wastes are encompassed by the SDGs.

Figure 2.
The 17 sustainable development goals [11].

The recent report, 21st Century Agriculture Renaissance: Solutions from the Land, [12] offers a vision for strategies encompassing climate smart agriculture through 1) sustainable intensification of production, 2) adaptive management and 3) greenhouse gas reduction. The report highlights projects from across the globe and across various farming systems with a focus on indicating which of the SDGs goals are addressed by these respective projects.

3. An evolving food and agriculture system (FAS) to 2050

Domestication of plants and animals can be traced as far back as around 11,000–9,000 BC [13]. From its origin, a fundamental element of the FAS has been a land-based (soil-based) agricultural production system. Today we have experienced the evolution of a highly advanced FAS through emergence of the science and technology of DigitalAg, artificial intelligence, sensors everywhere, internet of things (IoT), genomics (including CRISPR), drones and robots with the one consistent factor being the use of land, the soil. However, 'new' emerging subsystems are developing, based largely on nonland-based, even soilless-based indoor facilities. It is the purpose of this chapter to briefly trace this evolution of the FAS.

First, impressive highlights of key science and technological innovations in the 'conventional' FAS are discussed. Then, science and technological innovations in an 'alternative', largely soilless and indoor system are highlighted. The author recognizes that to do this in one chapter is beyond an ability to cover many significant science and technology innovations, particularly in the 'conventional' FAS. However, the choice has been to focus on science and technologies that are perceived to have potentially a large impact going forward to 2050, and are, in a sense, 'guesses' about a future FAS. A further caveat is there is an over emphasis on the natural sciences with a largely inadequate effort to address the very important social and cognitive sciences.

3.1 Innovations in 'conventional' FAS

The use of 'conventional' is intended to represent soil-based systems. For the purposes of this chapter, science and technological innovations are focused on: (i) computing and information science (Digital Ag), (ii) nanotechnology, (iii) biotechnology, (iv) renewable energy, (v) electrification, (vi) regenerative agriculture, and (vii) circular economy.

3.1.1 Computing and information science (Digital Ag)

The FAS has increasingly embraced computer and information science at many levels from large farmers to poorest farmers in developing countries through an integration of sensors, satellites, tablets and cell phones. Research, teaching and extension (outreach) programs in Digital Ag have been developed in many land grant universities in the U.S. and universities around the world. Like sustainability, Digital Ag is defined or described somewhat differently by various proponents. A description of Digital Ag by the Cornell Institute for Digital Agriculture (CIDA) is given in **Figure 3**, including linkages to basic elements of innovation, discovery, and analytics with broad applications to areas within the FAS. A key element of CIDA is its ability to bring diverse people together from the Colleges of Agriculture and Life Science, Engineering, Business, Veterinary Medicine, and Computer and Information Science at Cornell University.

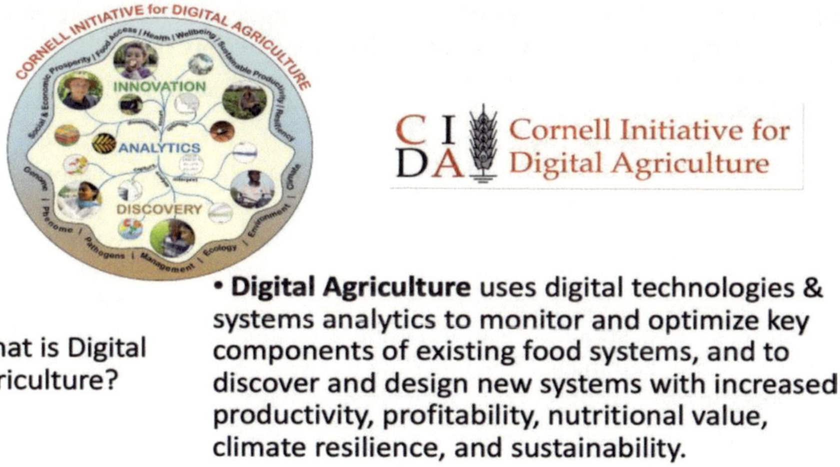

Figure 3.
The Cornell Institute for Digital Agriculture.

The capability of Digital Ag ultimately depends on an integration of critical elements for a successful system:

- Sensors (including drones, robotics, artificial intelligence) to initiate data acquisition in the field,

- Autonomous transfer of data from sensors [likely many, an Internet of Things agriculture (IoTA)] by wireless communication with digital devices (computers, tablets, and smart phones),

- Analytical devices with software capability (machine learning, artificial intelligence and handling of 'big' data) for storage, analysis, synthesis and reporting results, and

- Organizations (startups, consolidations and market developments) to apply recommendations to practice in the field.

Given that digitization is spreading through all aspects of food and agriculture one might ask, what is the difference between precision agriculture, smart agriculture and digital agriculture? There is not a unanimous definition, but for purposes of this chapter, the general descriptions offered by [14] are adapted to suggest that precision agriculture seeks to optimize conditions by means of sensory analysis and precise application technology, smart agriculture is a further development of precision agriculture to support decision making, and digital agriculture integrates concepts of both precision and smart agriculture to create value from data.

3.1.1.1 Sensors

It all begins with sensors and with the great advancements in sensor development, it is possible to study plant and animal physiology beyond the laboratory to measure, monitor and activate actions in plant, animal, and microbial production systems. The addition of the Internet of Things, Agricultural (IoTA), big data analysis and artificial intelligence is promoting a high-tech agriculture driven by data. Especially in the application of nanoscale science and technology, sensors and biosensors have been a major area of research and development. In a following section on nanotechnology, numerous examples of sensors in various applications are addressed.

3.1.1.2 Robotics

Robots have clearly moved from many industrial applications to become a significant new technology in the FAS. In labor-intensive crops and in specific identity applications, robots have assumed an important role. A few examples are: (i) identify weeds and implement weed control (e.g., mechanically remove weeds, employ microwave technology to kill weeds, and other methods); (ii) spot onset of plant diseases or pests and deliver intervention schemes (e.g., citrus greening, early potato blight, and many more); (iii) deliver fertilizer, pesticides, and herbicides at specific sites; (iv) spot controlled spray delivery in vineyards and orchards (including pollinator applications); (v) robotic 'duck' in rice fields to control weeds without pesticides; (vi) robots to pick fruit (e.g., apples, citrus, strawberries, raspberries, and more), **Figure 4** illustrates an autonomous fruit picking robot in a development stage; (vii) robot for transplanting; (viii) soil robots for soil testing and determine water-use effectiveness; (ix) in-food processing plants, robots to size, sort and package produce; and (x) autonomous robotic vehicles (including tractors, some electric) to perform field operations that could reduce soil compaction and simultaneously track data.

Robots have entered the dairy farm to milk and feed cows. Cows enter a special stall and are milked while feed is available, during milking based on production. Access to the milking stall is based on n times milking per day as a function of milk production. Cow identity is transmitted by an electronic animal tag and sensors within the teat cup provide data on temperature, milk conductivity, and milk quality. A highly desirable future biosensor would detect progesterone levels that could provide key data on reproductive status (estrus). A single robot station can handle about 40–50 cows per day which makes the system compatible with small farms as well as large farms. The milking robot has been adopted on small farms to address challenges of available human labor, freedom from the commitment of twice daily milking (minimum) permitting a more normal life, and because the cow can be milked more often, increased production has been experienced. **Figure 5** illustrates 2-robots on a small New York dairy with 100 cows. Also, a few large rotating milking parlors with robotic milking units have been installed across the world.

Figure 4.
Flying autonomous robot in development for picking fruit, (https://www.tevel-tech.com).

Figure 5.
Two robots on a New York farm of 100 cows.

The development and production of field and harvest robots is a global business. *Future Farming* [15] has produced a robot catalog with 35 field and harvest robots from sixteen countries. In this first edition, seven of the robots are manufactured in the U.S. and six from the Netherlands. It is anticipated that numbers will continue to increase significantly going forward.

3.1.1.3 Drones (unmanned aerial vehicles, UAV)

While drones (unmanned aerial vehicles) have been widely employed in military missions and for intelligence gathering, their use in agriculture is exploding. Relatively inexpensive and reasonably simple to operate, drones can be equipped with sensors, cameras and specialized hardware to perform a large array of functions in agriculture. Equipped with appropriate devices, drones are: (i) used to develop high-definition maps of fields that provide an ability to create prescriptive-defined application of sprays, fertilizer, pesticides, and herbicides, **Figure 6**; (ii) used to count the number of plants, fruit and flowers to forecast yields; (iii) employed to distribute seeds for crop planting; (iv) used when equipped with multispectral, hyperspectral and thermal cameras to measure chlorophyll, crop biomass, and plant health, as well as determine ground temperature, plant numbers,

Figure 6.
Group of drones capable of performing functions of high-definition maps of fields to create prescriptive-defined application of sprays, fertilizer, pesticides, herbicides and seeding.

soil water content, and estimate crop yields; (v) a potential way to deliver contraceptives to manage wild horse and burro population; (vi) used to monitor a plant's water stress and control irrigation for efficient water use; (vii) used, in absence, or in case of an inadequate number of normal bee pollinators, 'nanobees' (miniature drones) to supplement the pollination process; (viii) use drones in outdoor livestock systems to monitor animals for estrus behavior as well as control and manage the herd, and (xi) employed to monitor and track animals in inaccessible areas in the natural environment.

3.1.2 Nanotechnology

Nanoscale science and engineering offers the potential to significantly revolutionize the FAS. It can play an important role at each point along the FAS supply chain from production through consumption and including management of food losses and wastes [16, 17]. In broad terms, nanotechnology can be a key element in: (i) "re-engineering" of crops, animals, microbes and other living systems at the genetic and cellular level; (ii) development of efficient, "smart" and self-replicating production technologies and inputs; (iii) development of tools and systems for identification, tracking and monitoring; and (iv) manufacture of new materials and modify crops, animals and food products.

The major advancement of applications of nanotechnology in the FAS has occurred largely since 2000. A national research grants program at the USDA/NIFA (United States Department of Agriculture/National Institute of Food and Agriculture) initiated in 2002 has been an important driver of the research in the FAS over the past two decades. The areas of applications have included food quality and safety, animal health monitoring and management, plant systems, environmental systems, and assessment of societal impacts. Just a few applications are: (i) nanomaterials for crop and animal disease detection and detection of residues, trace chemicals, viruses, antibiotics and pathogens; (ii) enhance plant nutrient uptake, nutrient use efficiency, and fertilizer efficiency by controlled release of agrochemicals; (iii) seed coatings with nano-based chemicals to promote seed gemination and deliver long-term disease and pathogen resistance; (iv) DNA-based genetic materials using DNA-based nanobarcodes with a multi-probe sensor to detect pathogens (in plants, animals and environmental contaminants); (v) enhance water-use efficiency in crops by improving water retention and develop 'smart plants' to provide information to meet water needs and manage irrigation; and (vi) wide-spread advances in food packaging and food-contact materials for quality and increased shelf life (eliminate/reduce refrigeration).

Against this significant list of successful developments, the vision for the future of nanotechnology is impressive [17–21]: (i) enhanced sensitivity, selectivity, robustness, ease of use, cost-effective and longevity of nanosensors as key components of the field-distributed, intelligent sensor network for monitoring and control as part of the Internet of Agricultural Things (IoTA), **Figure** 7; (ii) use of common field crops (e.g., corn, soybean, and grains) and trees to make sustainable chemicals; (iii) design nitrogen-producing microbiome and seed coatings that promote crops to produce their own nitrogen fertilizer; (iv) tracking system for integrity of food (plant and animal) from production, transport, and storage to consumer consumption; (v) unique sensors: ingestible to monitor gut health, tooth sensor to measure food properties and chopsticks to detect food characteristics including nutrients; (vi) DNA life-like materials from agricultural biomass ranging from biosensors to biomanufacturing (replace petrochemicals) to development of value-added products including plastics that are biodegradable.

Figure 7.
Graphene sensors on plant leaf to sense water transpiration and measure plant water to control irrigation.

3.1.3 Biotechnology

The impacts of crop biotechnology has been studied over a 22-year period (1996–2018) on farm income and production [22] and on the environment [23]. Significant economic benefits at the farm level globally are estimated at $18.9 billion in 2018 and $225.1 billion (in nominal terms) for the 22-year-period. These gains are attributed at 52% to farmers in developing counties and 48% in developed countries with 72% of the gains based on yield and production increases and 28% from cost savings [22]. Returns on the investment in GM (genetically modified) crop seeds were calculated as an average of $4.41 per dollar invested in developing countries and an average of $3.24 per dollar invested in developed countries.

Assessments of environmental impact on GM crops estimate reduced global crop protection products use by 8.6% over the 22 years. Reduced GHG emissions through adoption of reduced tillage, that reduces fuel usage and improves soil carbon retention, are estimated to have an environmental impact reduction of 19% [23].

The annual report of the International Service for the Acquisition of Agri-biotech Applications (ISAAA) provides a yearly global update on adoption and distribution of biotech crops [24]. The 2019 report shows that GM crops increased to 29 countries with 190.4 billion hectares. A total of 72 countries have adopted biotech crops with 29 having planted crops and 43 additional countries importing biotech crops for food, feed, and processing.

The biological world in 2020 was marked by recognition of CRISPR (clustered regularly interspersed short palindromic repeats) with a Nobel Prize in Chemistry awarded to its inventors. Simply stated, CRISPR is a unique technology used to edit select genes by finding a specific bit of DNA inside a cell and then altering that piece of DNA. Already applied in human health, it is being used in plant science for traits that can prevent disease, create pest resistance, increase resiliency, and improve crop yields.

Animal biotechnology has contributed greatly to increasing livestock productivity through increased production, reproductive efficiency, genetic improvement, animal nutrition, and animal health [25]. Specifically recombinant bovine somatotropin (rBST) has been shown to increase feed conversion and milk yield. Major advances in animal reproduction has been experienced with biotechnology applied to genetics and breeding. The U.S. Food and Drug Administration approved in December 2020, a first-of-its-kind, intentional genomic alteration (IGA) in domestic pigs for food or human therapeutics [26].

Thus, if we are to create new crop varieties and increased yields and improved animal breeds, it is important to utilize the science of biotechnologies to advance benefits for both large and small farmers. The impressive potential of biotechnology should not be ignored and left underutilized.

3.1.4 Renewable energy

3.1.4.1 Solar energy

The challenges of meeting the needs of food, energy and water (frequently called a nexus) in the face of climate change have stimulated some innovative novel systems to co-locate agriculture and solar photovoltaics (PV), termed 'agrivoltaics'. The concept originally suggested in 1982 [27] has been further developed and analyzed by [28–31]. At present solar PV is being employed by large utility-grid systems and on rooftops but the opportunity to develop an integrated system with coupled application of PV and crop production on the same land maximizes land use without sacrificing crop land. In fact, a study of collocation in drylands [31] has shown synergistic benefits of reduced plant stress, improved yields and reduced PV panel heat stress. Development of enhanced semi-transparent PV panels would further advance collocation of PV panels and crop land. A conceptual rendering of the concept is illustrated in **Figure 8**.

Although **Figure 8** illustrates the solar PV panel elevated ('on stilts') to allow animals and equipment to move beneath the panels, another option could be ground mounted PV panels separated by an area between panels for farming [28]. At this point, the number of crops which have been evaluated under PV panels is limited. Also, the impact of PV panels on the microclimate of air temperature, wind speed and relative humidity needs significant study to assess plant response. Some studies have shown benefits for crops like tomatoes, and lettuce [30].

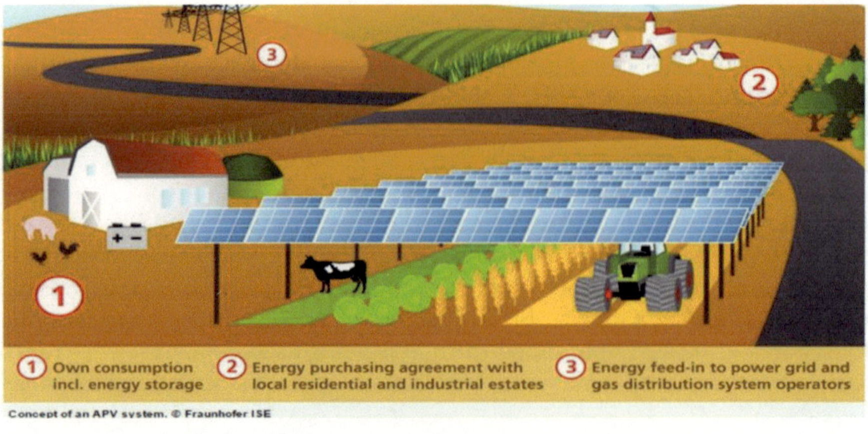

Figure 8.
A conceptual presentation of collocated solar PV and agricultural land with crops and animal production.

3.1.4.2 Wind energy

Much has changed from the early 1900's when many farmers used wind power to pump water and generate power from relatively small windmills. Today large wind turbines with generating capacity well above 1 MW are common on agricultural land, particularly in the West and Midwest, although many wind farms are found in other areas as well [32]. Like for solar PV, collocation of wind turbines on agricultural land has become common place. Farmers can lease land to wind developers, own turbines to generate power for their farm or as a farmer or group of farmers become a wind developer. Many farmers have found wind turbines on their land to be an important source of income. Typically, large turbines use a half-acre or less, including the access road, while allowing farming operations for cropping and grazing of livestock up to the base of turbines, **Figure 9**. As one farmer has been known to say, "it is a lot easier to milk a wind turbine than cows".

3.1.4.3 Bioenergy

3.1.4.3.1 Biofuels

At a time when the U.S. was dependent on imported transportation fuels, a Renewable Fuel Standard (RFS) was enacted by the Congress to create annual mandates for production of conventional biofuels and advanced biofuels. Corn ethanol became the predominant conventional biofuel and cellulosic-derived fuels as advanced biofuels. Significant controversary surrounded corn ethanol because of concerns of effects on food/feed prices, distortion of land use, increased cropland prices, and uncertainties about claims of environmental benefits. A report of the National Research Council (NRC, [33]) presented two findings: (1) the RFS may be an ineffective policy to reduce GHG emissions because the amount depends on how biofuels are produced, including changes to effects on land use, and (2) barriers to achieving the RFS due to high costs in producing cellulosic biofuels and market uncertainties. The U.S. Energy Information Administration [34], in projections to 2050, projects that the % of biofuels (ethanol, biodiesel and other biomass) may increase a bit between 10 and 13% depending on the scenario for oil prices.

Figure 9.
Large wind turbines integrated into agricultural crop land.

3.1.4.3.2 Biogas

A process using microorganisms, specifically a suite of bacteria, have been utilized to convert organic materials into biogas, primarily methane (~60–70%) and carbon dioxide (~30–40%), with small quantities of trace gases. Anaerobic digestion using methanogenic bacteria, in the absence of oxygen in airtight structures, has been used for many years, ranging from small home-owned digesters in China to advanced systems with increasingly large commercial tanks, highly instrumented in the U.S. and Europe. These large systems have been developed to manage animal manures together with food and other organic wastes by co-digestion to provide energy options of combined heat and power (CHP) and pipeline and transportation fuels) following processing and compression [35].

Anaerobic digesters number more than 250 in the U.S. while the number in Germany is now about 9,000. Germany has been particularly adept at using biogas to power bioenergy villages illustrating the potential for distributed energy generation in rural communities. It has become clear for economic success of anaerobic digesters that they operate as co-digesters by adding other organics in addition to animal manures [36]. Beyond direct products of biogas and digestate options, anaerobic digestion offers manure management opportunities for environmental benefits of reduced odors, reduced pollution and reduced GHG emissions. Opportunities for collaboration between businesses and farm digesters exist and are increasing. A recent example of this collaboration is a venture with founding members, Unilever, Starbucks and Dairy Farmers of America.

3.1.5 Electrification on the farm

The Rural Electrification Act of 1936 revolutionized rural America. Electric vehicles are revolutionizing the transportation sector. This revolution is also taking place in agriculture at an early stage with numerous equipment manufacturers launching, or working to develop, autonomous electric tractors [37]. Tractor companies, Monarch, Solectrac, Kubota, AGCO and John Deere to mention a few, are investing heavily in electric tractors and are in various stages in their development with the potential of limited availability as early as 2021. These tractors are equipped with autonomous hardware replete with many sensors and machine learning for data collection and tractor control.

At this point development of the electric tractor has been focused in the 30–40 hp. (horsepower) range (25 – 30 kW), largely due to the size and weight of batteries. An advantage of smaller equipment is potential for reduced soil compaction. **Figure 10** illustrates two experimental paths of John Deere, a battery-driven tractor and an electrically connected tractor (a long extension cord!).

First perceptions are that this high technology would be only applicable and affordable in 'industrialized' agriculture. However, the possibility of developing

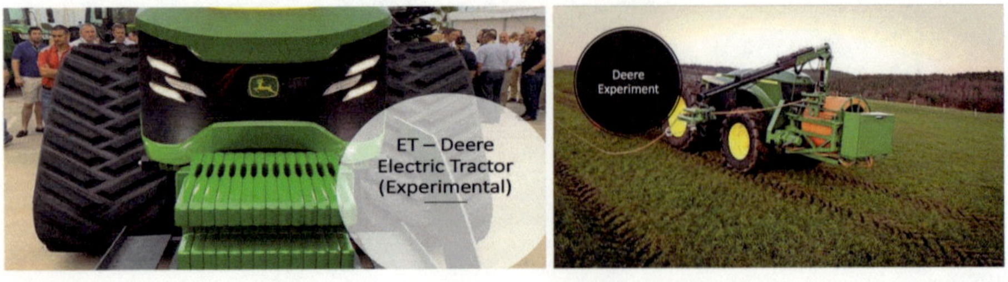

Figure 10.
John Deere's battery-electric tractor and connected electric tractor.

electric-driven tractors and equipment is certainly conceivable in the developing world because smaller tractors and machines are well adapted to the small land holdings. The author envisions the co-development of solar PV for charging batteries to power electric equipment. Rapidly developing advancements in battery technologies and decreasing cost will be keys to adoption in the developing world. The unique idea of a cord-connected electric tractor, while not likely to be an option in U.S. agriculture, might well be an excellent way to connect solar PV to power electric equipment for the small farmer.

3.1.6 Regenerative agriculture

As previously noted, agriculture is coming under increasing scrutiny because of GHG emissions and negative effects on the environment. Drawing much attention recently is the practice referred to as 'regenerative agriculture". The term has no universal definition but is frequently used to describe practices to promote soil health by increasing soil organic carbon [38]. Practices commonly perceived to advance regenerative agriculture are no-till farming, cover crops, diverse crop rotations, rotating livestock grazing, and lessened use of fertilizers, pesticides and herbicides. Cropping system diversification has been shown to reduce negative environmental impacts of soil erosion and nutrient runoff, and reduced cropping inputs while maintaining crop yields [39, 40].

While there is general agreement that regenerative agriculture practices improve soil health and provide environmental benefits, some researchers [38] report that regenerative agriculture practices have limited potential to increase soil carbon sequestration. Nevertheless, some corporations have set up a carbon sequestration market (Bayer) and a carbon credit for soil carbon sequestered (Land O'Lakes) for farmers. In addition, Cargill, McDonald's, and Walmart Foundation are collaborating with the World Wildlife Foundation on regenerative practices to improve grasslands of the Northern Great Plains.

It is suggested that going forward farmers will be paid for soil carbon storage. However, this requires an ability to measure soil carbon and quantify change in the field accurately over time to assess the effects of differing practices. Thus, future research is needed to find new ways of soil carbon sequestration and develop the data through measurement of soil carbon content.

3.1.7 Circular economy in FAS

The concept of has been recently introduced in the FAS. The guiding principle is to: (1) design out waste and pollution, (2) keep products and materials in use and (3) regenerate natural systems within the FAS [41]. The Ellen Macarthur Foundation [41] articulates three ambitions for a healthy urban food system as: (1) source food regeneratively, and locally where appropriate, (2) design and market healthier food products, and (3) make the most of food. For too long the FAS has been primarily a linear system from production, postharvest, processing, distribution and consumption without regard to wastes incurred along the value chain.

Although recently introduced in the FAS [42], the fundamental concept has been applied and described by terms like 'industrial ecology' or 'industrial symbiosis' in numerous areas. It has been employed to mean that a waste from an entity (business, for example) would become an input to another entity (business), thereby circulating materials and keeping them in use within the larger system, essentially the concept of an 'ecosystem' [43]. This concept has not been adopted widely, although the concept of a 'Food Eco-Industrial Park' would be intriguing.

Specifically, technologies and systems applied along the value chain are needed to reduce food losses and wastes which are estimated to be as much as 30–50% globally and remarkedly similar across regions.

3.2 Innovations in controlled environment food and agricultural systems (CEFAS)

Emerging technological innovations, particularly over the past two decades have developed, based largely on a nonland-based or soilless-based indoor systems. These developing initiatives are captured by the term, 'controlled environment food and agricultural systems" (CEFAS) broadly, and more specifically by emerging subsystems, 'controlled biologically-based indoor food systems' (CBIFS).

CEFAS has evolved from a protected environment provided by greenhouses, originally with soil as the growth medium to advanced greenhouses with nutrient solutions to replace soil. Vertical farms have evolved further by using height (vertical) dimension to create intensification of the growing environment and greater yields per m^2 (production area). Both advanced greenhouses and vertical farms employ highly sophisticated measurements, controls and management. Sensors, computer control, artificial intelligence, machine learning and robots are common. These technologies are primarily devoted to growing fruits and vegetables and not practical for common field crops such as corn, soybeans, cereal grains and tuber crops, although recent research has studied the potential for wheat grown in vertical farms [44]. In addition, an increasingly significant area of a sustainable food future is aquaculture and specifically the development of the recirculating indoor system.

3.2.1 Recirculating aquaculture systems (RAS)

Fish, including finfish and shellfish, contribute about 17% of global animal-based protein for human consumption and particularly so in developing countries which consume more that 75% while producing greater than 80% of the global fish supply [45]. A major concern is that the annual number of fish caught in the wild, particularly in the oceans, has stagnated since the 1990s. As world consumption of fish has grown, aquaculture (fish farming) has developed, and almost half of the fish consumed comes from aquaculture. It is estimated that aquaculture production needs to double from approximately 67 million tons (MT) in (2012) to about 140 MT in 2050 [2].

Aquaculture, as described above, is based primarily on confined operations in a water environment, marine, such as 'cages' in oceans (along coasts predominately) or freshwater, outdoor ponds on land. The concept of a recirculating indoor aquaculture system (RAS) over the past several decades has emerged as an alternative system with advantages of greatly reduced land use and major reduction in water requirements compared to ponds. Simply stated the water is filtered from the growing tanks (confined environment) and recycled for reuse in the tanks, **Figure 11**. RAS have performed well relative to measures of productivity and environment parameters. A comprehensive treatment of recirculating aquaculture systems is provided by [46]. Challenges persist because of high capital costs, feed sources, concern for fish diseases, food safety, and consumer acceptance. Consumer concerns that farmed fish tend to have lower levels of omega-3 fatty acids than wild fish [47] and concerns about the highly intensive growing environment have limited acceptance.

Aquaponics can be an added element to an RAS by combining plants and fish. In an aquaponics system, fish provide waste that effectively fertilizes plants, thereby

Figure 11.
A recirculating aquaculture system.

creating a closed loop system (circular economy) [46]. Plants act essentially as a filter by taking up nitrates in the system. The benefits are little waste from the overall system and inputs are minimized.

Clearly, as noted, large projected increasing consumer interest for seafoods provide a need to advance aquaculture generally and RAS specifically. Thus, efforts to intensify aquaculture production by RAS need to be directed at approaches to mitigate negative issues of RAS.

3.2.2 Greenhouses

The concept of growing plants in environmentally controlled areas can be traced back to Roman times [48]. The concept of the greenhouse, as we have come to know it today, began in the Netherlands and then England in the 17th century. They have evolved from simple row covers to very large structures in the 1960s when materials such as polyethylene films, aluminum extrusions, special galvanized steel, and PVC tubing became available for various structural support frames.

3.2.2.1 Basic greenhouses

For purposes of this chapter the basic greenhouse is one where a plastic film (polyethylene) is supported by a light frame, often a hoop or A-frame in form. This type of greenhouse is primarily dependent on solar energy for heat in a cold environment, although some heating device may be employed in severe situations. Any ventilation is accomplished by natural ventilation with manual openings or slots to promote air flow for cooling. In times of high solar energy, shade coverings such as clothes are used. The plants are grown in soil at ground level or in raised beds. Movement of plants and materials are managed manually by humans with assistance of simple devices and equipment. Because the costs for the basic greenhouse are relatively low, they are used in small farm operations or in many urban settings.

3.2.2.2 Advanced greenhouses

The advanced greenhouse is defined here as a greenhouse with a highly controlled environment, high automation under computer control and using a soilless growing medium, hydroponic solution. The controlled environment for plant production consists of intensive assessment of the environment by numerous sensors to measure and monitor the parameters of: temperature, pH, relative humidity, dissolved O_2 in nutrient solution, electrical conductivity for dissolved salts in nutrient solution, CO_2 of inside air, and light intensity from the sun and supplemental lighting, and PAR (photosynthetically activated radiation) in μmol/m^2/s. Quality and optimum plant growth is dependent on plants getting an optimum daily quantity of PAR (mol/m^2/d). If the daily PAR is not provided by the sun, the computer will implement supplemental lighting to meet the desired value, **Figure 12**.

An advanced greenhouse consists of a complete system from germination of seeds to the finished product. Typically, the seed is planted in a fibrous material such as rockwool cube to germinate. Following germination, the cubes are inserted into a material (like Styrofoam) to float on the surface of the nutrient solution until fully mature.

Temperature will be controlled typically by mechanical fan ventilation under computer control of air flow by managing air intake openings. Where appropriate evaporative cooling may be used to provide cooling. Addition of CO_2 can be added to increase plant growth. Shading material can be used to reduce excessive solar energy and moveable insulation can be used to reduce heat loss at night respectively. Beyond the controlled thermal and growing environment, the advance greenhouse will include a significant automation for materials handling including robots [49].

3.2.2.3 Vertical farms (VFs)

Based on advances discussed for advanced greenhouses, the vertical farm uses the vertical dimension (**Figure 13**) to grow plants in stacked layers thereby increasing greatly the amount of product grown per unit area [50–53]. Like for the advanced greenhouse, the growing environment in a vertical farm is closely controlled for temperature, humidity, ventilation and the properties of the nutrient solution, including introduction of robotics. Five reasons to take vertical farms seriously are: avoid effect of weather and weather extremes, large reduction in water usage by as much as 95%, plant yields are high and the growing cycle is short, lower food losses, shorter supply chains because VFs can be located in urban areas, and products can be produced year-round [54].

Key challenges for VFs are high capital and energy costs. The issues of high energy consumption in VFs is due to full artificial lighting (LEDs) and for meeting cooling and humidification loads. More efficient LEDS and using LEDS tailored to the light spectrum for the specific crop, rather than the full spectrum, may save

Figure 12.
An advanced controlled environment greenhouse.

Figure 13.
A vertical farm.

electricity. Possibly the residual heat could be used in a surrounding case where a need for heat is needed. Clearly, because of large capital costs and energy requirements VFs will be a 'niche' system until these are resolved. In comparison with advanced greenhouses, where solar energy is utilized and where greenhouses can also be located in urban environments (rooftops and vacant lots for example), VFs would seem to offer uncertain benefits.

Efforts to conduct a Life Cycle Assessment of VFs and, in addition, approaches for an integration of VFs into cities are critical to assess the future of VFs. Numerous VFs have been developed and a substantial number, as well, are in the planning stages in the U.S. and Asia. Some of these are conceptualized to include solar energy directly, and inclusion of aquaculture and even livestock production [55].

3.3 Innovations in 'alternative', biologically-based indoor food systems (CBIFS)

Foods, like all materials, are an assemblage of molecules arranged in a specific structure and one is witnessing significant new biological/biochemistry efforts to create foods from plant or animal cells from the 'bottom up'. Three technologies (CBIFS) are characterized in this overview as: (1) plant-based alternative foods, (2) cell-cultured foods, and 3D printed foods. Because they use biochemical building blocks from proteins, carbohydrates, fats, and oils from plants and animals, CBIFS is a 'new' agriculture.

While much of the hype has been directed to burgers [56], there has been substantial advancement of other alternative foods, such as, for eggs, fish, shrimp, milk, yogurt, chicken nuggets, and chicken tenders to mention a few. The objective of CBIFS is to develop food products that mimic traditional foods with significant benefits. The benefits can be: (i) an environment unaffected by weather/extreme weather; (ii) year-round production; (iii) shortened growing cycles and higher yields; (iv) reduction in land, energy, and water use, (v) lower food loss and waste; (vi) shorter supply chains, local access compatible with urban settings; (vii) reduction or elimination of pesticides and antibiotics; (viii) reduction of GHG emissions; (ix) reduction in water pollution; (x) potential for enhanced micronutrients, and (xi) eliminate animal welfare concerns (growing conditions and slaughter).

However, there are potential uncertainties (questions), such as: (i) high capital cost; (ii) timeline to market; (iii) in some cases, high energy consumption; (iv) consumer acceptance; (v) concern about food quality and safety, particularly nutritional content and presence of growth hormones; (vi) price to consumers,

(vii) potential contamination; (viii) impact and possible detrimental effect for small farmers; (ix) proprietary nature of processes; (x) unproven technology, and (xi) whether the CBIFS benefits large-scale economies to the detriment of markets for small farmers [56–58].

Sustainability is critical to any future food system and is a driving force for CIBFS. In broad terms, CIBFS seeks to develop foods that impose less environmental impact, enhance human health, and reduce ethical implications of traditional animal-agriculture production, particularly for meat. Global meat consumption is estimated to increase 3% per year to 2040 [59, 60]. However, several groups [60, 61] forecast a major protein disruption in the conventional animal-agriculture system where engineered foods at the molecular level will lead to a reduction of as much as 50% or more from conventional meat and dairy by 2040.

In this section, a brief overview is presented for three types of CIBFS: plant-based alternative foods, cell-cultured foods and 3D printed foods. Through research and development each of these subsystems has a potential to enhance sustainability, availability, reliability, consumer acceptance, and quality and safety of food production.

It should be noted that food cost to the consumer is a critical issue for success of any new product. Over the past 5–10 years, numerous entrepreneurs, start-ups and food companies have created alternative foods that are already in the marketplace. In many cases, the price to consumers, at present, is higher than equivalent traditional foods, but the difference has decreased over time. As these emerging alternative products are improved, it is likely that cost to the consumer will be reduced to be comparable or even less. The food system is a high complex, adaptive system and like in other emerging technologies, there will be major changes with business failures and new players going forward.

3.3.1 Plant-based alternative foods

The development of plant-based alternative foods has become a very hot area, particularly for plant-based meat alternatives [58, 62–64]. Development of plant-based meat alternatives have attracted the greatest public attention. Environmental, human health, and concerns for animal welfare are prime factors. Among more than 50 manufacturers currently, (*Beyond Meat, Light Life and Impossible Foods*), are leading developers of plant-based burgers that are widely available in grocery stores, restaurants, and online. While burgers are a major food product, other products such as ground beef sausage, bacon and hotdogs are available.

Globally the food and agricultural system is estimated to generate as much as 34% of total GHG emissions with 71% from agriculture and related land use and land use change [65]. The opportunity for plant-based alternatives to substantially reduce environmental impacts was determined in a comparative study (Life Cycle Assessment-LCA) of the *Beyond Burger* and a U.S. beef burger (quarter pounder) by the Center for Sustainable Systems at the University of Michigan [66]. The selected parameters were GHG emissions, cumulative energy use, water use, and land use. The comparison was made to an LCA study by the National Cattleman's Beef Asssociation [67]. For the *Beyond Burger* system the results showed 90% less GHG emissions, with 46% less energy, 99% less water and 93% less land use. *Impossible Foods* also commissioned a study which found that their burger uses 96% less land, 87% less water and 86% less fossil fuel than a quarter pound beef burger. Independent LCA studies would be beneficial, given the rapidly changing ingredients being used to create plant-based meat alternatives.

Plant-based protein sources (legumes and cereal grains) are an important choice for both the vegetarian and traditional meat consumer. However, challenges remain for developers of plant-based proteins to deliver a healthy, nutritionally safe, tasty flavor, texture and appearance (color) comparable to traditional products. Because the development of plant-based meats involves complex processing of many ingredients, there have been concerns expresses about health benefits. One report [68] reviewed formulation and nutrient content of some common commercial products (burgers, hams and chicken nuggets) which contained as many as 20–30 ingredients. Comparisons yield a mixed story because plant-based meats provide about the same calories as traditional meat with more sodium, more potassium (helps eliminate sodium), no cholesterol, more iron, more B vitamins, more calcium and more saturated fat. Also, there have been reported concerns [69] about whether high-temperature cooking (grilling, frying, etc.) of protein foods could generate toxins and carcinogens. Thus, there is a need to assess whether plant-based meat protein would be any less safe or safer than traditional meat.

Consumer acceptance of plant-based meats and food products is key to the ultimate impact of these products. Substantial improvements have been made during the decade with plant-based burgers such that many people find them indistinguishable from a traditional burger. Nevertheless, the author suggests the need for future research and study in: (i) independent LCS studies to quantify environmental benefits; (ii) further evidence on health benefits, nutrition, and safety; (iii) development of new plant proteins sources; (iv) reduction in number of non-protein ingredients; and (v) reduction in cost to consumer.

3.3.2 Cell-cultured foods

Cell-cultured meat, also known as cultivated meat, has advanced at a rapid pace over the past 20 years. The concept, although relatively simple, uses animal cells nurtured within a bioreactor to produce food that is designed to mimic meat products [70]. Compared to plant-based protein where protein is extracted from plants, cell-based meat is created from cells extracted from animals and grown in a culture. Specifically, a small piece of fresh muscle, obtained by biopsy, from a living animal is disrupted by a combination of mechanical and enzymatic methods to produce stem cells [71].

Using culturing methods, the adult stem cell (called satellite cells), in the presence of relatively high serum concentrations, divide leading to multiplying populations. Tissue engineering methods are then used to differentiate these expanded cells into muscle and fat tissue, which lead to generation of a cultured meat product closely resembling conventional meat [72]. **Figure 14** illustrates, in broad terms, the process based on the starting point of obtaining adult muscle stem cells from an animals or pluripotent stem cells from an embryo [73]. However, a recent study suggests that it may be possible to grow cultured meat with much less dependence on animals by using a soy-based scaffold for support of muscle cells and form a meat-like 3D-cell structure [74].

Today, there is no cultured meat available for consumers to purchase at retail or food service outlets in the U.S., unlike plant-based meat, but there are 20 or more start-ups in the cultured meat space [60]. However, the Singapore Food Agency approved in December 2020 a cultured chicken product by Eat Just [75]. With an investment more than $100 million in global funding by billionaires and venture capitalists, there are significant efforts to develop the area [55].

It is very difficult to reproduce the diversity of meats from the numerous existing animal species, breeds and cuts so there is a great need to optimize cell culture technology [76]. A LCA [77] and a (TEA) techno-economic assessment [78] have modeled future large-scale cell-cultured meat production facilities and have shown reduced overall environmental impacts and the potential to be cost-competitive

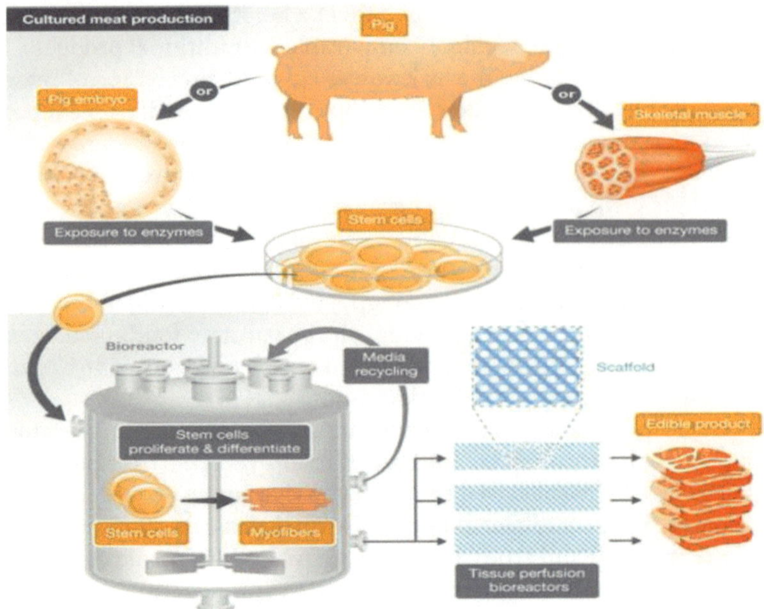

Figure 14.
The development process for cell-cultured meat [73].

with conventional meat by 2030. These are the first repots using data collected from active companies (more than 15) in the chain. The LCA shows cell-cultured meat is about 3.5 times more efficient (feed conversion ratio) than poultry which is the most efficient system of conventional meat production. The LCA in its comparisons with traditional meat includes the use of renewable energy in which case the reduction in GHG emissions shows a reduction of 17–92%, less land use of 63–95% and 51–78% less use of water depending on the respective conventional animal system.

However, studies by [76, 79] suggest that comparison is not that simple. They conclude that the relative comparisons with conventional meat depend on the type of systems for energy generation (i.e., decarbonized and renewable) and the specific production animal system. Thus, there is a need for independent and transparent LCA studies of cultured meat production.

A major challenge for cultured meat production is meeting consumer demands for flavor, texture, color, nutritional composition and cost [80, 81]. Conventional animal meat is high in protein with amino acids, vitamins and minerals. As development of cultured meat has advanced, the similarity to conventional meat has improved greatly, such that many people cannot distinguish a difference between the two [55]. An advantage that cultured meat can have is an ability to add texturizing ingredients, colorants, flavorings and nutrients to address sensorial and nutritional properties [80]. Challenges facing cell-cultured meat systems include: (i) replace animal serum with plant materials; (ii) reduce costs by material substitutions and advances in scalability; (iii) assessment of health effects, short and long term; (iv) conduct detailed, independent and transparent LCAs to quantify environment impacts; (v) eliminate growth hormone factors; (vi) develop cell lines that are more accessible; vii) safety assessment, particularly with respect to potential contaminants that might enter in the process; (viii) ability to replicate the diversity of conventional meats; (ix) address the issues of potential regulation, as well as labeling, and (x) develop a name or nomenclature for marketing (presently there is much variation and uncertainty about a common name).

3.3.3 3D printed foods

The combination of robotics and software has entered the realm of food manufacturing in the form of 3D printing [81–84]. 3D printing technology is a novel approach which can create complex geometries, tailored textures, and nutritional contents. The 3D technology can provide a 'customized food' to meet special dietary needs as well as mass customization. NovaMeat, a Spanish company, and Redefine Meat, an Israeli company, have utilized 3D technology to produce beef steaks and other meats that resemble animal meat.

In the 3D printing process, food ingredients are placed in cartridges, and the product is created layer by layer by a controlled robotic process, similar to 3D printing of non-food items. The technology has been employed to use tissue engineering to create meat and other food alternatives. Also, the 3D technology has been employed at the home scale to create 'designer 'foods. Depending on the specific food, ingredients can range from processed components (sauces, dough, etc.) to more elemental ingredients such as sugars, proteins, fats, and carbohydrates [82]. Some foods may require further processing, such as some form of cooking or storage. A significant challenge is linking material properties and structure to the printing process variables to get a desired 3D printed product. The parameters of control are those relating to the printer and those controlling the food-relevant parameters.

In Switzerland, Jungbunzlauer AG [85] is providing 'recipe cards' to guide consumers to use bio-based ingredients to create one's own dairy and meat alternatives. Recipe cards are available online to create foods such as non-dairy ice cream, cream cheese and yogurt as well as plant-based burgers and bratwurst. The respective recipe card provides a detailed list of ingredients, suppliers, quantities, together with directions to create the specific food product and with nutrition information as well. Thus, it seems not to be a great stretch that this information could lead to 3D printing of designer and specialized food products.

The 3D printing process compresses the value chain to a highly local system of inputs (ingredients), a single controlled process (the 3D printer) and a single output (the food product).

4. Conclusions

Meeting the demand for food of the growing world population will require both conventional land-based agriculture systems and controlled environment agriculture and food systems, including emerging controlled biology-based indoor food systems (CBIFS).

Already during the 21st century the impact of innovations in 'conventional' agriculture and food systems has been impressive:

- Digital Ag, driven by computing and information science, has progressed rapidly to offer technically advanced solutions to support an efficient FAS with decreases in food loss and waste with greater productivity, prosperity, and sustainability, with more to come,

- Sensors, robots and drones are, and will become even more ubiquitous moving forward,

- Nanoscale science and engineering, and biotechnologies (particularly CRISP) will continue to drive innovations in many areas of the FAS,

- Renewable energies of solar, wind and bioenergy are increasingly integrated into the FAS through co-location of sources on the land, as well as providing sustainable energy for rural communities, domestically and internationally,

- Electricity will be an increasing driving force in FAS for motive power of agricultural machines, including tractors and autonomous systems,

- Emphasis on the soil health is likely to increase substantially as a part of the initiatives in regenerative agriculture practices if it proves profitable and successful in sequestering significant carbon, and

- Circularity in the FAS is likely to be a major initiative in an effort to design out waste and pollution, keep materials in play and to maintain natural systems.

In acknowledging the fact that traditional FAS uses substantial land and creates considerable GHG emissions, an emerging area of food production has developed. Controlled Biologically-based Indoor Food Systems (CBIFS) can greatly reduce land area for agriculture production by largely soilless methods and produce foods with less environmental impact, enhance human health and avoid ethical concerns about traditional methods, particularly for meat. CBIFS described herein: recirculating aquaculture, advanced greenhouses, vertical farms, plant-based alternative foods, cell-cultured foods and 3D printing of foods are in various stages of development but are poised to grow in the future. Principal keys to growth are costs to the consumer and consumer acceptance. Nevertheless, numerous products are in the marketplace and already enjoy modest growth.

CBIFS use basic biochemical building blocks of proteins, amino acids, sugars, fats, carbohydrates, and oils from plants and animals. Thus, controlled environment agriculture and food systems, including CBIFS, should be viewed as complementary to conventional agriculture and typical supply chains. Further development of CBIFS foods that are as healthy, nutritious, safe and appealing as conventional foods will significantly contribute to the sustainability, resilience and circularity of food and agriculture systems.

Finally, there is a need to transcend the debate between numerous constituencies, rooted in ideological attitudes, to invoke and encourage an array of different approaches to meet the challenge of a food and agriculture system that is robust, safe, and sustainable in meeting the 17 Sustainable Development Goals for a sustainable planet.

Author details

Norman R. Scott
Department of Biological and Environmental Engineering, Cornell University, Ithaca, New York, USA

*Address all correspondence to: nrs5@cornell.edu

References

[1] National Research Council, A framework for assessing effects of the food system. 2015; The National Academies Press. Washington D.C.

[2] World Resources Report, Creating a sustainable food future: A menu of solutions to feed nearly 10 billion people by 2025. Final report July 2019; https://research,wri,org/sites/default/files/2019-07WRR_Food_Full_Report_0.pdf.

[3] Cornell Atkinson Center for Sustainability/Nature Sustainability, Socio-technical innovation bundles for agri-food systems transformation. December 2020; Nat Sustain 3, 973. https://doi.org/10.1038/s41893-020-00672-5

[4] EAT-Lancet Commission, Food Planet Health: Healthy diets from sustainable food systems. 2019; https://eatforum.org/content/uploads/2019/07EAT-Lancet_Commission_Summary_Report.pdf

[5] Walker, L., Buhler, D. Catalyzing holistic agriculture innovation through industrial biotechnology. 2020; Industrial Biotechnology 16(4): 189-208. DOI: 10.1089/ind.2020.29222.lpw

[6] NASEM, Innovations in the food system: Exploring the future of food: Proceedings of a workshop. 2019; The National Academies Press, Washington D.C.

[7] Friedman, T., Thank you for being late: An optimist's guide to thriving in the age of accelerations. Farrar, Straus and Giroux; 2016. 486 p. NY, NY

[8] Brundtland, G. World commission on environment and development: Our common future. 1987. https://sustainabledevelopment.unorg/content/documents.5987our-common-future.pdf

[9] National Research Council. Toward sustainable agricultural systems in the 21st century. 2010. 599p. National Academies Press, Washington D.C. DOI 10.17226/12832

[10] Weston, R. Sustainable Development: Definition and implementation strategies. 1992. Roy Weston, West Chester, PA

[11] United Nations Foundation. Sustainable development goals. unfoundation.org

[12] Solutions from the Lands. 21st Century agriculture renaissance: Solution for the lands. 2021. https://solutionsfromtheland.org/wp-content/uploads/2021/02/AgRenaissanceRpeort.pdf

[13] Zeder, M. The origins of agriculture in the near east. 2011. Current Anthropology. 52:221-235. https://www.jstor.org/stable/10.1086/659307

[14] agrocares. What is the difference between precision, digital and smart farming? 2018. agrocares.com

[15] Future Farming. World's first robotic catalogue with 35 propositions. 2020. futurefarming.com

[16] Scott, N., Chen, H. Nanoscale science and engineering for agriculture and food systems. 2012. Industrial Biotechnology 8(6): 340- 343 https://doi.org.10.1089/ind.2012.1549 (532-540) https://doi.org/10.1038/s41565-09-0439-5

[17] Scott, N., Chen, H., Cui, H. Nanotechnology applications and implications of agrochemicals toward sustainable agriculture and food systems. 2018. J. Agric. Food Chem. 66 (26): 6451-6456. DOI: 10.1021/acs.jafc.8b00964

[18] Giraldo, J., Wu, H., Newkirk, G. Kruss, S. Nanobiotechnology approaches for engineering smart plant sensors. 2019. Nature Biotechnology. 14(541-553). https://doi.org/10.1038/s41565-019-0470-6

[19] Lew, T., Sarojam, R., Jang, I, Park, B., Naqvi, N., Wong, M., Singh, G., Ram, R., Shoseyov, O., Saito, K., Chua, N., Strano, M. Species-independent analytical tools for next generation agriculture. 2020. Nature Plants. 6(1408-1417) https://doi.org/10.1038/s41477-020--00808-7

[20] Gilbertson, L., Pourzahedi, L., Laughton, S., Gao, X., Zimmerman, J., Theis, T., Westerhoff, P. Lowry, G. Guiding the design space for nanotechnology to advance sustainable crop production. 2020. Nature Nanotechnology. https://doi.org/10.1038/s41565-020-0706-5

[21] Kah, M., Tufenkji, N., White, J. Nano-enabled strategies to enhance crop nutrition and protection. 2019. Nature Nanotechnology 14(532-540) https://doi.org/10/1038/s41565-019-0439-5

[22] Brookes, G., Barfoot, P. GM crop technology use 1996-2018: farm income and production impacts. 2020. GM Crops and Foods 11(4). https://doi.org/10.1080/21645698.2020.1779574

[23] Brookes, G., Barfoot, P. Environmental impacts of genetically modified (GM) crop use 1996-2018: impacts on pesticide use and carbon emissions. 2020. GM Crops and Foods. 11(4). https://doi.org/10.1080/21645698.2020.1773198

[24] International Service for the Acquisition of Agri-Biotech Applications (ISAAA). ISAAA Brief 55-2019: Global status of biotech crops, 2020. www.isaaa.org

[25] Tonamo, A., Review status of animal biotechnology and options for improving animal production in developing countries. 2015. J. of Biology, Agriculture and Healthcare. 5(19): 21- 31. ISSN 2225-093X (online)

[26] FDA. Press Release December 14, 2020. Approves First-of-its-Kind Intentional Genomic Alteration in Line of Domestic Pigs for Both Human Food, Potential Therapeutic Uses

[27] Goetzberger, A., Zastrow, A., On the coexistence of solar-energy conversion and plant cultivation. 1982. Int. J. of Solar Energy. 1(1):55-69. https://doi.org/10.1080/01425918208909875

[28] Dinesh, H., Pearce, J., The potential of agrivoltaic systems. 2018. Renewable and Energy Reviews. 54(299-308). https://dx.doi.org/10.1016/j.rser.2015.10.024

[29] Dupraz, C., Marrou, H., Dufour, L., Nogier, A., Ferard. Y. Combining solar photovoltaic panels and food crops for optimizing land use: Toward new agrivoltaic schemes. 2011. Renewable Energy. 36(2725-2732). doi: 10.1016/j.renene.2011.03.005

[30] Adeh, E., Good, S., Calaf, M. Higgins, C. Solar PV power potential is greatest over croplands. 2019. natureresearch, scientific reports. 9:1142 https://doi.org/10.1038/s41598-019-47803-3

[31] Baron-Gafford, G., Pavao-Zuckerman, M., Minor, R., Sutter, L., Barnett-Moreno, I., Blackett, R., Thompson, M., Dimond, K., Gerlak, A., Nabhan, G., Macknick, E. 2019. Nature sustainability. 2(848-855) https://doi.org/10.1038/s41893-019-0364-9

[32] NREL (National Renewable Energy Laboratory). A clear vision for wind enhancement. www.nrel.gov

[33] National Research Council, Renewable fuel standard; potential economic and environmental effects of

U.S. biofuels policy. 2011. 416 p. The National Academies Press. https://doi.org/10.17226/13105

[34] EIA (Energy Information Administration. EIA projects U.S. biofuels production to slowly increase to 2025. 2021. www.eia.gov

[35] EPA. AgStar Development Handbook. 2020. 132 p. https://www.epa.gov/sites/production/files/2104-12/documents/agstar-handbook.pdf

[36] Labatut, R., Angenet, L. Scott, N., Biochemical methane potential and biodegradability of complex organic substrates. 2011. Bioresource Technology. 102(2255-2264) doi: 10.1016/j.biotech.2010.10.035

[37] Future farming. a website with continuing information and frequent updates in tracking electric autonomous equipment, including tractors. www.futurefarming.com

[38] World Resources Report, Creating a sustainable food future: A menu of solutions to feed nearly 10 billion people by 2025. Final report July 2019; Chapters 20-21. https://research,wri,org/sites/default/files/2019-07WRR_Food_Full_Report_0.pdf

[39] Hunt, N., Liebman, M., Thakrar, S., Hill, J. Fossil energy use, climate change impacts, and air quality-related human health damages of conventional and diversified cropping systems in Iowa, USA. 2020. Environ. Sci. Technol. 54(11002-11014) https://dx.doi.org/10.1021/acs.est.9b06929

[40] Tamburini, G., Bommarco, R., Wanger, T., Kremen, C., van der Heijden, M., Liebman, M., Hallin, S. Agricultural diversification promotes multiple ecosystems services without compromising yield. 2020. Sci. Adv. 6:eaba1715

[41] Ellen Macarthur Foundation. 2019. Cities and Circular Economy for Food, 66p. https://www.ellenmacarthurfoundation.org/assets/downloads/Cities-and- circular-economy-for-Food_280110.pdf

[42] ASABE. Resource: engineering and technology for a sustainable world. 2021. Special issue, March/April. www.asabe.org

[43] Valentine, S.V. 2016. Kalundborg Symbiosis: fostering progressive innovation in environmental networks. Journal of Cleaner Production 118: 65-77 http://dx.doi.org/10.1016/j.jclepro.2016.01.061

[44] Asseng, S., J.R. Guarin, M. Raman, O. Monje, G. Kiss, D. Despommier, F. M. Meggers, and P.P.G. Gauthier. 2020. Wheat Yield Potential in Controlled Environment Vertical Farms. PNAS 117 (32) 19131-19135. www.pnas.org/cgi/doi/10.1073/pnas.2002655117.

[45] OECD-FAO. 2017. Meat-Agricultural Outlook 2018-2027. Chapter 6. www.fao.org/3/i9166e/i9166e_chapter6_meat.pdf.

[46] Timmons, M., Guerdat, T., Vinci, B. Recirculating Aquaculture, 4th edition. 2018. Ithaca Publishing Company, LLC. ISBN 978-0971264670

[47] World Resources Report, Creating a sustainable food future: A menu of solutions to feed nearly 10 billion people by 2025. Final report. July 2019; Chapter 23 https://research,wri,org/sites/default/files/2019-07WRR_Food_Full_Report_0.pdf

[48] Janik, J., Paris, H., Parish, D. The cucurbits of Mediterranean Antiquity: Identification of Taxa from Ancient Images and descriptions. 2007. Annals of Botany 100(7): 1441-1457. doi.10.1093/aob/mcm242

[49] Ting, K., Lin, T., Davidson, P. 2016. Integrated urban controlled environment agricultural systems.

In: Kozai T, editor. LED lighting for urban agriculture. Springer-Science+Business Media, Singapore. p. 18-36 DOI 10.1007/978-981-10-1848-0_2. ch2

[50] Despommier, D. The Vertical Farm: Feeding the World in 21st Century. Martin's Press. NY, NY. 2010. 293 p.

[51] Benke, K., Tomkins, B. 2017. Future food-production systems: Vertical farming and controlled environment agriculture. Sustainability: Science, Practice and Policy 13(1): 13-26. https://doi.org/10.1080/15487733.2017.1394054

[52] Kozai, T. (Editor). 2018. Smart plant factory: The next generation indoor Vertical farms. Singapore: Springer.

[53] Kozai, T., Fujiwara K., Runkle, E. (Editors). 2016. Plant Factory and Greenhouse with LED Lighting. Singapore: Springer.

[54] Pinstrup-Andersen, P. Is It Time to Take Vertical Farming Seriously? 2017. Global Food Security. https://dx.doi.org/10.1016/j.gfs.2017.09.002

[55] Kalantari, F., Tahir, O., Lahijani, A., Kalantari, S. A review of vertical Farming technology: A guide for implementation of building integrated agriculture in cities. 2017. Advanced Engineering Forum 24 (76-91) doi.10.4028/www.scientific.net/AEF.24.76

[56] Purdy, C. Billion Dollar Burger. 2020 Penguin Random House. 252 p.

[57] NASEM. Innovations in the Food System: Exploring the Future of Food. Proceedings of a Workshop. 2019, Nation Academies Press. http://nationalacademies.org/hmd/Activities/Nutrition/FoodForum/2019-AUG-07

[58] He, J., Evans, N., Huaizhi, L., Shao, S. A Review on Plant-based Meat Alternatives: Driving Forces, History, Manufacturing, and Consumer Attitudes. 2020. Compr. Rev. Food Saf. 19(5): 2639-2656. https://doi:10.1111/1541-4337.12610

[59] FAO, Global food losses and food waste: Extent, causes, and prevention. 2011. Rome, Italy: United Nations FAO. http://www.fao.org/3/mb060e/mb060e00.htm

[60] A.T. Kearney. How Will Cultured Meat and Meat Alternatives Disrupt The Agricultural and Food Industry? 2020. Chicago, IL: A.T. Kearney Analysis. Industry[11613].pdf [74].pdf

[61] Tubb, C., Seba, T. Rethinking food and agriculture 2020-2030: The second domestication of plants and animals, the disruption of the cow, and the collapse of industrial livestock farming. www.rethinkx.com

[62] Sha, L., Xiong, Y. Plant protein-based alternatives of reconstructed Meat: Science, Technology and Challenges. 2020. Trnfs in Food Science & Technology. 102 (2020) 51-61. https://doi.org/10.1061/j.tifs.2020.05.022

[63] Curtain, F., Grafenauer, S. Plant-based meat substitutes in the flexitarian age: An audit of products on supermarket shelves. 2019. Nutrients (11) 11: 2603. doi.10.3390/nu11112603

[64] Friend, T. Value meal: Impossible Foods wants to save the world by inventing a better burger. 2019. THE NEW YORKER. September 30, 42-55. https:/www.newyorker.com/magazine/2019/09/30/can-a-burger-help-solve- climate-change

[65] Crippa, M., Solazzo, E., Guizzardi, D., Monforti-Ferrario, F., Tubiello, F., Leip, A. Food systems are responsible for a third of global anthropogenetic GHG Emissions. 2021. Nature Food. https://doi.org/10.1038/s43016-021-00225-9

[66] Heller, M., Keoleian, G. Beyond Meat's beyond burger life cycle assessment: A detailed comparison between a plant-based and animal-based protein source. 2018. Report No. CSS18-10. Center for Sustainable Systems, University of Michigan, Ann Arbor 1-38.

[67] Thoma, G., Putman, B., Matlock, M., Popp, J., English, L. Sustainability Assessment of U.S. Beef Production Systems. 2017. University of Arkansas Resiliency Center. https://scholarworks.uark.edu/rescentfs/3

[68] Bohrer, B. Review: nutrient density and nutritional value of meat products and non-meat foods high in protein. 2017. Trends in Food Sci. &Tech. 65(103-112) http://dx.doi.org/10.1016/j.tifs.2017.04.016.

[69] Barzegar, F., Kamankesh, M., Mohammadi, A. Heterocyclic aromatic amines in cooked food: A review on formation, health risk-toxicology and their analytical techniques. 2019. Food Chemistry. 280: 240-254. https://doi.org/10.1016/j.foodchem.2018.12.058

[70] Boler, D., Martin, J., Kim, M., J. Krieger, Milkowski, A., Mozdziak, P., Sylvester, B. 2020. Producing food products from cultured animal tissues. www.cast-science.org/wp-content/uploads/2020/04/QTA2020-1-Cultured-Tissues-1.pdf.

[71] Post, M. Cultured beef: Medical technology to produce food. 2013. J. Food and Agriculture. 94(6):1039-1041. Doi:10.1002/jsfa.6474

[72] Melzener, L., Verzijden, K., Buijs, A., Post, M., Flack., J. Cultured beef: From small biopsy to substantial quantity. 2020. J. Sci. Food and Agric. 101(1):7-14 https:// doi.org/10.1002/jsfa.10663.

[73] Tuomisto, H. The eco-friendly burger: could cultivated meat improve the environmental sustainability of meat products? 2019. EMBO Reports. 20(1). article e47395. https://doi.org/10.15252/embr.201847395

[74] Young J., Skivergaard, S. Cultured meat on a plant-based frame. 2020. Nature Food 1, 195.https://doi.org/10.1038/s43016-020-0053-6

[75] Ives, M. Singapore approves a lab grown meat product, a global first. 2020. New York Times. December 2, 2020

[76] Chriki, S., Hocquette, J. The myth of cultured meat: A review. 2020. Frontiers in Nutrition. https://doi.org/10.3389/fnut.2020.00007

[77] CE Delft. LCA of cultivated meat: Future projections for different scenarios. 2021. https://www.cedelft.eu.en/publications/2610/lca-of-cultivated-meat-

[78] CE Delft. TEA of cultivated meat: Future projections of different scenarios. 2021. https://www.cedelft.eu.en/publications/2609/tea-of-culivated-meat-future

[79] Lynch, J., Pierrehumpert, P. Climate impacts of cultured meat and beef cattle. 2019. Frontiers Sustain. Food Syst. doi.org/10.3389/fsufs.2019.00005

[80] Fraeye, I., Kratka, M. Sensorial and nutritional aspects of cultivated meat in comparison to traditional meat: Much to be inferred. 2020. Front. Nutr. https://doi.org/10.3389/fnut.2020.00035

[81] Dankar, I., Haddarah, A.,Omar, F., Sepulcre, F., Pujola, M. 3D Printing technology: The new era for food customization and elaboration. 2018. Trends in Food Science & Technology. 75(231-242). https://doi.org/10.1016/j.tifs.2018.03.018.

[82] Severini, C., Derossi, A., Azzollini, D. Variables affecting the printability of foods: Preliminary tests on cereal-based

products. 2016. Innovative Food Science and Emerging Technologies. 38(281-291). http://dx.doi.org/10.1016/j.ifset.2016.10.001

[83] Yang, F., Zhang, M., Bhandari, B. Recent developments in 3D food printing. 2017. Critical Reviews in Food Science and Nutrition, 57:14, 3145-3153. 10.1080/10408398.2015.1094732

[84] He, C., Zhang, M., Fang, Z. 3D Printing of food: Pretreatment and post- treatment of materials. 2019. Critical Reviews in Food Science and Nutrition, 60(14):2379-2392 https://doi.org/10.1080/10408398.2019.1641065.

[85] Jungbunzlauer. Natural Ingredients to Enhance Your Dairy & Meat Alternatives. 2020. www.jungbunzlauer.com

Section 3

Energy Systems Engineering

Chapter 19

Adsorption-Based Atmospheric Water Harvesting: Technology Fundamentals and Energy-Efficient Adsorbents

Muhammad Sultan, Muhammad Bilal, Takahiko Miyazaki, Uzair Sajjad and Fiaz Ahmad

Abstract

Nowadays, atmospheric water harvesting (AWH) became very essential to provide fresh potable water. This technique is in practice since 1900 (US661944A) by Edger S. Belden. Atmospheric water is a source of freshwater with 13000 trillion liters availability of water at any time and can be utilized in overcoming water shortage, especially in arid and rural areas. It holds up the water molecules in the form of vapors and accounts for adding 10% of all freshwater present on the earth. Mainly, the two most common methods have been used for the extraction of atmospheric water. First, the ambient air is cooled below the dew point temperature, and second in which the moisture in atmospheric air is adsorbed/absorbed using desiccant materials. Conventional vapor compression, thermoelectric cooling, dew, and fog water harvesting based systems/technologies possess some limits in terms of energy requirements, less efficiency, and high cost. However, the adsorption based AWH technology is relatively cheaper, environment friendly, and can be operated by a low-grade thermal energy source. The limited availability of commercial instruments to harvest atmospheric water using adsorbents indicates a lack of fundamental studies. The fundamental research on water adsorption, adsorption kinetics, regeneration conditions, and water collecting surface designs has not gained as much interest as required in the field of atmospheric water harvesting. In this regard, this book chapter discusses and presents the progress in the field of adsorbent materials and system designs along with the future directions to accelerate the commercialization of this technology.

Keywords: adsorption, desiccant dehumidification, atmospheric water harvesting, energy-efficient adsorbents, thermal energy, condensation

1. Introduction

Globally, water scarcity is considered one of the prime issues in the upcoming decades. Almost 2.1 billion people are lacking access to clean and fresh water [1]. **Figure 1** shows the water-stressed areas in the world. Middle East, Asia, South America, and some parts of Africa face water scarcity. Therefore, many studies

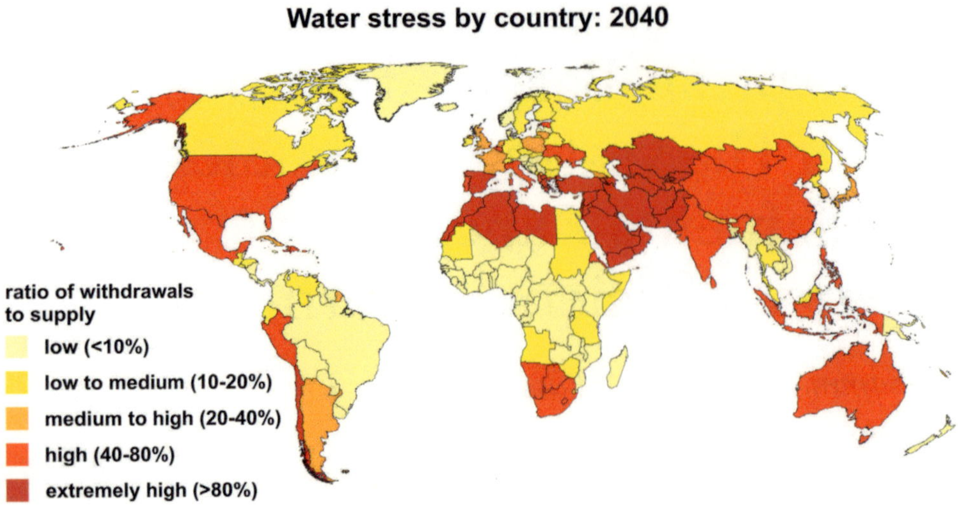

Figure 1.
World map showing the water-stressed areas by 2040 reproduced from [2].

have been investigated to supply enough water with cheaper, and portable methods [3, 4]. These methods include desalination, wastewater treatment, sewage recycling, and water harvesting from the atmosphere. The energy consumption in desalination systems is very high that is almost 50% of the cost, and make this technology inappropriate in most situations [5]. Also, seawater desalination is not suitable for remote areas and has many environmental problems. Thus, portable systems with less energy consumption are needed. Atmospheric water harvesting can be considered as a potential resource of fresh water in remote areas [6]. For this purpose, many researchers have introduced innovations for water production from humid air technology. This can be done by many ways i.e., using vapor compression cycle (VCC) [7–10], thermoelectric cooling (TEC) [11, 12],

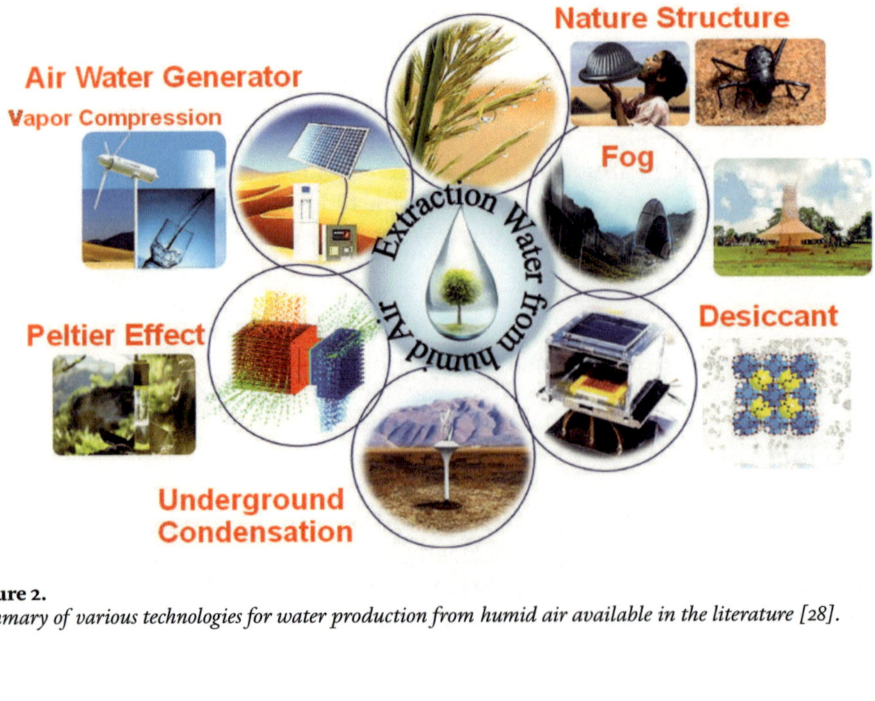

Figure 2.
Summary of various technologies for water production from humid air available in the literature [28].

absorption/adsorption refrigeration [13, 14], wind power with VCC [15], using solar chimneys [16, 17], using membranes [18, 19], and using adsorbent materials [20–27]. **Figure 2** summarized the various technologies investigated in the literature for producing water from the humid air. The purpose of all these technologies is to produce water and are using worldwide depending on the conditions and the requirements. Among all these technologies, desiccant based atmospheric water harvesting (AWH) shows a great potential to extract enough amount of drinkable water with less energy consumption [6]. The adsorption based AWH is possible in dry and desert regions with the lowest relative humidity. This technology utilizes renewable energy sources (solar, wind, and low-grade biomass) which ultimately lead towards the cheapest and most efficient systems. This chapter focuses on the fundamentals and principles of adsorption based AWH. The progress and perspectives and associated adsorption based AWH systems are also discussed in this study. Moreover, energy-efficient desiccant materials along with the recently developed new generation MOFs for AWH are also highlighted in this study. The main purpose of this chapter is to introduce the importance of AWH by employing various efficient desiccant materials.

2. Atmospheric water harvesting

2.1 Conventional AWH

Atmospheric water harvesting could be considered as a huge renewable source of water that can provide enough amount of water, but unfortunately is ignored [29]. Conventional water harvesting was started first when a Russian forester built a stone condenser during 1905 and 1912 and was considered as the early Greek dew condenser [30, 31]. Ziebold tested with this type of condenser and named as "the aerial wells", but unfortunately, this project was failed and the expected amount of water was not produced due to the low thermal conductivity and low heat capacity [32]. In 1957, a review was carried out on the absorption of water by the plants [33]. Since then several studies have been carried out focusing on the fog and dew harvesting by the plants and animals [34].

2.2 Modern AWH

Modern AWH shifted towards the innovations, methods, and technologies that can provide a significant amount of water in remote areas [35]. As mentioned earlier in the introduction section, various new methods have been proposed for AWH i.e., VCC, TEC, using membrane and adsorbent materials. Among these, the fog water was first collected with the help of nets in 1956 [36]. Shi et al. replaced these traditional meshes with vertically arranged wires to avoid the problems of clogging [37]. Dew water collection considered as the alternative approach because it is not majorly affected by climatic conditions and can provide water in most of the ambient environment [38]. A lot of advancements have been done in the designs of active condensers after the commercialization of mechanical refrigerators in the 1980s. The desiccant based dew water harvesting was taken into consideration in the Nineteenth century, in which the various desiccant materials capture the moisture from the atmosphere during the night, and then releases the moisture in vapor form during the day. This method has been proved the most energy-efficient and reliable technology because it employs solar energy and can provide water anywhere and anytime in the world.

3. Water vapor parameters in atmospheric air

Atmospheric air is a mixture of nitrogen, oxygen, and argon gas, and water vapors with varying contents. The relative humidity (Φ), absolute humidity (ω), and the dew point temperature (T_d) are considered as the most essential parameters of the air which can be used as the source of water. The relative humidity (Φ) represents the ratio of the partial pressure of water vapor (P_w) to the saturation pressure (P_s), while the absolute humidity (ω) represents the maximum amount of water that can be extracted from the air. The relative humidity can be expressed using (Eq. (1)) found in the literature [39, 40].

$$\Phi = \frac{P_w}{P_s(T)} \tag{1}$$

where, Φ represents the relative humidity, P_w denotes the partial pressure of water vapor, and P_s represents the saturation pressure. The relation between relative humidity, absolute humidity, temperature, and total air pressure can be described using (Eq. (2)) found in the literature [40].

$$\Phi = \frac{\omega P}{(0.622 + \omega) P_s(T)} \tag{2}$$

The dew point temperature (T_d) can be determined from (Eq. (2)) by solving for T at $\Phi = 1$ for given air pressure and absolute humidity. The water vapor saturation pressure (P_s) at any temperature (T) can be described using (Eq. (3)), while the total air pressure (P) can be described using (Eq. (4)) found in the literature [40].

$$P_s(T) = 610.94 \exp\left(\frac{17.625T}{243.04 + T}\right) \tag{3}$$

$$P = P_a + P_w \tag{4}$$

Total air pressure (P) is the sum of the partial pressure of dry air (P_a) and the vapor pressure of water in the air (P_w). The moist air enthalpy can be described using (Eqs. (5)–(7)) given in the literature [40].

$$H = H_a + \omega H_{wv} \tag{5}$$

$$H_a = C_{p,a} T \tag{6}$$

$$H_{wv} = H_{wv}(0°C) + C_{p,wv} T \tag{7}$$

where H_a term represents the enthalpy of dry air, H_{wv} term represents the enthalpy of the presence of water vapor, and $C_{p,a}$ denotes the heat capacity of air (kJ kg^{-1} C).

4. Principles of adsorption based AWH

Adsorption based AWH is unique in its way that it utilizes the desiccant materials to capture water vapors from the air and shows higher thermal efficiencies as compared to the traditional AWH systems. The main advantage is that the desiccant materials can be regenerated by solar thermal energy and the condensation process can occur at ambient conditions [41, 42]. **Figure 3** shows the adsorption based AWH process which consists of two stages. In the first stage, the desiccant material in contact with the ambient air at night which adsorbs the water vapors. In the second stage, the desiccant material is packed into a closed system where a significant amount of heat is provided to regenerate the desiccant material. Due to the regeneration process, the material desorbs the water vapors, and the collected vapors will be condensed into liquid form. With this approach, the AWH can be possible in low relative humidity areas. A lot of advancement has been done in the material designs, and system developments. **Figure 4** shows the dual-stage AWH device mechanism and prototype introduced in the literature [44]. A novelty in this device was that two adsorbent layers were used to improve the water production per day. The latent of condensation from the upper stage was used for the desorption purpose of the bottom stage. With this approach, the thermal efficiency can be improved, and this system can become more suitable for daily purposes. AQSOA Z01, zeolite material was experimentally tested and showed that a prototype can harvest up to 0.77 $L/m^2/day$ with an 18% increase as compared to the single-stage AWH device [44]. The results found that a temperature of 90°C on the solar absorber area can give a maximum water production for AQSOA Z01.

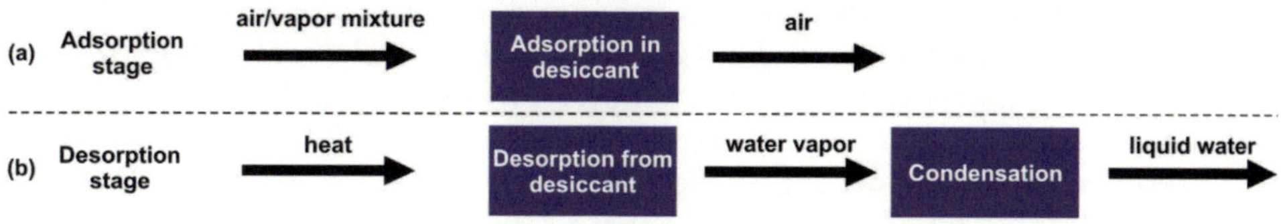

Figure 3.
Adsorption-based AWH process consists of two stages. (a) Adsorption stage (water vapors from the ambient air adsorbed in the adsorbent). (b) Desorption stage (water vapors desorbed from the adsorbent and condensed into liquid form) [43].

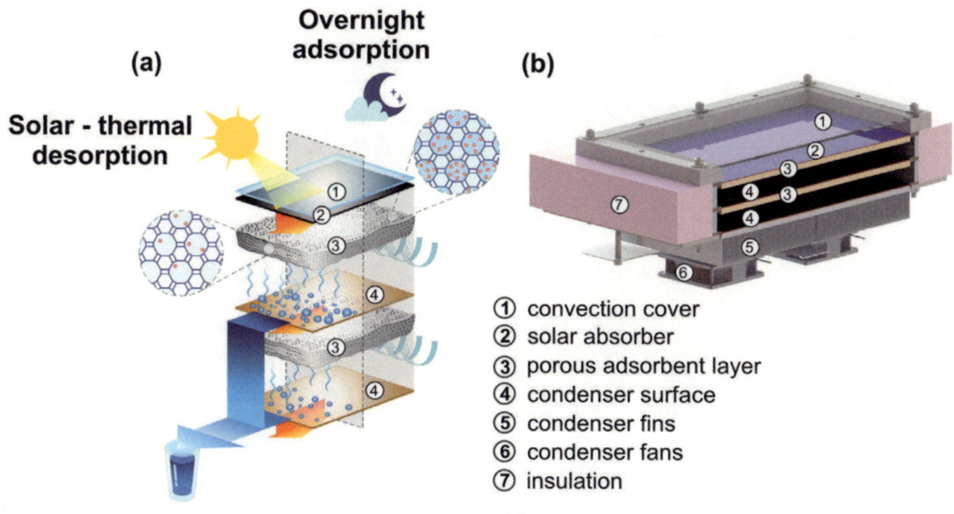

Figure 4.
Illustration of dual stage adsorption based AWH device. (a) Mechanism of dual stage adsorption based AWH in which adsorption process occurs during night when ambient air in contact with the adsorbent layer, while the desorption process occurs during day when device is closed, and heat is supplied to regenerate the adsorbent layer. (b) Dual stage AWH prototype consists of convection cover, solar absorber, adsorbent layer, and condenser [44].

5. Progress and perspectives in adsorption based AWH

Adsorption based AWH is a vital technology that can provide cost-effective water in arid areas. The vapor concentration in this technology can be achieved through desiccant materials which ad/absorb and desorb the water vapors from the air [45]. In this context, efficient desiccant materials are a key research priority and various materials have been developed. First, it was believed that the solid desiccant AWH systems can extract enough amount of water but requires a large amount of material which makes these systems very expensive [46]. Also, the operating costs of air blowers to circulate the air for both adsorption and desorption purposes make this system less attractive. However, the development of next-generation MOFs, nano-porous organic materials, and various composite desiccant materials shows great potential for AWH systems. **Figure 5** shows the recent progress in adsorption based AWH systems. Ideal desiccant material should possess the required properties of stability, hydrophilicity, and pore diameter. Adsorption capacities and densities are of great importance in any practical application [51, 52]. The desiccant materials with type IV and type V isotherms are most suitable for this application [43]. During the adsorption process, the materials adsorption capacity should linearly increase with relative humidity, while in the desorption process, the materials desorption capacity should drop steeply with the increased temperature. In this regard, progress has been made and Kallenberger et al. developed a composite material by incorporating the calcium chloride into an alginate-derived matrix [53]. The water uptake capacity of this material was almost linear with relative humidity and when adsorption temperature increases to 65°C, the water uptake capacity drops which shows that the desiccant material can be regenerated at low temperatures. Also, recently developed MOFs show this type of flexibility to harvest enough amount of water at the lowest relative humidity conditions [27, 54]. After the desorption process, the inlet air of the condenser is the outlet air of the desorber. It is worth noting that both the desorption and condensation temperature

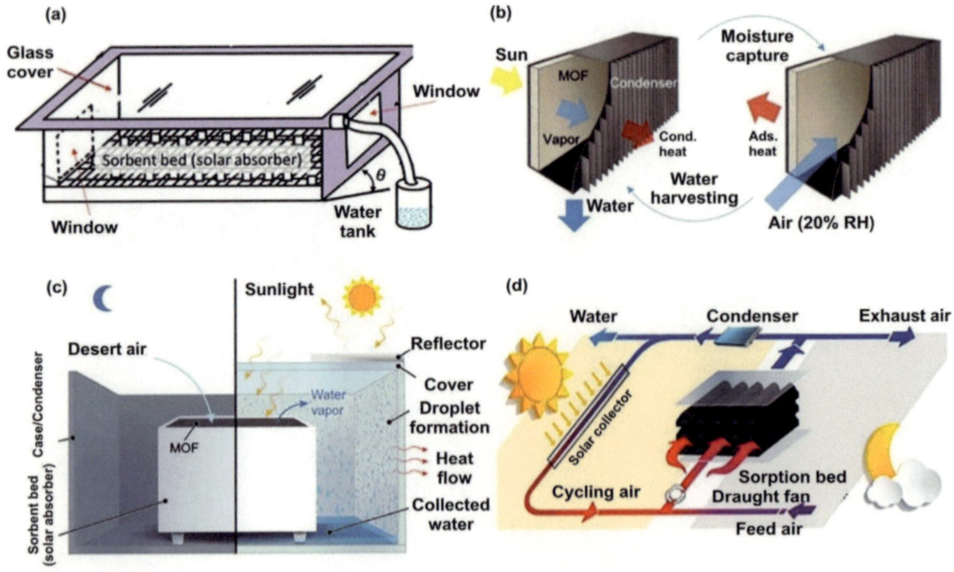

Figure 5.
Adsorption based AWH systems published in the literature. (a) Solar glass desiccant box type system [47]. (b and c) MOFs based AWH systems [48, 49]. (d) Packed columns desiccant matrix based AWH system mechanism [50].

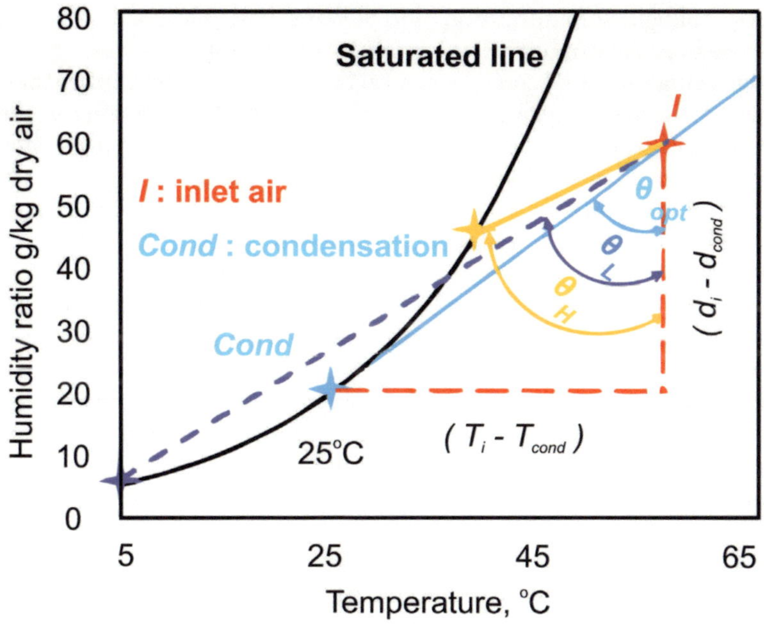

Figure 6.
The optimal condensing temperature on the psychrometric chart developed in the literature [46].

should be carefully chosen to balance the specific water production per day per unit collector area (SWP), and the specific energy consumption per unit mass water production (SEC). In this regard, Tu et al. developed a powerful tool to determine the proper desorption and condensation temperature [46]. **Figure 6** shows the optimal condensing temperature on the psychrometric chart in which the inlet air of the condenser is denoted by I (Ti, di), and the condensation states of the humid air are denoted by the stars on the saturated line. The tangent of the angle (θ) and SEC can be described using (Eqs. (8) and (9)) given in the literature [46].

$$\tan\theta = \frac{T_i - T_{cond}}{d_i - d_{cond}} \tag{8}$$

$$SEC = C_p\left(\frac{\varepsilon_T}{\varepsilon_d}\right) \times \tan\theta + h_{fg} \tag{9}$$

It is noted that when the line through the point I (Ti, di) is tangent to the saturated line, then the angle (θ) is at the smallest value, and therefore the condensing temperature at given inlet conditions for a minimum value of SEC can be obtained at the tangent point. An appropriate heat source and airflow rate can be chosen to find the optimum outlet air conditions of the desorber by using this tool.

6. Energy-efficient desiccant materials for AWH

6.1 Silica gel and hygroscopic salts based AWH materials

Energy efficient materials must have high capacity of adsorbing and desorbing water from the air [55]. Heidari et al. investigated a novel desiccant based evaporative cooling system for production of water [56]. The results showed that the silica gel-based system can harvest up to 585 L of water during a week. Milani et al. investigated a small scale air cooled silica gel based wheel dehumidifier for extraction of water from the atmosphere [57]. It was found that the system can generate more than 5.2 L of water per day in the ambient conditions of Sydney. A simulation model on TRNSYS was also built and found that the system can generate a cumulative of 18.5kL of water in the ambient conditions of Abu Dhabi, 10kL of water in London, and 13.8kL for the ambient conditions of Sydney. Similarly, various desiccant materials based on the hygroscopic salts were also investigated to produce water from humid air. **Table 1** shows the silica gel and hygroscopic salts-based materials used in atmospheric water harvesting systems. Hamed et al. investigated a system based on sandy bed impregnated with calcium chloride for atmospheric water harvesting [59]. The system was exposed to ambient air to absorbs the water vapors in the night and the desiccant material was covered with the glass layer where regeneration process will occur, and water vapors condensed into liquid form. It was found that the system can provide 1 L per m^2 of water per day. Wang et al. investigated a semi open system with a novel composite sorbent of LiCl with active carbon felt (ACF) for water production from humid air [61]. The system was tested at different experimental conditions and found that 14.7 kg, 13.6 kg, and 12.5 kg of water was obtained at conditions of 85%, 75%, and 65% relative humidity, respectively.

6.2 Zeolites based AWH materials

Zeolites are the family of porous crystalline and hydrated aluminosilicates that are widely used as the adsorbents in many applications. These materials can extract water from air at low relative pressures due to their affinity with water [62]. As zeolite materials have a framework structure, a high temperature is required to regenerate and desorb the water vapors. **Table 2** shows the summary of some potential zeolite materials with efficient adsorption capacities. Furukawa et al. studied the zeolite 13X and found that it can harvest water up to 0.40 g/g at low relative pressures [63]. The adsorption properties of Li-X zeolite and Na-X were investigated and found that these materials can be employed to extract water from air [64]. The results found that Li-X and Na-X can extract up to 0.244 g/g and 0.192 g/g respectively. The kind of ion in this type of zeolites not only influences the amount of adsorbed water but also the energy densities and heat of adsorption. Despite the high performance, the energy requirements for desorption purpose restricts the zeolite materials to be used in AWH systems [52].

Adsorbent	Material	Quantity	Water harvesting capacity	Reference
Silica gel	Desiccant wheel		585 L during a week	[56]
Hygroscopic salts	$CaCl_2$/cloth		1.5 L/m^2 day	[58]
	$CaCl_2$/cloth sand	1 kg	2.32 L/m^2 day	[26]
	$CaCl_2$/sand	1 kg	1 L/m^2 day	[59]
	LiCl/sand		90 mL/day, 115 mL/day	[24]
	$CaCl_2$/saw wood/ vermiculite		40–140 mL/kg/day	[60]
	LiCl/active carbon felt	40.8 kg	14.7 L	[22]

Table 1.
Summary of various silica gel and hygroscopic salts based desiccant materials for atmospheric water harvesting found in the literature.

Adsorbent	Material	Water harvesting capacity	Reference
Zeolite	Zeolite 13X	0.40 g water/g zeolite	[63]
	Li-X-Zeolite	0.244 g water/g zeolite	[64]
	Na-X	0.192 g water/g zeolite	[64]
	AQSOA type zeolites	0.1–0.3 kg water/kg zeolite	[65]

Table 2.
Zeolite based desiccant materials for atmospheric water harvesting found in the literature.

6.3 MOFs based AWH materials

MOFs have been researched for their water capture properties and they were found to be highly promising and energy efficient materials. Several members of the MOF family showed unprecedented water uptake property [63]. Specifically, zirconium MOFs made from Zr6O4(OH)4(-CO2)n secondary building units and carboxylate organic linkers showed very interesting properties in water adsorption [66]. MOF-841 was investigated and showed the maximum water uptake and maintained its structure over 80 adsorption–desorption cycles [63]. A similar trend was observed in other zirconium MOF named as MOF-801 which showed a water uptake at 10% relative humidity. Motivated by these results, MOF-801 based device was built and tested in Arizona, desert [49]. The device was consisted of two boxes, the inner box was open and holds the MOF material while, the outer box has a lid. The outer lid was open at night to allow the MOF-801 to in contact with ambient air and hold the water molecules in its pores and then the lid was closed in day and device was exposed to sunlight to regenerate the MOF material. This device was delivered 200–300 mL of water/kg of MOF/day at 20-40oC temperature and 5–40% relative humidity. This device showed remarkable results and proved as a first device in the history to extract water from the desert air. **Table 3** shows the water harvesting capacities of potential MOFs. It can be seen that Co_2Cl_2BTDD material delivered 0.82 g of water/g of MOF under 5–30% relative humidity conditions [68]. It was found that the pore diameter of this material was above the critical diameter for water capillary action which enabled water uptake at the limit of reversibility. **Figure 7** shows the framework structures of some potential MOFs used in AWH systems. The key in all MOFs is the framework structure which allow to trap water from low relative humidity conditions. The water harvesting through MOFs was moved to next level after the development of MOF-303 based device which showed extraordinary results at low relative humidity conditions and also exhibit adsorption and desorption cycles each on the scale of minutes [54]. This device was first tested in a laboratory and later in Mojave Desert at conditions of 10% relative humidity and 27°C and it delivered 0.7–1.0 L of water/kg of MOF/day [54]. It is clear from the discussion that MOFs can be considered as the potential and energy efficient materials for AWH. With these MOFs based AWH systems, not only clean water can be harvested in any climate but also to make this concept more mobile and dispensed [66].

6.4 Other AWH materials

Other adsorbent materials for AWH that have been interested and investigated in the last decade are nano porous super gels and super hygroscopic gels [72–74]. The main factors of these type of materials include the effective capturing of water molecules, high efficiency storage, and fast water desorption abilities under different climatic conditions [28]. **Figure 8** shows the nano-porous super hygroscopic hydrogel employed to harvest water from highly humid atmosphere zones [73]. This hydrogel was made up of Zn and O atoms in a unique ratio of 1:1.1.

Adsorbent	Material	Relative humidity (%)	Water harvesting capacity	Reference
Metal–Organic Framework	MOF-801	20	2.8 L	[27]
	MOF-303		0.175 L/kg	[67]
	MOF-841	5–35	44 wt%	[63]
	Co_2Cl_2BTDD	5–30	0.82 g/g	[68]
	UiO-66	40	0.052 g/g	[69]
	Banasorb-22		0.08 g/g	[70]
	Cr-soc-MOF-1	70	1.95 g/g	[71]
	HSO_3-UiO-66		0.038 g/g	[69]
	IRMOF-1		0.11 g/g	[70]

Table 3.
Metal–organic framework based desiccant materials for atmospheric water harvesting found in the literature.

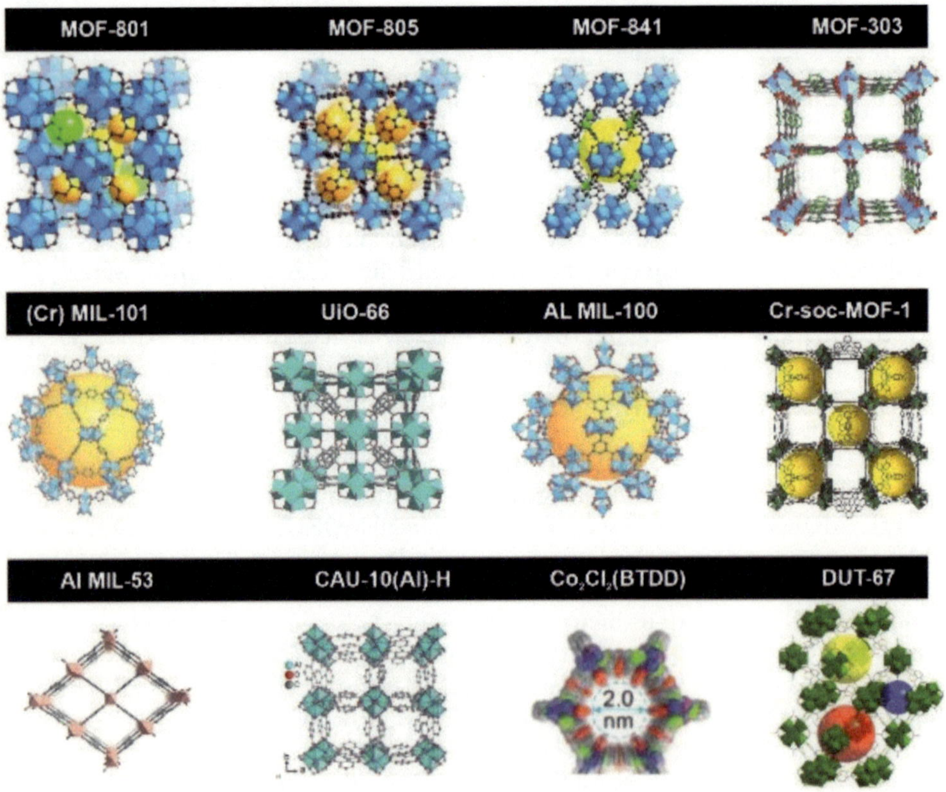

Figure 7.
Illustration of MOFs structures used in AWH systems.

It was found that this synthesized hydrogel has a high-water uptake of over 420% of its own weight. A steep increase in water absorption at high relative humidity of over 80% was shown by the hydrogel which makes it suitable for extraction of water from the humid air. The hydrogel showed the excellent stability for more than 1000 absorption/desorption cycles. It was concluded from the calculations that the absorption cycles of 15 min and desorption of 5 min could give the

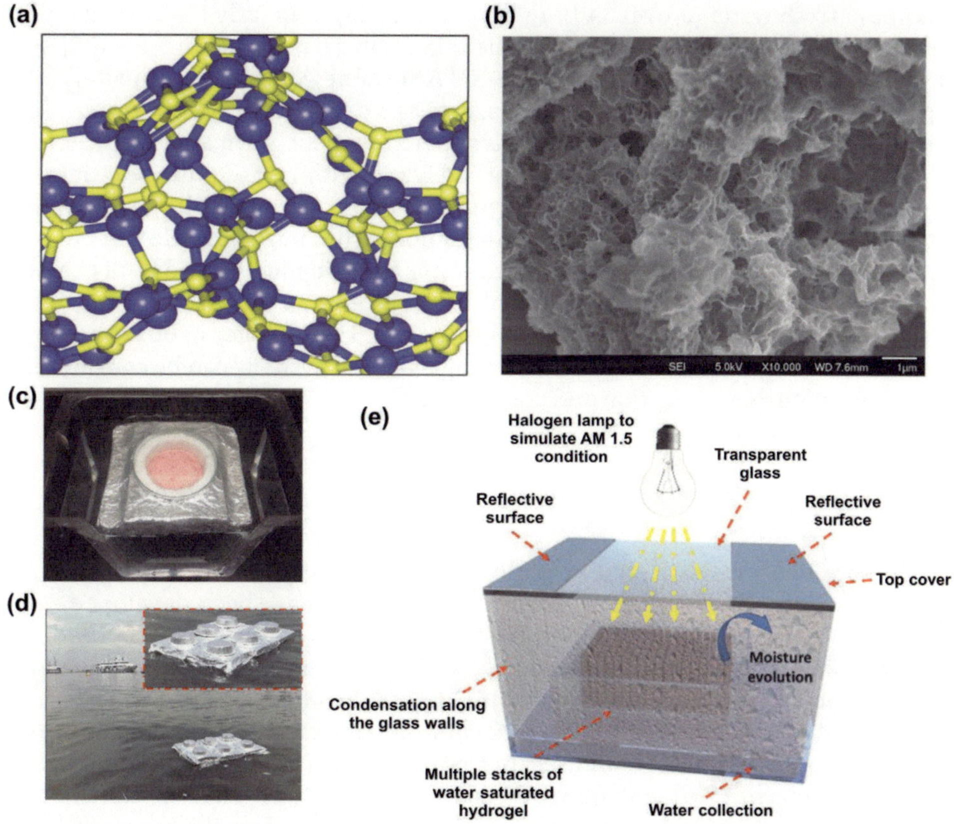

Figure 8.
Nano-porous super hygroscopic hydrogel-based AWH. (a) The hydrogel is made up of Zn:O ratio 1:1.1 (blue balls for zinc atoms, yellows balls for oxygen atoms). (b) SEM image of the hydrogel showing porous network. (c-d) The prototype developed for the absorption characteristics of the hydrogel by floating on the sea surface. (e) Schematic of the AWH system based on super hygroscopic hydrogel reproduced from [73].

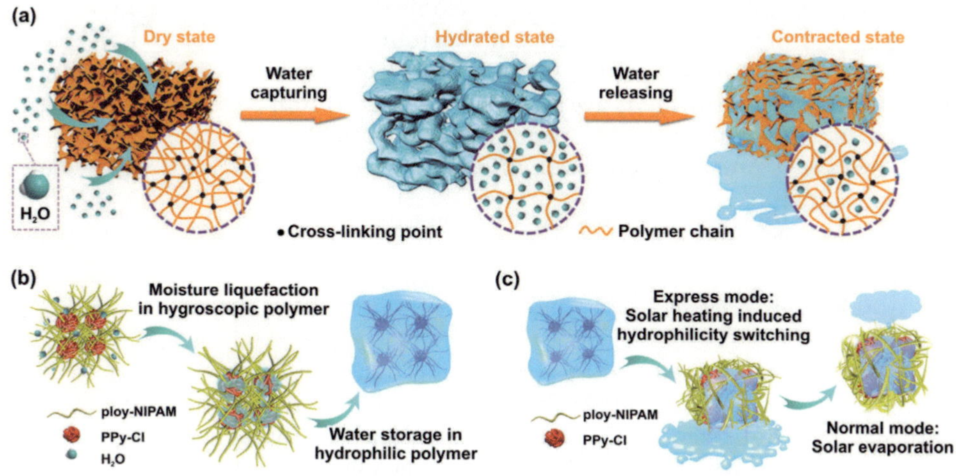

Figure 9.
Super moisture-absorbent gel-based AWH. (a) Illustration of the AWH process (moisture captured by the SMAG and releases water under room temperature). (b) Schematic showing the moisture absorption enabled by the SMAGs. (c) Schematic showing the express and normal modes for water harvesting reproduced from [72].

maximum fresh water of over 14 L/kg of hydrogel/day. Similarly **Figure 9** shows the illustration of super moisture-absorbent gel water harvesting process [72]. The super absorbent gel consists of poly-NIPAM framework which ultimately expands the internal area of the gel and serves as a pathway for water during desorption process. It was found that this super absorbent gel in saturated condition can directly release 50% of the absorbed water within 15–20 min once it is slightly heated to 40oC (denoted the "express mode"). After this phase, the water can be collected via condensation process (denoted the "normal mode"). The super moisture absorbent gel showed two water releasing modes and both can be powered by solar radiation. The super gel-based prototype was also investigated, and it was found that it can produced about 20 and 55 L of water in 60% and 90% relative humidity, respectively. These hydrogels-based systems can be considered as the low energy consumption and cost-effective.

7. Conclusions

The supply of freshwater to a rapidly growing world population is a great societal challenge. In this regard, several technologies have been developed and currently in use worldwide, but the advancements of additional methods for freshwater generation is very crucial to effectively address the global water scarcity. For this purpose, this chapter highlights the importance of adsorption based AWH which utilizes the desiccant materials to capture water vapors from the atmosphere and condenses into liquid form. The important water vapor parameters in ambient air are discussed in this study. The fundamental principles of adsorption based AWH are reviewed, moreover, the progress and perspectives in this technology also explained from the viewpoints of newly developed desiccant materials and the modified AWH systems designs. The study explores the energy efficient desiccant materials which are already employed in AWH systems. From the literature, it was found that the recently developed MOFs are promising due to their flexible nature and tailorable architectures and can harvest water from the atmospheric air at low relative humidity conditions. Some newly developed hygroscopic gels are also showing great potential to utilized in AWH systems. It was found that the temporal and spatial restrictions for AWH and as well as the energy requirements can also be reduced if the appropriate adsorbents are selected. The adsorption based AWH systems ensures no bulky equipment, more environment-friendly and cost effective. Thus, this study presents a comprehensive knowledge on AWH through adsorbent materials.

Acknowledgements

This work is carried out in the Department of Agricultural Engineering, Bahauddin Zakariya University, Multan-Pakistan. The work is financially supported by the Director Research/ORIC grants of Bahauddin Zakariya University (Multan), awarded to Principal Investigator Dr. Muhammad Sultan.

Conflict of interest

The authors declare no conflict of interest.

Author details

Muhammad Sultan[1*†], Muhammad Bilal[1†], Takahiko Miyazaki[2,3], Uzair Sajjad[4] and Fiaz Ahmad[1]

1 Department of Agricultural Engineering, Bahauddin Zakariya University, Multan, Pakistan

2 Department of Advanced Environmental Science and Engineering, Faculty of Engineering Sciences, Kyushu University, Fukuoka, Japan

3 International Institute for Carbon-Neutral Energy Research (WPI-I2CNER), Kyushu University, Fukuoka, Japan

4 Department of Mechanical Engineering, National Chiao Tung University, Hsinchu, Taiwan

*Address all correspondence to: muhammadsultan@bzu.edu.pk

† These authors contributed equally to this work.

References

[1] United Nations. (2018). n.d. https://doi.org/http://www.un.org/en/sections/issues-depth/water/.

[2] Maddocks A, Young RS, Reig P. 2015 Ranking the world's most water-stressed countries in 2040. World Resour Inst 26.

[3] Lundqvist J, Andersson A, Johannisson A, Lavonen E, Mandava G, Kylin H, et al. 2019 Innovative drinking water treatment techniques reduce the disinfection-induced oxidative stress and genotoxic activity. Water Res 155:182-92.

[4] Srithar K, Rajaseenivasan T. 2018 Recent fresh water augmentation techniques in solar still and HDH desalination–A review. Renew Sustain Energy Rev 82:629-44.

[5] Ahmed FE, Hashaikeh R, Hilal N. 2019 Solar powered desalination–Technology, energy and future outlook. Desalination 453:54-76.

[6] Bilal M, Sultan M, Miyazaki T. 2020 Spatiotemporal Investigation of Atmospheric Water Harvesting Potential Using Response Surface Methodology for Multan (Pakistan) and Fukuoka (Japan).

[7] Zhang RX, Zang RQ, Liu JX. 2016 Study on properties of water extraction from cooled air system under all operating conditions. Cryog Supercond 44:51-5.

[8] Zolfagharkhani S, Zamen M, Shahmardan MM. 2018 Thermodynamic analysis and evaluation of a gas compression refrigeration cycle for fresh water production from atmospheric air. Energy Convers Manag 170:97-107. https://doi.org/10.1016/j.enconman.2018.05.016.

[9] Habeebullah BA. 2009 Potential use of evaporator coils for water extraction in hot and humid areas. Desalination 237:330-45. https://doi.org/10.1016/j.desal.2008.01.025.

[10] Al-Farayedhi AA, Ibrahim NI, Gandhidasan P. 2014 Condensate as a water source from vapor compression systems in hot and humid regions. Desalination 349:60-7.

[11] Cao D, Zou Y. 2016 System Optimization of Water Exaction from Air by Semiconductor Cooling Air [J]. Build Energy Environ 35:71-3.

[12] Eslami M, Tajeddini F, Etaati N. 2018 Thermal analysis and optimization of a system for water harvesting from humid air using thermoelectric coolers. Energy Convers Manag 174:417-29. https://doi.org/10.1016/j.enconman.2018.08.045.

[13] Scrivani A, Bardi U. 2008 A study of the use of solar concentrating plants for the atmospheric water vapour extraction from ambient air in the Middle East and Northern Africa region. Desalination 220:592-9.

[14] Ibrahim NI, Al-Sulaiman FA, Saidur R. 2016 Performance assessment of water production from solar cooling system in humid climate. Energy Convers Manag 127:647-55.

[15] Solís-Chaves JS, Rocha-Osorio CM, Murari ALL, Lira VM, Sguarezi Filho AJ. 2018 Extracting potable water from humid air plus electric wind generation: A possible application for a Brazilian prototype. Renew Energy 121:102-15.

[16] Alexander B. 2011 Production of Freshwater and Energy from Earth's Atmosphere. Smart Grid Renew Energy 2011.

[17] Kashiwa BA, Kashiwa CB. 2008 The solar cyclone: A solar chimney for harvesting atmospheric water. Energy 33:331-9. https://doi.org/10.1016/j.energy.2007.06.003.

[18] Bergmair D, Metz SJ, De Lange HC, van Steenhoven AA. 2014 System

analysis of membrane facilitated water generation from air humidity. Desalination 339:26-33. https://doi.org/10.1016/j.desal.2014.02.007.

[19] Zhao B, Wang LY, Chung TS. 2019 Enhanced membrane systems to harvest water and provide comfortable air via dehumidification & moisture condensation. Sep Purif Technol 220:136-44. https://doi.org/10.1016/j.seppur.2019.03.034.

[20] Liu JY, Wang JY, Wang LW, Qi Y. 2016 Performance test of sorption air-to-water device. CIESC J 67:46-50.

[21] Ji JG, Wang RZ, Li LX. 2007 New composite adsorbent for solar-driven fresh water production from the atmosphere. Desalination 212:176-82. https://doi.org/10.1016/j.desal.2006.10.008.

[22] Wang JY, Liu JY, Wang RZ, Wang LW. 2017 Experimental research of composite solid sorbents for fresh water production driven by solar energy. Appl Therm Eng 121:941-50. https://doi.org/10.1016/j.applthermaleng.2017.04.161.

[23] Kabeel AE. 2007 Water production from air using multi-shelves solar glass pyramid system. Renew Energy 32:157-72.

[24] Srivastava S, Yadav A. 2018 Water generation from atmospheric air by using composite desiccant material through fixed focus concentrating solar thermal power. Sol Energy 169:302-15.

[25] Talaat MA, Awad MM, Zeidan EB, Hamed AM. 2018 Solar-powered portable apparatus for extracting water from air using desiccant solution. Renew Energy 119:662-74. https://doi.org/10.1016/j.renene.2017.12.050.

[26] William GE, Mohamed MH, Fatouh M. 2015 Desiccant system for water production from humid air using solar energy. Energy 90:1707-20. https://doi.org/10.1016/j.energy.2015.06.125.

[27] Kim H, Rao SR, Kapustin EA, Zhao L, Yang S, Yaghi OM, et al. 2018 Adsorption-based atmospheric water harvesting device for arid climates. Nat Commun 9:1-8. https://doi.org/10.1038/s41467-018-03162-7.

[28] Salehi AA, Ghannadi-Maragheh M, Torab-Mostaedi M, Torkaman R, Asadollahzadeh M. 2020 A review on the water-energy nexus for drinking water production from humid air. Renew Sustain Energy Rev 120:109627. https://doi.org/10.1016/j.rser.2019.109627.

[29] Wahlgren R V. 2001 Atmospheric water vapour processor designs for potable water production: a review. Water Res 35:1-22.

[30] Nikolayev V, Beysens D, Gioda A, Milimouk I, Katiushin E, Morel J-P. 1996 Water recovery from dew.

[31] Kogan B, Trahtman A. 2003 The moisture from the air as water resource in arid region: hopes, doubts and facts. J Arid Environ 53:231-40.

[32] Beysens D, Milimouk I. 2000 The case for alternative fresh water sources. Pour Les Resour Altern En Eau, Secher 11:1-16.

[33] Stone EC. 1957 Dew as an ecological factor: I. A review of the literature. Ecology 38:407-13.

[34] Malik FT, Clement RM, Gethin DT, Krawszik W, Parker AR. 2014 Nature's moisture harvesters: A comparative review. Bioinspiration and Biomimetics 9. https://doi.org/10.1088/1748-3182/9/3/031002.

[35] Shabir F, Sultan M, Miyazaki T, Saha BB, Askalany A, Ali I, et al. 2020 Recent updates on the adsorption capacities of adsorbent-adsorbate pairs for heat transformation applications. Renew Sustain Energy Rev 119:109630. https://doi.org/10.1016/j.rser.2019.109630.

[36] Gultepe I, Tardif R, Michaelides SC, Cermak J, Bott A, Bendix J, et al. 2007 Fog research: A review of past achievements and future perspectives. Pure Appl Geophys 164:1121-59.

[37] Shi W, Anderson MJ, Tulkoff JB, Kennedy BS, Boreyko JB. 2018 Fog harvesting with harps. ACS Appl Mater Interfaces 10:11979-86.

[38] Khalil B, Adamowski J, Shabbir A, Jang C, Rojas M, Reilly K, et al. 2016 A review: dew water collection from radiative passive collectors to recent developments of active collectors. Sustain Water Resour Manag 2:71-86.

[39] Gabler R, Petersen J, Trapasso L, Sack D. 2008 Physical geography. Nelson Education; .

[40] Cengel YA, Boles MA. 2007 Thermodynamics: An Engineering Approach 6th Editon (SI Units). The McGraw-Hill Companies, Inc., New York; .

[41] Sultan M, Miyazaki T, Saha BB, Koyama S. 2016 Steady-state investigation of water vapor adsorption for thermally driven adsorption based greenhouse air-conditioning system. Renew Energy 86:785-95. https://doi.org/10.1016/j.renene.2015.09.015.

[42] Sultan M, El-Sharkawy II, Miyazaki T, Saha BB, Koyama S, Maruyama T, et al. 2015 Insights of water vapor sorption onto polymer based sorbents. Adsorption 21:205-15.

[43] LaPotin A, Kim H, Rao SR, Wang EN. 2019 Adsorption-Based Atmospheric Water Harvesting: Impact of Material and Component Properties on System-Level Performance. Acc Chem Res 52:1588-97. https://doi.org/10.1021/acs.accounts.9b00062.

[44] LaPotin A, Zhong Y, Zhang L, Zhao L, Leroy A, Kim H, et al. 2020 Dual-Stage Atmospheric Water Harvesting Device for Scalable Solar-Driven Water Production. Joule:1-17. https://doi.org/10.1016/j.joule.2020.09.008.

[45] Shabir F, Sultan M, Niaz Y, Usman M, Ibrahim SM, Feng Y, et al. 2020 Steady-state investigation of carbon-based adsorbent-adsorbate pairs for heat transformation application. Sustain 12. https://doi.org/10.3390/su12177040.

[46] Tu Y, Wang R, Zhang Y, Wang J. 2018 Progress and Expectation of Atmospheric Water Harvesting. Joule 2:1452-75. https://doi.org/10.1016/j.joule.2018.07.015.

[47] Kumar M, Yadav A. 2015 Experimental investigation of design parameters of solar glass desiccant box type system for water production from atmospheric air. J Renew Sustain Energy 7. https://doi.org/10.1063/1.4922142.

[48] Kim H, Yang S, Rao SR, Narayanan S, Kapustin EA, Furukawa H, et al. 2017 Water harvesting from air with metal-organic frameworks powered by natural sunlight. Science (80-) 356:430.

[49] Fathieh F, Kalmutzki MJ, Kapustin EA, Waller PJ, Yang J, Yaghi OM. 2018 Practical water production from desert air. Sci Adv 4:1-10. https://doi.org/10.1126/sciadv.aat3198.

[50] Wang JY, Wang RZ, Tu YD, Wang LW. 2018 Universal scalable sorption-based atmosphere water harvesting. Energy 165:387-95. https://doi.org/10.1016/j.energy.2018.09.106.

[51] Sultan M, Miyazaki T, Koyama S, Khan ZM. 2018 Performance evaluation of hydrophilic organic polymer sorbents for desiccant air-conditioning applications. Adsorpt Sci Technol 36:311-26. https://doi.org/10.1177/0263617417692338.

[52] Sultan M, El-Sharkawy II, Miyazaki T, Saha BB, Koyama S. 2015 An overview of solid desiccant dehumidification and air conditioning systems. Renew Sustain Energy Rev 46:16-29. https://doi.org/10.1016/j.rser.2015.02.038.

[53] Kallenberger PA, Fröba M. 2018 Water harvesting from air with a hygroscopic salt in a hydrogel–derived matrix. Commun Chem 1:6-11. https://doi.org/10.1038/s42004-018-0028-9.

[54] Hanikel N, Prévot MS, Fathieh F, Kapustin EA, Lyu H, Wang H, et al. 2019 Rapid Cycling and Exceptional Yield in a Metal-Organic Framework Water Harvester. ACS Cent Sci 5:1699-706. https://doi.org/10.1021/acscentsci.9b00745.

[55] Sultan M, El-Sharkawy II, Miyazaki T, Saha BB, Koyama S, Maruyama T, et al. 2016 Water vapor sorption kinetics of polymer based sorbents: Theory and experiments. Appl Therm Eng 106:192-202.

[56] Heidari A, Roshandel R, Vakiloroaya V. 2019 An innovative solar assisted desiccant-based evaporative cooling system for co-production of water and cooling in hot and humid climates. Energy Convers Manag 185:396-409.

[57] Milani D, Qadir A, Vassallo A, Chiesa M, Abbas A. 2014 Experimentally validated model for atmospheric water generation using a solar assisted desiccant dehumidification system. Energy Build 77:236-46. https://doi.org/10.1016/j.enbuild.2014.03.041.

[58] Gad HE, Hamed AM, El-Sharkawy II. 2001 Application of a solar desiccant/collector system for water recovery from atmospheric air. Renew Energy 22:541-56.

[59] Hamed AM, Aly AA, Zeidan E-SB. 2011 Application of solar energy for recovery of water from atmospheric air in climatic zones of Saudi Arabia. Nat Resour 2:8.

[60] Kumar M, Yadav A, Mehla N. 2019 Water generation from atmospheric air by using different composite desiccant materials. Int J Ambient Energy 40:343-9.

[61] Wang JY, Wang RZ, Wang LW, Liu JY. 2017 A high efficient semi-open system for fresh water production from atmosphere. Energy 138:542-51. https://doi.org/10.1016/j.energy.2017.07.106.

[62] Hanif S, Sultan M, Miyazaki T. 2019 Effect of relative humidity on thermal conductivity of zeolite-based adsorbents: Theory and experiments. Appl Therm Eng 150:11-8.

[63] Furukawa H, Gándara F, Zhang YB, Jiang J, Queen WL, Hudson MR, et al. 2014 Water adsorption in porous metal-organic frameworks and related materials. J Am Chem Soc 136:4369-81. https://doi.org/10.1021/ja500330a.

[64] Jänchen J, Ackermann D, Stach H, Brösicke W. 2004 Studies of the water adsorption on zeolites and modified mesoporous materials for seasonal storage of solar heat. Sol Energy 76:339-44.

[65] Teo HWB, Chakraborty A, Fan W. 2017 Improved adsorption characteristics data for AQSOA types zeolites and water systems under static and dynamic conditions. Microporous Mesoporous Mater 242:109-17.

[66] Xu W, Yaghi OM. 2020 Metal – Organic Frameworks for Water Harvesting from Air, Anywhere, Anytime. https://doi.org/10.1021/acscentsci.0c00678.

[67] Fathieh F, Kalmutzki MJ, Kapustin EA, Waller PJ, Yang J, Yaghi OM. 2018 Practical water production from desert air. Sci Adv 4:1-10. https://doi.org/10.1126/sciadv.aat3198.

[68] Rieth AJ, Yang S, Wang EN, Dincă M. 2017 Record atmospheric fresh water capture and heat transfer with a material operating at the water uptake reversibility limit. ACS Cent Sci 3:668-72.

[69] Trapani F, Polyzoidis A, Loebbecke S, Piscopo CG. 2016 On the general water harvesting capability of metal-organic frameworks under well-defined climatic conditions. Microporous Mesoporous Mater 230:20-4.

[70] Wu T, Shen L, Luebbers M, Hu C, Chen Q, Ni Z, et al. 2010 Enhancing the stability of metal–organic frameworks in humid air by incorporating water repellent functional groups. Chem Commun 46:6120-2.

[71] Abtab SMT, Alezi D, Bhatt PM, Shkurenko A, Belmabkhout Y, Aggarwal H, et al. 2018 Reticular chemistry in action: A hydrolytically stable MOF capturing twice its weight in adsorbed water. Chem 4:94-105.

[72] Zhao F, Zhou X, Liu Y, Shi Y, Dai Y, Yu G. 2019 Super Moisture-Absorbent Gels for All-Weather Atmospheric Water Harvesting. Adv Mater 31:1806446.

[73] Nandakumar DK, Zhang Y, Ravi SK, Guo N, Zhang C, Tan SC. 2019 Solar energy triggered clean water harvesting from humid air existing above sea surface enabled by a hydrogel with ultrahigh hygroscopicity. Adv Mater 31:1806730.

[74] Nandakumar DK, Ravi SK, Zhang Y, Guo N, Zhang C, Tan SC. 2018 A super hygroscopic hydrogel for harnessing ambient humidity for energy conservation and harvesting. Energy Environ Sci 11:2179-87. https://doi.org/10.1039/c8ee00902c.

Chapter 20

Solar Technology in Agriculture

Ghulam Hasnain Tariq, Muhammad Ashraf and Umar Sohaib Hasnain

Abstract

Promotion of sustainable agriculture is one of the most priority development goal set by United Nations for achieving the food security to meet the ever-increasing global population food demand. Because of extreme importance of agriculture sector, significant technological developments have been made that played pivotal role for sustainable agriculture by value addition in agricultural products and meeting energy demands for machinery and irrigation. These developments include improved cultivation practices, processing units for agricultural products and operation of machinery and irrigation systems based on solar energy. Moreover, the emergence of new technologies and climate smart solutions with reduced carbon footprints have significantly addressed the ever-increasing fuel costs and changing climate needs. PV based solar irrigation pumps and agricultural machinery is typical example of this. Because, awareness of these technological development is essential to overcome energy issues, availability of energy to perform agricultural activities for sustainable agriculture at farm level and socioeconomic uplift of farming community to meet food requirements needs in the future. Therefore, this chapter attempts at providing the introduction of technologies for direct and indirect use of solar energy in the agriculture sector. The typical examples of direct use of solar energy like greenhouses or tunnel farming for cultivation of crops and vegetables and use of solar dryers for drying agricultural products have been comprehensively discussed. Similarly, the solar powered tubewells, tractors, and lights, etc. are few important examples of indirect use of solar energy and have also been discussed in this chapter. The indirect use is made possible by converting solar energy into electrical energy with the help of photovoltaic devices, called "solar cells". Also radio frequency (RF)-controlled seed sowing and spreading machines are discussed, which provide an eco-friendly method. Moreover, comprehensive discussion is made on solar based technologies in general as well regional context in view of their potential to scale-up and to address anticipated issues. The use of photovoltaics in agriculture is expected to be significant contribution in the near future that require urgent planning for the potential benefits and efficient use at the farm level. Therefore, the co-existence of "agrovoltaics" will be essential for the developments of agriculture and agroindustry.

Keywords: Sustainable agriculture, Solar Energy, Agricultural Machinery, Solar Irrigation, Greenhouse, Solar dryers, Agrovoltaics, Agroindustry

1. Introduction

The demand for energy in agriculture has increased significantly to meet the needs of growing population and increasing demand for food. For which not only the already available sources of energy are inadequate and have dwindled because their reserves are nearing to depletion. Therefore, along with other aspects for development in the field of agriculture, the field of research and exploration of new sources of energy is also the focus of interest of agro-researchers. Sun is an eternal center of energy, where solar fuel is being converted into solar energy by the fusion process since the birth of solar system. The use of solar energy is of central importance to meet energy demands. Fortunately, the blessings of Almighty Allah are that the solar energy has many features, which can be used directly and indirectly. For ensuring a sustainable future and addressing the increasingly serious impacts of climate change, especially global warming, developing countries are urgently seeking to switch from traditional energy to renewable energy [1]. Solar energy is abundant, free, and non-polluting; hence, it is considered one of the most competitive choices of all the renewable energy choices [2]. The agricultural sector also uses different methods to take advantage of these different features of solar energy for different applications. For example, the thermal properties of solar energy are used to dry foodstuffs, vegetables, crops, and meat, etc., which is a direct use of it. Drying of these goods is done by direct use of solar energy, but it needs long time which is a waste of time, also it is more likely to be contaminated with dust, malnutrition, food, insects and flies. In addition, unpredictable climate changes, such as wind and rain, can cause serious damages. In modern times, a variety of solar dryers are used for such direct use of solar energy. For the last few decades, solar energy has been used in various ways after converting it to other forms of energy such as chemical energy and especially electrical energy for various services and research has been given much importance for improvement of the conversion methods to capture solar energy. The conversion of solar energy into electrical energy "soletrical energy" has greatly increased the use in various spheres of life. Much research is being done in the field of agriculture for use of soletrical energy. And its use is sure to not only alleviate energy shortages for a variety of purposes, but is also a cheap, easy, unlimited and widely available source of energy on the whole earth throughout the year. The use of this soletrical energy for water pumping, lighting, pesticides spray, and various types of machinery such as tractors, etc., is being innovated day by day in agriculture. But utilization of solar energy in agriculture in this way is still limited, lot of awareness and research is required to be beneficiary of this blessings and hope of future energy requirements.

This chapter includes the awareness of solar energy and potential role of solar energy in the development of the agricultural sector and agroindustry. To avail the benefits of solar energy and consume it to perform various agro-affairs through different applications are discussed in this chapter. Moreover, research done so far to improve the agricultural sector through its use in various ways is also covered in this study. This study will provide coordination between energy researcher and farmers to utilize solar energy with its different characteristics.

1.1 Solar energy

The solar energy is a solar or sun fuel generating at the sun spreading everywhere in the universe and all planets of solar system rely on it. This is also named as clean energy, green energy, alternative energy or sustainable energy. This is the origin of most of the energy sources on earth. The solar energy coming from the sun is in the form of radiations of a range of values. Most of solar energy is captured in the interstellar space and only a small part of solar energy reaches on the earth. But this small quantity of solar energy reaching on earth surface in only one hour is still higher than the energy generated by all other available sources including hydro, nuclear and fossil fuels etc. At the sun about 4,000,000 tons of solar fuel is converted into energy per second, which is so huge comparatively to the conversion ability of a

1000-MW nuclear power station on earth having the capacity of converting only 0.130 Kg of nuclear fuel into energy in one year. The earth receives about 1366 watts per square meter from the sun, generally which varies with latitude [3]. All the accumulated energy in any form in the earth is because of solar energy, i.e., fossil fuels consisting of natural gas, oil and coal depends directly or indirectly on it. Moreover, all energy reserves are nearly equal to solar energy got from sun only in 20 days. Solar energy is such a fuel which will be lost with the universe. Utilization of solar energy is not a new concept or thinking, human being is utilizing this energy since its birth. Solar energy consists on a spectrum of range of wavelengths of radiations having different energies but most of the solar energy reaches on the earth surface consists on visible light and infrared light as shown in **Figure 1**. Although ultra violet part of this solar energy spectrum is higher in energy strength but lower in intensity. The more intensive part of this spectrum lies in visible part ranging from ~400 nm to 700 nm. Each part of this spectrum has its importance related to applications, i.e., white light for visible purpose lies in the part of solar spectrum 400 nm to 700 nm.

1.2 Assessment of photovoltaic power potential

The assessment of solar energy available in a particular region of the earth is necessary to further harness the source. Because, sustainable and affordable energy supply has strong correlation with the socioeconomic development of any country [5, 6]. Therefore, G20 countries that includes Argentina, Australia, Brazil, Canada, China, France, Germany, India, Indonesia, Italy, Japan, Republic of Korea, Mexico, Russia, Saudi Arabia, South Africa, Turkey, the United Kingdom, the United States, and the European Union consumes about 80% of the total energy. Most of the global energy requirements are meet from nonrenewable fossils fuels such as coal, oil and gas. Only, 9% energy requirements are meet by wind and solar energy globally for electricity generation. The global power mix trends of the year 2019 reveals that the increase in solar energy among other renewable sources is 24% i.e., about double than the addition of wind energy in a particular year [7].

The estimation of solar energy potential depends on many factors among the land cover is a major factor in the selection of a suitable area for solar PV generation installation. Direct solar resource is either estimated based on the Diffuse Horizontal Irradiance (GHI) or the Direct Normal Irradiance (DNI). However,

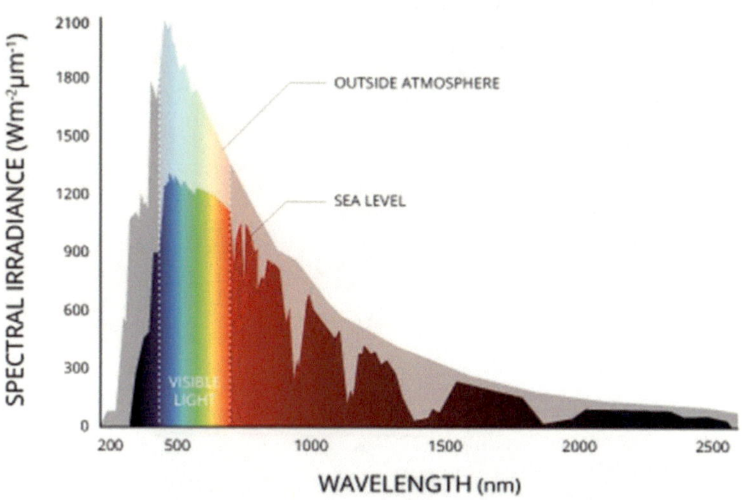

Figure 1.
Solar radiation spectrum [4].

actual solar potential for a region should be assessed by considering geographic, technological and economic potential. Because all the energy reaching to the earth surface cannot be harnessed due to geographically restricted areas, technological limitations due to limited efficiency of solar modules and energy production cost. For example, technological development directly determines the efficiency of the solar power transition. Initially, the PV modules efficiency of monocrystalline solar cells was 15% in 1950 which has now increased to 28% and polycrystalline reached 19.8% [8]. Similarly, governmental policy plays an important role in solar PV generation operation. Therefore, for a comprehensive solar energy potential analysis technological potential, economic potential, and other factors should be considered in addition to the solar energy resource. Researchers are assessing the global solar energy potential by considering these factors. For assessment of the solar potential of 147 countries, the data of Global horizontal irradiance (GHI) air temperature, PV power production potential, Index of seasonal, levelized cost of electricity and economic was used in GIS environment. In addition, some auxiliary data like terrain characteristics, built-up areas, population clusters, tree cover density, land cover and water bodies etc. data was also used to assess the technical potential for solar energy.

The Global Solar Atlas is prepared by Solargis that provides the easy access to solar resource and photovoltaic power potential data globally. Global Solar Atlas 2.0, is a free, web-based application developed and operated by the company Solargis s.r.o. on behalf of the World Bank Group, utilizing Solargis data, with funding provided by the Energy Sector Management Assistance Program (ESMAP). Maps and GIS data are available for 147 countries on online resources (**Figure 2**).

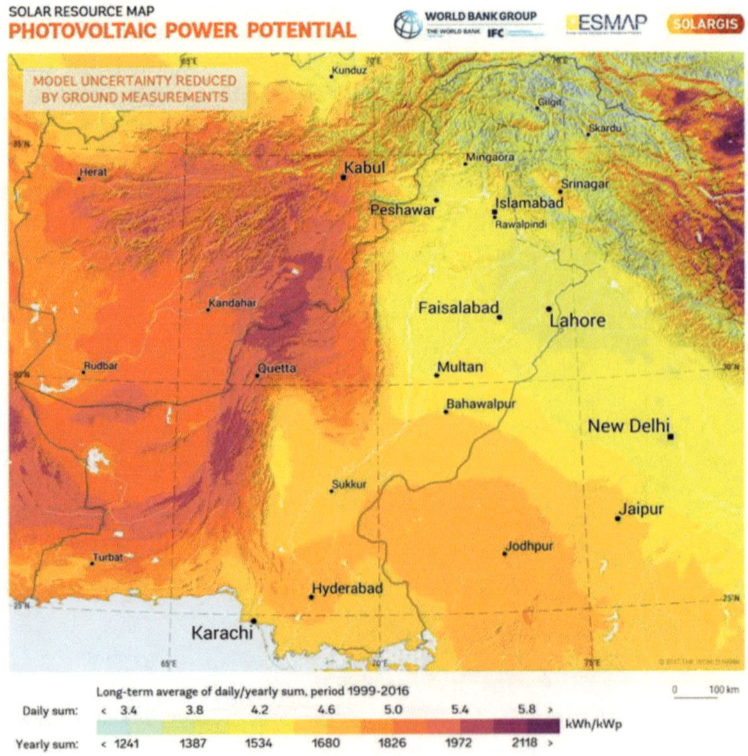

Figure 2.
Solar power potential of Pakistan, https://globalsolaratlas.info/map [7].

The direct solar radiation, having potential of concentrated Solar Power (CSP) and photovoltaic (PV), ranges 5–5.5 KWH/m^2 /day for more than 300 days a year in Southern Punjab. The range in almost all areas of Punjab is 4–6.5 KWH/m^2 /day [9].

2. Scope of solar energy

Climate change is caused by the human's activities relating to energy uses, as carbon dioxide emission is increasing 1.3% annually for the duration of 2014–2019 [10]. Meanwhile, the energy sector taking the responsibility by supporting the policies in technologies and renewable technologies are leading the energy market globally for new energy generation capacity [10]. The year 2020 was a best year for photovoltaic and wind energy market with almost 115GW and 71 GW were added respectively [11, 12]. However, the pace of world's energy transition from traditional fossil fuels to these renewable technologies is far from alignment with Paris Agreement [10]. Although 90% of total electricity energy will be generated with renewable supply by 2050, for which 63% of total electricity needs will be supplied by wind and solar photovoltaics [10]. Solar photovoltaic installed power generation would reach to 14000 GW by 2050 [10]. Solar energy and solar photovoltaic are attractive candidates to fulfill the electricity needs for domestic utility and to run electric vehicles, also cooling and heating requirements.

2.1 Solar technologies

Solar technologies are in common use in simple forms like drying in sun and basking in sunshine since the birth of earth, and people are using some other simple solar technologies including solar water heating and solar cookers by consuming direct sunshine or solar energy. The global solar PV market has rapidly grown by 50% over the past decade [13]. During 2011, more than 29 Giga Watt (GW) new solar PV industry was installed worldwide which was 70% increase compared to the year 2010. Global PV capacity exceeded 69 GW with 70% installed in European countries. During 2017, close to 73 GW of solar capacity added worldwide [7]. Since last few decades, solar energy is being used by converting it into electrical energy with the help of devices called solar cells or photovoltaic devices. These devices are now set up on the hope to fulfill the energy needs and becoming a technology ladder. Another energy converting device is thermocouple which consists on a pair of semiconducting wire with one end connected and other ends are free and when connected end side heated with solar energy than a potential difference is appeared across free ends. Under ordinary sun light efficiency of thermocouples is very low but concentrated sun energy can increase the efficiency of thermocouples. Solar cells convert directly sunlight energy to electricity while thermocouple convert heat from sunlight into electricity [3]. A schematic flow chart of solar energy utilization via different ways is shown in **Figure 3**.

2.2 Solar Technologies in Agriculture

Technology at agricultural farms is changing and improving rapidly. These developments are improving the farm machinery and equipment, farms facilities and buildings, both for crops and animals at farms. As we all know solar energy is the largest and cheapest energy resource on earth. Solar energy can easily fulfill energy provision and supply at agriculture farms. Various solar energy absorbing devices and systems have been developed and are in work for agricultural

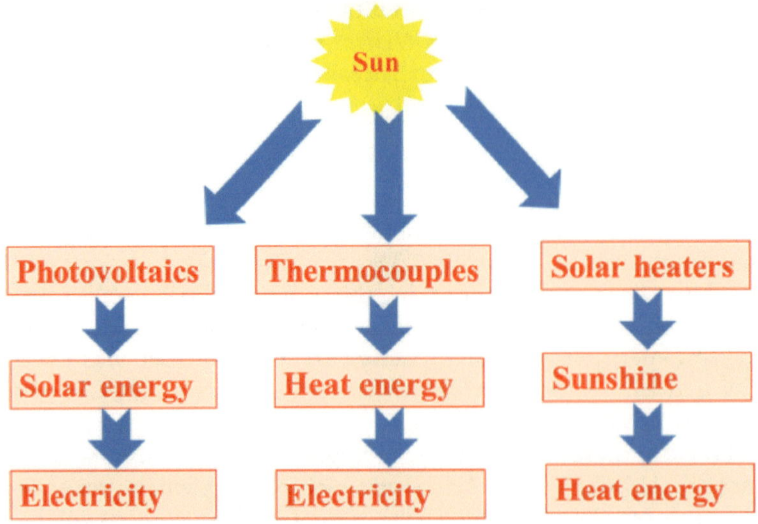

Figure 3.
Utilization of solar energy via different ways.

applications. This includes solar thermal and electric devices such as solar spraying machine, solar greenhouse heating, solar crop dryers, solar water pumps, ventilation for livestock, solar aeration pumps, solar electricity etc.

- **Solar PV operated water lifting/pumping system:**

Solar PV pumping systems are quite helpful to operate the pressurized irrigation system. Specifically, solar pumps may be useful as water lifting devices in irrigation canals and also to evenly distribute water in those areas where traditional water systems could not have access, such as in the elevated hilly lands.

- **Solar spraying and seed sowing machines:**

The solar pesticides sprayer machine is designed for small farmers to improve their productivity. They can easily carry and handle these machines with rechargeable batteries and direct solar illumination options. Mostly pesticide spraying activity is done in the day time, so these spray machines could be used by directly capturing solar energy, which prevents the installation of batteries in these machines. Also, solar powered seed spreading and sowing machines introduce a simple and convenient way of seeds spreading and sowing to small fields, and also in those areas where traditional machinery could not be available. It will be more useful for small farmers and agrarian society. Thus, solar-powered automatic pesticide sprayer and seed sowing machines will facilitate farmers to leave the heavy-duty machines and also provide easy access to work in remote areas of the countryside where general machinery is not readily available [14]. Today radio frequency controlled solar sowing machines are also designed to provide farmers eco-friendly sowing and spreading of seeds. These RF solar controlled sowing machines work with the help of blue tooth, which sow the seeds at controlled depth and distance between seeds [15].

- **Solar crop drying:**

One of the applications of solar energy in agriculture is a solar drying system which is based on variety of options. Solar dryers are available different shapes and structures. Different types of solar dryer are available for various applications, which is used for drying of agricultural products like potatoes, grains, carrots and mushrooms. Depending upon heating arrangement active dryers and passive dryers are two main types. In active solar dryers, external means are used for solar energy heat transfer, like pumps and fans are used for solar energy flow from solar energy collector to crops drying beds, while passive dryer heat is circulated in natural way by wind pressure or buoyancy force or with the combination of these both [16].

- **Solar greenhouse heating:**
  Generally, greenhouses around the world use sunlight to meet their lighting needs for photosynthesis, but they are not ready to use the sun for heat. Rather, they rely on conventional energy sources, such as oil or gas, to produce greenhouse temperatures for winter plant growth. However, solar-powered greenhouses (SGHs) are built to use solar energy for both heating and lighting. Also, these greenhouses reduce the damage caused by excess solar energy from the ambient to the greenhouse during hot sunny periods. A controlled environment is available in these SGHs.

- **Solar powered tractors:**
  Tractor is a fundamental machinery in agriculture, which made the farming much easier and increased the crops yield and production. Tractor converted the agriculture farming into agroindustry by performing lot of functions with the help of variety of tools and equipment. Usually, tractors consume oil to run and work, which increases the budget of farming also cause the pollution in atmosphere by producing carbon dioxide during combustion. Solar powered tractors became good option which could work directly under the sun by consuming solar energy through PV system in day time and also could continue working in night time with the help of utilizing energy stored in batteries. Although solar powered tractors are in preliminary stage of development but results are hopeful for bright agriculture future [17].

2.2.1 Solar machinery and tractors

Tractor is a most important and central technology and machinery at any agricultural farm. A tractor provides power to perform many tasks, including plowing, seeding, planting, fertilizing, spraying, cultivating, and harvesting crops at farms. Tractor are also used for transporting crops and materials at farms and market. Modern agricultural developments and to increase production to accomplish the needs of human being best farming can be done by using multifunctional compact tractors. Tractors have great social and economic impact on agricultural activities.

Commonly tractors use diesel oil as an energy source. Solar machinery and tractors use solar energy converted in to electricity. One way of using solar energy in form of electrical energy is by using solar panels fixed on machinery or tractors, a schematic diagram is shown in **Figure 4**.

Another way of using solar energy is converting it into electricity at solar power station and charging the batteries of tractors. But in this way energy stored in batteries of a solar electric tractor is very small and a tractor could not work for a long time with a single charging of batteries at solar power station. A challenge for solar electric tractors working in the fields is that the energy density of batteries is low which reduce the working efficiency of tractors. Also charging time of batteries is comparatively is large so exchangeable batteries idea could be used to run tractors for long time [18].

2.2.2 Solar irrigation

Irrigation is a basic need for the crops to grow that play to meet the global food demand. Irrigation demands for crops can be meet by three different sources

Figure 4.
Schematic diagram of solar powered tractor [18].

categorized as green water, blue water and non-renewable groundwater. Green water refers the use of effective precipitation for crop growth that is stored in the soil root zone and blue water to the surface freshwater available in rivers, lakes, reservoirs and the groundwater. Agriculture sector is the major water consumers in the world and accounts for approximately 70% consumption of fresh water [19]. An estimated 67% of the world's crop production still comes from rainfed agriculture [20], where crops requirements are fulfilled from the water held in the root zone of soil. Moreover, the large solar energy potential i.e., more than 6 kWh/m^2 and existence of underground water potential make the solar irrigation well suited for arid and semi-arid regions.

In Asia, especially Pakistan, China, India, and the United States account for 68% of fresh water withdrawals for irrigated agriculture, out of which ~34% is consumed by India only. In Pakistan and India, about 37 million electric and diesel tubewells have been installed in the irrigated area. Therefore, there is great potential to convert these tubewells on solar energy. In Pakistan, there is a 2,900,000 MW solar energy potential due to its geographical location with more than 300 sunshine days, 26-28°C average annual temperature and 1900–2200 kWh/m^2 annual global irradiance [9]. The southern part of Pakistan where annual Direct Normal Irradiance (DNI) is above 5 kWh/m^2/day which is ideally suitable for photovoltaic technologies for irrigation. In Pakistan, about 1.1 million tubewells exist out of which 0.8 million are diesel operated and 0.3 million are electric. The use of tubewells have increased in Pakistan because the surface water supplies are not sufficient to meet the irrigation requirements. Therefore, significant withdrawal is done from the groundwater resources that ranks Pakistan at 4th in the world. Overall, at global scale, estimated groundwater abstraction ranges between 600 and 1100 km^3 yr.$^{-1}$ [21]. For the year 2000 the reported abstraction rate and estimated groundwater depletion per country with range of uncertainty of India, United States, China and Pakistan is given in **Table 1**.

Country	Abstraction (km^3 yr^{-1})	Depletion (km^3 yr^{-1})	D/A (%)
India	190 (±37)	71 (±21)	37 (±19)
United States	115 (±14)	32 (±7)	28 (±9)
China	97 (±14)	22 (±5)	22 (±9)
Pakistan	55 (±17)	37 (±12)	69 (±48)

Table 1.
Reported groundwater abstraction rate and estimated groundwater depletion per country with ranges of uncertainty for the year 2000 [21].

Significant withdrawal of groundwater shows the importance and the potential of solar energy in irrigation as a substitute of fossil fuels and ultimately providing an environmentally sustainable solution to address the climate changes. Therefore, solar based irrigation can provide a sustainable solution for groundwater pumping which otherwise requires expensive and unreliable energy. Solar powered tubewells have several advantages over traditional systems. For example, diesel or propane engines require not only expensive fuels but also create noise and air pollution. Moreover, the overall initial cost, operation and maintenance cost, and replacement of a diesel pump are 2–4 times higher than a solar photovoltaic (PV) pump. Therefore, solar water pumping system is a cost effective, environment friendly and have low maintenance solution that makes it ideal system for pumping groundwater particularly for remote locations.

Solar energy can also be used for pumping water from the storage ponds to irrigate the crops. However, solar irrigation is coupled with the High Efficiency irrigation Systems (HEIS) for potential use of available water. Because, it is believed that an economics of solar-powered pumping systems can only be justified, if it is properly designed and linked with high-efficiency irrigation systems such as drip, bubbler, sprinkler or bed and furrow irrigation methods. For example, recently, in Pakistan, solar coupled drip irrigation systems have been installed on 21,255 acres during three years (2016–2017 to 2018–2019) [22]. Moreover, promotion of high value Agriculture through HEIS envisages installation of solar systems on 20,000 acres, especially the water scares and saline groundwater areas. Therefore, there is great potential to adopt the innovative solution for the areas where the solar systems have been installed due to limited water availability and saline areas. Moreover, there is increasing trend in farmers that can be observed to use these solar pumps for surface irrigation in the plain areas. Moreover, these solar pumps are used to irrigate limited lands of farmers. Therefore, after fulfilling the irrigation requirements the energy can be used for other purposes at farm level. However, there is little evidence to use this available energy where option to connect with the grid is not available. Grid connected solar pumping system is being considered economically viable in the rural areas. For example, a study shows that Levelized Energy Cost (LEC) of the grid-connected SWPS through Life Cycle Cost (LCC) is 4–54% less than the off-grid system depending on the size of the pump [23]. Therefore, it is necessary to provide the alternate utilization of the available energy of solar pumping system for better capacity utilization and economic viability, especially for larger solar pumping units.

Solar water pumping is based on photovoltaic (PV) technology that converts sunlight into electricity to pump water. The PV panels are connected to a motor (DC or AC) which converts electrical energy into mechanical energy. This mechanical energy is used to operate a pump to pump out the water from the ground. The capacity of a solar pumping system to pump water is a determined on the basis of head, flow, and power to the pump. The water pump will draw a certain power which a PV array needs to supply. A typical solar pumping system comprise of a pumping unit, solar panels, inverter, PV mounting structure and foot valves etc. The details of the solar pumping system components and its design can be found in literature [24, 25]. Solar water pumps may be categorized as submersible, surface, and floating water pumps. Submersible pumps are preferred to extract the required quantity of water from deeper depths. However, surface pumps are useful to extract water from the shallow groundwater aquifers. The temperature beyond 25°C decreases the solar output. The dust accumulation also decreases the PV panels efficiency. If a sprinkler cleaner/cooler is not installed then it requires the additional 25–30% PV panels to accommodate the dirt and temperature effects. However, it depends on the air quality conditions of the region. The use of a sprinkler for dust removal and reducing the temperature effects has been found to improve PV solar panel performance by 7–9%. Moreover, solar powered pumping systems efficiency can be increased up to 20% by manually tracking the solar panels. The use of automatic sun tracking improves the pump efficiency but increase the system cost considerably [25].

2.2.3 Solar dryer

Preservation of crops to keep them without rotting and decomposition for long time is essential activity in agriculture. It is required to keep them fresh and nutritious to carry them from fields to consumers. This process of preservation may be from domestic to industrial level depending upon farm size and crops distribution strategies. Different preservation methods include freezing, canning, drying and dehydration. Among these, drying of crops and food is simple and easy method which can work at any temperature and environment. Drying is an easy way to remove moisture from crops and food products in order to keep them with desired content of moisture. It also extends the storage life and enhancement of quality for long time. Basically, drying involves some heating process to vaporize moisture from crops and food products kept in dryers. In earlier time, drying was done by putting crops in open sun, but this method was more likely to be contaminated with dust, malnutrition, food, insects and flies. Thus, from last few decades, many sophisticated dryers are used to remove moisture from foods and crops. Main parameter to control is the temperature of crops which is done by providing certain amount of flow of heat. This heat can be provided by hot air blow through the crops, which may be very costly set up. Fortunately, solar radiations are better source of heat, and solar thermal energy can be used for drying purpose to dry crops, foods, vegetables, grains and any other crops' products. These solar dryers are made in different shapes, sizes and structures to enhance their activity. In these solar dry Different types of solar dryers are in practice for various applications depending on method of heat transfer, their geometry and structure, such as [16];

i. Active dryers

ii. Passive dryers

iii. Integrated dryer

iv. Distributed dryer

v. Mixed mode dryer

vi. Solar cabinet dryer

vii. Green house dryer

Most of these solar drying systems either active or passive can be identified in further three sub-classes of solar dryers [26];

- Direct (integral) type
- Indirect (distributed) type
- Hybrid (mixed mode) type

A most common solar dryer is based on racks design attached with a solar collector, which can collect solar energy in higher amount and can achieve higher drying temperature in result. Solar collector could be a simple black box managed with a transparent cover. Natural convection or an ordinary solar fan could be used to flow the hot air from solar collector to the crops placed on the racks as shown in the **Figure 5**. In agroindustry for large scale applications mechanized solar dryer is used, which is an active dryer type, in which solar heated boilers are used to heat the air, and forced to by fans to approach the crops' beds [28].

2.2.4 Solar fertilization

Fertilizers have central role in the modern agriculture to increase the yield of crops. For fertilizers production ammonia is one of the most important chemicals, which is produced through a well-known Haber-Bosch thermochemical process. By this process 140 million tons of ammonia is being produced per year. This ammonia production consumes large amount of energy nearly 2.5 exajoule per year. To run the process hydrogen is obtained from methane which results 340 million tons of CO_2 per year [29]. Due to huge costs for establishment of plants, centralized production of ammonia with <100 plants worldwide are in function. For better utilization of fertilizers decentralization of traditional fertilizers is compulsory. To overcome these hardens solar energy-based fertilizations is a good option. Solar energy can convert dinitrogen into such nitrogen products which became nutrients for crops. Such

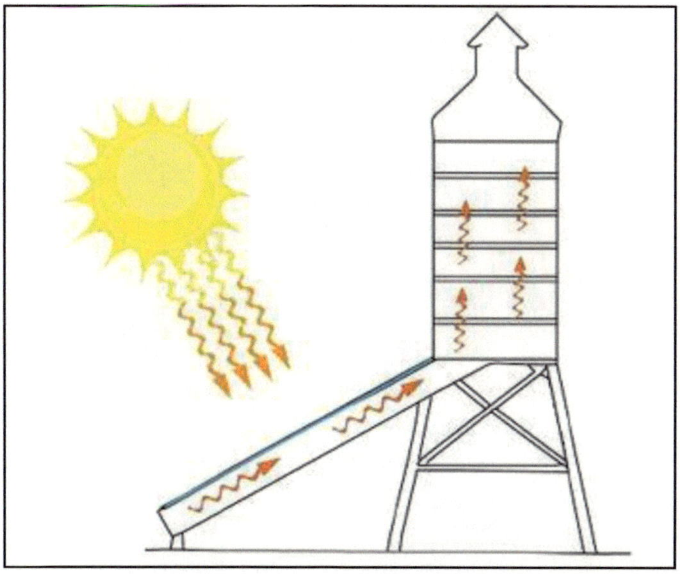

Figure 5.
Indirect solar dryer based on solar collector, racks and chimney [27].

nitrogen products produced by the solar energy are called solar fertilizers. The possibility of producing solar fertilization at country's level may be able to reduce cost of nitrogen based nutrient production by minimizing costs of transportation across the international borders. Also, it will provide employment to jobless workers at country level. Organizing solar fertilizers in developing countries will improve agriculture in remote areas of each country and farmers could become comfortable and satisfied. Above all solar fertilizers will reduce and cut off methane consumption and carbon associated threats to environment. Solar fertilization production is simply based on solar energy, water and nitrogen from air to produce nitrogen-based fertilizers near or at the farms, which also an eco-economic advantage. Management of these solar fertilizers will reduce ammonia use. A study revealed that 250 petajoules of energy/year could be saved by reducing10% use of ammonia or urea-based fertilizers [30].

The developments of solar fertilization need good and reliable strategy for dinitrogen fixation at ambient temperature. These developments can be made by the help of bioengineering, and catalysis research under precise conditions and approach [31, 32]. Such fixation of nitrogen in solar fertilizers can be accomplished by efficient electrochemical and photochemical natural process, which are expected to have significant lower concentration of nitrogen. These solar fertilization with lower concentration is characteristically safer and enable better nutrient managing [33]. The solar fertilizer production is similar in some aspects to the solar hydrogenation production, as light absorption, catalysts' reaction and energy transfer from absorbent material are involved in both processes. However, solar fertilizers would be integrated with agriculture farm infrastructure and for different application. Some of the key aspects of such processes required for production of solar fertilizers include capture or absorption of solar energy, catalysis reaction and separation process for production of solar fertilizers [34–38]. In this whole process of solar fertilizer production sun energy from sun light or solar fuel is absorbed by solar cells and/or photocatalytic particles which provide a potential to initiate an electrochemical reaction to convert dinitrogen, oxygen and water in to nitrogen products like nitrates and including ammonia in aqueous solution schematically shown in the **Figure 6**.

- **Absorption of solar energy**

  As production of solar fertilizers is based on absorption of utilization of solar fuel by solar energy from sun and converting it to chemical energy by two

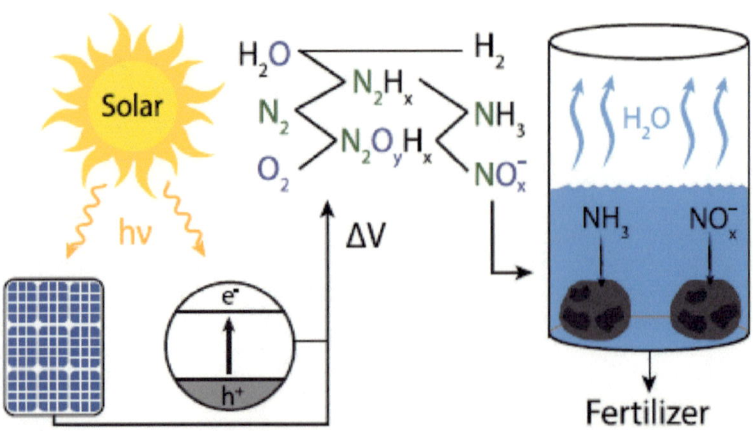

Figure 6.
Schematic diagram of solar absorption, catalysis reaction and separation process for solar fertilizers' production [29].

ways; *i, direct absorption* of sun light in a photocatalysis process (photochemistry), and *ii, indirect absorption* of sun light in a PV-electrolysis (photovoltaics and electrochemistry). A third hybrid approach (direct + indirect) is photoelectrochemistry in which electrical biasing is required absorption of sun light [37, 38]. These solar fuel technologies have good motivations for production and utilization at decentralized remote locations or at agricultural sites as compared to centralized huge industrial production.

- **Catalysis reaction**

  After absorption of the solar energy the conversion of molecular dinitrogen, oxygen and water is the central process of production of solar fertilization. For this a catalyst is required to dissociate triple bond of dinitrogen at favorable temperatures. Most approaches for this nitrogen dissociation have focused on chemical reduction of nitrogen to produce ammonia. For nitrogen reduction one of the best catalysts is based on carbon which shows an efficiency of 5% for electrical-to-ammonia in an aqueous solution [39].

- **Separation process for production**

  The chemical separation process for generation of reactants and convert the effluents to a fertilizer is an important step of solar fertilization technology. Because nitrates, ammonia and urea are water soluble which make a challenge for separation and concentration of products. Aqueous electrolytes are used in many electrochemical techniques for this separation process. Generally, these separation process require sophisticated techniques and processes for particular catalyst. This separation can be moderated with supported catalysts [40].

2.2.5 Solar dairy farms

Milk value chain from small dairy forms to market could be improved by using solar cooling technology. Milk cooling technology is costly and mostly small dairy farm (SDF) owners have lake of facilities for this purpose. Usually, these SDF owners are associated in dairy cooperatives which are responsible for managing to collect the milk from member owners and then supply collected milk to market or dairy plants. Lake of facilities of milk cooling in hot weather under warm climate conditions can lead to high bacterial contamination in milk. Solar dairy farming is based on solar technology.

- **Solar freezers or refrigerators at dairy farms**

  An emergency and simple way of saving milk is by using ice or freezers for cooling purpose. But, most of SDF exists in remote areas where transmission lines are not possible. In these areas solar powered freezers is a good option. Ice produced in these freezers could be used in milk cans for a better and effective cooling. Different institutions are working for developments of solar dairy farms specially for milk cooling. At Institute of Agricultural Engineering of the University of Hohenheim a solar milk cooling system has been designed which is based on the utilizations of ordinary milk-cans in Tunisia. In these designed solar dairy farms solar freezers are being used to produce ice for milk cooling. These milk canes can preserve milk for six to sixteen hours depending on amount of ice put in milk cans [41]. These solar dairy farms have great potential to improve dairy values and more efficient in remote and off grid areas by using environment friendly and clean energy. **Figure** 7 dairy farmers' comments and observations on the impacts that those farmers experienced due to use of solar technology [41].

- **Solar heating for steam generation**

  Sterilization process is an important activity at dairy farm for which low temperature steam is used. Parabolic trough collectors are commonly used to generate steam and other high temperature applications. At dairy farms solar water heater could be installed to raise the water temperature from 27–67°C [42]. A lot of furnace oil and other fuels could be saved by using solar heating at dairy farms.

2.2.6. Solar greenhouse production

All crops at agriculture farms needs proper environment including moister in air, temperature and light intensity. These parameters have great impact on for crops growth and yield, but we have not any control on them. All these parameters are controlled and determined by the nature, which are never remain constant and all time favorable. A lot of variations exist in environment and weather, sometime favorable and sometime very bad for crops. For continuous production at agriculture farms a favorable environment and conditions are required. Such a proper environment and promising conditions could be provided at solar greenhouse. Solar greenhouse is a covered structure where crops and vegetables are grown under favorable climate conditions and proper environment for the growth and production of plants. In greenhouse a controlled sunlight is managed for photosynthesis and also an adequate temperature is maintained suitable for plants whether outside is hot or cold. Vegetables could be grown throughout the year in these solar

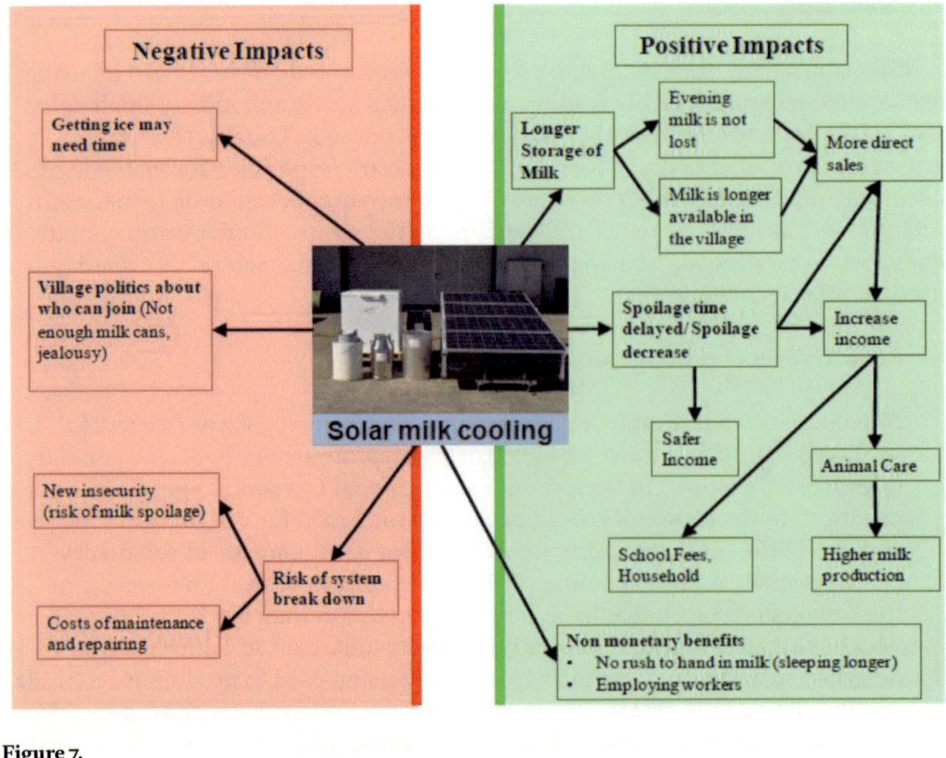

Figure 7.
Effect of small-scale solar milk cooling [41].

greenhouses. In these greenhouses solar energy is collected and stored in many ways and therefore they differ in designs. There are many parameters that effect the growth of plants in solar greenhouses. Among these parameters' intensity of sunlight, temperature of greenhouse, temperature of surroundings, humidity of greenhouse and surroundings, nutrients and carbon dioxide etc. Greenhouses provide such an environment to plants that they can grow in controlled conditions and optimized values of all these parameters.

Sunlight intensity

Sunlight, water and carbon dioxide are essential ingredients to produce carbohydrate and oxygen in photosynthesis process occurred in the chlorophyll of chloroplasts of plant cells. Initially chloroplast is responsible of absorption of sunlight and then for following chemical reaction.

$$H_2O + CO_2 + Sunlight \rightarrow Oxygen + Carbohydrate \tag{1}$$

These carbohydrates are used in the growth of the plants. In the respiratory process the energy is released which is used for the growth of plants and fruits. Better control of sunlight is responsible of efficient photosynthesis process and carbohydrate production. Sunlight intensity varies from beginning of day to time of noon from 0 to 150000 lux respectively. It also varies for weather difference like in cloudy days light intensity goes lower and some types of plants could not grow appropriately. For low and high sunlight intensity level, the photosynthesis process very much effected and plant's growth and yield are limited. Sunlight intensity is different required for photosynthesis in different plants like cucumber can grow in high intensity of sunlight, while tomato, lettuce and carrot need lower intensity of sunlight. Light intensity can be increased in the regions where light intensity is lower by different methods like by painting the walls and roof of greenhouses. Moreover, additional lighting may be required in the darken days to increase the light intensity as well its duration. For this additional lighting different types of lamps are used which are powered by solar cells.

- **Temperature of greenhouse**

Other than sunlight temperature is another parameter which should be optimum for biochemical reactions in the different types of plants. Temperature of plants surroundings and soil is very much dependent on sunlight intensity, humidity, air velocity and carbon oxide in the greenhouse. Temperature may affect different activities like food and water in root system, transportation of minerals in stems and leaves, and photosynthesis process. Also, for different stages of development of plants like germination, growing, flowering, fruit beginning and fruit reap or maturation, different temperature is required as shown in **Figure 8**.

- **Humidity in greenhouse**

Humidity in greenhouse environment plays a vital role in plants' growth and health, as relative humidity ranging from 30 to 70 percent is perfect for plants' growth, while comparatively higher relative humidity i.e., more than 90 percent is harmful for plants' health as it provides a suitable environment to pathogenic organisms' growth. Solar greenhouses provide controlled humidity in the environment and surroundings of plants growing within the greenhouse, where generally relative humidity between 55 to 65 percent and environment temperature between 20 to 25°C could be controlled.

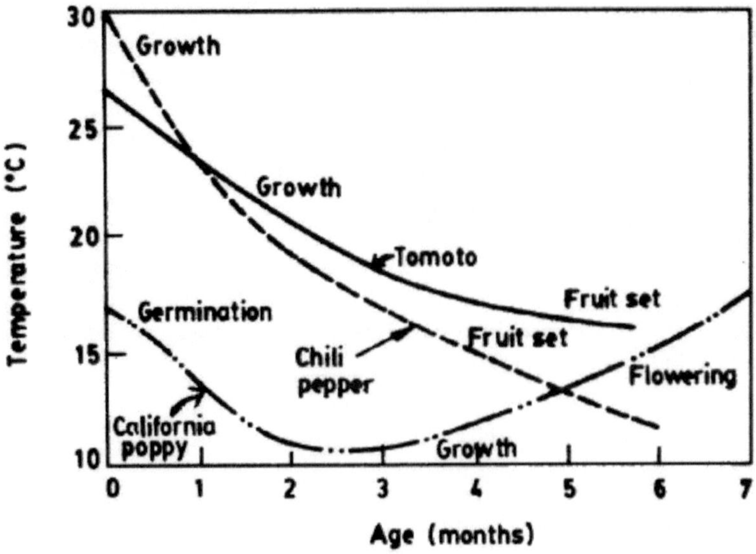

Figure 8.
Optimum temperature at night for growth and production [43].

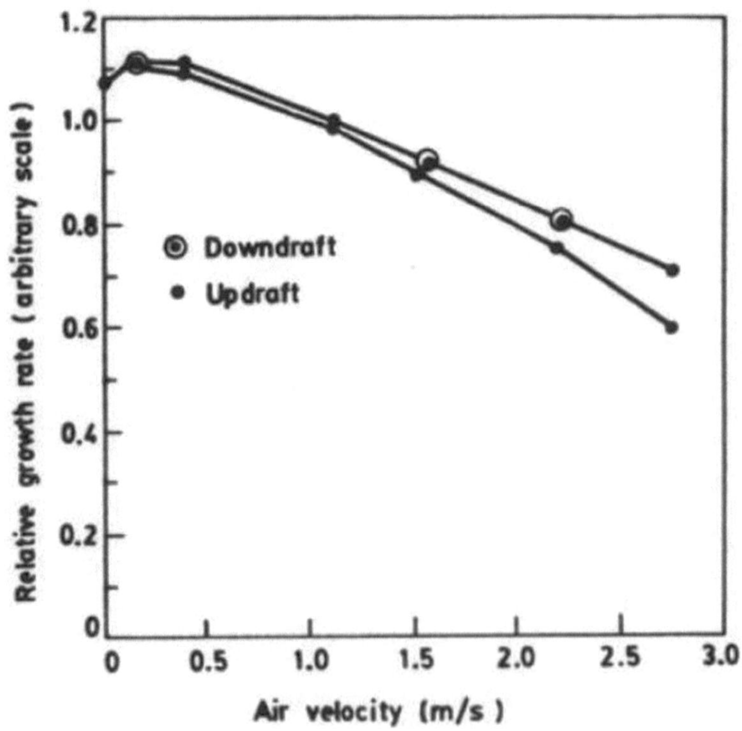

Figure 9.
Effect of air speed on leaf's growth [43].

- **Air transport and carbon dioxide**

Air transport affect the evaporation of water, availability of CO_2, cooling effects etc. so the growth of plants is affected. The air speed the plant's transpiration and

Figure 10.
(A) Shed type solar greenhouse; (B) Quonset solar greenhouse [44].

water vapors from plant to outside air, movement of CO_2 for photosynthesis. The air speed effects the plant growth as shown in **Figure 9** that the leaf's growth is effected by increase in air speed [43].

- **Solar powered greenhouse design**

The design of a solar powered greenhouse is different from an ordinary greenhouse in following few aspects;

- Glazing should be oriented in such a way that it can receive maximum solar energy.
- Use different heat storage materials to hold solar heat in winter.
- Use such glazing materials which minimize heat loss.
- Natural ventilation used for cooling in summer.

Two primary solar greenhouse designs are; i, Shed Type, & ii, Quonset Hut [44]. The orientation of shed type solar greenhouse is based on its length side along east to west direction as shown in **Figure 10A** [44]. Its north wall is painted or covered with some reflective material. The Quonset huts do not have any covered or insulated wall. Their structures are so that absorption of solar energy and distribution of solar heat is enhanced. Although insulation of solar greenhouse walls is required to minimize the solar heat losses, as shown in **Figure 10B** [44].

3. Conclusions

Technologies at agricultural farms are improving rapidly to facilitate farmers and bringing innovations in farming business. But this rapid increase of technology dependent agriculture farming required lot of energy resources. Also, the energy consumption increases the production cost of agriculture products. To overcome these energy and cost issues cheaper, easily and abundantly available energy sources are required. Fortunately, sun is a huge source of energy with abundant solar fuel on it, which can last till the life of earth. Thus, the solar energy is the largest and cheapest energy resource available on earth. Solar energy can easily fulfill energy need and supply at agriculture farms. Solar energy-based agriculture farms can easily accomplish energy requirements and reduce cost production. Utilization of solar energy at agricultural farms includes different types of machinery and equipment depending on task to accomplish by using different characteristics of solar energy like heating or converted in some other form of energy, such electrical or chemical. These applications include solar thermal and electric devices such as solar spraying machine, solar greenhouse heating, solar crop dryers, solar water pumps, ventilation for livestock, solar irrigation pumps, solar electricity etc. These solar energy equipped machineries also include radio frequency solar controlled sowing and spreading of seeds. Solar energy is a trustful and reliable source to compensate all requirements of energy for future.

Acknowledgements

This work was supported by Higher Education Commission Pakistan (HEC) through "National Research Program for Universities (NRPU)" project No: 10304/Punjab/ NRPU/R&D/HEC/ 2017 HEC is gratefully acknowledged for this support.

Author details

Ghulam Hasnain Tariq[1*], Muhammad Ashraf[2] and Umar Sohaib Hasnain[3]

1 Department of Physics, Khwaja Fareed University of Engineering and Information Technology, Rahim Yar Khan, Pakistan

2 Department of Agricultural Engineering, Khwaja Fareed University of Engineering and Information Technology, Rahim Yar Khan, Pakistan

3 Department of Electrical Engineering, Attock Campus, COMSATS University Islamabad, Pakistan

*Address all correspondence to: hasnain.tariq@kfueit.edu.pk

References

[1] T. Muneer, M. Asif, S. Munawwar, Sustainable production of solar electricity with particular reference to the Indian economy, Renewable and Sustainable Energy Reviews, doi:10.1016/j.rser.2004.03.004.397, **9** (2005) 444-473.

[2] H. Fang, J. Li, W. Song, Sustainable site selection for photovoltaic power plant: An integrated approach based on prospect theory, Energy Conversion and Management, **174** (2018) 755-768. doi:10.1016/J. ENCONMAN.2018.08.092. 406

[3] F.H. Cocks, "Energy Demand and Climate Change", WILEY-VCH Verlag GmbH & Co. KGaA, Weinheim, Germany, 2009.

[4] Fondriest Environmental, Inc. "Solar Radiation and Photosynethically Active Radiation." Fundamentals of Environmental Measurements. 21 Mar. 2014.

[5] K. Fatai, L. Oxley, F.G. Scrimgeour, "Modelling the causal relationship between energy consumption and GDP in New Zealand, Australia, India, Indonesia, The Philippines and Thailand", Mathematics and Computers in Simulation, **64**(3) (2004) 431-445.

[6] T. Muneer, S. Mauble, M. Asif, "Prospects of solar water heating for textile industry in Pakistan", Renewable and Sustainable Energy Reviews, **10**(1) (2006) 1-23.

[7] https://globalsolaratlas.info/map, solar radiation and meteorological data.

[8] V.V. Tyagi, N.A.A. Rahim, N.A. Rahim, J.A.L. Selvaraj, "Progress in solar PV 505 technology", Research and achievement, Renewable and Sustainable Energy 506 Reviews, **20** (2013) 443-461. doi:10.1016/j.rser.2012.09.028.

[9] S.M.N. Nawaz, S. Alvi, "Energy security for socioeconomic and environmental sustainability in Pakistan", e00854, Heliyon, **4** (2018) 1-21.

[10] Preview of IRENA (2021), World Energy Transitions Outlook: 1.5°C Pathway, International Renewable Energy Agency, Abu Dhabi.

[11] PV-Magazine (2020), "Wood Mackenzie expects 115 GW of solar this year", 28 October 2020, www.pv-magazine.com/2020/10/28/wood-mackenzie-expects-115-gw-of-solar-this-year/ (accessed 15 January 2021).

[12] ESI Africa (2020), "GWEC: Wind power market set for record growth despite COVID-19", 6 November 2020, www.esi-africa.com/industry-sectors/renewable-energy/gwec-windpower-market-set-for-record-growth-despite-covid-19/, (accessed 30 January 2021).

[13] IRENA (2019), Future of Solar Photovoltaic: Deployment, investment, technology, grid integration and socio-economic aspects (A Global Energy Transformation: paper), International Renewable Energy Agency, Abu Dhabi.

[14] K. Singhal, G. Prajapati, "Solar Powered Seed Sowing Machine", International Journal of Applied Engineering Research, **13**(6) (2018) 259-262.

[15] T. Devaraj, Sakthiya R. S. Raja, M. Janarthanan, "RF Controlled Solar Seed Sowing Machine", IOP Conf. Series: Materials Science and Engineering, 955 (2020) 012105, doi:10.1088/1757-899X/955/1/012105.

[16] A. Balasuadhakar, T. Fisseha, A. Atenafu, B. Bino, "Review on Passive Solar Dryers for Agricultural Products", IJIRST –International Journal for Innovative Research in Science & Technology, **3** (01) (2016).

[17] S.M. Sutar, M.C. Butale, "Solar Powered Multi-Functioned Electric

Tractor for Agriculture Usage", International Journal of Research in Engineering, Science and Management, **3** (3) (2020).

[18] A.S. Vaidya, "A Study of Solar Electric Tractor for Small Scale Farming", International Journal of Science and Research (IJSR), **8**(3), (2019).

[19] I.A. Shiklomanov, (2000), World water resources and water use: Present assessment and outlook for 2025, in World Water Scenarios Analyses, edited by F. R. Rijsberman, Earthscan, London.

[20] F.T. Portmann, S. Siebert, and P. Döll, "MIRCA2000—Global monthly irrigated and rainfed crop are as around the year 2000: A new high-resolution data set for agricultural and hydrological modelling", global Biogeochemical Cycles, **24** (2010) 1-24.

[21] Y. Wada, L.P.H. van Beek, and M.F.P. Bierkens, "Nonsustainable groundwater sustaining irrigation: A global assessment", WATER RESOURCES RESEARCH, **48** (2012) 1-18.

[22] Three years (2016-17 to 2018-19) On Farm Water Management (OFWM), Govt of the Punjab, Pakistan, 2019.

[23] S.R. Mantri, R.S. Kasibhatla and V.K.B. Chennapragada, Grid-connected vs. off-grid solar water pumping systems for agriculture in India: a comparative study, Energy Sources, Part A: Recovery, Utilization, and Environmental Effects, DOI: 10.1080/15567036.2020.1745957, (2020).

[24] M. Saleem, M. Ashraf, and J. Fripp, "Solar-Powered Pumping Systems for High Efficiency Irrigation: Design Manual for Practitioners". A joint Report published by PARC, PCRWR and ICARDA-Pakistan under USDA project "Watershed Rehabilitation and Irrigation Improvement in Pakistan", (2015) 42.

[25] S.S.Chandel, M.N. Naik, R. Chandel, 'Review of solar photovoltaic water pumping system technology for irrigation and community drinking water supplies", Renewable and Sustainable Energy Reviews, **49** (2015) 1084-1099.

[26] G. L. Visavale, Principles, Classification and Selection of Solar Dryers. In Solar drying: Fundamentals, Applications and Innovations, Ed. Hii, C.L., Ong, S.P., Jangam, S.V. and Mujumdar, A.S., ISBN - 978-981-07-3336-0, Published in Singapore, (2012) 1-50.

[27] B. Freischlad, "solar energy applications in agriculture in Senegal", DOI:10.13140/RG.2.2.27048. 47365, (2017).

[28] M. Wakjira, "Solar drying of fruits and windows of opportunities in Ethiopia", African Journal of Food Science, **4**(13) (2010) 790 – 802.

[29] B.M. Comer, P. Fuentes, C.O. Dimkpa, Yu-Hsuan Liu, C.A. Fernandez, P. Arora, M. Realff, U. Singh, M.C. Hatzell, and A.J. Medford, "Prospects and Challenges for Solar Fertilizers", Joule, **3** (2019), 1578-1605.

[30] P.G. Levi, and J.M. Cullen, "Mapping global flows of chemicals: from fossil fuel feedstocks to chemical products", Environ. Sci. Technol., **52** (2018) 1725-1734.

[31] E.J. Vicente, and D.R. Dean, "Keeping the nitrogen-fixation dream alive", Proc. Natl. Acad. Sci. USA, **114** (2017) 3009-3011.

[32] S.L. Foster, S.I. Bakovic, R.D. Duda, S. Maheshwari, R.D. Milton, S.D. Minteer, M.J. Janik, J.N. Renner, and L.F. Greenlee, "Catalysts for nitrogen reduction to ammonia", Nature Catal., **1** (2018) 490-500.

[33] R. Schlogl, "Catalytic synthesis of ammonia-a "never-ending story", Angew. Chem. Int. Ed. **42**, (2003) 2004-2008.

[34] B.A. Pinaud, J.D. Benck, L.C. Seitz, A.J. Forman, Z. Chen, T.G. Deutsch, B.D.

James, K.N. Baum, G.N. Baum, S. Ardo, et al. Technical and economic feasibility of centralized facilities for solar hydrogen production via photocatalysis and photoelectrochemistry. Energy Environ. Sci., **6** (2013)1983.

[35] M.R. Shaner, H.A. Atwater, N.S. Lewis, and E.W. McFarland, "A comparative technoeconomic analysis of renewable hydrogen production using solar energy", Energy Environ. Sci., **9** (2016) 2354-2371.

[36] J.H. Montoya, L.C. Seitz, P. Chakthranont, A. Vojvodic, T.F. Jaramillo, and J.K. Nørskov, "Materials for solar fuels and chemicals", Nat. Mater., **16** (2016) 70-81.

[37] N.D. McDaniel, and S. Bernhard, Solar fuels: thermodynamics, candidates, tactics, and figures of merit. Dalton Trans., **39** (2010) 10021-10030.

[38] J. Highfield, "Advances and recent trends in heterogeneous photo(electro)-catalysis for solar fuels and chemicals", Molecules, **20** (2015) 6739-6793.

[39] Y. Song, D. Johnson, R. Peng, D.K. Hensley, P.V. Bonnesen, L. Liang, J. Huang, F. Yang, F. Zhang, R. Qiao, et al. A physical catalyst for the electrolysis of nitrogen to ammonia. Sci. Adv., **4** (2018) e1700336.

[40] I.V. Gursel, T. Noe¨, Q. Wang, and V. Hessel, "Separation/recycling methods for homogeneous transition metal catalysts in continuous flow", Green Chem., **17** (2015) 2012-2026.

[41] A.S. Rojas, V.T. Toledo, F. Mrabet, and J. Müller, "Improving milk value chains through solar milk cooling", ZEF Working Paper Series, ISSN 1864-6638, Center for Development Research, University of Bonn, (2018).

[42] D.D. Desai, J.B. Raol, S. Patel, I. Chauhan, "Application of Solar energy for sustainable Dairy Development", European Journal of Sustainable Development, **2** (4) (2013) 131-140.

[43] H.P. Garg, "*Advances in Solar Energy Technology*", D. Reidel Publishing Company, Dordrecht, Holland 1987.

[44] S. Gorjian, T.T. Hashjin, B. Ghobadian, "Solar Powered Green-houses", SET2011, 10th International Conference on Sustainable Energy Technologies, Istanbul, Turkye, (2011).

Chapter 21

Experimentally Investigated the Development and Performance of a Parabolic Trough Solar Water Distillation Unit Concerning Angle-Wise

Fahim Ullah

Abstract

The PTC performance was evaluated at four (i.e., 25o, 35o, 45o, and 55o) different adjusting Angles and it clearly showed that the adjusting Angles is highly significant, affecting the efficiency of the collector. The PTC received mean solar radiation 513 kJ.m-2.hr-1 with the absorbing temperature of the absorber in PTC was noted 123oC, 115oC, and 113oC consecutively the months of the year with the adjusting angles of 25o, 35o, and 45o respectively. Distilled water from the solar water distillation unit was found to improve the laboratory's quality and wash equipment in the hospital. PTC's efficiency noted 26.9%, 26.3%, and 26.1% with the distilled water up to 217, 313, and 343 ml.m-2.day-1 for the adjusting Angles of 25o, 45o, and 35o respectively. From the result, it concluded that to obtain maximum distilled water, the PTC should be set on adjusting Angles of 25o, 35o, and 45o. The average unit price of distillate from the solar still is assessed as Rs. 2.64/L-m2 with a payback period is 365 days. The unit distillate cost is seen to reduce significantly from Rs. 4.92/L to Rs. 1.57/L. It concluded from results that the distilled water of PTC relatively decent quality.

Keywords: Solar energy, Parabolic Trough Collector, Efficiency, Distilled water and Adjusting Angles

1. Introduction

1.1 Background of the study

Most developing countries are in a vital energy crisis [1, 2]. The demand for energy has increased over the years because of the increasing world population and expansion of global industries, especially food and feed. Most of the energy consumption is from power generation, transportation, and industry community sectors [3, 4]. Moreover, most utility energy is taken from fossil oil, gas, and coal. Many developed countries have their policies to find alternative energy.

Energy plays an essential role in the industrialization and economic development of a country. A country will be prosperous if it has sufficient energy resources to fulfill its needs [1]. Besides the available energy resources, countries must work hard to explore and conserving renewable energy resources. The total solar energy received from the atmosphere and absorbed by Earth's oceans and landmasses is approximately 3.85 x 1024 W.yr-1 [2]. The total solar energy coming from the sun is so vast in one year, which is twice about the energy produced from the resource of the Earth's non-renewable, i.e., coal, oil, and natural gas [3, 4]. Pakistan is being located between 23.8o to 36.7o north latitude and 61.1o to 75.8o East Longitude. It is rich in renewable energy resources.

Ahsan et al., [5] attempted to find suitable resources to produce alternative energy such as biomass, solar energy, geothermal, hydropower, wind energy and ocean energy. Nanjing is a city found in Jiangsu, China. It is located 32.06 latitude and 118.78 longitude and it is situated at elevation 22 meters above sea level. It is rich in renewable energy resources. Solar energy has brilliant prospectus at a latitude of 32o which can be utilized for making electricity by photovoltaic (PV) cells, drying of products by solar collectors, water heating and water distillation systems [6, 7]. A total of 174,000 terawatts (TW) of energy reaches the earth at the upper atmosphere to form the incoming solar radiation (insulation) [8, 9]. From this total energy approx. 30% reflected to the atmosphere, and the remaining energy is absorbed by the clouds, oceans and land masses [10, 11]. **Figure 1** shows the incoming, absorbed and reflected solar radiation from the atmosphere. It shows the spectrum of solar light/radiation at the surface of the earth, which mostly spreads across the visible range and near the infrared ranges. Most of the population lives in the areas where the land received the insolation levels in the range of 150–300 watt/m2 per day or 3.5–7.0 kWh/m^2 per day [11–13].

Hydroelectric and thermal solar energy has enormous prospective sources of renewable energy. Energy plays a vital role in the industrialization and economic development of a country [14, 15]. A country will be prosperous if it has sufficient energy resources to fulfill its needs. In addition to the available energy resources, countries must work hard to explore and conserve renewable energy resources. The solar energy coming from the sun is so vast in one year that it amounts to twice the

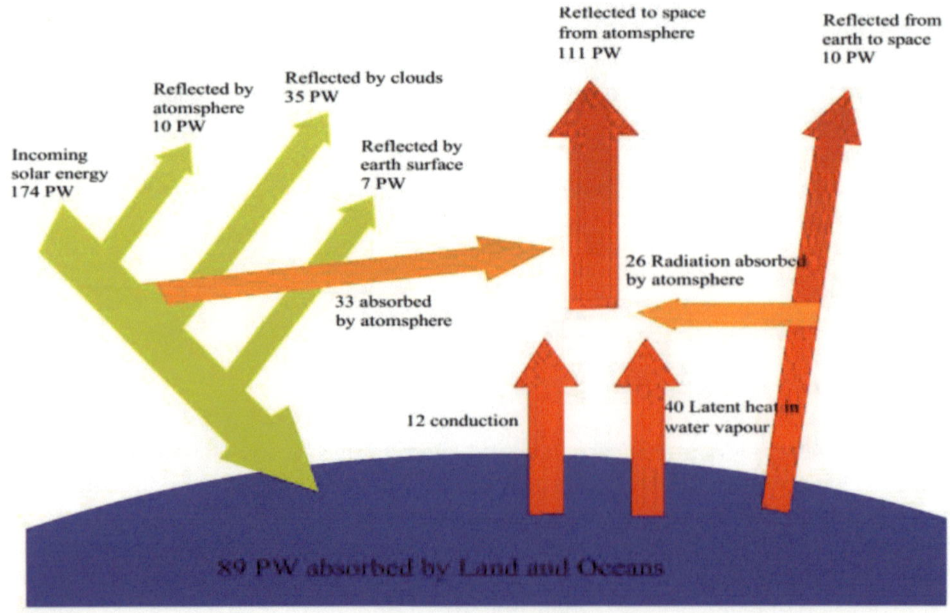

Figure 1.
Schematic diagram of the incoming, absorbed and reflected solar radiation intensity.

energy which is produced from the resources of the earth's non-renewable energy such as coal, oil, natural gas [16, 17]. Renewable energy provides a clean and nontoxic energy source. The key sources of energy are the sun, wind, biomass, waves and geothermal energy [18, 19]. Solar energy can be exploited in the form of thermal energy by using different kinds of solar collectors for different purposes, i.e., dehydration of fruits & vegetables, water distillation and producing electricity [19, 20].

Energy is an elementary need for agriculture and other industries [21, 22]. Different resources, like wood, coal, fossil fuels and nuclear chemicals were used as foundations for energy, but all these sources are getting rare [23, 24]. By using resources like wood, coal and fossil fuels for energy utilization, we are adding significant agents producing while environmental pollution and global warming. Due to high prices and shortages in the future, scientists of the world have established other energy resources called renewable energy resources including solar, tidal, wind and biomass [25, 26]. Wind and tidal energy are present in small areas of the globe while solar energy is present universally. The sun is the eventual source of energy for the earth. Energy from the sun is interminable and green as it does not create pollution and global warming [27, 28]. The sun gives us electromagnetic particle emission called solar energy and this energy can be consumed for different purposes like the drying of agricultural products, heating buildings, for irrigation purpose and for producing electricity [29, 30]. In the fourth century B.C., different methods were used for getting dried fruits & vegetables, which were very difficult to be performed. The dehydration of fruits & vegetables and other crops dried by open-air sun drying was not satisfactory, because the products became infected with bacteria, rodents, and insects, and worsen quickly due to the high ambient temperatures and relative humidity [31, 32].

1.2 Status of solar energy usage

Figure 2 showed the status of different sources of energy usage in the year 2016 in China. Solar energy is the most promising technology in the world [33]. Energy plays an essential role in the industrialization and economic development of a country. A country will be prosperous if it has sufficient energy resources to fulfill its needs [34]. Besides the available energy resources, countries must work hard for

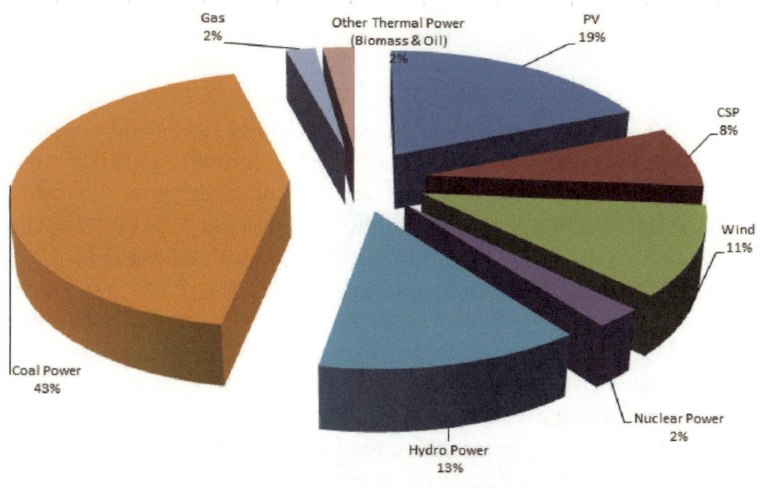

Figure 2.
Status of different sources of energy usage.

exploring and conserving renewable energy resources. The total solar energy received from the atmosphere and absorbed by the earth, oceans and land masses is approximately 3.85 x 1024 W.yr-1 [35]. The total solar energy coming from the sun is so vast that in one year it is twice the energy produced from the resource of the earth's non-renewable, i.e., coal, oil, natural gas [36, 37].

The concept of alternative energy is to develop other resources as a substitute for petroleum and to reduce the central issue of global warming. China and Pakistan import fossil fuels annually, equivalent to 40% of all total imports to fulfill the energy requirements of the country while spending 7 billion dollars. From the survey, it is clearly shown that by the year 2050, energy needs are expected to be three times the current needs in China and Pakistan while supplies are less than inspiring. For the utilization of this incoming solar energy, different kinds of solar collectors were used for various purposes, i.e., dehydration of fruits and water distillation.

1.3 Review of the literature

Renewable and sustainable energy resources are the best substitute for conventional fuels and energy sources in a country's energy security and sustainable developed as well as its minimal environmental impact. China and Pakistan are making attempts to promote and support the utilization of alternative energy and to improvement in energy efficiency. Different researchers have conducted experiments on the drying of different fruits and vegetables, and the desalination process using different solar collectors. The researchers [38–40] found that hot air drying reduces the risk of the development of Alfa toxins in fruits. They experimented with a pilot airflow cabinet dryer with the greatest loss in ascorbic acid. They also concluded that pretreated fruits take a shorter time to dry as compared to controlled fruit. **Figure 2** shows the Status of different sources of energy usage in 2016.

Most of the energy consumption is from power generation, transportation, and the industrial community sectors. Moreover, the most utility energy is taken from fossil oil, gases, and coal. Many developed countries have policies to find alternative energy. Many researchers have attempted to find suitable resources to produce alternative energy such as biomass, solar energy, geothermal energy, hydropower, wind energy and ocean energy. The concept of alternative energy is to develop other resources as a substitute for petroleum and to reduce the central issue of global warming.

1.4 Solar collectors

Solar energy is a well-known process used for drying fruits and vegetables, while it is also usable for other purposes, i.e., water distillation and ventilation, etc. [41, 42]. Different types of collectors are used for collecting energy from the sun, but the flat plate solar collector and parabolic trough solar collector are the most appropriate for getting more tracking sunlight for the dehydration of fruits and vegetables and water distillation [43, 44]. Other researchers [45, 46] have reported that drying is the most dynamic process for better quality of fruits and vegetable. The researchers [47, 48] said that the flat plate solar collector is the best method for the heating of water with convective heat flow having an efficiency of 35–45%. Several years of research showed that the flat plate solar collector is better for the use of heating of farm shops, dairy buildings [49, 50]. Efficiency is the important parameter of flat plate solar collector for the heating of water and dehydration of different kinds of agricultural fruits and vegetables [51, 52]. The ability of the collector depends on an optimum combination of temperature and flow rate [53].

Solar collectors can be utilized for different purposes such as the purification and distillation of liquids, the drying of products, the heating of water for various purposes, for lighting at night and for water pumping [38]. The researchers [39, 40, 54] found that solar energy is one of the promising techniques in renewable energy for getting the pure and clean water from potable water resources. There are so many techniques which are used for heated water to produce clean and pure water i. e. solar collector, solar photovoltaic, etc. [55–60]. In this research project, we have designed two solar collectors, i.e., flat plate solar collector and parabolic trough solar collector. Both were used for the dehydration of fruits, i.e., apples, apricots, and loquats, etc., and also for water distillation purposes with the development of a single-axis tracking control system.

Parabolic trough solar collector.

This type of collector is used in solar power plants [61]. A trough-shaped parabolic reflector is used to concentrate sunlight on an insulated tube (Dewar tube) or heat pipe, placed at the focal point, containing coolant which transfers heat from the collectors to the boilers in the power station [62]. In a parabolic dish collector, one or more parabolic dishes focus solar energy at a single focal point, similar to the way reflecting telescopes focuses starlight [63]. The shape of a parabola means that arriving light rays which correspond to the dish's axis will return toward the focus [61]. Light from the sun reaches the earth's surface almost entirely parallel, and the plate aligned with its axis pointing at the sun permits almost all incoming radiation to replicated toward the focal point of the plate [64]. Most damages in such collectors are due to deficiencies in the parabolic shape and lacking reflection. Losses due to atmospheric trickle are minimal [65–111]. However, on a hazy or foggy day, light is diffused in all directions through the atmosphere, which significantly reduces the efficiency of a parabolic dish [1].

Solar energy has brilliant prospects at the latitude of 34o, which can be utilized for making electricity by Photovoltaic (PV) cells, drying of products by solar collectors, water heating, and water distillation systems [5]. The wick type solar collector with load and no-load at the adjusting Angles of 10o, 20o, 30o, and 40o tested in summer and winter. The average yield of distilled water was 2300 ml/m2/day1 in winter and 3400 ml/m2/day1 in summer reported by [6, 7]. The distilled water production was 182 ml/m2.hr. with the difference between glass and sea-water, and the solar system's efficiency was noted 21.3%. They concluded that in 48-hour, the distilled water production increased from 3000 ml/m2 to 3200 ml/m2 by [8, 9]. The solar efficiency still determined with different inclination adjusting Angles of 15o, 30o, and 45o to compare various conventional solar distillation systems' energy determination. The use of polyvinyl chloride material in collectors have increased the efficiency, studied by [10]. The collector's daily output was in the range of 2 to 4 l/m2/day, and efficiency was calculated by 27% tested the flat plate collector at the inclination adjusting Angles of 45o with the horizontal facing due south. They studied the collector from 08:00 AM to 05:00 PM during sunlight hours, and output was increased by 31% evaluated by [11]. The product's output was increased from 2240 ml/day to 3510 ml/day during October to December using the flat plate solar collector with the adjusting Angles of 35o with different parameters conducted by [12].

The basin's efficiency- type solar still was highest in Jun, July, and August up to 75% with solar irradiance, and the output of the basin was 7000 ml/m2.day. It was further studied that the output of a solar still was decreased without using the condenser collector [13]. The solar still plant studied with different tilt adjusting Angles of 15o, 25o, 35o, 45o, and 50o and reported that maximum output was obtained by adjusting Angles of 35o during May. They noted the maximum absorber temperature at 01:0 PM to 02:0 PM [14, 15]. Distilled water is used in various industries, nuclear-powered ships as a coolant, various beverages, Lead-acid batteries, automotive cooling systems, steam irons for pressing clothes and surgical instruments washing, etc. An electric water distillation plant commonly prepares distilled water, but it has a high initial cost and requires electricity.

On the other hand, solar energy can prepare distilled water [16, 17]. Different designs of solar collectors are available that can be used for water distillation. Solar distillation plants can work by the natural water cycle, and it can receive the solar energy to warm the water so that the water boils and evaporates. The vapors are then condensed in distilled water forms as it cools down reported by [18]. The solar collector can be utilized for different purposes such as purification and distillation, dried water heating, heating of water for different purposes, lighting at night, and water pumping [19]. In the solar desalination system, water is converted to steam using the sun's energy, and then these vapors condense as pure water. After the condensing of vapors, it's free of salts and other impurities.

The solar distillation water plant is a cheap and straightforward method to distill or purify water reported by [20]. This plant required solar radiation as heat that can convert water into the vapors form. Therefore, for solar distillation, the 2260 kj.kg-1 energy is required to evaporate the water evaluated by [21].

1.5 Significance of the study

There is a shift record in the adoption of solar technology due to a shortage of electric power. Energy is the primary need nowadays and to fulfill the requirement of people to use solar thermal collectors to overcome the lack of solar energy. Solar thermal collectors convert solar radiance to heat and then this heat is given to a fluid which utilizes this heat to produce distilled water form tape. It was also used for the warm purposes in the buildings to convert water to steam. The performance of solar collectors was a keen factor to use them efficiently for dehydration and water distillation purposes. Energy is the input or the heat given to the collectors that are available from the solar radiation daily and to apply for some useful purpose, i.e. water distillation. Efficiencies based on the first law and second law of thermodynamics.

To overcome and maintain the problem of water distillation, researchers indicate the prefer ability of solar collectors' i.e. flat plate solar collector and parabolic trough solar collector are suggested for increasing the yield during the water distillation. The water distilling is the most important parameters during the distillation with the process of solar collectors. Most of the studies focused on the distillation process with the using of flat plate solar collector and corrugated plate solar collector, but few were focused on the design of parabolic trough solar collector and concentrating parabolic collector. Producing of distilled water can be done using solar energy, but there is a need for sophisticated technology for distillation without affecting the quality of produced distilled water. The multistage water distillation process is a well-known process used by many researchers for tap water distillation.

Although a lot of research has been done in this field, there is a gap regarding energy and cost-efficient use of water distillation. A considerable amount of energy is consumed in order to maintain the water distillation process. To fill the gaps in the data, a research program was carried out to experimentally investigated the development and performance of a parabolic trough solar water distillation unit concerning angle-wise; we aimed to get the necessary data to rate commonly used solar collector designs and to identify the required modifications.

1.6 Objectives of the study

The research project was carried out to study the **experimentally investigated the development and performance of a parabolic trough solar water distillation**

unit concerning angle-wise. In the present research study, solar distilled water unit was developed in the form of PTC in the Department of Agricultural Mechanization, FCPS, The University of Agriculture, Peshawar- KPK, Pakistan with the primary objectives of the study are as follows:

- To fabricate the parabolic trough collector (PTC).
- To investigate the development of parabolic trough solar water distillation unit concerning Angle-wise.
- To study the performance of PTC with Angle-wise (Efficiency).
- To compare the quality of prepared solar distilled water with the available distilled water in laboratories.

2. Materials and methods

2.1 Parabolic trough collector

Parabolic trough solar water distillation unit consists of a parabolic reflector, as shown in **Figure 3**. The reflector was made of a Galvanized iron sheet. The sun rays strike on the reflector sheet and then reflect the absorber's one focus point (used for distillation). For constructing the PTC, the focal length was calculated by using two methods. One is the software used to find the focal point, which name as a Parabolic Calculator 2.0 version, as shown in **Figure 4**, and secondly, the Eq. (1), studied by [22], is used to find the focal point of PTC. The cross-sectional area of PTC was calculated using Eq. (2) reported by [23].

$$Y = \frac{x^2}{16\ f} \text{ or } f = \frac{x^2}{16\ y} \tag{1}$$

$$Art = Wrt \text{ x } Lrt \tag{2}$$

The absorber consists of a black-painted pipe, which received water from a storage tank and then heated up and converts to vapors form with solar radiation

Figure 3.
General view of parabolic trough solar water distillation unit.

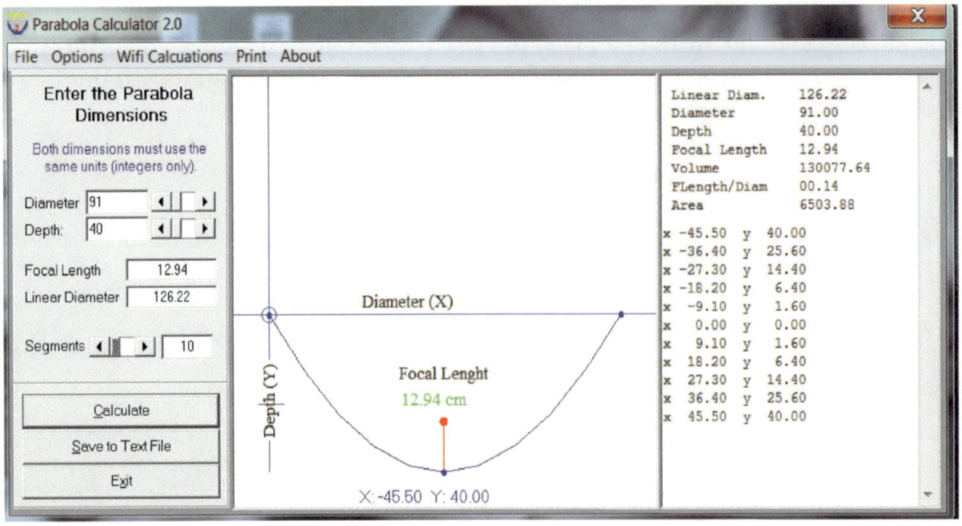

Figure 4.
The dimension of parabolic trough.

intensity absorbance. The absorber area and volume were calculated by the following Eqs. (3) and (4).

$$A_{ab} = \pi \times D_{ab} \times L_{ab} \quad (3)$$

$$V_{ab} = \pi \times r^2 \times L_{ab} \quad (4)$$

Two storage tanks were used in the experiment. One for inlet water, which contained tap water, and the other used for outlet distilled water. The collector is oriented along the east–west axis along the longitude and the altitude of the experimental area. The tilt adjusting Angles of the collector adjusted with an adjustable stand to collect maximum solar radiation [24].

2.2 Solar radiation intensity

The solar radiation intensity is the amount of energy received from the sun per unit time per unit area on the Earth. The SRI was recorded daily, weekly, and monthly with a Mechanical Pyranometer and recorded in Mechanical Pyranograph. Eq. (5) is used to determine solar radiation reported by [25].

$$I_s = 368 \times V_c \quad (5)$$

2.3 Performance of parabolic trough collector

The Performance of the PTC assisted for the solar water distillation unit was evaluated in terms of the quantity of distilled water obtained during a laboratory experiment. The distillation unit (condenser) is attached to the absorber, and it received water vapors from the absorber through the outlet opening-jet. The vapors cooled down to low temperature in the distillation unit to become liquid form. For sea-water distillation, we used this system, and it works well. Still, the scaling effect came after the sea-water passing through the absorber, and also through the distillation unit, it was blocked both systems with the microbes and some other type of micro-organism. Still, we clean both the system after three days using the chemical of concentrated nitric acid to clean that system for all the scaling effect cause.

Efficiency is the ratio of heat available to the collector (input) and distilled water (output). The PTC's performance was evaluated at different adjusting Angles, i.e., 250, 350, 450, and 550 without sun tracking system in a whole day for three consecutive months of the year, 2012, i.e., June to October. The temperature data was also recorded in this experiment. Eq. (6) was used for the efficiency of PTC studied by [26].

$$\eta\ (\%) = \frac{\text{Mass of distilled water (kg)} \times 2260 \left(\frac{\text{kJ}}{\text{kg}}\right)}{\text{Solar energy (kJ)}} \times 100 \tag{6}$$

2.4 Testing of water quality

The purity of distilled water was tested with the help of E.C meter (Model No: 4310). Before using the E.C meter, it was calibrated with the standard 0.1 and 0.01KCL solutions, and the S.I unit of E.C meter is expressed in Siemens [27]. When the E.C meter reading is in the range of 0-30μS.cm-1, the distilled water is free from impurities, i.e., Ca, Mg, Zn, and Na. While the reading greater than 30μS.cm-1 means that the distilled water contains impurities in the form ions reported in the literature [28].

2.5 Economic analysis

The procedure described by [29] is utilized for economic analysis of the solar still, and the main factors used in the analysis of the desalination unit are described as; annual fixed Cost (AFC), sinking fund factor (SFF), salvage cost (S), annual salvage cost (ASC), Annual maintenance cost (AMC), Total annual Cost (TAC) and Cost per liter (CPL) and Md is the annual average productivity.

$$\text{AFC} = (\text{CRP})\text{P} \tag{7}$$

$$\text{SFF} = \frac{\text{i}}{(\text{i}+1)^{\text{n}} - 1} \tag{8}$$

$$\text{S} = 0.2\text{P} \tag{9}$$

$$\text{ASC} = (\text{SFF})\text{S} \tag{10}$$

$$\text{AMC} = 0.1\ \text{AFC} \tag{11}$$

$$\text{TAC} = \text{AFC} - \text{ASC} + \text{AMC} \tag{12}$$

$$\text{CPL} = \frac{\text{TAC}}{\text{M}_\text{d}} \tag{13}$$

3. Results and discussion

3.1 Solar radiation intensity

Solar radiation intensity data were recorded every week with a Mechanical Pyranometer during the consecutive months of the year. Mean solar radiation intensity data were calculated during the daytime from 07:00 AM to 04:00 PM, as showed in **Figure 5**.

The graph line shows the highest mean value of solar radiation intensity recorded up to 625.5 kJ.m-2.hr-1 at 01:00 PM. The data trend shows that solar radiation intensity starts gradually increasing from the daytime 07:00 AM to

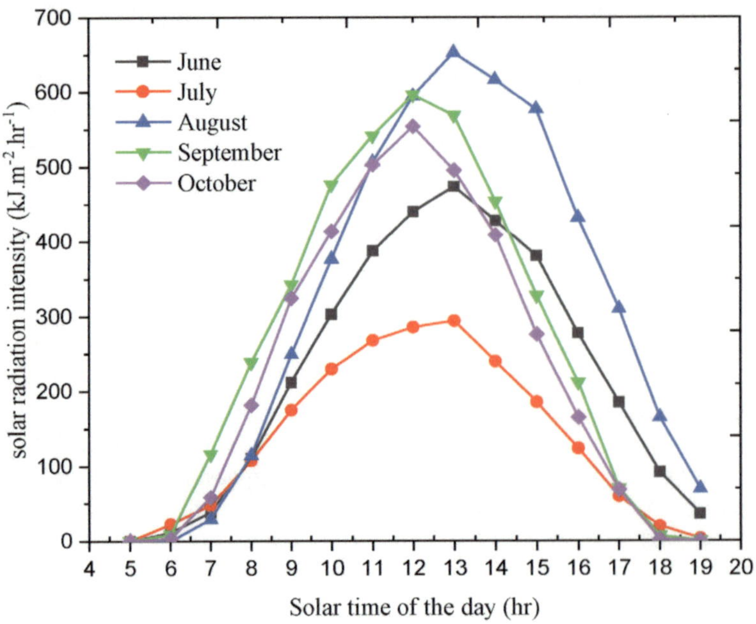

Figure 5.
Mean solar radiation intensity for the consecutive months of the year, 2012.

01:00 PM and then started gradually decreasing from 01:00 PM to 04:00 PM during the experiment. The results agree with the finding of [30, 31], who reported the solar radiation intensity was in the range of 500 W-m-2 using the solar water desalination system with different plates. The data show that the highest solar radiation intensity was noted at 01:00 PM due to higher radiation. Because in the morning, the sun was clear, and radiation was highest; after the daytime 02:00 PM, the sun was covered with the light clouds, so that's why the radiation was decreasing. The results agree with the finding of [32], who reported that the solar radiation intensity was 368.00 kJ.m-2.hr-1 during October 2012. The results are in agreement with the finding of [33], who reported that the solar radiation intensity was 368.00 kJ.m-2.hr-1 during October 2010, because in the morning time radiation was least due to light clouds and air blowing in October, while the daytime from 10:00 AM the radiation was increased with the clear sky.

3.2 The temperature of the parabolic trough collector

The mean range of absorber temperature of the PTC during the time of the day, i.e., 07:00 AM to 04:00 PM, was recorded from the consecutive months of the year at different adjusting Angles, i.e., 25o, 35o, 45o, and 55o are presented in **Table 1**.

The mean highest value of absorber temperature was recorded 123oC at adjusting Angles of 45o, similar to the finding results [4], who reported the air stream temperature 120oC. Because in the morning, the sun was not clear, so the temperature was the lowest while after the daytime at 10:00 AM, the temperature was increasing with the increasing solar radiation. The data results indicated that the mean highest absorber temperature was recorded 113oC at the adjusting Angles of 25o and 35o. The results agree with the finding by [34], who reported that the absorber temperature of the PTC for the time of the day, i.e., 07:00 AM to 04:00 PM, was 18oC to 110oC during September 2011. Results are similar to the finding by [35], who reported the air stream temperature in the range of 80oC to 120oC. Similarly, the results contradict the finding of (HP 1985). It was reported

Adjusting angles	Mean ambient temperature	Mean absorber temperature
25o	25	120
35o	28	115
45o	22	123
55o	20	110

Table 1.
Adjusting angles wise range of mean temperature on PTC.

that the absorber temperature of parabolic trough solar collectors for the whole day was 69oC to 91oC in October 2011.

3.3 Performance of parabolic trough collector

The mean highest output of distilled water was ranged from 472 ml.m-2.day-1 to 782 ml.m-2.day-1 for the different adjusting Angles are shown in **Figure 6**. The mean maximum output of distilled water was recorded 782 ml.m-2.day-1 for the adjusting angles of 45o, followed by 734 ml.m-2.day-1 and 718 ml.m-2.day-1 with the adjusting angles of 35o and 25o respectively. Similarly, the mean minimum output of PTC's distilled water was noted 472 ml.m-2.day-1 with the adjusting angle of 55o because the sun path was at the range of 80o to 85o adjusting Angles for the PTC, and we collect the date up to 17 days. Results are similar to the finding [36, 37], who observed that the distilled water production increased up to 600 ml. m-2.day-1 with the solar chimney power generation-sea water desalination of the synthetic system. The results contradict the result [21, 38], who observed that the average yield of distilled water was 2300 ml. m-2.day-1 in winter on the single solar wick type distillation plant.

Figure 6 shows that the average aggregate distillation yield of solar distillation units corresponds to the average annual condition. Based on meteorological data (solar radiation) obtained from the website's information, it is assumed that the

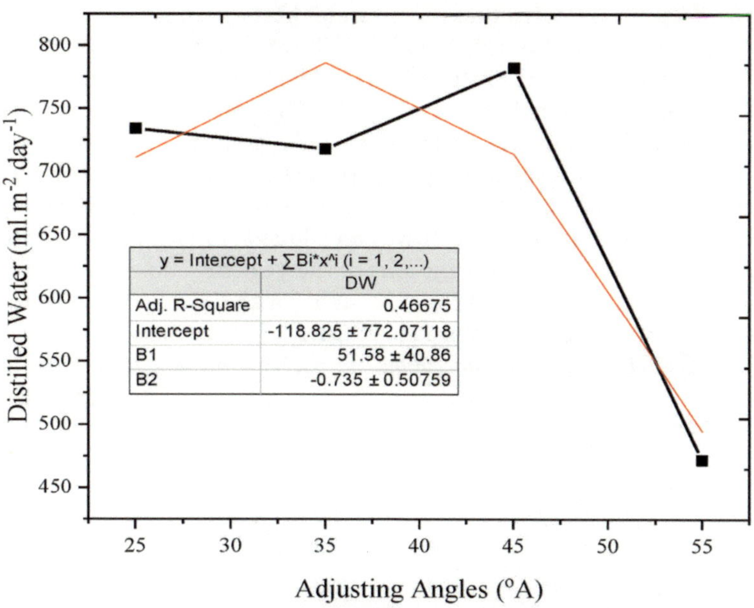

Figure 6.
Mean output of distilled water for the months of the year.

average annual condition is equal to the average condition attained during the study [50, 51]. From the results of the study, 625.5 kJ.m-2.hr-1 has been noted as the average daily solar radiation. The 550.2 kJ.m-2.hr-1 has been noted as the average yearly solar radiation in the area (as described on the website). It is reasonable to consider the test period equivalent to the average annual condition since the average annual obtainability of solar radiation in the area is appropriate to be adjacent to the experiment.

Likewise, the results are not in line with [39], who reported that water production was 6000 ml.day-1 with desalination process low temperature. The standard error bars are applied to distilled water data in the graph, which showed the standard error between the consecutive months of the year and adjusting Angles. The results are similar to the finding [40], who observed that the distilled water production increased up to 600 ml. m-2.day-1 with the solar chimney power generation-sea water desalination of the synthetic system. Results are not in line with the result [41], who reported that water production was 6000 ml. day-1 with desalination process low temperature. The results are similar to the founding of [42, 43], who observed that the distilled water production increased up to 600 ml. m-2.day-1 with the solar chimney power generation-sea water desalination of the synthetic system. However, the results are not similar to the result [18], who observed that the average yield of distilled water was 2300 ml. m-2.day-1 in winter on the single solar wick type distillation plant. Likewise, the results are not in line with the result (WD & Tamme, 2008), who reported that water production was 600 ml.day-1 with a low-temperature desalination process. PTC assisted for solar distilled water at tilt adjusting Angles of 35o and 45o worked efficiently for the maximum output of distilled water for the three consecutive months of the year.

3.4 The efficiency of parabolic trough collector

The efficiency of PTC at different adjusting Angles 25o, 35o, 45oand 55o during the consecutive months of the year are shown in **Table 2**. PTC's mean efficiency per day at different adjusting angles varied from 17.9% to 26.9% in **Table 2**.

From the data, it was noted that the mean highest efficiency of 26.9% was found at the adjusting Angles of 25o, followed by 26.3% and 26.1% was noted at the adjusting angles of 45o and 35o respectively, while the mean minimum efficiency was noted 17.9% at the adjusting Angles of 55o. The data shows that the PTC was performing well at the adjusting angles of 25o, 35o, and 45o compared to other adjusting Angles. PTC's low efficiency may be due to cloudy days at the adjusting Angles of 65o so that the absorber's temperature was not reached to the required amount for the distillation of water. It was concluded from the result of the mean efficiency of the three consecutive months of the year that the PTC is efficiently working at the adjusting angles of 25o, 35o, 45o, for the distillation of water. The reason may be a rapid change of sun rays striking on the collector, which affected the absorber's focal line at adjusting angles of 25o, 35o, and 45o compared to 15o, 55o

Adjusting angles	SRI	AT`	Ei	Dw	Eo	%η
25o	1480.1	1.39	2057.3	244.6	552.7	26.9
35o	1489.2	1.39	2070.0	239.1	540.3	26.1
45o	1612.6	1.39	2241.6	260.6	588.9	26.3
55o	1429.4	1.39	1986.9	157.2	355.2	17.9

Table 2.
Mean efficiency of PTC assisted for solar distilled water.

65o, respectively. Results are near the findings by [18], who reported that solar efficiency still was 16.1%. Likewise, results are similar to the finding [44]. They reported that the efficiency increases by 9.2%, with the increasing absorber area from 0.51 m2 to 0.62 m2.

3.5 Description of water quality analysis

The water twisted by the parabolic slot solar collector is estimated for quality from numerous characteristic points of view. The water management laboratory department at Peshawar Agricultural University in Pakistan verified feed water and distillate samples from various angles, i.e., 25o, 35o, 45o, and 55o. The samples werc verified for pH, electrical conductivity, Alkalinity, total dissolved solids (TDS), and chloride content. **Table 3** reports the characteristic seats and average values of the three random samples composed of feed water and distillate samples from different days. The table also includes acceptable limits for available properties from the studies reported [45]. pH represents the acidity of the water sample, determined at a gauge of 0 to 14. A sample with a pH of 7.0 designate neutral values, slower than 7.0 designates acidity, and above 7.0 is considered to be essential solutions. The study results noted the range from 7.26 and 8.18 pH values of feed water samples, while for distilled water, the pH was noted with an average of 7.46.

Conductivity (E.C) (m/s) is the capability of an ingredient to conduct current, which is proportional to the absorption of numerous melted salts (obtainable in the form of ions (cations and anions)). The average value of 901.20 m/s was noted for the inlet's electrical conductivity from the study result simple. It was found that the conductivity of the distilled sample was very little up to 19.75 and 28.52 m/s associated with the inlet. Similarly, Alkalinity (mg/L) is the capability of water to counteract acids. From the present results of the study, the alkalinity value of feed water samples was detected in 400 to 412 mg/L, while 14 to 24 mg/L values were recorded for the distilled water samples. As a result, significant differences in the Alkalinity of feed water (406 mg/L) and distilled water (18.80 mg/L) samples were detected.

Total Dissolved Solids (TDS) (mg/L) assessments are indicators for assessing the overall quality of water. Therefore, the TDS test provides a qualitative measure, although it does not approximate approximates in the sample. It was detected that in feed water samples, TDS values recorded between 463 mg/L and 470 mg/L and 2.69 mg/L to 13.88 mg/L was noted for the TDS of distillate samples, which demonstrating enhancements in water quality achieved from solar energy. Similarly, the concentration of chloride in water raises electrical conductivity and, consequently, its corrosive character. The present results of the study indicated that the range of saline water was 55.00–71.90 mg/L, while for the distillate sample, the value was noted in the range of 10.90–13.40 mg/L. Therefore, it can be inferred that the quality of water obtained from solar energy is still suggestively better-quality; in addition to the above products, the production of distilled water tasteless, tasteless, and colorless. Thus, distilled water produced from parabolic trough solar collector is potable. The results agree with the finding [46, 47]; they reported that the adjusting Angles of 35o and 45o is the best for the PTC for producing maximum distilled water. The performance of PTC was best at the adjusting Angles of 35o for the maximum output of distilled water reported by [48].

3.6 Comparison of D. W regarding e. C with available distilled water IN the agriculture university Peshawar (AUP)

Distilled water obtained from PTC in the Department of Agricultural Mechanization was compared with the other distilled water, prepared with the EDU in

Property	Feedwater				Distillate water					Acceptable limit [45]
	Sample 1	Sample 2	Sample 3	Average	25o	35 o	45 o	55 o	Average	
pH	7.61	7.34	8.18	7.70	7.40	7.41	7.50	7.45	7.44	6.5–8.5
E. C (m/s)	901.00	910.00	896.00	893.33	20.12	24.40	25.80	25.10	23.85	NA
Alkalinity (mg/L)	410.00	400.00	412.00	407.33	24.00	20.00	16.00	14.00	18.5	200
TDS (mg/L)	466.00	470.00	463.00	466.33	9.55	11.49	2.69	11.58	8.82	500
Chloride (mg/L)	57.40	71.90	57.40	62.23	12.40	10.90	11.40	11.90	11.65	250

Table 3.
Properties of feed water and distilled water angle wise

different Departments of the University regarding E.C is shown in **Figure 7**. The E. C of distilled water which was prepared in the Department of Agricultural Mechanization through PTC was 18μS.cm-1, which is in the range of the E.C of distilled water 15μS cm-1, 16μS cm-1, 19μS cm-1, and 20μS cm-1, which was collected from different departments of the University (AUP) which they prepared through EDU. From the E.C meter of the distilled water, it is clear from **Figure 7** that the E.C of distilled water prepared by PTC is similar to the prepared distilled water through EDU. Standard error bars are applied to the E.C data of distilled water of different department wise, which showed how much error is present in the data. For Peshawar-Pakistan climatic conditions, the annual average daily yield from the parabolic trough solar distillation unit can be assumed to be 782 ml.m-2.day-1 [49].

Nevertheless, the economic assistances analysis described below will highlight the effects of design parameters, i.e., adjusting parabolic trough solar distillation unit angles. The solar distillation unit is expected to function for an ordinary of 153 days per year (established on the yearly sunshine period in North Peshawar-Pakistan). **Table 4** encapsulates the outcomes of the economic investigation.

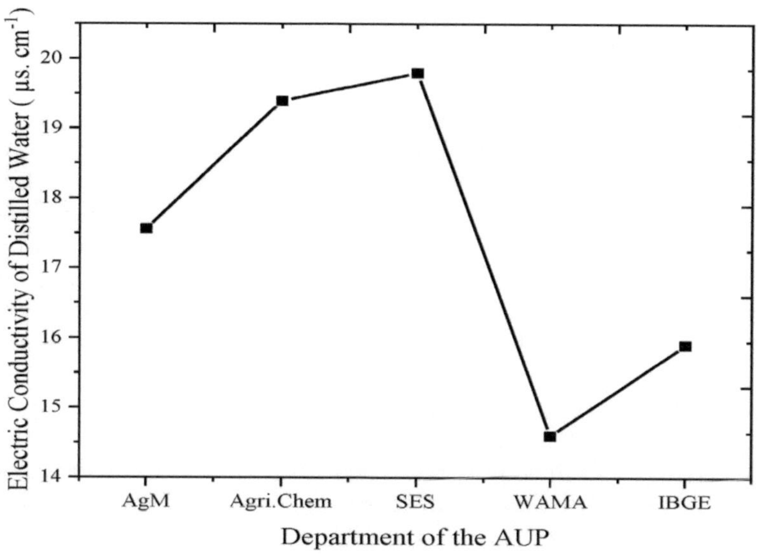

Figure 7.
Electric conductivity of distilled water in departments of the university.

Economic parameters	Value
Fixed Cost	Rs, 10,000
Annual salvage cost	Rs, 110 per annum
Annual fixed cost	Rs, 2000 per annum
Annual maintenance cost	Rs, 200 per annum
Annual water productivity	281.520 L.m-2
The market price of distilled water	Rs, 13/L
Total annual cost	Rs, 1800/annum
Cost of water produced	Rs, 2.64/L-day
Payback period	365 days

Table 4.
Economic analysis of solar distillation unit.

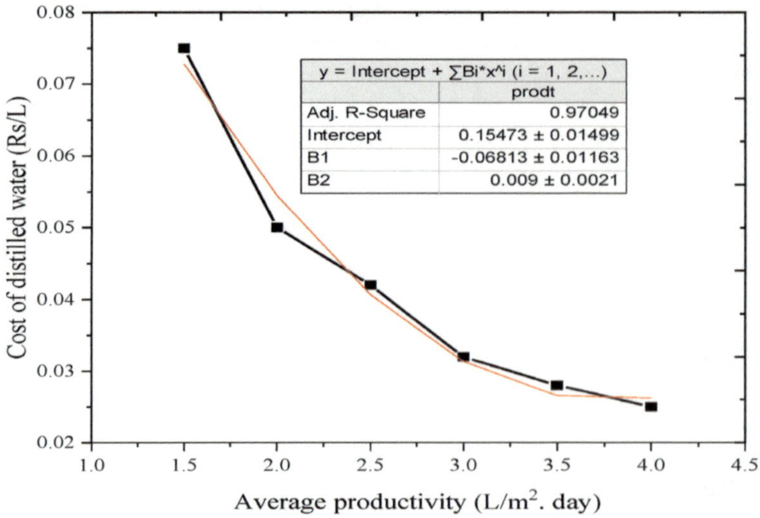

Figure 8.
Cost of average productivity of distilled water.

Based on cost analysis, the assessed unit cost of fractions is Rs 2.64/L2. Correspondingly, the payback period is 365 days (Because a sunny day of the year is supposed to be 153 days). The decrease in the angle of alteration of the solar distillation device has little effect on the desalination device's production rate. Nevertheless, in this study, the angle adjustment has been an essential factor affecting solar distillation devices' daily productivity. Therefore, the impact of distillate production on water costs was investigated, taking into account the increase in distillate production in the four cases considered under the present study.

Assess the consistent rate per liter of distilled water, as shown in **Figure 8**. The study results concluded that the cost per unit of distillation had been suggestively decreased from Rs, 4.92/L to Rs, 1.57/L. As a result, the cost is significantly reduced (approximately 68%) distilled water achieved with the increase in distillation components, resulting from improvements in the solar distillation design, with adjustment angles, i.e., 25o, 35o, 45o, and 55o.

4. Conclusion

Water and energy are the basic needs for us to lead an everyday life on Earth. Solar energy technologies and their usage are the most important and useful for developing countries to sustain their energy needs. For the distillation process, the use of solar energy is one of the essential techniques. From the results of the laboratory experiment, it was concluded that:

- A PTC can be used for the production of a reasonable amount of distilled water.

- For the maximum amount of distilled water for PTC, the adjusting angels were recommended 25o, 35o, and 45o.

- From the results, it was cleared that PTC's maximum efficiency was noted 26.9%, 26.3%, and 26.1% for the adjusting Angles 25o, 45o, and 35o, respectively.

- The cost per distiller has been expressively compact from Rs 4.92/L to Rs 1.57/L.

- The average unit price of solar distillate is still evaluated at Rs 2.64/L-m2, with a recovery period of 365 days.

After careful considering of the experiment, the following suggestion drawn from the results of the study:

- The PTC is a cheap source of making solar distilled water as compared to the EDU plant.
- The collector should be placed in open space for absorbing maximum solar radiation to get maximum distillation.
- The collector's focal point should be re-adjustable, and the reflecting sheet should be used as good quality.
- For increasing collector efficiency, a sensor revolving system should be developed concerning direct reflection of sun rays.
- To repeat the experiment for the whole year at different adjusting Angles and day timings.

Acknowledgements

This work was edited for proper English language, grammar, punctuation, spelling, and overall style by native English speaking editors at American Journal Experts (AJE).

Declarations

Availability of data and materials: The datasets used and analyzed during the current study are available from the corresponding author on reasonable request.

Competing interests: The authors declare that they have no competing interests" in this section.

Funding: The author(s) received no financial support for the research, authorship, and publication of this article.

Author Contributions: The research article with several authors has its contributions to work. Fahim Ullah conducted the experimental work, methodology, formal analysis, and data curation, writing-original draft preparation. At the same time, Mansoor Khan Khattak reviewed and editing as well as supervised.

Nomenclature

Y (cm)	Depth of Parabola
X (cm)	Diameter of Parabola
f (cm)	The focal point of Parabola
Art (cm2)	The cross-sectional area of Parabolic Trough
Wrt (cm)	Width of parabolic trough
Lrt (cm)	Length of parabolic trough
Aab (cm2)	The cross-sectional area of the Absorber pipe
π (3.14)	Constant term

Dab (cm)	The diameter of the Absorber pipe
Lab (cm)	Length of Absorber pipe
Is (SRI)	Solar Radiation Intensity (kJ.m-2.hr-1)
Vc	Mechanical Pyranograph value (Cal.m-2.sec-1)
η	Efficiency
2260	Vaporization requirement (kJ-kg-1)
AT	Area of Parabolic Trough (m2)
Ei	Energy Input (kJ.m-2.hr-1)
DW	Distilled water (m.hr-1)
Eo	Energy output (kJ.m-2.hr-1)
Vab	The volume of the Absorber pipe (cm3)
R	The radius of the Absorber pipe

Acronyms

PTC	Parabolic Trough Collector
SDW	Solar Distilled Water
E.C	Electrical Conductivity
S.I	System International Unit
KCL	Potassium Chloride Solution
Ca, Mg, Zn, and Na	Calcium, Magnesium, Zinc, and Sodium
EDM	Electrical Conductivity Meter

Author details

Fahim Ullah[1,2]

1 Joint International Research Laboratory of Information Display and Visualization, School of Electronic Science and Engineering, Southeast University, Nanjing, P.R. China

2 Department of Agricultural Mechanization, FCPS, The University of Agriculture, Peshawar, KP, Pakistan

*Address all correspondence to: fahimullah320@seu.edu.cn

References

[1] A. Ahsan, I. KMS, T. Fukuhara, and G. AH, "Experimental study on evaporation, condensation, and production of a new tubular solar still," Desalination, vol. 260, pp. 172–179, 2010.

[2] A. Firoz, "Solar Radiation Studies at Karachi Pakistan," Karachi, Pakistan, 1989.

[3] R. FR, O. MG, F. Gordillo, and M. López-Martínez, "Application of new control strategy for sun tracking," Energy Convers Manag., vol. 48, pp. 2174–2184, 2007.

[4] R. Forristall, "Heat transfer analysis and modeling of a parabolic trough solar receiver implemented in engineering equation solver," NREL is a U. S. Dep. Energy Lab., pp. 2–18, 2003.

[5] Ghaffar MA, "The energy supply situation in the rural sector of Pakistan and the potential of renewable energy technologies," Renew. Energy, vol. 6, no. 8, pp. 941–976, 1995.

[6] Ghoreishi SM, Kamali H, Ghaziaskar HS, and D. AA, "Optimization of supercritical extraction of linalyl acetate from lavender via box–Behnken design," Chem. Eng. Technol., vol. 35, no. 9, pp. 1641–1648, 2012.

[7] A. Giannuzzi, E. Diolaiti, M. Lombini, D. R. A, B. Marano, and G. Bregoli, "Enhancing the efficiency of solar concentrators by controlled optical aberrations: method and photovoltaic application," Appl Energy, vol. 145, pp. 211–222, 2015.

[8] D. İbrahim and K. Fergun, "Drying and Rehydration Behaviors of Convection Drying of Green Peas," Dry. Technol., vol. 29, 2011.

[9] Igual M, Garcia ME, Martín-Esparza ME, and Martínez-Navarrete N, "Effect of processing on the drying kinetics and functional value of dried apricot," J. Food Res. Int., no. 47, pp. 284–290, 2012.

[10] R. V. Padilla, G. Demirkaya, D. Y. Goswami, E. Stefanakos, and M. M. Rahman, "Heat transfer analysis of parabolic trough solar receiver," Appl. Energy, vol. 88, no. 12, pp. 5097–5110, 2011.

[11] Jamil AA, Osama A, and Ahmed AS, "Design and Performance Assessment of a Parabolic Trough Collector ," Jordan J. Mech. Ind. Eng., no. 8, p. 1, 2014.

[12] Jegadheeswaran S and P. SD., "Performance enhancement in latent heat thermal storage system: A review," Renew Sust Energ Rev, vol. 13, no. 9, pp. 2225–2244, 2009.

[13] Z. JG et al., "Preparation of paraffin/expanded graphite phase change composites for thermal storage," New Carbon Mater, vol. 24, pp. 114–118, 2009.

[14] S. JH, D. Shum, H. Djaballah, and S. DA, "A high-throughput screen for alpha particle radiation protectants," Assay Drug Dev Technol, vol. 5, 2010.

[15] P. JC, A. Martin, and M. Camacho, "Resazurin microtiter assay plate: A simple and inexpensive method for detection of drug resistance in mycobacterium tuberculosis," Antimicrob Agents Chemother, vol. 46, pp. 2720–2722, 2002.

[16] J. John, M. Joy, and A. EK, "Inhibitory effects of tamarind (Tamarindus indica L.) on polypathogenic fungi," Allelopath. J, vol. 14, pp. 43–49, 2004.

[17] K. JM and H. SF, "Nanoparticle-enhanced phase change materials (NEPCM) with great potential for improved thermal energy storage," Int

Commun Heat Mass Transf., vol. 34, pp. 534–543, 2007.

[18] F. Ullah and M. Kang, "Performance Evaluation of Parabolic Trough Solar Collector with Solar Tracking Tilt Sensor for Water Distillation," Energy Environ., 2019.

[19] D. Wu and W. Liu, "A new fault diagnosis approach for the pitch system of wind turbines," Adv. Mech. Eng., vol. 9, no. 5, p. 168781401770335, 2017.

[20] Tiwari S, Tiwari GN, and Al-Helal IM, "Performance analysis of photovoltaic thermal (PVT) mixed mode greenhouse solar dryer," Sol. Energy, no. 133, pp. 421–28, 2016.

[21] F. Ullah and M. Kang, "Performance evaluation of parabolic trough solar collector with solar tracking tilt sensor for water distillation," Energy Environ., p. 0958305X1983927, 2019.

[22] B. AA, A.-H. AA, S. IAE, and O. MZ, "A solar still augmented with a flate plate collector," Desalination, vol. 172, pp. 227–234, 2005.

[23] E.-S. AA, A.-G. AA, A.-H. FS, and A. SF, "'Thermal performance of a single basin solar still with PCM as a storage medium,'" Appl. Energy, vol. 86, pp. 1187–1195, 2009.

[24] E.-S. AA and M. Al-Dossari, "A mathematical model of single basin solar still with an external reflector," Desalin Water Treat, vol. 26, pp. 250–259, 2011.

[25] A. Abadi Musyafa' Soeprijanto, A.,, I, "Type-2 fuzzy logic Controller based pv passive two-axis solar tracking system," Int. Rev. Electr. Eng., vol. 10, pp. 390–398, 2015.

[26] Abu-Zour AM, Riffat SB, and M. Gillott, "New design of solar collector integrated into solar louvers for efficient heat transfer," Appl Therm Eng, vol. 26, no. 16, pp. 1876–1882, 2006.

[27] Agha KR, "The thermal characteristics and economic analysis of a solar pond coupled low temperature multi stage desalination plant," Sol. Energy, vol. 83, no. 4, pp. 501–510, 2009.

[28] Agboola OP, Atikol U, and Assefi H, "Feasibility assessment of basin solar stills," Int. J. Green Energy, vol. 12, pp. 139–147, 2015.

[29] Arasu AV and Sornakumar T, "Design manufacture and testing of fiberglass reinforced parabola trough for parabolic trough solar collectors," Sol. enegy, vol. 81, no. 1, pp. 1273–1279, 2007.

[30] Fahim U et al., "Impact of drying method of figs with small scale flat plate solar collector," World J. Eng., vol. 08, no. 0054, 2016.

[31] Fahim U, Lubna H, Muhammad SM, and Kamran H, "Effect of irradiation on the quality of dried fig," Res. J. Appl. Sci. Eng. Technol., vol. 12, no. 10, pp. 1007–1010, 2016.

[32] Faqir MA, Saeed A, and Maqam D, "Storage effect on physiochemical and sensor characteristics of dried apricot jam," Pak. J. Food. Sci, vol. 14, no. 1–2, pp. 43–47, 2004.

[33] W. FD, "Solar collector energy storage and materials," the MIT press, vol. Master the. Cambridge, 1991.

[34] Fudholi A, Sopian K, Ruslan M, Alghoul M, and Sulaiman M, "Review of solar dryers for agricultural and marine products," Renew. Sustain. Energy Rev., no. 14, pp. 1–30, 2010.

[35] Z. FY, M. Han, and J. Feng, "Design of sun tracker used in the solar battery," J. Control Eng. China, vol. 16, p. 109 - 111, 2009.

[36] Soteris K, "Use of parabolic trough solar energy collectors for sea-water

desalination," Appl. Energy, vol. 60, pp. 65–88, 1998.

[37] M. Souliotis and Y. Tripanagnostopoulos, "Experimental study of CPC type ICS solar systems," Sol. Energy, vol. 76, pp. 389–408, 2004.

[38] F. Ullah, M. K. Khattak, M. Khan, L. Hassan, and K. Hasrat, "The impact of the dehydration process on guava by using of the parabolic trough solar collector," PAK. J. FOOD SCI, vol. 26, pp. 92–97, 2016.

[39] Varun S, Sharma A, and Sharma N, "Construction and performance analysis of an indirect solar dryer integrated with solar air heater," Procedia Eng., no. 38, pp. 3260–69, 2012.

[40] Vijayavenkataraman S, Iniyan S, and G. R, "A review of solar drying technologies," Rene. Sustain. Energy Revi, pp. 2652–2670, 2012.

[41] Wang BH and Liu GH, "Improvement in the astronomical parameters computation for solar radiation observation," J. Acta Energiae Solaris Sin., vol. 12, no. 1, p. 27-32, 1991.

[42] Wang R, Hong GF, and B. Han, "Transcript abundance of rml1, encoding a putative GT1-like factor in rice, is up-regulated by Magnaporthe grisea and down-regulated by light," Gene, no. 423, pp. 105–115, 2004.

[43] H. Wang and W. Fang, "Solar irradiance absolute radiometer with the ability of automatic solar tracking," J. Chinese Opt., vol. 4, p. 252-258, 2011.

[44] S. Wolf-Dieter and M. Eck, "Buffer storage for direct steam generation," Sol. Energy, pp. 1277–1282, 2006.

[45] Yousefi MA, "Experimental study of natural convection heat transfer from vertical array of isothermal horizontal elliptic cylinders," Exp. Therm. Fluid Sci., vol. 32, pp. 240–248, 2007.

[46] Zhang L, Sheng Y, and Hong X, "Experimental study on condensation heat transfer characteristics of steam on horizontal twisted elliptical tubes," Appl. Energy, vol. 97, pp. 881–887, 2012.

[47] Zhang P, Ma F, and Xiao X, "Thermal energy storage and retrieval characteristics of a molten-salt latent heat thermal energy storage system," Appl. Energy, vol. 173, pp. 255–271, 2016.

[48] Zhang Y, "Solar thermal power technology," J. High-technology Ind., vol. 7, p. 28-30, 2009.

[49] Zhang Y, Li M, and Ji X, "The impact of the condenser mirror's parameters of solar trough reflector on the concentrating characteristic," J. Yunnan Norm. Univ. Nat. Sci. Ed., vol. 33, no. 4, p. 14-19, 2013.

[50] Zhang Z, Yan X, and Xiao ZJ, "Experimental research on single phase flow and convection heat transfer in heat-generating porous media," Nucl. Power Eng, vol. 33, pp. 1–4, 2012.

[51] J. Zhang et al., "High-performance ultrathin organic−inorganic hybrid silicon solar cells via solution-processed Interface modification," ACS Appl. Mater. Interfaces, vol. 9, p. 21723−21729, 2017.

[52] Abo-Elkasem M, Osama EM, and Fatouh M. Development of design optimized simulation tool for water desalination system. [J] Desalination, 2016, 398, 157–164.

[53] Abu-Khadera MM, Badranb OO, and Abdallah S. Evaluating multi-axes sun-tracking system at different modes of operation in Jordan. [J] Renew Sust Energy Rev, 2008, 12, 864–873.

[54] Agha KR. The thermal characteristics and economic analysis of a solar pond coupled low temperature multi-stage desalination plant. [J] Sol. Energy, 2009, 83, 501–510.

[55] Ahmed HM. Seasonal performance evaluation of solar stills connected to passive external condensers. [J] Sci. Res Essays 2012, 7, 1444–1460.

[56] Aljane F, Toumi I, and Ferchichi A. HPLC determination of sugars and atomic absorption analysis of mineral salts in fresh figs of Tunisian cultivars. [J] African Journal of Biotechnology, 2007, 6, 599.

[57] Alkhudhiri A, Darwish N, and Hilal N. Membrane distillation: A comprehensive review. [J] Desalination, 2012, 287, 2–18.

[58] Anant S, Buddhi D, and Sawhney RL. Solar water heaters with phase change material thermal energy storage medium. A review. [J] Renewable and Sustainable Energy Reviews, 2009, 13, 2119–2125.

[59] Arunkumar T, David D, Velraj R, Ravishankar S, Hiroshi T, and Vinothkumar K. Experimental study on a parabolic collector assisted solar desalting system. [J] Energy Conversion and Management, 2015, 105, 665–674.

[60] Arunkumar T, Velraj R, D. D., and Ravishankar S. Influence of crescent-shaped absorber in water desalting system. [J] Desalination, 2016, 398, 208–213.

[61] Balaji D, Jayaraj K, Phani SVSK, and Ramana MVM. Water quality improvement studies in LTTD plant. [J] Journal Desalination and Water Treatment 2016, 57, 24705-24715.

[62] Bennett CL. Optimal heat collection element shapes for parabolic trough concentrators. [J] Journal of Solar Energy Engineering-Transactions of the ASME, 2008, 2.

[63] Bhardwaj R, Ten MVK, and Mudde RF. Maximized production of water by increasing area of condensation surface for solar distillation. [J] Applied Energy 2015, 154, 480–490.

[64] Bouchekima B. A solar desalination plant for domestic water needs in arid areas of South Algeria. [J] Desalination, 2002, 153, 65 - 69.

[65] Chafik E. A new sea water desalination process using solar energy. [J] Desalination, 2002, 153, 25 – 37.

[66] Chaudhry MA, Raza R, and Hayat SA. Renewable energy technologies in Pakistan: Prospects and challenges. [J] Renewable Sustainable Energy Reviews, 2008, 13, 1657 - 1662.

[67] Chong KK, and Wong CW. General formula for on-axis sun-tracking system and its application in improving tracking accuracy of solar collector. [J] Solar Energy 2009, 83, 298–305.

[68] Chow TT. A review on photovoltaic /thermal hybrid solar technology. [J] Applied Energy, 2010, 87, 365-379.

[69] Du S, and Xia X. Numerical investigation on effects of nonparallelism of solar rays on concentrating solar power. [J] Acta Energiae Solaris Sinica, 2006, 27, 388–395.

[70] Duffie J, and Beckman W. Solar Engineering of Thermal Processes. (3rd Edn). [J] J. Wiley & Sons, Inc 2006.

[71] El-Bahi A, and Inan D. Solar still with minimum inclination, coupled to an outside condenser. [J] Desalination, 1999, 123, 79 - 83.

[72] El-Sebaii AA, and El-Bialy E. Advanced designs of solar desalination systems: A review [J] Renewable and Sustainable Energy Reviews 2015, 49 1198–1212.

[73] Eltawil MA, and Omara ZM. Enhancing the solar still performance using solar photovoltaic, flat plate collector and hot air. [J] Desalination 2014, 349, 1–9.

[74] Fahad AA, Feridun H, and Ibrahim D. Performance assessment of a novel system using parabolic trough solar collectors for combined cooling, heating, and power production. [J] Renewable Energy 2012, 48, 161-172.

[75] Shanmugasundaram K, and Janarthanan B. Performance analysis of the single basin double slope solar still integrated with shallow solar pond. [J] Int. J. Innov. Res. Sci. Eng. Technol, 2013, 2, 5786–5800.

[76] Sinha S, Kumar S, and Tiwari GN. Active solar distillation system-an investment alternative to a solar hot water system. [J] Energy Convers. Mgmt, 1994, 35, 583 - 588.

[77] Sudhir C, Nimesh P, and Hitesh NP. A Critical Review on Solar Water Heater. [J] International journal of advanced engineering technology, 2013, 50-52.

[78] Sampath KK, Arjunan TV, Pitchandi P, and Kumar PS. Active solar distillation—A detailed review. [J] Renewable and Sustainable Energy Reviews, 2010, 14, 1503–1526.

[79] Sethi AK, and Dwivedi VK. Design, fabrication, and performance evaluation of double slope active solar still under forced circulation mode. [J] Int. J. Therm. Environ. Eng 2013, 6 27–34.

[80] Shanmugasundaram K, and Janarthanan B. Performance analysis of the single basin double slope solar still integrated with shallow solar pond. [J] Int. J. Innov. Res. Sci. Eng. Technol 2013, 2 5786–5800.

[81] Sharshir SW, Yang N, Guilong P, and Kabeel AE. Factors affecting solar stills productivity and improvement techniques: A detailed review. [J] Applied thermal engineering 2015, DOI: http://dx.doi.org/doi: 10.1016/j.applthermaleng.1011.1041.

[82] Shashikanth M, Binod K, Yennam L, Mohan PSK, Nikhila A, and Sonika V. Solar water distillation using energy storage material. [J] Procedia Earth and Planetary Science 2015, 11.

[83] Shiva G, Barat G, Teymour TH, and Ahmad B. Experimental performance evaluation of a stand-alone point-focus parabolic solar still. [J] Desalination 2014, 352, 1–17.

[84] Singh DB, and Tiwari GN. Effect of energy matrices on life-cycle cost analysis of partially covered photovoltaic compound parabolic collector active solar distillation system. [J] Desalination 2016, 397, 75–91.

[85] Soteris K. Use of parabolic trough solar energy collectors for sea-water desalination. [J] Applied Energy 1998, 60, 65–88.

[86] Vivar M, Fuentes M, Norton M, Makrides G, and de-Bustamante I. Estimation of sunshine duration from the global irradiance measured by a photovoltaic silicon solar cell. [J] Renewable and Sustainable Energy Reviews 2014, 36, 26–33.

[87] Rubio FR, Ortega MG, Gordillo F, and López-Martínez M. Application of new control strategy for sun tracking. [J] Energy Convers Manage, 2007, 48, 2174–2184.

[88] Saleh A, Qudeiri JA, and Al-Nimr MA. Performance investigation of a salt gradient solar pond coupled with desalination facility near the Dead Sea. [J] Energy, 2011, 36, 922–931.

[89] Sampathkumar K, Truman TV, Pitchandi P, and Senthilkumar P. Active solar distillation—A detailed review. [J] Renew. Sust. Energy. Rev, 2010, 14, 1503–1526.

[90] Scrivani A, El-Asmar T, and Bardi U. Solar trough concentration for fresh water production and wastewater

treatment. [J] Desalination, 2007, 206, 485–493.

[91] Sengar SH, Mohod AG, Khandetod YP, Modak SP, and Gupta DK. Design and development of wick type solar distillation system. [J] Journal of Soil Science and Environmental Management, 2011, 2, 125 - 133.

[92] Sethi AK, and Dwivedi VK. Design, fabrication, and performance evaluation of double slope active solar still under forced circulation mode. [J] Int. J. Therm. Environ. Eng, 2013, 6, 27–34.

[93] Singh RV, Dev R, Hasan MM, and Tiwari GN. Comparative energy and exergy analysis of various passive solar distillation systems. [J] World renewable energy Congress, 2011, 3929-3936.

[94] Sümeyye A, Meltem T, Şeref T, and Mehmet Ö. Effects of different sorbic acid and moisture levels on chemical and microbial qualities of sun-dried apricots during storage. [J] Food Chemistry, 2015, 356–364.

[95] Tchanche BF, Lambrinos Gr, Frangoudakis A, and Papadakis G. Exergy analysis of micro-organic Rankine power cycles for a small scale solar driven reverse osmosis desalination system. [J] Applied Energy, 2010, 87, 1295e1306.

[96] Tomson T. Discrete two-positional tracking of solar collectors. [J] Renew Energy, 2008, 33, 400–405.

[97] Omara, Z., Eltawil, M., and Elnashar, E. A new hybrid desalination system using wicks/solar still and evacuated solar water heater. [J] Desalination 2013, 325, 56–64.

[98] Veera GG, and Nirmalakhanda N. Desalination at low- temperature and low pressure. [J] J. Science Direct. Desalination, 2009, 344, 239 – 247.

[99] Ravi G, Naga SS, Devender V, and Hima BB. Solar water distillation using three different phase change materials. [J] Materials Today: Proceedings, 2015, 2, 1868 - 1875.

[100] Weidong H, Peng H, and Zeshao C. Performance simulation of a parabolic trough solar collector. [J] Solar Energy, 2012, 86, 746–755.

[101] Velmurugan V, and Srithar K. Performance analysis of solar stills based on various factors affecting the productivity—A review. [J] Renew. Sust. Energy. Rev, 2011, 15, 1294–1304.

[102] Vivar M, Fuentes M, Norton M, Makrides G, and Bustamante I. de. Estimation of sunshine duration from the global irradiance measured by a photovoltaic silicon solar cell. [J] Renewable and Sustainable Energy Reviews, 2014, 36, 26–33.

[103] Younas MS, Masood SB, Imran P, and Muhammad S. Development of zinc fortified chitosan and alginate based. [J] Pak. J. Agri. Sci, 2014, 4, 1033-1039.

[104] Zheng B, and Weng YW. A combined power and ejector refrigeration cycle for low temperature heat sources. [J] Solar Energy, 2010, 84, 784e791.

[105] Naveen KG, Arun KT, and 1Subrata KG. Heat transfer mechanisms in heat pipes using nanofluids – A review. [J] Experimental Thermal and Fluid Science, 2018, 90, 84-100.

[106] Abadi, I., Musyafa', A., Soeprijanto, A., . Type-2 fuzzy logic Controller based pv passive two-axis solar tracking system. [J] International Review of Electrical Engineering, 2015, 10, 390-398.

[107] C. S. Chin, A. B., W. Mcbride, . Design, Modeling and Testing of A Standalone Single Axis Active Solar Tracking Using MATLAB/Simulink [J]

Renewable Energy, 2011, 36, 3075–3090.

[108] Taher. M, S. E. A., Sassi. B. N. Performance modeling and Investigation of Fixed, Single and Dual Axis Tracking Photovoltaic Panel in Monastir City, Tunisia. [J] Renewable and Sustainable Energy Reviews, 2011, 15, 4053–4066.

[109] Sunggur C. Multi axes sun tracking system wit PLC control for photovoltaic panels in Turkey. [J] Renewable Energy, 2009, 34, 1119-1125.

[110] Hossein. M, A. K., Arzhang. J, Karen. A, Ahmad. S, . A Review of Principle and Sun Tracking Methods for Maximizing Solar System Output. [J] Renewable and Sustainable Energy Review, 2009, 13, 1800-1818.

[111] Yan. Z, J. Z. Application of Fuzzy Logic Control Approach in a Microcontroller-Based Sun Tracking System. [J] WASE International Conference on Information Engineering, 2010, 2, 161–164.

Section 4

Computer Applications in Agriculture

Chapter 22

Precision in Agriculture Decision Making Based on Machine Learning

Ratnmala Bhimanpallewar and M.R. Narasinga Rao

Abstract

Farming is the one of the major occupations in India. Increase in population is increasing the demand of food, whereas soil degradation causing decrease in yield. Technology is contributing in agriculture domain through software and hardware enhancement. One of the software-based contribution is for predicting the suitable crop. Same field can be suitable for one crop and not for another one, so it is better to choose the one which can lead to better yield. There are many predictive algorithms available. Algorithms which can work for suitability analysis need to test and choose the best one. Such predictive algorithms need dataset in appropriate format. Once the quality data is available correct predictions can be made. Data mining, machine learning are the branches comprise of algorithms, which can be trained based on dataset. Here we are introducing algorithms for decision making based on field data.

Keywords: machine-learning, agriculture, suitability, yield, dataset, algorithm, fragmented land, crop

1. Introduction

Agriculture is the main occupation in India and no doubt experienced farmers are the expert at doing everything. They can take the decisions; some decisions may fail so need to do some analysis for good results. One of the decisions we are considering here is "which crop to be adopted to have better yield?". It needs to analyze all the features on which yield is dependent and predict the suitable crop/crops. The basic feature i.e. agriculture land is the natural feature, which cannot prepare artificially and that too increase in the soil degradation is one of the serious issues farmers are facing [1]. There are multiple reasons affecting on soil quality one of them is, use of fertilizers without knowledge. For increasing yield farmers are using unbalanced quantity of fertilizers [2], without knowing the effect of it.

Even young farmers cannot take decision about which crop is suitable for current situation (soil quality, environment etc.). Farmers are adopting some crop and using chemicals for better yield, we have a good example for this from case study at Kolhapur district [3] which was resulted in causing cancer. To avoid such worst impacts we are thinking about recommending the suitable crop/crops. To get the genuine decision we are using historical data. Once the pattern of historical results identified successfully, predictions can be done appropriately.

There are many algorithmic approaches available for predictive analysis, machine learning is the branch which makes computer to learn from the dataset available. We are discussing about the 1) Role of machine learning in precise decision making about agriculture, 2) How to classify suitability level of crop/crops and feature values.

2. Agriculture decision making: Current scenario

Since old days farmers are taking farming decisions about agriculture by experience. Here we are discussing about the decision of choosing the crop to be adopted. Though agriculture is a major sector in India, same trend is going on. Automated and advanced systems in the form of software and physical machines are invented and available for use in developed countries, some of them are available in developing countries like India as well. Available facilities are either unaffordable or not easily approachable. Some systems are developed for some particular geographical area, it cannot be adopted for India as it is. Software techniques available in developed countries are not applicable due to fragmented land, because those systems are developed considering unfragmented land.

In India Krishi Vigyan Kendra (KVK) are the centers made available for farmer's guidance. It is responsible to spread technical knowledge among farmers. They conduct training, awareness programs to achieve some goals set as below:

1. Develop advisory services for farmers.

2. Conduct training program on different trends.

3. Conduct training programs for different level of people working in agriculture.

4. Agriculture field testing

5. Demonstration for recent innovations

It is playing an important role for needy farmers. Multiple KVK centers are there, so they can work according to the location to develop location specific solutions. It also has contribution in providing quality products like planting material, seed, organic material, livestock related products etc. As above mentioned, all agriculture related centers are available to facilitate people working in agriculture sector.

Despite all the knowledge centers, testing facilities available, farmers are continuing with traditional methods. The major change they have adopted is after Green revolution. Green revolution has shown drastic increase on yield [4, 5]. Other than using high-yielding varieties of seeds and the improved quality of fertilizers farmers are not interested to adopt new technics. There are many reasons behind it, as discussed further.

The one who work in farm are not well educated, not aware about the ongoing innovations. Even if they become aware, they cannot afford it. Economic condition of most of the farmers in India is not sufficient to purchase the automated systems available in the market. At another end, the educated, economically well, and aware people not much involved in actual farming and related occupations. The traditional farming methods are dependent on natural parameters. Uncertainty in the nature directly effects on agriculture production, so not getting yield as expected.

Thus, since long ago there is no improvement in the economic condition of small/marginal scale farmers in India.

3. Existing systems

Decision making system available in other countries. One of the popular decision making system is Agriculture Land Suitability Evaluator (ALSE), is the crop specific evaluator at Peninsular Malaysia [6]. Here crop specific means, it works for mango, citrus, guava, papaya and banana as well. Base for decision making is cultivation history, cultivation knowledge, land characteristics, climate features such as annual precipitation, dry season length per month, land slope, nutrient availability, and retention. Some other features are used for land availability and suitability evaluation in Tuban Regency, Java Island [7]. Spatial multi-criteria analysis has been done based on parameters like, land elevation, slope, slope direction, land use/land cover, land capability, integrating soil order, climate, and accessibility. Outcome prepared was land use plan by the spatial pattern. This is the area having most fertile land in the country. The weights are assigned to above mentioned criteria with the help of eight experts involved for sub-criterions. Criteria wise scores are assigned in the range of 0–10, according to involvement of sub criterions under each criterion. Weighted sum overlay method used those weights to prepare suitability map. Suitability analysis for crop Soybean in Indonesia [8], to satisfy the local need. Regular domestic consumption of soybean was more so the, need were not fulfilled. The research was conducted in Karawang Regency, West Java, Indonesia to identify suitable area for soyabean plantations in paddy fields and prepared plan for it. The suitability classes defined according to FAO categories from suitable (S2) to not suitable (N). For wheat crop suitability analysis conducted in North Carolina [9]. The case study under taken was rain-fed wheat. Five criterions considered were soil-fertility, climate, soil-features, soil-organic-matter, soil-quality, soil-chemistry and seventeen sub-criteria under that. This system also considers geographic information systems (GIS) as base and the square root method is used, called multi-criteria analysis. Percentage of land suitability for organic wheat was highly suitable- 18.6% and moderately suitable- 76.8% in Duplin country. Existing yield simulation method was also based on Moderate Resolution Imaging Spectroradiometer [10]. A case study for corn and soybean yield simulated within a certain scope of area, predictions are given by the United States Department of Agriculture (USAD)- National Agricultural Statistics Service (NASS). All above discussed systems are suitable for unfragmented land not for fragmented lands I India.

Decision making system available in India

As discussed above still the decision making in India mainly for small/marginal scale fields is by traditional way. So, farmers are not getting expected yield due to many reasons like decreasing quality of soil, uncertainty in nature. Till the moment automated machines are not used by small/marginal scale Indian farmers and are dependent on labors. There are many reasons which makes them to be dependent on labors as below:

1. Poor economic condition, so unable to purchase machines.

2. In fragmented lands, it is difficult to use the big machineries like tractor and tractor accessories.

3. Unaffordable cost of automated systems.

4. Unaware due to lack of interest.

Though the farming is the main occupation in India, due to urbanization people are moving to urban areas in search of jobs and it causes labor deficiency. So, labor cost is increasing which add on to farming expenses and again it is lowers the economic condition of farmers.

4. Introduction to technology in agriculture decision making

Decision making is in the context of choosing the appropriate crop which give good outcome in available natural conditions like soil, environment. The reason behind it is soil and environment cannot be controlled. So, first we need to study about the technologies available to analyze the data collected through monitoring and recording the soil and environmental features. There are existing prediction techniques we can study to choose the suitable method and we can do the crop suitability analysis based on historical data. Suitability analysis will give the level of suitability for crop/crops and then user can take the decision accordingly. The crop/crops having fair suitability level can be adopted for better yield.

We have divided this advisory system into sub parts as below:

4.1 Collecting data

4.1.1 Identifying the features

The first important task is to identify the features affecting on crop yield. As per the guidelines available at country level [11], state-wise [12], local [13] through variety of sources. We have listed the fallowing recurrent features from variety of sources and suitable ranges of the possible features.

a. Topography

b. Temperature

c. Soil type

d. Soil Quality

e. Rainfall

f. Soil moisture

g. pH

h. EC

i. Soil nutrients: Nitrogen, Phosphorus, Potassium

j. Soil micronutrients

All above features and some features can be added or removed as per requirement and availability of data like humidity, soil texture etc. Some of the features remain constant for the period of a year or more than that ex. Topography of land does not change for years. Some features need to measure for the period of a season or less than that ex. rainfall required at the beginning that is seeding phase of the crop is different than the growing phase of the crop. As usual we are considering two seasons of cropping i.e. Kharif and Rabi. There are three main phases of the crop seeding, growing and maturity. Each phase has different requirement for rainfall, soil-moisture, temperature, nitrogen level, phosphorus level and potassium level. We will discuss about the influence of features on crop growth one by one. To consider these values for analysis purpose we need to normalize the feature values. Some of the feature values are considered in the numeric form and some need to convert into categorical form. Textual values cannot be considered as it is, it needs to map to the numeric categories. For better understanding of the technique let us take an example of the crop wheat.

a. Topography

Topography is nothing but the slope of the crop land. The field having slope less than 10 degree is good to use for cropping. It is a natural feature which can be controlled artificially. With or without analysis we can say this is vital feature contributing towards decision making for crop suitability. Sloppy fields are not good at holding the water at higher ends and may have clayey soil at lower end this uneven nature is not suitable for crop growth. Some of the crops with less water requirement can be adopted provided the slope is less. To use the topography values for algorithm purpose we have categorized it. Plane surface is always most suitable for every crop so, the categorization is done as 1-Plane, 2- slope less than 10 degree and 3-slope more than 10 degree.

b. Temperature

It is again a natural parameter. Geographical location of India is such that enough solar energy is available. This feature values need to consider phase wise. Suitable ranges of temperature is discussed in **Table 1**. If temperature goes below or above the range crop yield get reduced. The values of temperature can be considered as it is in the numeric form with unit °C (degree Celsius).

c. Soil type

Type of the soil plays a key role in crop yield. Water holding capacity is dependent on the soil type. For some of the crops clayey soil is good and for

Sr. No.	Temperature	Soil quality	Suitability
1	24°C	Good	Level1
2	22°C	Good	Level1
3	20 °C	Good	Level1

Table 1.
Sample tuple in dataset.

some loamy. Few crops can be grown in sandy soil as well. Soil type is textual value, cannot be considered as it is. Here we are categorizing it according to water holding capacity. First category is 1- Loamy, 2- Clayey, 3- Slity, 4- Sandy.

d.Soil quality

Quality of the soil is dependent on multiple features like soil nutrients, micronutrients, texture, water holding capacity and it may varies according to the chemical used, erosion occurred etc. Krishi Vigyan Kendra helps to know the quality of soil in terms of levels. With reference to that we are considering the soil quality levels as 1- Good, 2- Moderate, 3- Marginal 4- Low.

e. Rainfall

Rainfall is the natural feature on which other features dependent. Water requirement for different crop is different. Precipitation has some annual pattern, sometimes it may vary. Here we can directly consider the numeric values or range of it for analysis purpose. This feature needs to measure according to phases of the crop.

f. Soil moisture

Water holding capacity of the soil, supplied water decides the soil moisture level. Good quality and type of soils has enough moisture content. Not only rainfall but irrigation sources also contribute to decide the soil moisture values. Here we are taking its numerical values as it is phase wise.

g. pH

It shows acidity or alkalinity of soil and is measured in pH units. Different crops can bear different level of acidity in soil and water. This numerical value is considered in the range of 0 to 14.

h.EC

Soil electrical conductivity is measure of amount of salinity, one of the indicators of soil health. Excess salinity levels occur in arid and semiarid regions. For this feature as well numeric values considered directly. The range of it is from 0.611 to 25.9 dS m − 1.

i. Soil nutrients: Nitrogen, Phosphorus, Potassium

Major soil nutrients are nitrogen, phosphorus, potassium. This feature is artificially manageable. Deficiency can be resolved by adding the fertilizers available in the market but excess amount available can not reduced in any way. If farmer get their soil tested, they come to know the existing level of nutrients and amount of fertilizers to be added. But they do not approach for it and add the fertilizers without knowing the requirement. This is leading to soil degradation. Once we know the suitable crop and existing amount of nutrients, it is easy to recommend the appropriate amount of fertilizers to add and avoid the soil degradation up to certain extent.

j. Soil micronutrients

 There are subcategories of soil micronutrients. Soil cab be tested into laboratory to know the availability of micronutrients, so that we can understand one aspect of soil quality and take future decision.

These are some basic feature we have considered here. We need to keep all the information, while recording these feature values: 1) what was the location? 2) what were the date and time? 3) what was the cropping season? 4) what was the crop/crops adopted? and 5) what was the final yield? There are some other features affecting on yield but either has less effect or values are not available easily. If more factors are considered, results get quite improved.

4.1.2 Collecting data using feature values

There are two different sources from which dataset can be collected. Either we can have sensor-based device which will sense actual field features and record the dataset. According to features we want to monitor; device can be integrated with respective sensors (available in market) and microcontroller to control the data recording [14] and storing process. This method can be referred as monitoring feature values. Another method is to prepare dataset from available online or offline sources called as gathering feature values.

4.1.2.1 Monitoring feature values

Environmental data is available with the meteorological department since long ago. Parameters required for agriculture are with the different scope ex. Rainfall measured by meteorological department is area wise, city wise whereas for agriculture purpose rainfall need to measure at specific location. Soil features also vary one farm to another farm as per the crop adopted and fertilizers used. So the global reports prepared cannot be referred as it is for deciding suitable crop. Crop specific monitoring system can be used as described below to measure the field specific features.

A monitoring system with microcontroller and accessories [14] to sense the nearby features like.

a. Rainfall sensor

b. Moisture sensor

c. pH sensor

d. EC sensor

e. Environmental sensor

f. NPK measuring kit

All the accessories like above can be used to monitor the actual field. There is lot of variety in the accessories available in the market. So according to the device used, scope of monitoring that feature get varies. Depending on the area of field one or more than one system needs to plant in the same field, so that whole farm will get monitored to get more accuracy in the feature value. According to components used cost of this monitoring system get varies. The frequency of feature monitoring can be every second, minute or hour as per the code written, thus it is always editable as per user need. Data can be gathered in the tabular form, so that it can easily converted to SQL database for further analysis. Backup of data can be fetched from device as an when required. As we know microcontroller like raspberry-pi has own memory in the form SD card from which data can be fetched or we can make the provision for data transfer by using some network protocol, as raspberry-pi has default support for Wi-Fi, we just need to do the configuration and coding accordingly. This kind of data backup need only once in a season. It is better to maintain the data along with date and time (time is optional in some cases) and exact place of the data recording.

4.1.2.2 Gathering feature values

Metrological department already has the system to monitor the environmental features accurately. Also, Krishi Vigyan Kendra available at different places providing the values of soil and water features after soil and water testing. The drawback of this method is we do not get the data belongs to same location for which we want to do the suitability analysis. Metrological department has their setup at certain places only not at every place we might use for suitability analysis. KVK reports are also from the different fields, we cannot guarantee that all the soil samples collected are belongs to same location for which we want to do the suitability analysis. Here definitely we can filter out and choose the dataset belongs to same region which will give appropriate prediction. So, we collect the available and authorized data from metrology department, Krishi Vigyan Kendra and the offices working under agriculture universities and government maintaining data related to farming. This data is compiled further to fetch the values of identified features under Section 4.1.1 and crop yield value. Data need to maintain in tabular format so it can easily map to SQL database for further analysis. It is mandatory to maintain the data along with date and time (time is optional in some cases) and exact place of the data recording.

4.1.3 Preparing dataset from feature data

The frequency of monitored features can be changed by averaging the value for required interval. Same data can be customized for different purposes like weather prediction, crop suitability analysis, to understand the pattern of field features etc. Here we are talking about crop suitability analysis. Cropping seasons play important role in the crop specific suitability analysis. So, the interval for data analysis is also a season. Under season every crop goes through the three main phases seeding(s), growing(g), maturity(m). Each phase needs some basic requirement like in the seeding phase favorable temperature for wheat is 20°C -25°C, in the growing phase it is 15–30°C and in maturity phase it is 14°C -15°C [12] at state Punjab in India whereas the it can tolerate the temperature in the range of 3.5–35°C. It means if the crop gets that feature values in expected range in all the phases it will lead to better yield (i.e. high suitability level S1 or S2) otherwise yield may get reduced than expected (i.e. low suitability level S3 to N2). As per the crop season and the phase, interval of data collected is decided. Some of the features need to consider phase wise (ex. Rainfall) and for some features single value need to consider (ex. nitrogen, phosphorus, potassium (NPK). Along with the above identified features under Section 4.1.1 we also need to get to know the season, yield and crop/crops adopted for the fields considered through data gathering.

4.2 Analyze data for decision making

Collected data is further analyzed to understand the crop specific suitability for variety of crop for particular season, the one which is more suitable is advised to adopt. Here the logic is- based on historical data collected. From environmental data we understand the pattern of environmental features for current season and current phase of crop and predict that aggregate/common values in the respective season. Some of the features like NPK, pH are available as the current value of that feature, so it can be taken as it is not any prediction required here. Standard feature values/ranges required for crop is already well known to the farmers who are in farming business itself. For new farmers, the guidelines are available for cropping, feature values suitable are mentioned crop wise and phase wise under the guidelines available [12, 13]. Guidelines are available in the form of books, reports, online resources, also offline centers like KVK are always available to guide. The information provided by such resources is static information.

Here we are discussing about analyzing the suitability level of crop based on current situation dynamically. For such decision making, historical data with few current feature values can make available as discussed under Section 4.1. The suitable methodologies need to choose and customize for agriculture application. Suitability results need to compare with the existing results available, that is called as testing in machine learning. Variety of methods can be applied and then the methodology which gives more accuracy can be used further. Once the data is available, the methods based on data analysis are more suitable. Extracting the required information from available dataset is called as Data mining.

Data analysis means, identifying something based on current and historical data. Broadly it is categorized into two categories qualitative and quantitative analysis. If you want to get the answer for why, what, or how we need to go for qualitative analysis. If you want to know some statistical or categorical value, then go for quantitative analysis. Here we are discussing about suitability level analysis. According to Food and Agriculture Organization of the United Nations, suitability level is a categorical value [15], so quantitative approach has been used. Quantitative approaches further categorized as text analysis, statistical analysis, diagnostic analysis, predictive analysis, and prescriptive analysis. Here we want to predict the suitability based on historical and some current values, so need predictive analysis. If we have the historical database of the farm features along with the yield, yield of the crop/crops can decide respective suitability level of the crop for storing into dataset.

Ex. For crop C in particular farm area for season S, expected highest yield is Y_H Tones/Hector and lowest expected yield is Y_L Tones/Hector (it can be zero Tone in worst condition).

Then, interval between two adjacent levels Y_i calculated as,

$$Y_i = Y_H - Y_L / 4 \tag{1}$$

Five ranges of yield from higher to lower, are computed as,

$$\text{Level1 is} >= Y_H \tag{2}$$

$$\text{Level2 is} \langle Y_H \text{ and} \rangle = Y_H - Y_i \tag{3}$$

$$\text{Level3 is}\langle\text{Level2 and}\rangle = \text{Level2} - Y_i \tag{4}$$

$$\text{Level4 is}\langle\text{Level3 and}\rangle = \text{Level3} - Y_i \tag{5}$$

$$\text{Level5 is} < \text{Level4} \tag{6}$$

Example,
Expected high yield Y_H = 20 Tones/Hector and lowest yield Y_L = 3 Tones/Hector.
So,
Interval.

$$Y_i = Y_H - Y_L / 4 = 20 - 3 = 17 / 4 = 4.25$$

$$\text{Level1 is} >= 20\,\text{Tones / Hector}$$

$$\text{Level2 is}\langle 20\,\text{and}\rangle = Y_H - Y_i = 20 - 4.25\,(15.75\,\text{Tones / Hector})$$

$$\text{Level3 is}\langle 15.75\,\text{and}\rangle = 15.75 - 4.25\,(11.5\,\text{Tones / Hector})$$

$$\text{Level4 is}\langle 11.5\,\text{and}\rangle = 11.5 - 4.25\,(7.25\,\text{Tones / Hector})$$

$$\text{Level5 is} < 7.25\,\text{Tones / Hector}$$

There are five suitability levels defined by FAO [15] as below,
S1- suitable.
S2- moderately suitable.
S3- marginally suitable.
N1- not suitable due to physical reasons.
N2- not suitable due to economic reasons.
Now if we know actual yield Y then, suitability level decided as,

$$\text{If Yield falls under Level1 then, suitability level} -> \text{S1} \tag{7}$$

$$\text{If Yield falls under Level2 then, suitability level} -> \text{S2} \tag{8}$$

$$\text{If Yield falls under Level3 then, suitability level} -> \text{S3} \tag{9}$$

$$\text{If Yield falls under Level4 then, suitability level} -> \text{N1} \tag{10}$$

$$\text{If Yield falls under Level5 then, suitability level} -> \text{N2} \tag{11}$$

Thus, we can get the values for suitability level for historical dataset.

5. Machine learning: Methods and applications

Machine learning is the analysis technique which enables computer to learn from experience. Here experience is in the form of dataset. More the accuracy in dataset more accurate the results are. Here accuracy has the different dimensions. It depends on multiple things like 1)Duration of the dataset captured (ex. Data gathered from last 10 years is better than the data captured for last 1–2 years), 2) Features considered (It's better to consider maximum characteristics on which crop yield is dependent), 3)Frequency of the data recording (ex. The values of the features in field varies after certain time, so it's better if we could record every variation in dataset), 4)Duration of the data considered for analysis (ex. To do predictive analysis for the crop in kharif season we need to consider data recorded in kharif season only, the data recorded in rabi season will work as noise). 5)Missing data (if missing data is more in proportion then accuracy in data decreased). We can prepare the dataset from historical data available and the monitoring device. Learning from the historical data and comprehending for current situation is known as supervised machine learning, one of the best predictive analysis techniques. Supervised machine learning is further categorized into regression, classification, naive Bayesian model, random forest model, neural networks, support vector machines.

Classification is the method where we divide dataset into defined classes. Class is nothing but category of the instance. Ex. In a school student are categorized into different classes. The class is decided based on features of the student like age, result of the previous year, date of birth etc. Based on marks of the students, further they can be categorized into pass class and fail class.

6. Role of machine learning in agriculture decision making

Some of methods are already used in agriculture, those are decision making methods based on some mathematical computations. More or less those are based on basic features of machine learning. Few examples we will discuss here. A land evaluation is done based on features belong to climate and site-soil [16]. It is developed using statistics and neural network model techniques. This is bit static method not user friendly, so not used widely. Decision support system [6] is based on soil features like topography, nutrients, history of cultivation, precipitation etc. Static categorization of the feature values is done to evaluate suitability level. The drawback of this system is also a static nature. Konstantinos G Liakos has discussed in detail how machine learning played role in agriculture precision through crop management, yield prediction, disease detection, soil and water management [17]. Patricio and Rieder [18] mentioned that artificial intelligence plays important role in improving accuracy in agriculture. During 2013–2017 data captured by camera was analyzed using support vector machine classifier. Uddin, Mohammad Shorif has discussed contribution of machine learning and computer vision in agriculture [19]. Machine learning helps in decision making, for better productivity and more precise systems. In developed countries machine learning is introduced in agriculture too early for different purpose like farming prediction is done using classification [20, 21], Artificial Neural Network technique used for crop yield prediction [22]. Even for study regarding plant disease statistical and machine learning approaches has been used [23–27].

Decision making about suitable crop

As we have discussed under Section 4, we can get the feature values for a particular crop in a particular season as per the list of features identified under Section 4.1.1 and also, we can compute the level of suitability of the same crop using the methodology discussed under Section 4.2. Here, crops specific suitability is considered as output class. Suitability level is further divided into five classes [15], so classification method of machine learning is chosen. According to the discussion under Section 5, we can say that supervised technique classification is suitable for crop specific suitability analysis. We can treat the computed suitability levels based on yield as class-value for that particular record/tuple and all other feature-values as input-feature-values of the same tuple. Any classification technique, which can classify the records into more than one classes based on input features called as multi-class classification technique. Any multi-class classification technique can be used and further customized [28] to get the appropriate suitability analysis. Similar to output classes computed using Eqs. (7)–(11) input classes can be computed as below.

If input feature is categorical value then no need to compute the levels, it can be directly mapped. Example soil quality is one of the features need to consider having three different having categorical values good, moderate and average. Then it will be mapped to input levels as below:

Level1	Soil_quality = Good
Level2	Soil_quality = Moderate
Level3	Soil_quality = Average

Let us consider the input X_i is environment feature temperature at Punjab state in India. Lowest temperature is considered as 0°C and highest temperature observed is 50°C. We know in the growing phase of the crop wheat for high growth rate the favorable temperature is 20–25°C. Wheat cannot tolerate the temperature below 3.5°C so, Level5 is less than or equal to 3.5°C. It cannot tolerate the temperature above 35°C so, Level1 is above 35°C.

Thus,
Highest value is 35°C
Lowest value is 3.5°C
So, as per Eq. (1) interval is calculated as,
$X_i = (35–3.5)/4 = 7.75$.

X_1 is	35
X_2 is	$X_1 - X_i = 35–7.75 = 19.5$
X_3 is	$X_2 - X_i = 27–8 = 11.75$
X_4 is	$X_3 - X_i = 21–8 = 4$

To convert into input class levels, the dynamic levels will be computed. If the available dataset has tuples as shown in **Table 1**.

To simplify we will round up the values. The levels will be converted as below.

Level1 > =35°C.
Level2 < 35°C and > = 20°C.
Level3 < 20°C and > =12°C.
Level4 < 12°C and > =4°C.
Level5 < 4°C.

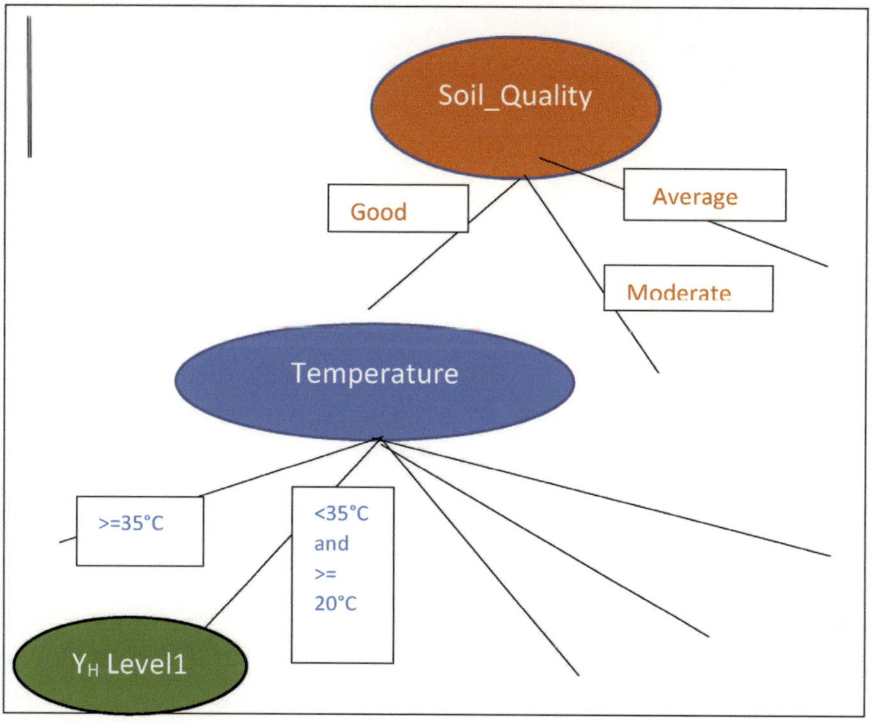

Figure 1.
Decision making tree.

Now, as per tuples in database for suitability level 1 of output the mapping input level in Level2, thus for decision tree classification intermediate result based on above data belongs **Table 1**, partial decision tree will be as shown in **Figure 1**.

Based on simple dataset available as per **Table 1**, simple decision tree has been formed. If more features will be considered levels of the tree will be increased. Always the last level of the tree will output class i.e. suitability/yield level. For n number of input features tree will have n + 1 level [29].

7. Conclusion

We can conclude that machine learning can be applied in agriculture decision making. More the balanced and appropriate dataset is available better the decision can be taken. Here the machine learning approach we used, called as decision tree classification. So, we can say quality of decision tree formation is dependent on quality of input dataset. Advanced decision tree approaches work on variety of data values like categorical, constant and discrete (numerical as well as text values with some preprocessing) values. We can consider all these variety of agriculture features for processing and decision making.

Author details

Ratnmala Bhimanpallewar[1*] and M.R. Narasinga Rao[2]

1 Vishwakarma Institute of Information Technology, Pune, India

2 K.L. University, Vijayawada, India

*Address all correspondence to: ratnmalab@gmail.com

References

[1] Food and Agriculture Organization of the United Nations (FAO). India Country Programming Framework. 2016; Available from: http://www.fao.org/3/a-bp575e.pdf

[2] FAO_Chapter 6 - Causes of land degradation [Internet]. United Nation: FAO; Available from: http://www.fao.org/docrep/v4360e/V4360E00.htm#Contents

[3] Timespoints TOI. Kolhapur Cancer news dated Aug 30 2018. Times of India [Internet]. 2018;1-18. Available from: http://timesofindia.indiatimes.com/articleshow/66977019.cms?utm_source=contentofinterest&utm_medium=text&utm_campaign=cppst

[4] Ameen A, Raza S. International Journal of Advances in Scientific Research Green Revolution: A Review QR Code *Correspondence Info. Int J Adv Sci Res [Internet]. 2017;3(12):129-37. Available from: https://doi.org/10.7439/ijasr.v3i12.4410

[5] Flachs A. Encyclopedia of Food and Agricultural Ethics. Encycl Food Agric Ethics. 2016;(November).

[6] Elsheikh R, Mohamed Shariff ARB, Amiri F, Ahmad NB, Balasundram SK, Soom MAM. Agriculture Land Suitability Evaluator (ALSE): A decision and planning support tool for tropical and subtropical crops. Comput Electron Agric [Internet]. 2013;93:98-110. Available from: http://dx.doi.org/10.1016/j.compag.2013.02.003

[7] Widiatmaka, Ambarwulan W, Setiawan Y, Walter C. Assessing the suitability and availability of land for agriculture in tuban regency, East Java, Indonesia. Appl Environ Soil Sci. 2016;2016.

[8] Widiatmaka W, Ambarwulan W, Firmansyah I, Munibah K, Santoso PBK, Sudarsono S. Land suitability and dynamic system modelling to define priority areas of soybean plantation in paddy fields in Karawang, West Java. Agrivita. 2014;36(3):235-48.

[9] Karimi F, Sultana S, Shirzadi Babakan A, Royall D. Land Suitability Evaluation for Organic Agriculture of Wheat Using GIS and Multicriteria Analysis. Pap Appl Geogr [Internet]. 2018 Jul 3;4(3):326-42. Available from: https://doi.org/10.1080/23754931.2018.1448715

[10] Doraiswamy PC, Hatfield JL, Jackson TJ, Akhmedov B, Prueger J, Stern a. Crop condition and yield simulations using Landsat and MODIS. Remote Sens Environ. 2004;92:548-59.

[11] Chandrasekaran B, Annadurai K, Somasundaram E. A Text book of Agranomy. 2010. 1-856 p.

[12] Services IA-I and. No Title [Internet]. Nagarjuna Education Trust, Hyderabad, India. 2020. Available from: http://www.ikisan.com

[13] Agriculture University. Krishi Darshani 2008. Pune; 2008.

[14] Ratnmala Bhimanpallewar, M. R. Narasinagrao, A prototype model for c ontinuous agriculture field monitoring and assessment, International Journal of Engineering & Technology, 2018; 7 (2.7): 179-182.

[15] FAO. Chapter 3 : Land suitability classifications. A Framew L Eval Soils Bull 32 [Internet]. 1976;1-9. Available from: http://www.fao.org/3/X5310E/x5310e00.htm

[16] D. De La Rosa, F. Mayol, E. Diaz-Pereira, *M. Fernandez*, and D. De La Rosa, A land evaluation decision support system (MicroLEIS DSS) for agricultural soil protection: With special

reference to the Mediterranean region, *Environ. Model. Softw.*, 2004; 19(10):929-942.

[17] Konstantinos G. Liakos, Machine Learning in Agriculture: A Review, Sensors, 2018; 18(8):2674:1-29.

[18] Jha K, Doshi A, Patel P, Shah M, Artificial Intelligence in Agriculture, 2019, 2: 1-12.

[19] Uddin, Mohammad Shorif, Computer Vision and Machine Learning in Agriculture, ISBN 978-981-336-424-0.

[20] Washington Okori, Joseph Obua,Machine Learning Classification Technique for Famine Prediction, Proceedings of the World Congress on Engineering, WCE 2011, London, U.K, 2011 Jul; II

[21] S.R.Rajeswari , Parth Khunteta, Subham Kumar,Amrit Raj Singh,Vaibhav Pandey, Smart Farming Prediction Using Machine Learning, International Journal of Innovative Technology and Exploring Engineering ISSN: 2278-3075, 2019 May; 8(7): 190-194

[22] Miss.Snehal, S.Dahikar, Dr.Sandeep V.Rode, "Agricultural Crop Yield Prediction Using Artificial Neural Network Approach". International Journal of Innovative Research in Electrical, Electronic, Instrumentation and Control Engineering, Vol. 2, Issue 1, January 2014

[23] Gutierrez D. D., Machine Learning and Data Science: An Introduction to Statistical Learning Methods with R. Basking Ridge, NJ: Technics Publications, 2015

[24] Mehra, Lucky K. et al., Predicting Pre-Planting Risk of Stagonospora Nodorum Blotch in Winter Wheat Using Machine Learning Models, Frontiers in Plant Science, 2016 Apr; PMC. Web. 16, 7 (2016): 390

[25] Kim KS, Beresford RM, Development of a disease risk prediction model for downy mildew (Peronospora sparsa) in boysenberry., Walter M Phytopathology. 2014 Jan; 104(1):50-6.

[26] Rumpf, T., A.-K. Mahlein, U. Steiner, E.-C. Oerke, H.-W. Dehne, and L. Plümer., Early Detection and Classification of Plant Diseases with Support Vector Machines Based on Hyperspectral Reflectance, Computers and Electronics in Agriculture 74.1 (2010): 91-99. Web.

[27] M.P.Raj, P,R,Swaminarayan, J.R.Saini, D.K.Parmar, Application Of Pattern Recognition Algorithm In Agriculture : A Review, Int.J.Advanced Networking and Application, 2015; 6(5).

[28] Ratnmala Bhimanpallewar, M. R. Narasinagrao, A Machine Learning Approach to Assess Crop Specific Suitability for Small/Marginal Scale Croplands, International Journal of Applied Engineering Research, 2017 Nov;12(23):13966-73.

[29] Ratnmala Bhimanpallewar, M. R. Narasinagrao, Evaluating the Influence of Soil and Environmental Parameters in Terms of Crop Suitability using Machine Learning, Indian Journal of Agricultural Research, 2021; DOI:10.18805/IJARe.A-4942

Chapter 23

Tomato Leaf Diseases Detection Using Deep Learning Technique

Muhammad E.H. Chowdhury, Tawsifur Rahman, Amith Khandakar, Nabil Ibtehaz, Aftab Ullah Khan, Muhammad Salman Khan, Nasser Al-Emadi, Mamun Bin Ibne Reaz, Mohammad Tariqul Islam and Sawal Hamid Md. Ali

Abstract

Plants are a major source of food for the world population. Plant diseases contribute to production loss, which can be tackled with continuous monitoring. Manual plant disease monitoring is both laborious and error-prone. Early detection of plant diseases using computer vision and artificial intelligence (AI) can help to reduce the adverse effects of diseases and also helps to overcome the shortcomings of continuous human monitoring. In this study, we have extensively studied the performance of the different state-of-the-art convolutional neural networks (CNNs) classification network architectures i.e. ResNet18, MobileNet, DenseNet201, and InceptionV3 on 18,162 plain tomato leaf images to classify tomato diseases. The comparative performance of the models for the binary classification (healthy and unhealthy leaves), six-class classification (healthy and various groups of diseased leaves), and ten-class classification (healthy and various types of unhealthy leaves) are also reported. InceptionV3 showed superior performance for the binary classification using plain leaf images with an accuracy of 99.2%. DenseNet201 also outperform for six-class classification with an accuracy of 97.99%. Finally, DenseNet201 achieved an accuracy of 98.05% for ten-class classification. It can be concluded that deep architectures performed better at classifying the diseases for the three experiments. The performance of each of the experimental studies reported in this work outperforms the existing literature.

Keywords: Smart agriculture, automatic plant disease detection, deep learning, CNN, classification

1. Introduction

Thousands of years ago, the development of agriculture led to the domestication of main food crops and animals today. One of the major global problems that humanity faces today is food insecurity [1] of which plant diseases are a major cause [2]. According to one estimate, plant diseases collectively account for a crop yield loss of around 16% globally [3]. The global potential loss from pests is estimated to be around 50% for wheat and 26–29% for soybean [3]. Plant pathogens are classified into major groups of fungi, fungus-like organisms, bacteria, virus, viroid, virus-like organism, nematodes, protozoa, algae, and parasitic plants. Artificial intelligence (AI), machine learning (ML), and computer vision have provided significant help in numerous applications including power prediction from renewable resources [4, 5] and biomedical applications [6, 7]. The application of AI has seen a great boost during the COVID-19 pandemic period for the detection of lung-related diseases [8–11] and other prognostic applications [12]. Similar advanced technology can be used to mitigate the adverse effects of plant diseases by their early-stage detection and diagnosis. Recently, the application of AI and computer vision to automatic detection and diagnosis of plant diseases is being extensively studied because manual plant disease monitoring is tedious, time-consuming, and labor-intensive. Sidharth et al. [13] applied a Bacterial Foraging Optimization-based Radial Basis Function Network (BRBFNN) to automatically identify and classify plant disease achieving the classification accuracy of 83.07%. Convolutional neural network (CNN) is a very popular neural network architecture that is used successfully for a variety of computer vision tasks in diverse fields [14]. CNN architecture and its different variants have been utilized by researchers for the classification and detection of plant diseases. Sunayana et al. [15] compared different CNN architectures for disease detection in potato and mango leaves achieving an accuracy of 98.33% for AlexNet and 90.85% for a shallow CNN model. Guan et al. [15, 16] used a pre-trained VGG16 model to estimate the disease severity in apple plants and achieved an accuracy of 90.40%. Jihen et al. [17] used LeNet [18] model to classify healthy and diseased banana leaves and achieved an accuracy of 99.72%.

Tomato is a major food crop across the globe with a per capita consumption of 20 kilograms per year and represents about 15% of average total vegetable consumption. North America is consuming 42 kilograms of tomatoes per capita per year while Europe is consuming 31 kilograms of tomatoes per capita per year [19, 20]. To meet the global demand for tomatoes, it is imperative to devise techniques for improving crop yield and early detection of pests, bacterial, and viral infections. Several works have been done in employing artificial intelligence-based techniques to improve tomato plants' survival by early detection of diseases and subsequent disease management. Manpreet et al. [21] used a pre-trained CNN-based architecture known as Residual Network or commonly called ResNet to classify seven tomato diseases with an accuracy of 98.8%. Rahman et al. [22] proposed a deep learning-based fully-connected network to classify Bacterial Spot, Late Blight, and Septorial Spot disease from tomato leaf images and achieved an accuracy of 99.25%. Fuentes et al. [23] to classify ten diseases from tomato leaves images considered three main families of detectors: Faster Region-based Convolutional Neural Network (Faster R-CNN), Region-based Fully Convolutional Network (R-FCN), and Single Shot Multibox Detector (SSD). These detectors were combined with different variants of deep feature extractors VGG16, ResNet50, and ResNet152 for Faster R-CNN, ResNet-50 for SSD, and ResNet-50 for R-FCN for real-time disease and pests' recognition and achieved the highest Average Precision of 83% with VGG16 on top of FRCNN. Agarwal et al. [24] proposed a Tomato Leaf Disease Detection (ToLeD) model, a CNN-based architecture for the classification of ten diseases from tomato leaves images achieving an accuracy of 91.2%. Durmuş et al. [25] evaluated AlexNet and SqueezeNet architectures for the classification of ten diseases from tomato leaves images and achieved an accuracy of 95.5%. Although the disease classification and detection in plant leaves are well-studied in tomatoes and other plants, the reliability of leaf images with varying back-ground on large image classes are not well-studied, since the real-world images can vary greatly in terms of lighting conditions, image quality, orientation, etc. [18].

This chapter has the following main contributions: (i) Investigation of the classification tasks for different variants of CNN architecture for binary and different multi-class classifications of tomato diseases. Several experiments employing different CNN architectures were conducted on raw images. Three different types of classifications were done in this work: (a) Binary classification of healthy and diseased leaves, (b) Six-class classification of healthy and four diseased leaves, and finally, (c) Ten-class classification with healthy and 9 different diseases classes. (ii) The performance achieved in this work outperforms the existing state-of-the-art works in this domain.

The rest of the chapter is organized in the following manner: Section 1 gives a brief introduction, literature review, and motivation for the study. Section 2 describes the different types of plant pathogens. Section 3 provides the methodology of the study with technical details such as the dataset description, pre-processing techniques, and details of the experiments. Section 4 reports the results of the studies followed by discussions in Section 5 and finally, the conclusion is provided in Section 6.

2. Pathogens of tomato leaves

Fungi is the predominant plant pathogens and it can cause multiple diseases including early blight, septoria leaf spot, target spot, and leaf mold. Fungi can attack plants through different sources such as infected soil and seeds. Fungal infections can spread from one plant to another through animals, humans, machinery, and soil contamination. The fungal attack vectors include plant pruning wounds, insects, leaf stomata, and others. The early blight disease of tomato plants is caused by the fungus, which affects the plant leaves. If it affects the seedlings' basal stems, adult plant's stem, and fruits, it is called collar rot, stem lesion, and fruit rot, respectively [26, 27]. Numerous methods have been devised for the control of early blight but the most effective methods are cultural control i.e. efficient soil, nutrients, and crop management to reduce infections and also with the use of fungicidal chemicals. Septoria leaf spot of tomato plants is caused by fungus [28, 29], which releases tomatinase enzyme that speeds up the degradation of tomato steroidal glycoalkaloids α-tomatine [30, 31]. The target spot disease of tomato plants is caused by the fungus [32, 33]. Symptoms of target spot disease in tomato plants are necrotic lesions of light brown color in the center [34, 35]. The lesions spread to a larger blighted leaf area and result in early defoliation [34, 35]. The target spot also damages the fruit directly by entering into the fruit pulp [34, 35]. The leaf mold disease of plants is caused by the fungus [36, 37]. It occurs during periods of extended leaf wetness. Bacteria is also a major plant pathogen. Bacteria enter plants through wounds such as insect bites, pruning, cuts, and also through natural openings such as stomata. Plant's surrounding environmental conditions such as temperature, humidity, soil conditions, availability of nutrients, weather conditions, and airflow are important factors in determining the bacterial growth on the plant and the consequent damage. Bacterial spot is a plant disease caused by bacteria [38, 39]. Molds are also a major cause of plant diseases. Late blight disease of tomato and potato plants is caused by mold [40, 41]. The appearance of dark uneven blemishes on leaves tips and plant stems are a few of the symptoms. Tomato yellow leaf curl virus (TYLCV) is a devastating virus causing tomato disease. This virus attacks the plant through another insect. Although tomato plants are unhealthy diseased leaves and iii) ten-class classification of healthy and various diseased leaves. In study II, different types of tomato leaf diseases are classified into the group of diseases while in study III, different classes of unhealthy and healthy leaf images were classified. Similar experiments the primary hosts for the virus, this viral infection has been reported in several other plants including beans and pepper, tobacco, potatoes, and eggplants [42, 43]. In the last few decades, due to the rapid spread of the disease, the research focus has been shifted to damage control of yellow leaf curl disease [44–47]. Another viral disease that specifically affects tomato plants is caused by Tomato mosaic virus (ToMV). This virus is found worldwide and affects not only tomatoes but other plants as well. Symptoms of ToMV infection include twisting and fern-like appearance of leaves, damaged fruit with yellow patches, and necrotic blemishes [48, 49].

3. Methodology

The overall methodology of the study of the paper is summarized in **Figure 1**. This study used tomato leaf data from the plant village dataset [50, 51], where tomato leaf images are provided. As explained earlier, the paper has three different studies: (i) binary classification of healthy and unhealthy leaves; (ii) six-class classification of healthy and different disease group leaves were conducted; and (iii) ten class of healthy and several different diseased leaves were carried out. The classification is done using pre-trained networks- ResNet18, MobilenetV2, InceptionV3, and DenseNet201 that have been comparatively successful in previous publications [8, 10, 11, 52–57].

3.1 Datasets description

In this study, plant village tomato leaf images dataset was used [50, 51], where 18,162 tomato leaf images are available. All images were divided into 10 different classes, where one class is healthy and the other nine classes are unhealthy (such as- bacterial spot, early blight, leaf mold, septoria leaf spot, target spot, two-spotted spider mite, late bright mold, mosaic virus, and yellow leaf curl virus), and 9 unhealthy classes are categorized into five subgroups (namely-bacterial, viral, fungal, mold and mite disease). Some sample tomato leaf images, for healthy and different unhealthy classes from plant village dataset are shown in **Figure 2**. Moreover, a detailed description of the number of images in the dataset is also shown in **Table 1**, which is useful for classification tasks discussed in detail in the next section.

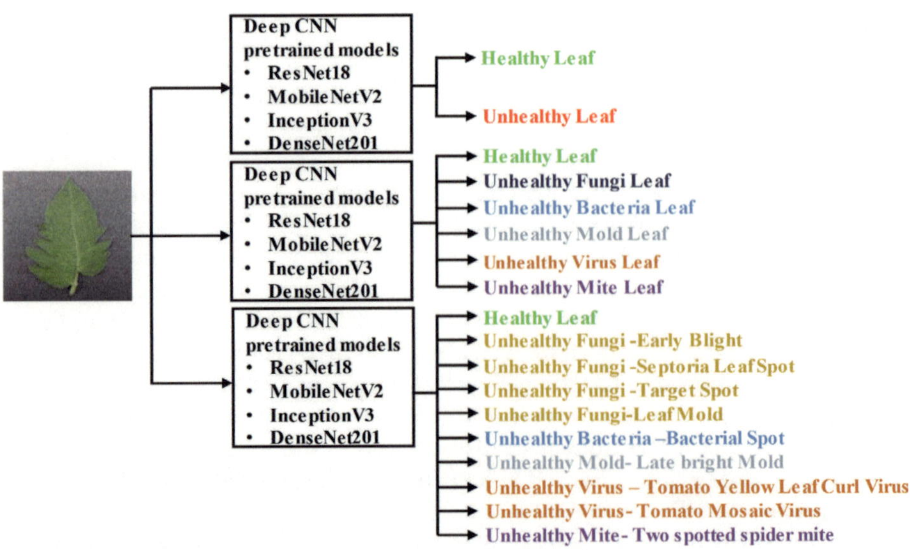

Figure 1.
Overall Methodology of the study.

Figure 2.
Sample images of healthy and different unhealthy tomato leaves from the plant village database [3].

Class	Unhealthy					Healthy
	Fungi	**Bacteria**	**Mold**	**Virus**	**Mite**	
Sub Class	Early blight (1000)	Bacterial spot (2127)	Late bright mold (1910)	Tomato Yellow Leaf Curl Virus (5357)	Two spotted spider mite (1676)	Healthy (1591)
	Septoria leaf spot (1771)			Tomato Mosaic Virus (373)		
	Target spot (1404)					
	Leaf mold(952)					
Total Tomato Leaf Images (18,162)						

Table 1.
The number of tomato leaf images for healthy and different unhealthy classes.

3.2 Preprocessing

3.2.1 Resizing and normalizing

The various CNN network has input image size requirements. Thus, the images were resized to 299 × 299 for Inceptionv3 and 224 × 224 for Resnet18, MobilenetV2, and DenseNet201. Using the mean and standard deviation of the images of the dataset, z-score normalization was used to normalize the images.

3.2.2 Augmentation

Training with an imbalanced dataset will result in a biased model because the dataset is not balanced and does not contain a comparable number of images for the various categories. As a result, data augmentation can aid in the creation of a similar number of images in each class, resulting in reliable results, as reported in numerous recent publications [6–11]. To align the training images, three augmentation techniques (rotation, scaling, and translation) were used. The images were rotated in a clockwise and counterclockwise direction with an angle of 5 to 15 degrees for image augmentation. The scaling process involves enlarging or shrinking the image's frame size, and 2.5 percent to 10% image magnifications were used in this analysis. Image translation was accomplished by converting images by 5–20% horizontally and vertically.

3.3 Experiments

Four pre-trained CNN models were investigated that were originally trained on ImageNet Database [58] to classify tomato leaf images. Three different classification experiments were carried out in this study. **Tables 2–4** summarize the details of the images in the experiments for three different classification of leaf images separately. Two of the four pre-trained networks are shallow (MobilenetV2, and ResNet18), while the other two are deep (Inceptionv3, and DenseNet201) to see whether shallow and deep networks are appropriate for this application. **Table 5** presents a summary of the parameters (Batch size (BS), Learning rate (LR), Epochs (E), Epochs patience (EP), Loss function (LF), Optimizer (OP)) for classification in experiments.

All of the studies were conducted on an Intel Xeon Processor E5–2697 v4, 2.3 GHz with sixty-four GB RAM and a sixteen GB NVIDIA GeForce GTX 1080 GPU using the PyTorch library and Python 3.7.

Database	Types	Total No. of images/class	For Both Segmented and Unsegmented experiment		
			Train set count/fold	Validation set count/fold	Test set count/fold
Plant village dataset	Healthy	1591	1147*10 = 11470	127	317
	Unhealthy (9 diseases)	16570	11930	1326	3314

Table 2.
Summary of the binary classification experiment.

Database	Types	Count of images/class	For Both Segmented and Unsegmented experiment		
			Train set count/fold	Validation set count/fold	Test set count/fold
Plant village dataset	Healthy	1591	1147*3 = 3441	127	317
	Fungi	5127	3692	410	1025
	Bacteria	2127	1532*2 = 3064	170	425
	Mold	1910	1375*3 = 4125	153	382
	Virus	5730	4126	458	1146
	Mite	1676	1207*3 = 3621	134	335

Table 3.
Summary of the six-class classification experiment.

3.4 Performance matrix

Important performance metrics for classification experiment is stated in Eqs. (1)–(5):

$$Accuracy = \frac{TP + TN}{(TP + FN) + (FP + TN)} \tag{1}$$

$$Sensitivity = \frac{(TP)}{(TP + FN)} \tag{2}$$

$$Specificity = \frac{(TN)}{(TN + FP)} \tag{3}$$

$$Precision = \frac{(TP)}{(TP + FP)} \tag{4}$$

Database	Types	Count of images/class	For Both Segmented and Unsegmented experiment		
			Train set count/ fold	Validation set count/ fold	Test set count/ fold
Plant village dataset	Healthy	1591	1147*3 = 3441	127	317
	Early blight	1000	720*5 = 3600	80	200
	Septoria leaf spot	1771	1275*3 = 3825	142	354
	Target spot	1404	1011*3 = 3033	112	281
	Leaf mold	952	686*5 = 3430	76	190
	Bacterial spot	2127	1532*2 = 3064	170	425
	Late bright mold	1910	1375*3 = 4125	153	382
	Tomato Yellow Leaf Curl Virus	5357	3857	429	1071
	Tomato Mosaic Virus	373	268*13 = 3484	30	75

Table 4.
Summary of the ten-class classification problem.

Parameters for classification model	
BS	**16**
LR	0.001
E	15
EP	6
SC	5
LF	BCE
OP	ADAM

Table 5.
Summary of parameters for classification experiments.

$$F1\_score = \frac{(2 * TP)}{(2 * TP + FN + FP)} \tag{5}$$

Here, true positive (TP) is the number of correctly classified healthy leaf images and true negative (TN) is the number of correctly classified unhealthy leaf images. False-positive (FP) and false-negative (FN) are the misclassified healthy and unhealthy leaf images, respectively.

4. Results

The performance of various networks in the different experiments is reported in this section.

In this study, three different experiments were conducted for tomato leaf images and the comparative performance for four different CNNs for the three classification schemes is shown in **Table 6**. It is apparent from **Table 6** that all the evaluated pre-trained models perform very well in classifying healthy and unhealthy tomato leaf images in two-class, six-class, and ten-class problems.

Among the networks trained with leaf images for two-class, six-class, and ten-class problems, Densenet201 outperformed other trained models except without segmented two-class and with segmented six class problems where InceptionV3 was the best-performing network. Moreover, shallow networks ResNet18, and MobilenetV2 both showed comparable performance to most of the deep networks for the classification of images.

DenseNet201 outperforms others and for six-class and ten-class problems showed accuracy, sensitivity, and specificity of 97.99%, 97.99%, 99.54% and 98.05%, 98.03%, 99.76%, respectively. On the other hand, InceptionV3 produced the best result with accuracy, sensitivity, and specificity of 99.2%, 99.2%, and 96%, respectively for the two-class problem. **Figure 3** clearly shows that the Receiver operating characteristic (ROC) curves for two-class, six-class, and ten-class

Classification	**Model**	**Overall**		**Weighted**		
Binary Classification		Accuracy	Precision	Sensitivity	F1-score	Specificity
	ResNet18	98.4	98.4	98.37	98.37	95.2
	MobileNet	98.42	98.42	98.38	98.33	95.45
	DenseNet201	98.9	98.85	98.66	98.76	95.56
	Inceptionv3	**99.2**	**99.23**	**99.2**	**99.25**	**96**
Six-Class Classification	ResNet18	96.86	96.84	96.84	96.84	99.18
	MobileNet	96.74	96.76	96.74	96.74	99.25
	DenseNet201	**97.99**	**97.99**	**97.99**	**97.98**	**99.54**
	Inceptionv3	97.65	97.67	97.65	97.63	99.41
Ten-Class Classification	ResNet18	96.75	96.77	96.79	96.76	99.65
	MobileNet	97.2	97.18	97.19	97.17	99.7
	DenseNet201	**98.05**	**98.03**	**98.03**	**98.03**	**99.76**
	Inceptionv3	97.35	97.34	97.35	97.34	99.69

Table 6.
Summary of the tomato leaf disease classification performance using original leaf images.

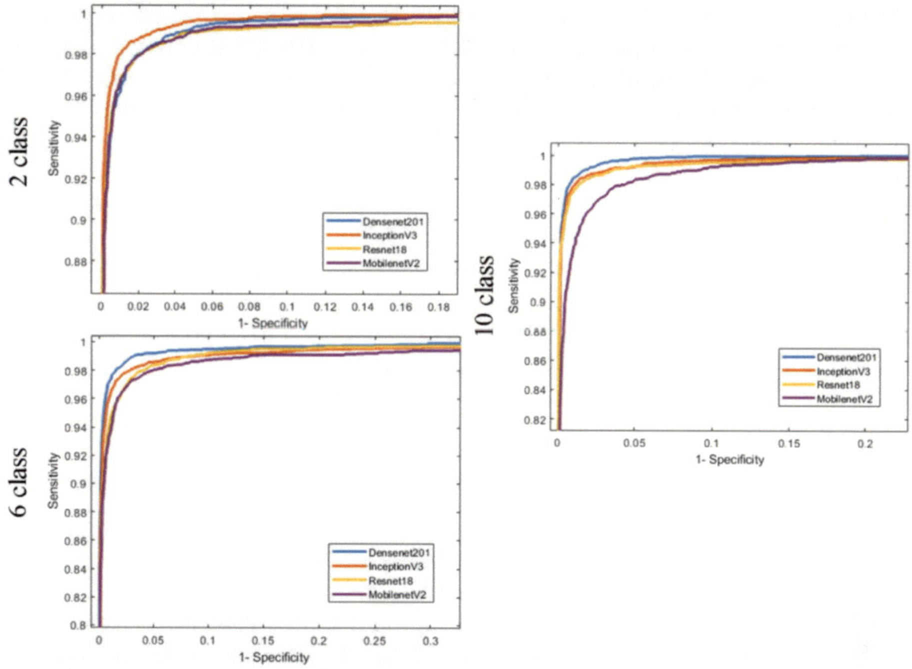

Figure 3.
Comparison of the ROC curves for healthy, and unhealthy tomato leaf image classification using CNN based models for two-class, six-class, and ten-class Classification.

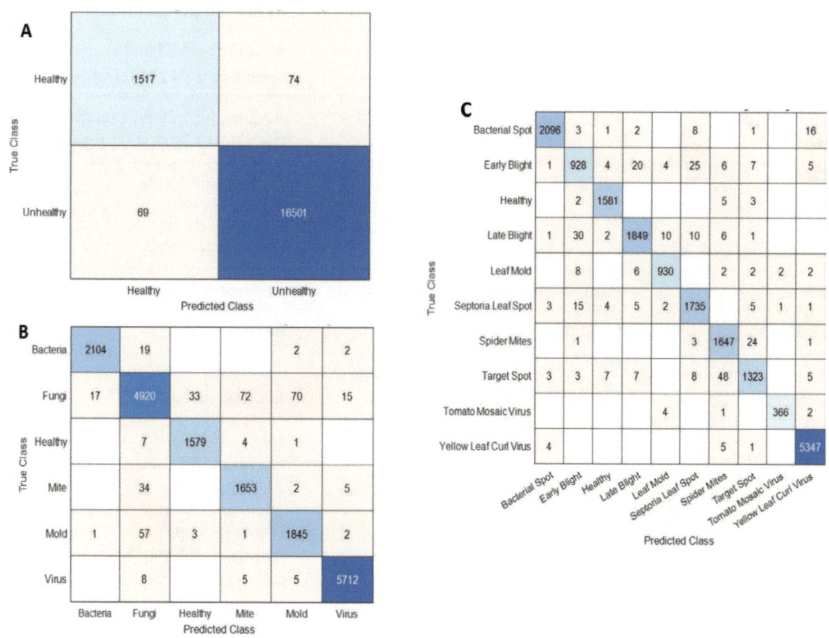

Figure 4.
Confusion Matrix for healthy, and unhealthy tomato leaf image classification using CNN based models for (A) Binary-class, (B) six-class, and (C) ten-class Classification.

problems of tomato leaf images. It is evident from **Figure 3** that network performances are comparable for 2-, 6- and 10-class problems. However, deep networks can provide better performance gain for 6- and 10-class problems.

The confusion matrix for the best performing networks for the different classification problems are shown in **Figure 4**. It can be noticed that even with the best performing network InceptionV3 for two-class tomato leaf images, 69 out of 16,570 unhealthy tomato leaf images were miss-classified as healthy and 74 out of 1,591 healthy tomato leaf images were miss-classified as unhealthy images.

For the six-class problem, which consisted of one healthy class and five different unhealthy classes, only 27 out of 1,591 healthy tomato leaf images were miss-classified as unhealthy images, and 385 out of 16,570 unhealthy tomato leaf images consisted of one healthy class and nine different unhealthy classes, only 32out of 1,591 healthy tomato leaf images were miss-classified as unhealthy images and 382 out of 16,570 unhealthy tomato leaf images of nine different categories were miss-classified as healthy or any other unhealthy classes.

5. Discussion

Plant diseases are a major threat to global food security. Latest technologies need to be applied to the agriculture sector to curb diseases. Artificial intelligence-based technologies are extensively investigated in plant disease detection. Computer vision-based disease detection systems are popular for their robustness, ease of acquiring data, and quick results. This research investigates how different CNN-based architectures perform on classification of tomato leaf images. The study was divided into 3 sub-studies of 2 class classification (Healthy, and Unhealthy), 6 class classification (Healthy, Fungi, Bacteria, Mold, Virus, and Mite), and 10 class

Paper	Database	Reported performance
P. Tm et al. [59] (2018)	Plant village dataset (10 classes)	Accuracy-94.85%, Precision-94.81%, Sensitivity-94.81%, F1 Score – 94.81%
Mohit et al. [24] (2020)	Plant village dataset (10 classes)	Accuracy-91.20%, Precision-90.90%, Sensitivity-92.90%, F1 Score- 91.60%
H. Durmuş et al. [25] (2017)	Plant village dataset (10 classes)	Accuracy-95.50%
Keke et al. [60] (2018)	Plant village dataset (2 classes)	Accuracy-97.20%
Belal et al. [61] (2018)	Plant village dataset (2 classes)	Accuracy-98.00%
Ouhami et al. [62] (2020)	Own dataset (6 classes, 666 images)	Accuracy-95.65%
Fuentes et al. [63] (2018)	Plant village dataset (9 classes)	Accuracy-96.00%
Madhavi et al. [64] (2020)	Own dataset (2 classes, 520 images)	Accuracy-80.00%
Proposed Study	Plant village dataset (2 classes,6 classes, and 10 classes) 18162 images	(Binary Class) Accuracy-99.2%, Precision- 99.23%, Sensitivity-99.2%, F1 Score- 99.25% (Six Class) Accuracy-97.99%, Precision- 97.99%, Sensitivity-97.98%, F-1 Score- 97.54% and (Ten Class) Accuracy-98.05%, Precision- 98.05%, Sensitivity-98.03%, F-1 Score- 98.03%

Table 7.
Results in the paper compared with other state of the art results.

classification (Healthy, Early blight, Septoria leaf spot, Target spot, Leaf mold, Bacterial spot, Late bright mold, Tomato Yellow Leaf Curl Virus, Tomato Mosaic Virus, and Two-spotted spider mite). Overall, the DenseNet201 model outperformed every other model except for binary classification, where the InceptionV3 model outperformed other models. In the binary classification of healthy and diseased tomato leaves, InceptionV3 showed an overall accuracy of 99.2%, while DensNet201 showed an overall accuracy of 99.67%. In 6 class classification, DenseNet201 showed an overall accuracy of 97.99%, while InceptionV3 showed an overall accuracy of 97.65%. In 10 class classification, DenseNet201 showed an overall accuracy of 98.05%, while InceptionV3 showed an overall accuracy of 97.35%. The results in the paper are comparable to the state-of-the-art results and are also summarized in **Table 7**. Although the Plant Village dataset used in this study contains images taken in diverse environmental conditions, the dataset is collected in a specific region and is of specific breeds of tomatoes. A study conducted using a dataset containing images of other breeds of tomato plants from different regions of the world may result in a more robust framework for early disease detection in tomato plants. Furthermore, the lighter architecture of CNN models with non-linearity in the feature extraction layers might be useful to investigate for portable solutions.

6. Conclusion

The stages of the process into the infinite possibilities of machine learning for agriculture applications, complete with case studies. ResNet, MobileNet, DenseNet201, and InceptionV3 are examples of state-of-the-art pre-trained CNN models that do an excellent work of classifying diseases from plant leaf images. When compared to other architectures, the DenseNet201 was found to be better at extracting discriminative features from images. The trained models can be used to detect plant diseases early and automatically. As a result, preventive actions can be adopted faster. This research could help with early and automated disease detection in tomato crops, due to the use of cutting-edge technology like smartphones, drone cameras, and robotic platforms. The proposed structure can be combined with a feedback system that provides appropriate insights, treatments, disease prevention, and control techniques, resulting in improved crop yields.

Acknowledgements

This work was made possible by HSREP02-1230-190019 from the Qatar National Research Fund, a member of Qatar Foundation, Doha, Qatar. The statements made herein are solely the responsibility of the authors.

Conflict of interest

There is no conflict of interest.

Author details

Muhammad E.H. Chowdhury[1*], Tawsifur Rahman[1], Amith Khandakar[1], Nabil Ibtehaz[2], Aftab Ullah Khan[3], Muhammad Salman Khan[3,4], Nasser Al-Emadi[1], Mamun Bin Ibne Reaz[5], Mohammad Tariqul Islam[5] and Sawal Hamid Md. Ali[5]

1 Department of Electrical Engineering, Qatar University, Doha, Qatar

2 Department of Computer Science and Engineering, Bangladesh University of Engineering and Technology, Dhaka, Bangladesh

3 AI in Healthcare, Intelligent Information Processing Laboratory, National Center for Artificial Intelligence, Pakistan

4 Department of Electrical Engineering (JC), University of Engineering and Technology, Peshawar, Pakistan

5 Department of Electrical, Electronic and Systems Engineering, Universiti Kebangsaan Malaysia, Bangi, Selangor, Malaysia

*Address all correspondence to: mchowdhury@qu.edu.qa

References

[1] M. E. Chowdhury, A. Khandakar, S. Ahmed, F. Al-Khuzaei, J. Hamdalla, F. Haque, *et al.*, "Design, construction and testing of iot based automated indoor vertical hydroponics farming test-bed in qatar," *Sensors,* vol. 20, p. 5637, 2020.

[2] R. N. Strange and P. R. Scott, "Plant disease: a threat to global food security," *Annual review of phytopathology,* vol. 43, 2005.

[3] E. Oerke, "Crop losses to pests," *The Journal of Agricultural Science,* vol. 144, p. 31, 2006.

[4] F. Touati, A. Khandakar, M. E. Chowdhury, *S. Antonio* Jr, C. K. Sorino, and K. Benhmed, "Photo-Voltaic (PV) Monitoring System, Performance Analysis and Power Prediction Models in Doha, Qatar," in *Renewable Energy*, ed: IntechOpen, 2020.

[5] A. Khandakar, M. EH Chowdhury, M. Khoda Kazi, K. Benhmed, F. Touati, M. Al-Hitmi, *et al.*, "Machine learning based photovoltaics (PV) power prediction using different environmental parameters of Qatar," *Energies,* vol. 12, p. 2782, 2019.

[6] M. H. Chowdhury, M. N. I. Shuzan, M. E. Chowdhury, Z. B. Mahbub, M. M. Uddin, A. Khandakar, *et al.*, "Estimating Blood Pressure from the Photoplethysmogram Signal and Demographic Features Using Machine Learning Techniques," *Sensors,* vol. 20, p. 3127, 2020.

[7] M. E. Chowdhury, A. Khandakar, K. Alzoubi, S. Mansoor, A. M Tahir, M. B. I. Reaz, *et al.*, "Real-Time Smart-Digital stethoscope system for heart diseases monitoring," *Sensors,* vol. 19, p. 2781, 2019.

[8] T. Rahman, A. Khandakar, M. A. Kadir, K. R. Islam, K. F. Islam, R. Mazhar, *et al.*, "Reliable tuberculosis detection using chest X-ray with deep learning, segmentation and visualization," *IEEE Access,* vol. 8, pp. 191586-191601, 2020.

[9] A. Tahir, Y. Qiblawey, A. Khandakar, T. Rahman, U. Khurshid, F. Musharavati, *et al.*, "Coronavirus: Comparing COVID-19, SARS and MERS in the eyes of AI," *arXiv preprint arXiv:2005.11524,* 2020.

[10] M. E. Chowdhury, T. Rahman, A. Khandakar, R. Mazhar, M. A. Kadir, Z. B. Mahbub, *et al.*, "Can AI help in screening viral and COVID-19 pneumonia?," *arXiv preprint arXiv:2003.13145,* 2020.

[11] T. Rahman, M. E. Chowdhury, A. Khandakar, K. R. Islam, K. F. Islam, Z. B. Mahbub, *et al.*, "Transfer Learning with Deep Convolutional Neural Network (CNN) for Pneumonia Detection using Chest X-ray," *Applied Sciences,* vol. 10, p. 3233, 2020.

[12] M. E. Chowdhury, T. Rahman, A. Khandakar, S. Al-Madeed, S. M. Zughaier, H. Hassen, *et al.*, "An early warning tool for predicting mortality risk of COVID-19 patients using machine learning," *arXiv preprint arXiv:2007.15559,* 2020.

[13] S. S. Chouhan, A. Kaul, U. P. Singh, and S. Jain, "Bacterial foraging optimization based radial basis function neural network (BRBFNN) for identification and classification of plant leaf diseases: An automatic approach towards plant pathology," *IEEE Access,* vol. 6, pp. 8852-8863, 2018.

[14] Y. LeCun, P. Haffner, L. Bottou, and Y. Bengio, "Object recognition with gradient-based learning," in *Shape, contour and grouping in computer vision*, ed: Springer, 1999, pp. 319-345.

[15] S. Arya and R. Singh, "A Comparative Study of CNN and AlexNet for Detection of Disease in Potato and Mango leaf," in *2019*

International Conference on Issues and Challenges in Intelligent Computing Techniques (ICICT), 2019, pp. 1-6.

[16] G. Wang, Y. Sun, and J. Wang, "Automatic image-based plant disease severity estimation using deep learning," *Computational intelligence and neuroscience,* vol. 2017, 2017.

[17] J. Amara, B. Bouaziz, and A. Algergawy, "A deep learning-based approach for banana leaf diseases classification," *Datenbanksysteme für Business, Technologie und Web (BTW 2017)-Workshopband,* 2017.

[18] O. Ronneberger, P. Fischer, and T. Brox, "U-net: Convolutional networks for biomedical image segmentation," in *International Conference on Medical image computing and computer-assisted intervention*, 2015, pp. 234-241.

[19] F. Statistics, "Food and Agriculture Organization of the United Nations," *Retrieved,* vol. 3, p. 2012, 2010.

[20] I. Adeoye, O. Aderibigbe, I. Amao, F. Egbekunle, and I. Bala, "TOMATO PRODUCTS'MARKET POTENTIAL AND CONSUMER PREFERENCE IN IBADAN, NIGERIA," *Scientific Papers: Management, Economic Engineering in Agriculture & Rural Development,* vol. 17, 2017.

[21] M. Kaur and R. Bhatia, "Development Of An Improved Tomato Leaf Disease Detection And Classification Method," in *2019 IEEE Conference on Information and Communication Technology*, 2019, pp. 1-5.

[22] M. A. Rahman, M. M. Islam, G. S. Mahdee, and M. W. U. Kabir, "Improved Segmentation Approach for Plant Disease Detection," in *2019 1st International Conference on Advances in Science, Engineering and Robotics Technology (ICASERT)*, 2019, pp. 1-5.

[23] A. Fuentes, S. Yoon, S. C. Kim, and D. S. Park, "A robust deep-learning-based detector for real-time tomato plant diseases and pests recognition," *Sensors,* vol. 17, p. 2022, 2017.

[24] M. Agarwal, A. Singh, S. Arjaria, A. Sinha, and S. Gupta, "ToLeD: Tomato Leaf Disease Detection using Convolution Neural Network," *Procedia Computer Science,* vol. 167, pp. 293-301, 2020.

[25] H. Durmuş, E. O. Güneş, and M. Kırcı, "Disease detection on the leaves of the tomato plants by using deep learning," in *2017 6th International Conference on Agro-Geoinformatics*, 2017, pp. 1-5.

[26] R. Chaerani and R. E. Voorrips, "Tomato early blight (Alternaria solani): the pathogen, genetics, and breeding for resistance," *Journal of general plant pathology,* vol. 72, pp. 335-347, 2006.

[27] Q. Wu, Y. Chen, and J. Meng, "DCGAN-based data augmentation for tomato leaf disease identification," *IEEE Access,* vol. 8, pp. 98716-98728, 2020.

[28] R. N. Cabral, W. A. Marouelli, D. A. Lage, and A. C. Café-Filho, "Septoria leaf spot in organic tomatoes under diverse irrigation systems and water management strategies," *Horticultura Brasileira,* vol. 31, pp. 392-400, 2013.

[29] A. C. Café-Filho, C. A. Lopes, and M. Rossato, "Management of plant disease epidemics with irrigation practices," *Irrigation in Agroecosystems,* p. 123, 2019.

[30] J. Zou, S. Rodriguez-Zas, M. Aldea, M. Li, J. Zhu, D. O. Gonzalez, *et al.*, "Expression profiling soybean response to Pseudomonas syringae reveals new defense-related genes and rapid HR-specific downregulation of photo-synthesis," *Molecular plant-microbe interactions,* vol. 18, pp. 1161-1174, 2005.

[31] G. Li, T. Chen, Z. Zhang, B. Li, and S. Tian, "Roles of Aquaporins in Plant-Pathogen Interaction," *Plants,* vol. 9, p. 1134, 2020.

[32] R. Schlub, L. Smith, L. Datnoff, and K. Pernezny, "An overview of target spot of tomato caused by Corynespora cassiicola," in *II International Symposium on Tomato Diseases 808*, 2007, pp. 25-28.

[33] J. Zhu, L. Zhang, H. Li, Y. Gao, W. Mu, and F. Liu, "Development of a LAMP method for detecting the N75S mutant in SDHI-resistant Corynespora cassiicola," *Analytical biochemistry,* vol. 597, p. 113687, 2020.

[34] K. Pernezny, P. Stoffella, J. Collins, *A. Carroll*, and A. Beaney, "Control of target spot of tomato with fungicides, systemic acquired resistance activators, and a biocontrol agent," *PLANT PROTECTION SCIENCE-PRAGUE-,* vol. 38, pp. 81-88, 2002.

[35] J. Abdulridha, Y. Ampatzidis, S. C. Kakarla, and P. Roberts, "Detection of target spot and bacterial spot diseases in tomato using UAV-based and benchtop-based hyperspectral imaging techniques," *Precision Agriculture,* vol. 21, pp. 955-978, 2020.

[36] C. F. de Jong, F. L. Takken, X. Cai, P. J. de Wit, and M. H. Joosten, "Attenuation of Cf-mediated defense responses at elevated temperatures correlates with a decrease in elicitor-binding sites," *Molecular plant-microbe interactions,* vol. 15, pp. 1040-1049, 2002.

[37] J. Calleja-Cabrera, M. Boter, L. Oñate-Sánchez, and M. Pernas, "Root growth adaptation to climate change in crops," *Frontiers in Plant Science,* vol. 11, p. 544, 2020.

[38] F. Louws, M. Wilson, H. Campbell, D. Cuppels, J. Jones, P. Shoemaker, *et al.*, "Field control of bacterial spot and bacterial speck of tomato using a plant activator," *Plant Disease,* vol. 85, pp. 481-488, 2001.

[39] K. Qiao, Q. Liu, Y. Huang, Y. Xia, and S. Zhang, "Management of bacterial spot of tomato caused by copper-resistant Xanthomonas perforans using a small molecule compound carvacrol," *Crop Protection,* vol. 132, p. 105114, 2020.

[40] M. Nowicki, M. R. Foolad, M. Nowakowska, and E. U. Kozik, "Potato and tomato late blight caused by *Phytophthora infestans*: an overview of pathology and resistance breeding," *Plant disease,* vol. 96, pp. 4-17, 2012.

[41] A. Buziashvili, L. Cherednichenko, S. Kropyvko, and A. Yemets, "Transgenic tomato lines expressing human lactoferrin show increased resistance to bacterial and fungal pathogens," *Biocatalysis and Agricultural Biotechnology,* vol. 25, p. 101602, 2020.

[42] E. Glick, Y. Levy, and Y. Gafni, "The viral etiology of tomato yellow leaf curl disease–a review," *Plant Protection Science,* vol. 45, pp. 81-97, 2009.

[43] M. Dhaliwal, S. Jindal, A. Sharma, and H. Prasanna, "Tomato yellow leaf curl virus disease of tomato and its management through resistance breeding: a review," *The Journal of Horticultural Science and Biotechnology,* vol. 95, pp. 425-444, 2020.

[44] M. Ghanim and H. Czosnek, "Tomato yellow leaf curl geminivirus (TYLCV-Is) is transmitted among whiteflies (*Bemisia tabaci*) in a sex-related manner," *Journal of Virology,* vol. 74, pp. 4738-4745, 2000.

[45] M. Ghanim, S. Morin, M. Zeidan, and H. Czosnek, "Evidence for transovarial transmission of tomato yellow leaf curl virus by its vector, the whitefly*Bemisia tabaci*," *Virology,* vol. 240, pp. 295-303, 1998.

[46] Y.-Z. He, Y.-M. Wang, T.-Y. Yin, E. Fiallo-Olivé, Y.-Q. Liu, L. Hanley-Bowdoin, *et al.*, "A plant DNA virus replicates in the salivary glands of its insect vector via recruitment of host DNA synthesis machinery," *Proceedings of the National Academy of Sciences,* vol. 117, pp. 16928-16937, 2020.

[47] H. Choi, Y. Jo, W. K. Cho, J. Yu, P.-T. Tran, L. Salaipeth, *et al.*, "Identification of Viruses and Viroids Infecting Tomato and Pepper Plants in Vietnam by Metatranscriptomics," *International Journal of Molecular Sciences,* vol. 21, p. 7565, 2020.

[48] L. Broadbent, "Epidemiology and control of tomato mosaic virus," *Annual review of Phytopathology,* vol. 14, pp. 75-96, 1976.

[49] Y. Xu, S. Zhang, J. Shen, Z. Wu, Z. Du, and F. Gao, "The phylogeographic history of tomato mosaic virus in Eurasia," *Virology,* vol. 554, pp. 42-47, 2 020.

[50] D. Hughes and M. Salathé, "An open access repository of images on plant health to enable the development of mobile disease diagnostics," *arXiv preprint arXiv:1511.08060,* 2015.

[51] 'SpMohanty/PlantVillage-Dataset'. (Accessed on 24 Jan 2021). [online]. Available: https://github.com/spMohanty/PlantVillage-Dataset

[52] H. Wang, Z. Wang, M. Du, F. Yang, Z. Zhang, S. Ding, *et al.*, "Score-CAM: Score-weighted visual explanations for convolutional neural networks," in *Proceedings of the IEEE/CVF Conference on Computer Vision and Pattern Recognition Workshops*, 2020, pp. 24-25.

[53] H. Wang, M. Du, F. Yang, and Z. Zhang, "Score-cam: Improved visual explanations via score-weighted class activation mapping," *arXiv preprint arXiv:1910.01279,* 2019.

[54] R. Azad, M. Asadi-Aghbolaghi, M. Fathy, and S. Escalera, "Bi-directional ConvLSTM U-net with Densley connected convolutions," in *Proceedings of the IEEE International Conference on Computer Vision Workshops*, 2019, pp. 0-0.

[55] (Accessed on:01 August 2020). [Online]. Available: https://github.com/imlab-uiip/readme

[56] T. Rahman, A. Khandakar, Y. Qiblawey, A. Tahir, S. Kiranyaz, S. B. A. Kashem, *et al.*, "Exploring the Effect of Image Enhancement Techniques on COVID-19 Detection using Chest X-rays Images," *arXiv preprint arXiv:2012.02238,* 2020.

[57] J. Schlemper, O. Oktay, M. Schaap, M. Heinrich, B. Kainz, B. Glocker, *et al.*, "Attention gated networks: Learning to leverage salient regions in medical images," *Medical image analysis,* vol. 53, pp. 197-207, 2019.

[58] J. Deng, W. Dong, R. Socher, L.-J. Li, K. Li, and L. Fei-Fei, "Imagenet: A large-scale hierarchical image database," in *2009 IEEE conference on computer vision and pattern recognition*, 2009, pp. 248-255.

[59] P. Tm, A. Pranathi, K. SaiAshritha, N. B. Chittaragi, and S. G. Koolagudi, "Tomato leaf disease detection using convolutional neural networks," in *2018 Eleventh International Conference on Contemporary Computing (IC3)*, 2018, pp. 1-5.

[60] K. Zhang, Q. Wu, A. Liu, and X. Meng, "Can deep learning identify tomato leaf disease?," *Advances in Multimedia,* vol. 2018, 2018.

[61] B. A. Ashqar and S. S. Abu-Naser, "Image-Based Tomato Leaves Diseases Detection Using Deep Learning," 2018.

[62] M. Ouhami, Y. Es-Saady, M. El Hajji, A. Hafiane, R. Canals, and M. El Yassa, "Deep Transfer Learning Models for Tomato Disease Detection," in *International Conference on Image and Signal Processing*, 2020, pp. 65-73.

[63] A. F. Fuentes, S. Yoon, J. Lee, and D. S. Park, "High-performance deep neural network-based tomato plant diseases and pests diagnosis system with refinement filter bank," *Frontiers in plant science,* vol. 9, p. 1162, 2018.

[64] G. Langar, P. Jain, and N. Panchal, "TOMATO LEAF DISEASE DETECTION USING ARTIFICIAL INTELLIGENCE AND MACHINE LEARNING," *INTERNATIONAL JOURNAL,* vol. 5, 2020.

Chapter 24

A Nonlinear Fuzzy Controller Design Using Lyapunov Functions for an Intelligent Greenhouse Management in Agriculture

Lukman Adewale Ajao, Emmanuel Adewale Adedokun, Joseph Ebosetale Okhaifoh and Habib Bello Salau

Abstract

The importance of agronomists in large-scale production of food crops under considerate environmental weather conditions cannot be overemphasized. However, emerging global warming is a threat to food security due to its effect on soil depletion and ecosystem degradation. In this work, the design of the proposed intelligent context is to observe, model and simulate greenhouse control system activity towards the management of the farm crop growth as the affected salient environmental parameters. Characteristically, temperature and humidity are the major factors that determine the crop yield in a greenhouse but the case of a dry air environment or beyond $30^0C - 35^0C$ of high air humidity will affect crop growth and productivity. A Mamdani technique of fuzzy logic controller with non-linear consequent is used for intelligent greenhouse design in the LABVIEW virtual environment. This approach is used to mimic the human thought process in the system control by setting some logical rules that guide the greenhouse functions. For the system stabilization achievement, a direct method of Lyapunov functions was proposed. The simulation model result shows that, the average temperature of 18.5^0C and humidity 65% is achieved for a decent environment of crop growth and development during winter. However, the average temperature and humidity achieved during summer is 27.5^0C&70% respectively. For every season that is beyond 30.5^0C *and* 75% of temperature and humidity will require automation of roof opening and water spilled.

Keywords: Agricultural technology, Artificial intelligence, Fuzzy logic, Greenhouse, Nonlinear system

1. Introduction

Agriculture is an important aspect of any nation's development which usually requires appropriate seasoning irrigation and fertilization to produce a quantity of food products [1]. The seasoning control application of fertigation (fertilizer and irrigation) techniques has proven efficient in plant growth, development, and yield large crop production [2]. Computers and electronics play a significance role in the development and mechanization of agricultural products through the recent applications of ubiquitous technology of Internet of Things (IoT). This advancement and dynamic methods of control theories application helps improving the agricultural equipment (mechanization) and the processes. The recent integration of artificial intelligence (AI) and computational intelligence (CI) into the agro-mechanical machine and mechatronics system (embedded sensors and robotic) shaped the agricultural technology and their commercialization.

So, the studies have indicated a strong link between agriculture and economic growth as the backbone of national sources of income and commercial development [3]. The increasing demand in food consumption nationwide as resulting from increased daily population explosion that necessitated the provision of precision agriculture monitoring [4] and to ease the farming process as well as abnormalities in the farm environment. Although, farming is an essential means of increasing food production, recently its cultivation is decreasing and becoming inversely proportional to the rising population. This is partly due to the phenomenon of global warming [5]. As a result of changes in climate conditions and its threat of conservatory, the need arises for an agricultural development control system to manage this condition for high yield of crop production [6–8].

The changes in climate condition increases the tropical storm intensity and frequency due to rising temperatures and climate pattern that change mutually. Whereas, the warming of temperatures ocean and sea levels rising escalate the disaster storms growth with excess heat trapped in the atmosphere. It is observed that dissolving of heat energy and excess of carbon dioxide gas has significant damage on the ocean. Like oceanic acidification that affect reproduction and formation of animal shells, oceanic heat waves affect coral reefs with the frustration of fish migration, and oceanic dead zones created as a result of deoxygenation process [9, 10].

Consequently, there is a need to prevent and manage common emission releasing into the atmosphere with effect on agricultural crop growth and environmental degradation effect. This threat of emission releasing contributes immensely to the effect of climate change which affects the successful cultivation of farm crops [11]. The United Nations' World Meteorological Order (WMO) confirmed that the world planet is about 1.1°C warmer, and is forecasting an increase from 4–5°C towards end of the century. Others factors of agronomy-house sustenance depend on the environmental weather condition which includes temperature, humidity, winds, light intensity, and solar radiation. The statistical overview of primary sources of greenhouse gas emission is given in **Figure 1**, which include industry, transportation, building, agriculture, forestry, electricity and heat production [12]. While the sources of releasing those gases, are Methane (NH_4), Nitrous Oxide (NO), and Carbon Dioxide (CO) from the industrial processes, fossil fuel, bush burning, forestry, sewage disposal and other land use [13].

A Greenhouse is a controlled place where plants are grown under control conditions of ambient temperature, humidity, water vapor, light intensity, and carbon (iv) oxide [14]. The environmental conditions for greenhouses can be varied according to the plants need to get most out of the plants and for high efficiency. Since the environmental conditions of the greenhouse need to be adjusted for optimal growth, the size and cost of labor increase proportionally to the size of the greenhouse and the number of plants [15, 16]. A greenhouse is a structure designed with glass walls or transparent material and a glass/translucent roof used to grow food crops and plant cultivation (such as tomatoes and tropical flowers under controlled environmental conditions [17, 18].

The efficient management and monitoring of greenhouse plants condition require the integration of an artificial intelligence system (AI) or automated control system (ACS) based on context-aware software design (CASD). Therefore, a

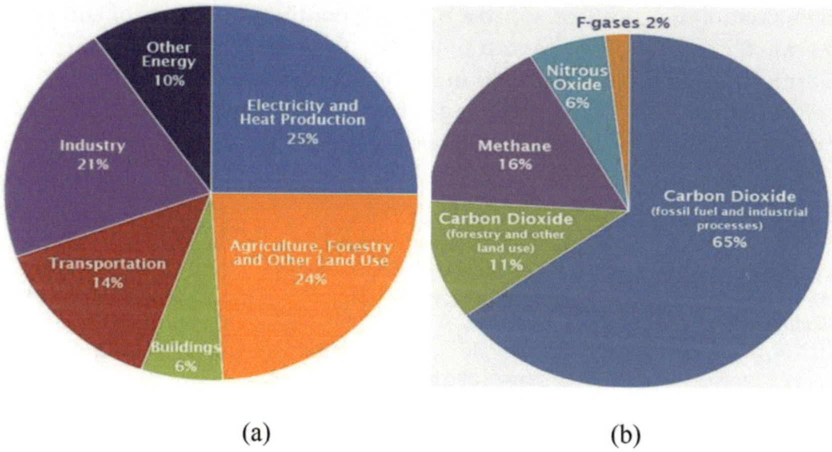

Figure 1.
The statistical overview of primary sources of greenhouse gas emission. (a) The consequential effect of rapid gaseous emission (greenhouse gas emission) to the atmosphere will be increases with time. (b) This includes carbon dioxide (CO) with percentage of 147%, methane (CH_4) with percentage of 259%, and nitrous oxide (N_2O) with percentage of 132%.

Greenhouse Development Rights (GDR) framework is proposed in [19] to safeguards the right of development as a possible global solution to the climate change challenges. It is shown that GDR approach is an international context for (China and the USA) which provide funding for the development mechanism of a greenhouse as an approach to address global climate change challenges. The GDR is a foundation for future evolution and industrialized developed countries. In another approach of control and keeping hothouse cool is the development of a smart controller for grid stabilization using optoelectronic system [20].

This book chapter contribution aims at presenting an intelligent greenhouse control based on the non-linear consequent fuzzy logic controller using LabVIEW for agricultural technology. The system helps to monitoring greenhouse parameters and acts based on the specified fuzzy rules to control the system environmental condition without or with little human intervention. Thus, there is a need for an intelligent greenhouse control system in agricultural technology that can reduce human labor costs, increase productivity, and reduce human intervention.

2. Related works

The global warming crisis necessitated the development of a real-time monitoring and control system for managing change in environmental temperature conditions. This temperature change plays an important role in the soil contents of farm crops. Therefore, the use of computer technology approach (such as embedded systems and AI) has been newly adopted in realizing the design of automation control and monitoring system of ambient temperature in greenhouse management. The greenhouse control system is developed using LabVIEW simulation software for data collection and analysis of conservation in [21]. The work mainly focuses on adjusting the temperature environment using a thermostat and sensor to detect and control the hotness of the greenhouse. The process was simulated and implemented in the developing system platform of Labview software. An optimized sprinkler irrigation system for predicting use of budding land based on soil features using fuzzy logic decision approach in [22]. The significance of adopting this fuzzy logic in land evaluation is a suitable approach for the continuous nature of soil properties and provide an accurate distribution index for predicting land use.

An optimized method of cultivation in the greenhouse automated system with smart environments using an embedded system development approach in [23]. This industrial automated greenhouse model is developed for plant experimentation at

Title	Strength	Limitation
Design of an Intelligent Management System for Agricultural greenhouses based on the Internet of Things [24].	Successfully developed a remote monitoring system for greenhouses using ZigBee protocols. Users can remotely control and manage greenhouse parameters such as temperature and humidity.	Absence of an intelligent technique. Although the control method is remote, it is also manual.
Smart greenhouse monitoring using Internet of Things [25].	A system capable of remotely monitoring greenhouse parameters via a web application.	No intelligent technique presents. Lack of control mechanism.
Research on the control system of the intelligent greenhouse of IoT based on ZigBee [26].	Successfully developed a ZigBee based system capable of remotely monitoring and controlling greenhouse parameters	Absence of intelligent technique. Control is manual.
Internet of Things based smart greenhouse: remote monitoring and automatic control [27].	Implemented a smart greenhouse using GSM/GPRS for remotely monitoring and controlling greenhouse parameters. The system is capable of automatically controlling the parameters if they are out of the specified range.	Absence of an intelligent technique for the control of parameters.
Intelligent greenhouse design based on Internet of Things (IoT) [28].	Developed an intelligent greenhouse using Cloud service for remotely monitoring greenhouse variables. The system is also capable of automatically controlling the parameters if they fall below or above specified values.	Absence of intelligent control technique.
Smart greenhouse using IoT and cloud computing [16].	Successfully developed a monitoring interface for greenhouse parameters using IoT and cloud computing	Absence of intelligence and control technique.
Design and implementation of a smart greenhouse [18].	Successfully developed a smart greenhouse control system to monitor and control the parameters in a tomato farm. The system automatically controlled actuators to regulate greenhouse variables.	Absence of intelligent technique.
Intelligent Monitoring Device for Agricultural Greenhouse Using IoT [29].	The author proposes a monitoring system for greenhouses using wireless sensor networks and IoT. The proposed system incorporates a microcontroller that transmits information that can be monitored with an Android Application.	Absence of intelligent technique. No control technique specified.

Table 1.
Summary of the related works.

the University of Alicante to control air-conditioning, soil condition, and irrigation in the system. The optimization services integrated into this system model designed help in the detection and prediction of agricultural production of smart environments. But the optimized smart environment greenhouse does not consider controlling the system conditions during rainfall, summer, and winter. Other authors that contribute to the development of automated and intelligence-based greenhouse control and monitoring system is analyzed in **Table 1**.

From these literatures, it is observed that the limitation is on the part of intelligence incorporated into the system with linearized fuzzy model improvement. Also, the season management of crop cultivated area in the greenhouse with automatic control technique are not studied. Hence, this book chapter aims to fill those gaps by implementing a non-linear consequent fuzzy logic controller system for the decision-making process and automatic control of the greenhouse system with an approach of context-aware software design ontology. This book chapter is organized into 5 sections. The introductory part discussed the general background of study in Section 1, Section 2 presented the related works. The research methodology is presented in Section 3, while sub-Section 3.1 mathematical modeling of the greenhouse control system in sub-Section 3.1, sub-Section 3.2 presented a linearize and non-linear consequent fuzzy controller design for greenhouse control. Sub-Section 3.3 contained Lyapunov function for stabilization of non-linear consequent fuzzy controller. Sub-Section 3.4 presented simulation and implementation of non-linear consequent fuzzy controller based-greenhouse design in LabVIEW. The results and discussion are presented in Section 4, sub-Section 4.1 contained Intelligent greenhouse management based nonlinear control simulation results. Sub-Section 4.2 presented simulation results of a Lyapunov stability of nonlinear control system. Section 5 gives the conclusion and recommendations for future works.

3. Methodology

Context-aware systems are software systems designed with the ability to sense (sensor) and adapt to the environmental conditions for the solution required to the problem design [30, 31] through a fuzzy controller. This design involves determining what the system needs to sense, make adaptations, and respond to sensor information. It requires sensing temperature and humidity, and then adapt to the environmental condition for the greenhouse system control and management using nonlinear fuzzy controller system with direct method of Lyapunov functions to achieved stabilization. The system modeling and design need a focus value or parameter to influence the designed value such that it can sense the elements and manipulate them in case of irregularities. So that it can make the element relevant to the purpose of the design and the designer focus. An overview of the approach

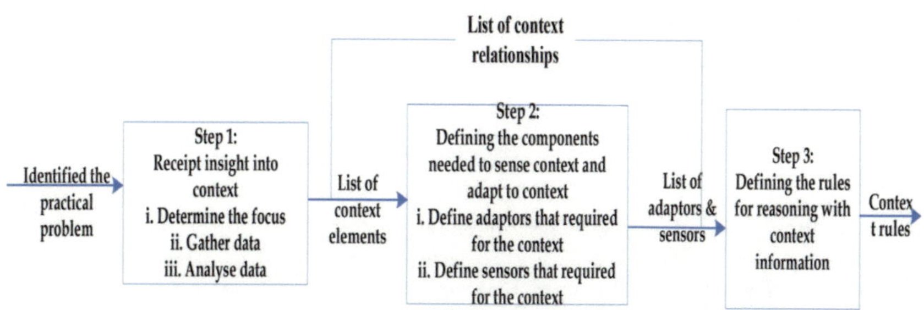

Figure 2.
An overview to the element approach in the CAS design.

design for an intelligent greenhouse control system includes practical problem identification, insight to the context, elements components required in the sense and adaptation, and logical reasoning rules for information as illustrated in **Figure 2**.

The fuzzy logic controller architecture [32] consisting of crisp input rules, Fuzzification (knowledge-based or linguistic rules), fuzzy inference engine (logic rules), and Defuzzification (output crisp values). The input of a fuzzy control system parameter can be adjusted to improve the system (fuzzy mechanism) performance using the Eqs. (1) and (2).

$$\theta^{(n)} = \alpha(P_0, P_1, P_2, P_3 \dots, P_n) \tag{1}$$

$$\theta^{(n)} = \alpha\left(\theta^{(n-1)}, P_n\right) \tag{2}$$

where, $\theta^{(n)}$ is define as a set of input parameter to adjust at time t, T_n and P_n is the parameter collected at a time T_n.

The non-homogeneity consequent of the fuzzy logic controller system technique is adopted in the design to sense the greenhouse environment and adapted for a unique solution of a design problem. A visual graphical programming system-design platform and software development environment called Laboratory Virtual Instrument Engineering Workbench (LabVIEW) was used to achieve the context-design. It is very efficient and commonly used in engineering as a context-aware system design for data acquisition, instrument control, and industrial automation system. It is a multi-threading and multiprocessing hardware system that is automatically engaged by the in-built scheduler during the execution flow structure (nodes) of a graphical block diagram. The connection wires will propagate the variables and execute the process immediately all its input data reachable.

This system is used to control the temperature and humidity of the greenhouse system using a non-homogeneity control system. The temperature and humidity inputs parameter are set and the system keeps both values constant regardless of the outside temperature of the controlled system. This is achieved using the combination technique of the linearized system with non-linear fuzzy, and adopt Lyapunov function to achieve system stability in the model. This model helps in controlling the opening greenhouse roof for rainfall and sunshine, and/or by turning on the sprinkler to reduce the temperature as presented in the algorithm of **Table 2**. The

Algorithm for greenhouse temperature management (Roof opening, closed and water spilled)
t is the time, t_l is minimum air temperature, t_h is the maximum air temperature
Procedure for greenhouse roofing control $(time, t_l, t_h)$
$t \leftarrow$ *air temperature value*
If time between 8 : 00*am and* 8 : 00*pm, then*
$t_{avg} \leftarrow$ *air temperature average*
if $t < t_l$, *then*
Control greenhouse roofing (*Closed*)
else if $t \geq t_h$
Control greenhouse roofing (*Open small*)
else
If $t_{avg} - t < t_l$, *then*
Control greenhouse roofing (*Closed, No water spill*)
else if $t_{avg} - t \geq t_h$, *then*
Control greenhouse roofing (*Open, Water spill*)

Table 2.
Greenhouse temperatures management and control

decision-making process of the system is achieved using a combination technique of linear and non-linear approach consequent of the fuzzy logic controller system. The sub-system irrigation and ventilation classification help the agronomist to manage the setpoints of the control input variables. This irrigation-ventilation model is an intelligence unit that is used for the senses and responds to immediate action by introducing the prediction and optimization facilities that are supervised by the agronomist as presented in **Figure 3**.

The calories required for heating the air in the greenhouse is calculated as expressed in Eq. (3). For the determining value of temperature, it requires average heat of 0.30Kcal to achieve a one-meter cube of air. It is observing that 1 kW heat can produce 860Kcal, and a heat source of 30 W can produce 25.8Kcal heat per hour, and equivalent to 0.43Kcal heat per minute [33].

$$\varnothing = \mathcal{M}\mathbb{C}\ell\daleth \tag{3}$$

where $\varnothing$ is the heat, $\mathcal{M}$ is the mass, $\mathbb{C}$ is the heating temperature (0.24 Kcal/kg), $\ell\daleth$ is the difference in temperature.

3.1 Mathematical modeling of greenhouse control system

The behavior of the greenhouse microclimate is dynamic and combinations of physical processes involve mass balance and energy transfer. The physical processes involved are used in estimating the greenhouse climate. The amount of energy leaving the greenhouse can be calculated as expressed in Eqs. (4) and (5).

$$E_{total} = -E_{loss} + E_{gain} \tag{4}$$

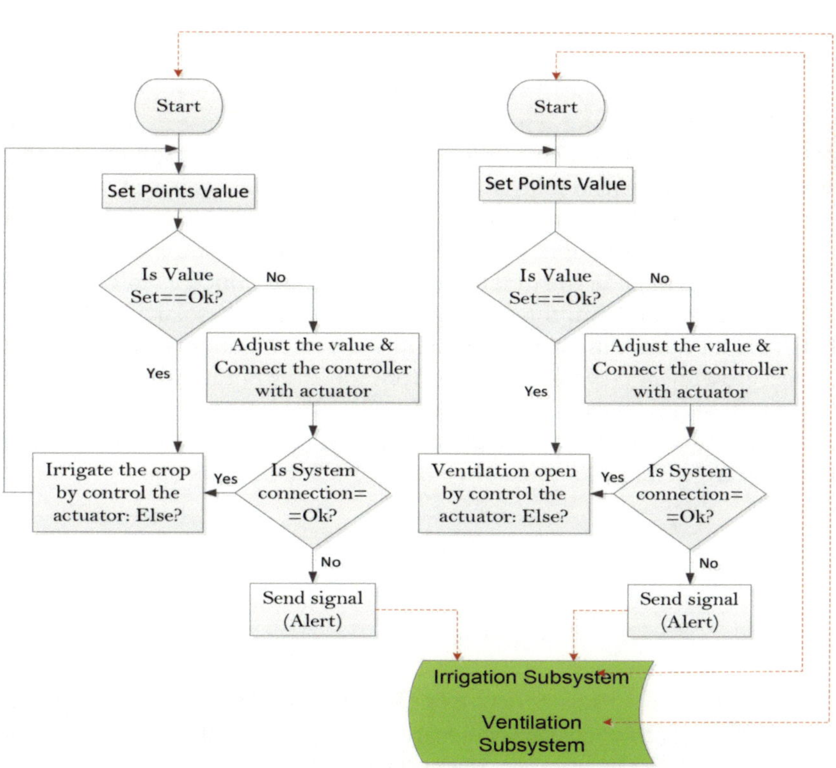

Figure 3.
Intelligent greenhouse monitoring and control flowchart.

$$E_{loss} = E_k + E_r + E_v + E_{inf} + E_{cond} \quad (5)$$

where, E_{total} is the total energy balance (W), E_{gain} is the amount of energy entering the greenhouse (W), E_{loss} is the amount of energy leaving the greenhouse (W), E_k is heat loss due to conductive heat loss (W), E_v is the heat transfer due to ventilation (W), E_{inf} is heat transfer due to infiltration (W), E_r is heat transfer due to the longwave radiation and E_{cond} is heat loss due to condensation (W).

The conductive loss encompasses all the heat transfers through the greenhouse cover from the internal to the external air, conductive heat transfer through the covering material and radiative heat transfer can be expressed as in Eq. (6). The thermal wave radiation exchange from the interior greenhouse to outside can be calculated as given in the non-linear Boltzmann relation in Eq. (7) and Eq. (8). Therefore, the ventilation of heat lost in the greenhouse is proportional to the rate of air exchange and the differences occur between the inside and outside air temperature [34], and the loss can be determined as in Eq. (9).

$$E_k = A_c h(\lambda_i - \lambda_0) \quad (6)$$

$$E_r = A_c \sigma \varepsilon(\lambda_i - \lambda_0) \quad (7)$$

$$E_v = G\rho C_p(\lambda_i - \lambda_0) \quad (8)$$

$$and, G = \omega r_v k_v A_v \quad (9)$$

where λ_o is outside air temperature (K), λ_i is inside air temperature (K), h is the conductive heat transfer coefficient (W/m^2), A is the area of greenhouse cover (m^2), Q_r is radiation loss, ε is the combining emissivity between the cover and sky, σ is Boltzmann constant, ρ is air density (kg/m^3), C is the specific heat of air (J/kg K), G is airflow due to ventilation (m^3/s), w is the wind speed (m/s), r_v is percent of the ventilator opening, k_v is the slope of the curve showing the ventilation flux divided by wind speed variation and A is area of the ventilator (m^2).

The heat energy is transfer within the intelligent greenhouse system as a result of infiltration of energy loss which is due to the exchange air through cracks occurs in the greenhouse and is considered. Since the infiltration rate is based on the volume of water vapor changed per unit cover area (roof and walls). This volume of water vapor is directly proportional to the wind velocity and the temperature difference from both inside to outside the greenhouse can be determined as in Eq. (10). Then, the sources of heat gain from the greenhouse model include solar radiation heat which is the most determinant of heat gain by the intelligent greenhouse system during crop growing and system heating from the environment [35]. So, the energy of the greenhouse can be calculated as in Eq. (11), the heat transfer from tubes to the greenhouse environment is expressed as in Eq. (12) and the internal temperature increases are within the range of (0.3–0.7) which 0.3 was chosen.

$$E_{inf} = 0.5VN(\lambda_i - \lambda_0) \quad (10)$$

$$E_r = A_g \gamma \tau I \quad (11)$$

$$Q_{hs} = mC_p(\lambda_{\omega i} - \lambda_{\omega 0}) \quad (12)$$

where, H_{inf} is the infiltration heat loss (W), λ_i is the temperature inside the greenhouse (K), λ_o is the outside temperature (K) of a greenhouse, V is greenhouse volume (m^3), and N is the number of air changes per hour (h^{-1}), E_r is solar energy radiate into the greenhouse environment (W), I is total external solar energy falling on a horizontal surface of the greenhouse (W/m^2), A is an area of greenhouse floor

(m^2), τ is radiation light transmission to the greenhouse cover, γ is constant of the proportion of solar radiation that radiates into the greenhouse. Q_{hs} is heat gain from the heating system (W), m is the heating water flow rate (kg/s); $\lambda_{\omega i}$ is heating water inlet temperature (°C), $\lambda_{\omega 0}$ is heating water outlet temperature (°C) and C_p is the specific heat capacity of water (J/kg K).

3.2 A linearize and non-linear consequent fuzzy controller design for greenhouse management

A closed-loop or called feedback controller transfer function is adopted since the output of the intelligent control system $\varphi(t)$ is fed back into the system through a sensory measurement device (sensor) γ. The comparison is for reference value $\tau(t)$, where the controller system α takes the error ε (difference) between the reference point or set values and the output to adjust the inputs μ feedback to the system under control β. From the perspective of implementation of the controller with a linear approach and time-invariant, the elements of the transfer function $\alpha(s)$, $\beta(s)$, and $\gamma(s)$ do not depend on time where α is controller, β is the system under controller (plant), and sensor measurement denotes γ [36–38].

We can analyze the systems using the Laplace transform on the variables as expressed in Eqs. (13)–(16).

$$\varphi(s) = \beta(s)\mu(s) \tag{13}$$

$$\mu(s) = \alpha(s)\varepsilon(s) \tag{14}$$

$$\varepsilon(s) = \tau(s) - \gamma(s) - \varphi(s) \tag{15}$$

By solving $\varphi(s)$ in terms of $\tau(s)$ can be expressed as given in Equation

$$\varphi(s) = \left(\frac{\beta(s)\alpha(s)}{1 + \beta(s)\alpha(s)\gamma(s)}\right)\tau(s) = \aleph(s)\tau(s) \tag{16}$$

The closed-loop or feedback transfer function of the greenhouse control system is expressed as $\aleph(s)$ in Eq. (17), where the numerator is identified as open-loop (forward gain) from τ input parameter to φ output values, and the denominator is a feedback loop that goes around the system called loop gain. So, if $|\beta(s)\alpha(s)| \gg 1$, that is, it has a standard model with each value of s, and if $|\gamma(s)| \approx 1$, then $\varphi(s)$ is approximately equal to $\tau(s)$ and the output system is close to the reference input.

$$\aleph(s) = \frac{\beta(s)\alpha(s)}{1 + \gamma(s)\alpha(s)\alpha(s)} \tag{17}$$

The flowchart technique for a linearized and non-linear fuzzy model for the optimization function of the greenhouse control and management model is illustrated in **Figure 4**. This mechanism operates as a reference model to the non-linear system and is connected in parallel in such a way that the linear system passes across the non-linear for better stability.

The state-space model for the non-linear fuzzy controller is given in Eq. (18), which increases the fuzzy rules quantity exponentially with non-linearities measures. The delayed in the state-space model for the fuzzy controller is given in Eq. (19). Where $\check{x}(t)$ is the state vector of $\check{x}(t) \in \Re^{n_x}$, $\upsilon(t)$ is an input vector for $\upsilon(t) \in \Re^{n_\upsilon}$, s is the number of rules, $\omega(t)$ is the available premise vector, $\delta_k(\omega)$ is the

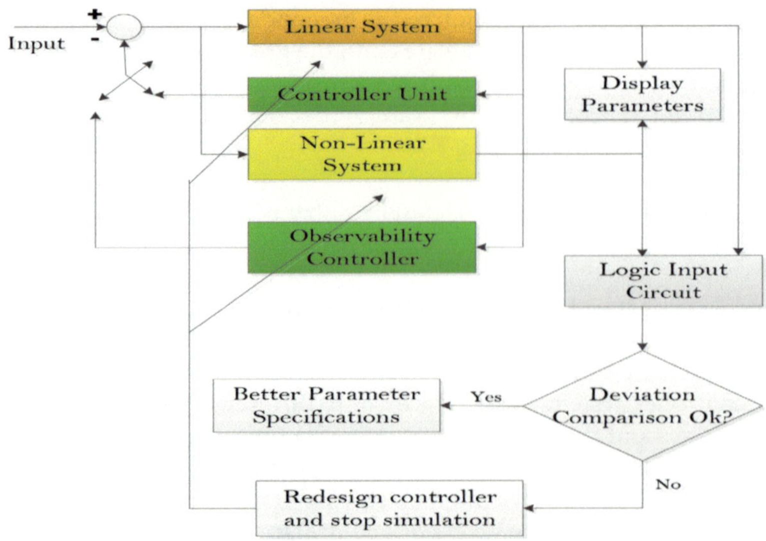

Figure 4.
Flowchart of combined linearized and non-linear fuzzy system.

membership function, α_k and β_k are the linear models, and the convex sum is given as: $\delta i(\omega) \in [0, 1], \sum_{k=1}^{s} \delta i(\omega) = 1$.

$$\breve{x}(t) = \sum_{k=1}^{s} \delta_k(\omega(t))(\alpha_k x(t) + \beta_k(t)) \tag{18}$$

This state-space model for the fuzzy system can be expanded to determined the time delay dependent as given in equation, where $\tau(t)$ is the delay time dependent, and $\delta_m(\omega(t - \tau(t)))$ is the delay states that dependent on fuzzy membership functions.

$$\breve{x}(t) = \sum_{k=1}^{s} \sum_{m=1}^{s} \delta_k(\omega(t))\delta_m(\omega(t - \tau(t)))(\alpha_{km} x(t) + \Im_{km} x(t - \tau(t)) + \beta_{km} \upsilon(t - \tau(t))) \tag{19}$$

The notation of $\tau(t) := \tau$ can be expressed as Eq. (20), and the closed-loop fuzzy model for non-linear time-dependent is in Eqs. (21) and (22). This is to reduce the number of fuzzy rules and to serve the purpose of measured-state and non-linearities unmeasured-state [34, 39, 40].

$$\Psi_{\omega\tau} = \sum_{k=1}^{s} \sum_{m=1}^{s} \delta_k(\omega(t))\delta_m(\omega(t - \tau(t)))\Psi_{km} \tag{20}$$

$$\breve{x}(t) = \alpha_{\omega\tau} x(t) + \Im_{\omega\tau} x(t - \tau) + \beta_{\omega\tau} \upsilon(t - \tau) \tag{21}$$

$$\breve{x}(t) = \alpha_{\omega\tau} x(t) + \Im_{\omega\tau} x(t - \tau) + \beta_{\omega\tau} G\psi(\xi x(t)) + \beta_{\omega\tau} \upsilon(t - \tau) \tag{22}$$

The $x(t - \tau)$ is the state vector for time-delayed, $\upsilon(t - \tau)$ is an input vector time-delayed, G is the system matrix, x(t) is a function of linear combination for each input to the model, and $\psi(\xi x(t))$ is a vector function. The boundary condition for the existence of vector function in the model can be expressed as $\psi_k, k = 1, \ldots, r$ *occur in* b_k such that; $0 \leq \frac{\psi_k(v)-\psi_k(w)}{v-w} \leq b_k$ as in Eq. (23).

$$\psi(!) = \begin{pmatrix} \psi_1(.) \\ \psi_2(.) \\ \psi_3(.) \\ \ldots \\ \psi_r(.) \end{pmatrix} \tag{23}$$

3.3 Lyapunov function for stabilization of non-linear consequent fuzzy controller based-greenhouse

For the stabilization and dynamical nature of the system, a Lyapunov non-linear function (LNF) is adopted to operate the system model as a linear with a limited range of function at every region. This approach of LNF helps the model to present auxiliary nonlinear feedback which can be operated as linear for control design purposes. Since a Lyapunov direct method of stability criterion for a linear system can be defined, suppose u = 0, and the exist two-point p > 0 and q > 0. Therefore, a linear system is asymptotically stable at the beginning for any given symmetric that existed given a unique solution that used for stability analysis as given in Eq. (24), with $\wp(\theta) = \sum_{j=1}^{m} \theta_j \wp_j,\ \wp_j > 0$, where $\eth < 0$. But the choice of q can be made arbitrarily which is mostly set as q = 1, an identity matrix p for all successive principal minors of p is positive using Sylvester theorem as expressed in Eq. (25).

$$\begin{aligned} \dot{\mathrm{x}}(\mathrm{t}) &= \alpha \mathrm{x}(\mathrm{t}) + \beta \mathrm{u}(\mathrm{t}) \\ y(t) &= \gamma x(t) + \delta u(t) \end{aligned} \tag{24}$$

$$P = \begin{matrix} P_{11} & P_{12} & P_{13} \\ P_{21} & P_{22} & P_{23} \\ P_{31} & P_{32} & P_{33} \end{matrix},\ P_{11} > 0. \text{ Therefore } \Delta[P] > 0 \tag{25}$$

However, the parameters for the fuzzy model-dependent can be given as in Eq. (26).

$$\begin{cases} \dot{x} = \alpha(\theta)x + \beta_d(\theta)d + \beta_u(\theta)u \\ \varepsilon = \gamma_\varepsilon(\theta)x + \delta_{\varepsilon d}(\theta)d + \delta_{\varepsilon u}(\theta)u \\ y = \gamma_\varepsilon(\theta)x + \delta_{\varepsilon d}(\theta)d, \end{cases} \tag{26}$$

where $\alpha(\theta) = \sum_{i=1}^{m} \theta_i(\sigma)\alpha_i$, $\beta_d(\theta) = \sum_{i=1}^{m} \theta_i(\sigma)\beta_{d,i}$, $\beta_u(\theta) = \sum_{i=1}^{m} \theta_i(\sigma)\beta_{u,i}$, $\gamma_\varepsilon(\theta) = \sum_{i=1}^{m} \theta_i(\sigma)\gamma_{\varepsilon,i}$, $\delta_{\mathrm{\varepsilon d}}(\theta) = \sum_{i=1}^{m} \theta_i(\sigma)\delta_{\mathrm{\varepsilon d},i}$, $\delta_{\mathrm{\varepsilon u}}(\theta) = \sum_{i=1}^{m} \theta_i(\sigma)\delta_{\mathrm{\varepsilon u},i}$, $\gamma_y(\theta) = \sum_{i=1}^{m} \theta_i(\sigma)\gamma_{y,i}$, $\delta_{yd}(\theta) = \sum_{i=1}^{m} \theta_i(\sigma)\delta_{yd,i}$, $\theta = [\theta_1(\sigma) \ldots \theta_m(\sigma)]^T$.

The associated weighting function of normalized fuzzy with the ith system are calculated through the degree of fuzzy membership functions $\theta_1(\sigma)$ and premise variable with a closed interval of $[0, 1]$ which must satisfy these properties in Eq. (27);

$$\theta \leq \theta_i(\sigma) \leq 1, \sum_{i=1}^{m} \theta_i(\sigma) = 1, \sum_{i=1}^{m} \theta_i(\sigma) = 0. \tag{27}$$

The state-space matrices function can be replaced in the derivation with a new introduce operator as expressed (Eqs. (28) and (29)), where α, P can be replaced with ω, and the subscript μ refers to all signals (x, d, ε) and η_μ is the dimension of signal μ.

$$P_\mu := \begin{bmatrix} \theta_2 I_{n\sigma} \\ \vdots \\ \theta_m I_{n\sigma} \end{bmatrix}, \omega^c := \begin{bmatrix} \omega_2 - \omega_1 \\ \vdots \\ \omega_m - I_1 \end{bmatrix}, \tag{28}$$

$$\omega^{\mathcal{R}} := [\, \omega_2 - \omega_1 \dots \omega_m - \omega_1] \tag{29}$$

The notation in Eq. (29) can be expressed as given in Eq. (30) using fuzzy weighting membership functions properties.

$$\omega(\theta) = \sum_{i=1}^{m} \theta_i \omega_i = \omega_1 + \omega^{\mathcal{R}} P_\mu = P_\mu^T \omega^c \tag{30}$$

Therefore, the Lyapunov fuzzy model function for the system stability is given with $\wp(\theta) = \sum_{j=1}^{m} \theta_j \wp_j,\ \wp_j > 0$, where $\eth < 0$ as expressed in Eq. (31)–(33).

$$\eth = \varkappa^T \wp(\theta) \varkappa \tag{31}$$

$$\eth = \varkappa^T \left(\sum_{j=1}^{m} \theta_j \wp_j + \alpha^T(\theta) \wp(\theta) + \wp(\theta) \alpha(\theta) \right) \varkappa \tag{32}$$

$$\eth = \varkappa^T \begin{bmatrix} I_{n\sigma} \\ \alpha(\theta) \end{bmatrix}^T \varkappa \begin{bmatrix} \sum_{j=1}^{m} \theta_j \wp_{j,} & \wp(\theta) \\ \wp(\theta) & 0 \end{bmatrix} \begin{bmatrix} I_{n\sigma} \\ \alpha(\theta) \end{bmatrix} \varkappa \tag{33}$$

From the expression given in Eqs. (27)–(29), these symbolizations can be achieved as given in Eq. (34), when the fundamental matrix will be represented as X, while Y is the out factor, and $Z := \left[I_{n\sigma} \ A^T(\theta) \right]^T$. Then, $YZ = \left[I_{n\sigma} \ P_\mu^T \ A^T(\theta) \right]^T$.

$$\begin{bmatrix} \sum_{j=1}^{m} \theta_j \wp_j & \wp(\theta) \\ \wp(\theta) & 0 \end{bmatrix} = \begin{bmatrix} I_{n\sigma} & 0 \\ P_\mu & 0 \\ 0 & I_{n\sigma} \end{bmatrix}^T \begin{bmatrix} \sum_{j=1}^{m} \theta_j \wp_j & 0 & \wp_1 \\ 0 & 0 & (\wp^C) \\ \wp_1 & (\wp^C)^T & 0 \end{bmatrix} \begin{bmatrix} I_{n\sigma} & 0 \\ P_\mu & 0 \\ 0 & I_{n\sigma} \end{bmatrix} \tag{34}$$

Therefore, the condition of Lyapunov stability expression in Eq. (34) is comparable with $YZ^T X(YZ) < 0$. So, the matrix YZ can be reform as given in Eq. (35).

$$YZ = P_\mu * \begin{bmatrix} 0 & P_\mu \\ 0 & P_\mu \\ \omega^{\mathcal{R}} & \alpha_1 \end{bmatrix} \tag{35}$$

But the variable X which depends on the fuzzy weighing function derivative can be solved using conservatism of LMI-based stabilization conditions as in Eq. (36). The constraint notation is $\sum_{j=1}^{m} \theta_j \wp_j = 0$ and $\wp_j + F - \wp_1 \geq 0, j \in \mathrm{I}[2, m]$.

$$\sum_{j=1}^{m} \dot{\theta}_j \wp_j = \dot{\theta}_1 F + \sum_{j=2}^{m} \phi_j \left(\wp_j + F - \wp_1 \right) \le \dot{\theta}_1 F \le \sum_{j=2}^{m} \dot{\theta}_j \left(\wp_j + F - \wp_1 \right) \tag{36}$$

So, if $\Phi_1 := \Phi_1 F + \sum_{j=2}^{m} \dot{\theta}_j \left(\wp_j + F - \wp_1 \right)$ and $\Phi_2 := -\Phi_1 F + \sum_{j=2}^{m} \dot{\theta}_j \left(\wp_j + F - \wp_1 \right)$, the stability of Lyapunov fuzzy weighing function is guaranteed by the expression given in Eq. (37).

$$(YZ)^T X_k (YZ) < 0 \; k = 1, 2, \tag{37}$$

where

$$X_k := \begin{bmatrix} \Phi_k & 0 & \wp_1 \\ 0 & 0 & \wp^c \\ \wp_1 & (\wp^c)^T & 0 \end{bmatrix}$$

For the fuzzy system controller to be asymptotically stabilized, then it is given that $u = U(\theta) Q^{-1}(\theta) x$ with $U(\theta) = \sum_{j=1}^{m} \theta_j U_j$ and $Q(\theta) = \sum_{j=1}^{m} \theta_j Q_j$. The Lyapunov fuzzy system function is given as $\forall = x^T Q^{-1}(\theta) x$ and the system controller can be expressed as $u = U(\theta) Q^{-1}(\theta) x$, and the condition for stabilization $d = 0$ can be finally described as in Eq. (38) with a similar derivation of LMI-based stabilization condition [41, 42].

$$= -\sum_{j=1}^{m} \theta_j Q_j + A(\theta) Q(\theta) + Q(\theta) A^T(\theta) + B_u(\theta) U(\theta) + U^T(\theta) B_u^T(\theta) > 0 \tag{38}$$

3.4 Implementation of non-linear consequent fuzzy controller based-greenhouse design in LabVIEW

The Fuzzy Inference System (FIS) consists of two inputs (temperature and humidity) and two outputs (electric roof and water spills). A Mamdani fuzzy logic technique was implemented in this study due to its wide acceptance and suitability for this application. The triangular membership functions were implemented for all inputs and outputs. The input 'Temperature' had membership function values of 'cold', 'normal', and 'warm', while the input 'Humidity' membership function had

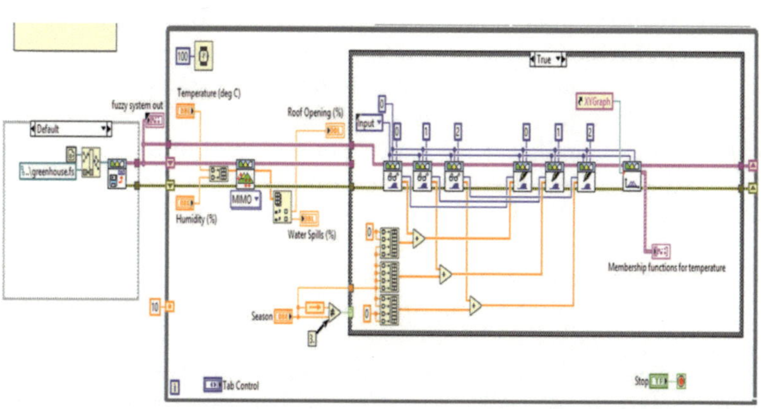

Figure 5.
Block diagram of an intelligent greenhouse control system.

values of 'dry', 'normal', and 'wet'. As for the outputs, the membership function for 'Electric Roof' signified the level of the opening for the roof. The output membership function parameter is 'closed', semi-open', and 'open'. The output 'Water Spills' represented the amount of water to be spilled by the sprinkler. This parameter has membership function values of 'low', 'moderate', and 'more'. Besides, the greenhouse control system was designed to consider each of the four major seasons (spring, summer, fall, and winter). As a result of this weather variation, each season has different membership function values for the weather conditions. The block diagram is illustrated in **Figure 5**.

4. Results and discussion

4.1 Simulation results of an intelligent greenhouse management based nonlinear control

The intelligent greenhouse control system was designed and simulated in LabVIEW using non-linear consequent for the controller. Two major interface environments were used to achieve the design of the system, the front panel, and the block diagram interfaces. The LabVIEW environment also provides a tool for fuzzy logic designs and the fuzzy logic designer has three interfaces, namely: Variables, Rules, and Test System. These interfaces respectively give the user an interface to specify the inputs and outputs of the system, provide the IF-THEN rules, and test the system to analyze the performance. In LabVIEW, an algorithm was implemented for the intelligent control of the greenhouse. This algorithm was implemented using a block diagram for the simulation of a nonlinear based intelligent greenhouse control system.

The interface has a knob that can be used to select a particular season. Also, the temperature and humidity can be altered to view various results. Selecting different values for temperature and humidity result in different outputs for roof opening and the water spills through system actuators. These outputs are determined by the fuzzy logic controller. Depending on the season selected, the outputs of the FIS will differ even with the same inputs. This is mainly because each season uses a different membership function for its decision-making. Considering these scenarios, experiments were conducted for each of the four seasons with the same input values. This was done to analyze varying results of the seasons and to examine the effectiveness of the control system. During this summer season, the dynamic sensor deployed to the environment is temperature and moisture sensors for monitoring the temperature and humidity of the greenhouse at constant temperature input of 25°C and the relative humidity of 85%. The membership function for temperature has three stages cold, normal, warm. It observed that temperature starts to normalizes from 22.5–32.5 degrees celcius to get constant temperature input. Therefore, the roof is open at 50%, and water spilled at the relative humidity of 40.1%. The simulation of fuzzy controller based intelligent greenhouse during the summer season was presented in **Figure 6**, and its fuzzy membership functions. The surface view of the dynamic system testing is presented in **Figure 7**.

In this work, a knob is designed to mimic the outside environment based on four possible weather conditions in a year (summer, spring, rainfall, winter). The constant temperature parameter set is 25°C and humidity at 65%. The membership function for temperature has three stages cold, normal, warm. During the summer, the temperature starts to normalize from 22.5–32.5 degrees Celsius to get a constant temperature value. Then, from the understanding of physics, an increase in temperature reduces humidity and relatively controls the sprinkler to turn ON and

DOI: http://dx.doi.org/10.5772/intechopen.98409

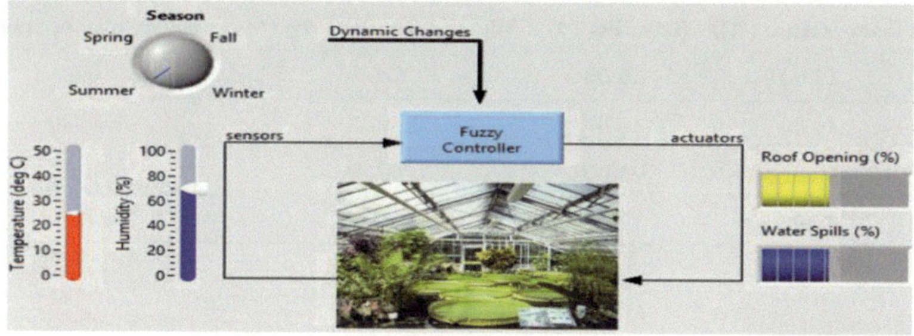

(a)

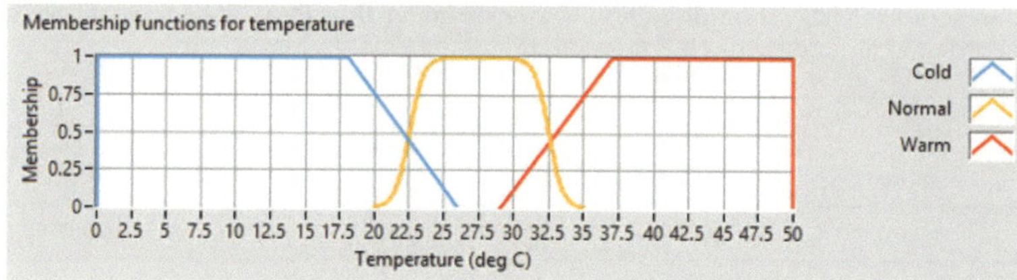

(b)

Figure 6.
A greenhouse simulation model for spring season and its membership function. (a) A greenhouse simulation model for the summer season. (b) The simulation model result of an intelligent greenhouse environment shows that the average temperature of (17:5°C) and humidity (55%) are conducive for crop growth and development without requiring roof opening or water spills.

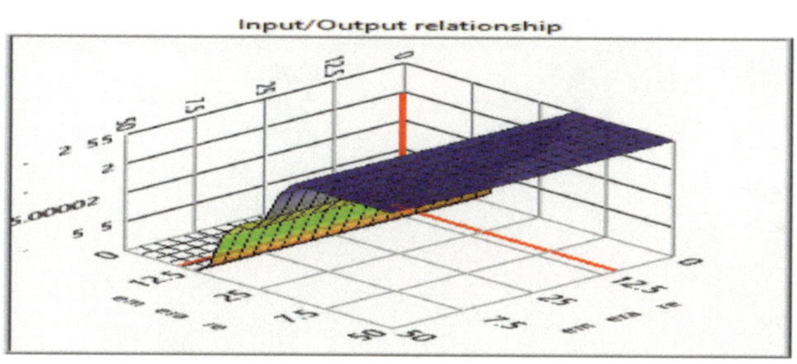

Figure 7.
Surface view of the dynamic system testing.

Series	Temperature (°C)	Humidity (%)	Water sprinkler flow (%)	Electric roof opening (%)
1	10.00	60.00	0.00	0.00
2	20.00	40.00	20.80	10.56
3	30.00	20.00	90.43	50.00
4	40.00	10.00	100.00	100.00

Table 3.
Results for the summer season.

cause the roof opening. All these calculations are handled logically by the fuzzy logic controller in the software-context based on the input and possible output variables. **Tables 3–6** presented the results obtained for summer, spring, winter, and fall seasons respectively, and the graphical representation of the results obtained is in **Figures 8–11**.

Series	Temperature (°C)	Humidity (%)	Water sprinkler flow (%)	Electric roof opening (%)
1	10.00	60.00	0.00	0.00
2	20.00	40.00	50.38	30.30
3	30.00	20.00	80.79	70.50
4	40.00	10.00	100.00	100.00

Table 4.
Results for the spring season.

Series	Temperature (°C)	Humidity (%)	Water sprinkler flow (%)	Electric roof opening (%)
1	10.00	60.00	50.00	50.00
2	20.00	40.00	80.80	70.63
3	30.00	20.00	90.70	70.63
4	40.00	10.00	100.00	100.00

Table 5.
Results for the winter season.

Series	Temperature (°C)	Humidity (%)	Water sprinkler flow (%)	Electric roof opening (%)
1	10.00	60.00	0.00	0.00
2	20.00	40.00	75.80	60.00
3	30.00	20.00	80.23	70.00
4	40.00	10.00	100.00	100.00

Table 6.
Results for rainfall season.

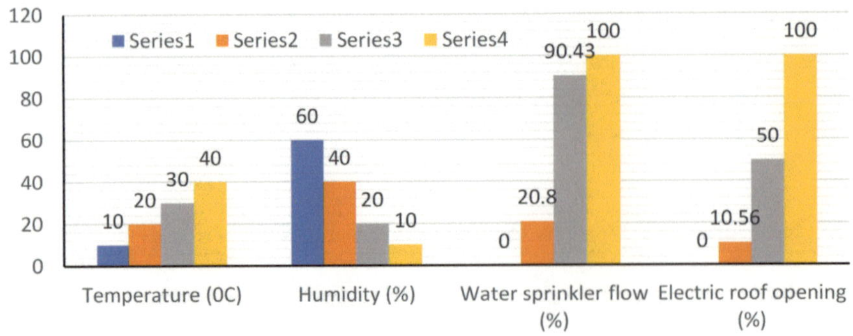

Figure 8.
Graphical representation of summer season parameters.

In this context, a knob is designed to mimic the outside environment based on four possible weather conditions in a year (summer, spring, rainfall, winter). The constant temperature parameter set is 25°C and humidity at 65%. The membership function for temperature has three stages cold, normal, warm. During the summer, the temperature starts to normalize from 22.5–32.5 degrees Celsius to get constant temperature value. Then, from the understanding of physics an increase in temperature reduces humidity and relatively controls the sprinkler to turn ON and the cause the roof opening. All these calculations are handled logically by the fuzzy

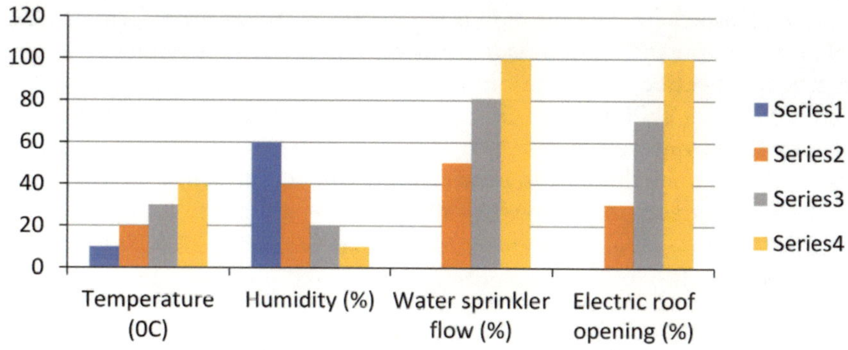

Figure 9.
Graphical output for simulation of spring season parameters.

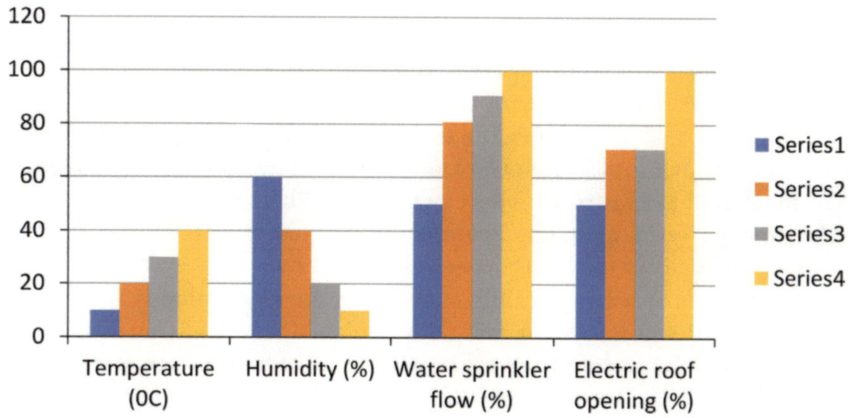

Figure 10.
Graphical output for simulation of winter season parameters.

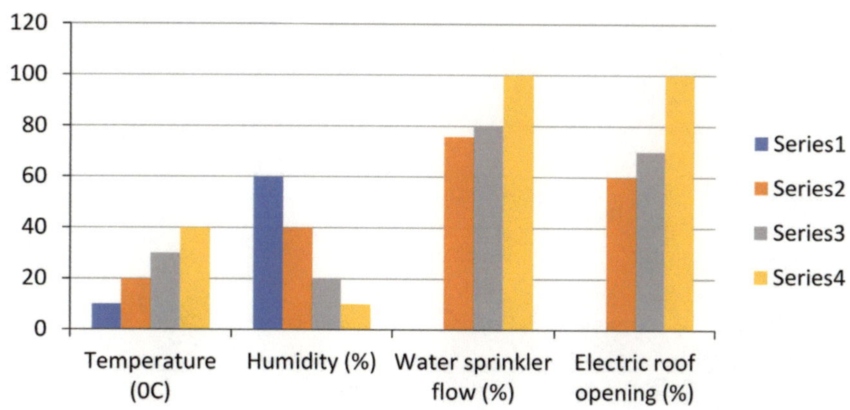

Figure 11.
Graphical output for simulation of rainfall season parameters.

logic controller in the software-context based on the input and possible output variables.

The results for each season depending on different environmental and season behavior which is processed by non-linear consequent fuzzy logic controller. This is achieved using different membership functions for each season. Since each season has its unique weather conditions and temperature requirements. The variation based on the season's implementation is to ensure an effective performance of controller during the different seasons. Furthermore, it can be observed from the results that irrespective of the season, higher temperatures lead to wide roof openings and high-water spill levels. This is done to reduce the temperature to the level specified by the farming environs. Also, low temperatures result in no roof openings or water spillage, since there is no need to lower the temperature further. But, during the summer season, the average temperature and humidity required is $(27.5^0C\&65\%)$ respectively. For every season that beyond $(30.5^0C\&75\%)$ of temperature and humidity will require automation of roof opening and water spilled.

4.2 Simulation results of a Lyapunov stability of nonlinear control system

The nonlinear fuzzy controller system for managing intelligent greenhouse was simulated in the MATLAB environment to achieved the stabilization of linearizing system, when its asymptotically stable using Lyapunov function. From the state-space of fuzzy model given in Eqs. (18), (25) and (26), the characteristics equation is derived as $|\ell\imath - \text{Å}|$, and the description is given [42]. If $f(\ell, \lambda) = |\ell\imath - \text{Å}|$, the system is universal and stable since the eigenvalues are positioned at the left-half side. Also, the eigenvalues (λ) follow a trend when plotted a multi-dimensional of $f(\ell, \lambda)$ as illustrated in **Figure 12**. This eigenvalue (λ) help to achieved a steady with better dynamic performances, good compensation quality and fast responses of the system as it moves closer to the trend of red spotted lines. The system controller undergoes processes to achieve stabilization when the eigenvalue is $\lambda = -1 * 10^2$ at periods of (0–0.50) seconds using Fast Fourier Transform (FFT) analysis.

For instance, we considered the continuous-time of nonlinear system to compute the equilibrium points and steadiness (stability) of the system as given in Eq. (39). The control system pathways based on the dynamic nature was verified in the MATLAB simulation environment for the chosen value of $g = 2$, $g = 3$, $g = 4$, and $g = 5$.

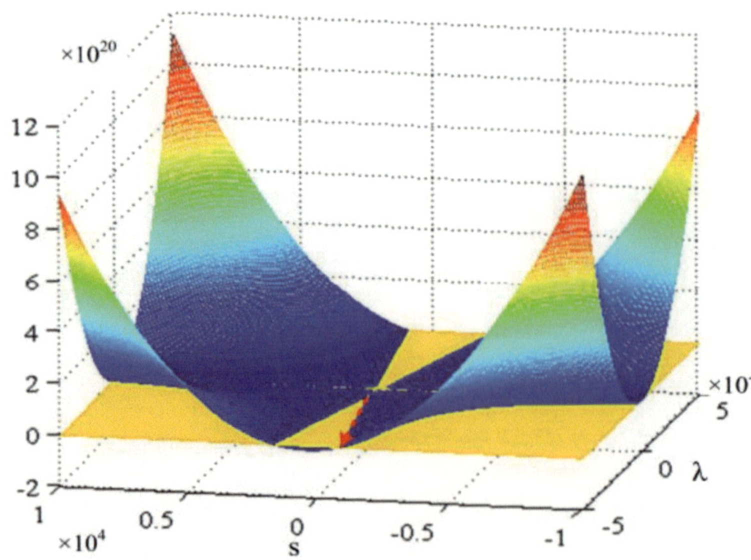

Figure 12.
A multi-dimensional design of eigenvalues $f(\ell, \lambda.)$

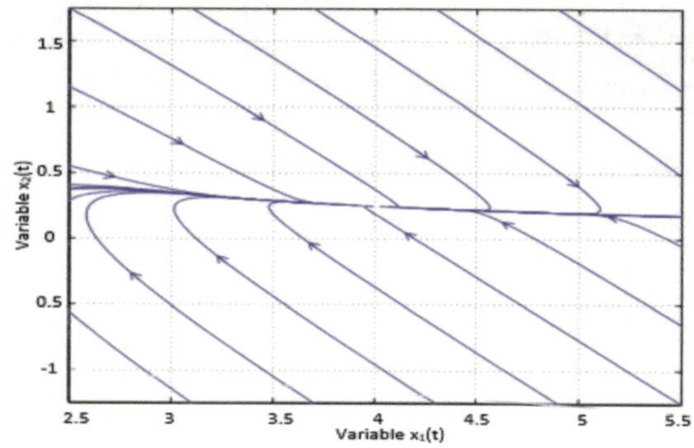

(a) $g = 2$

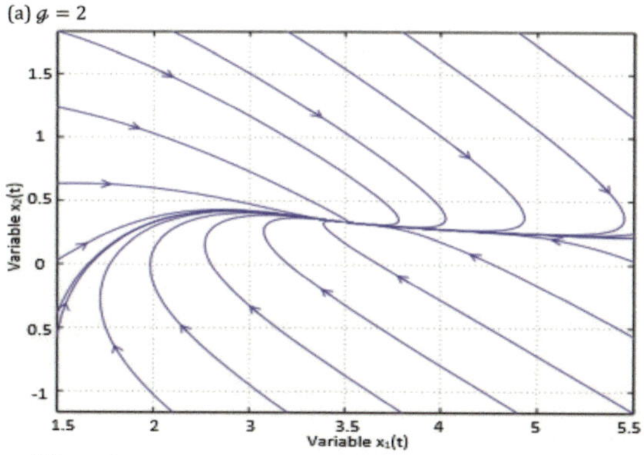

(b) $g = 3$

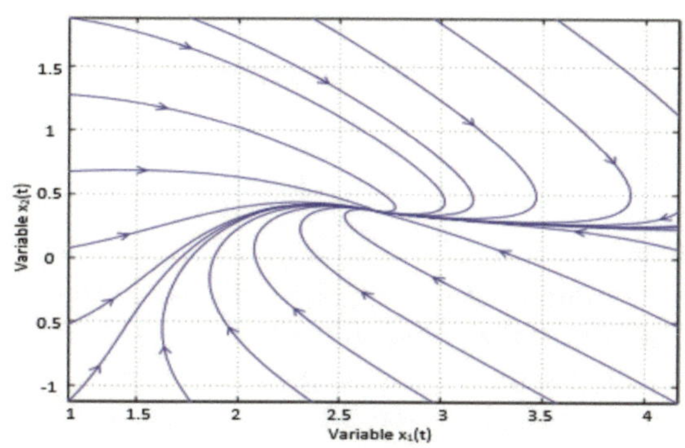

(c) $g = 4$

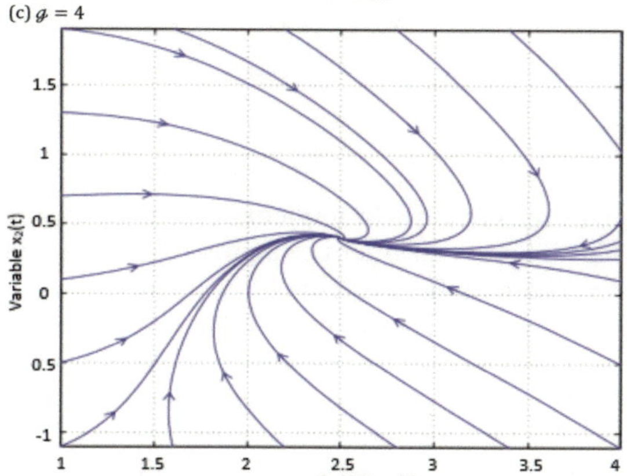

(d) $g = 5$

Figure 13.
Pathway line within the setpoint environment when $g = 2$, $g = 3$, $g = 4$, and $g = 5$.

$$\begin{cases} 0 = g(2 - x_1) + x_1^2 x_2 \\ 0 = x_1 - x_1^2 x_2 = x_1(1 - x_1 x_1), when\ x_1 = 0, and\ x_1 x_1 = 1 \end{cases} \tag{39}$$

Therefore, we substitute $x_1 x_1 = 1$ into the first equation, as $x_1 = 0$ does not satisfy the condition. It gives $2g - gx_1 + x_1 = 0 \rightarrow x_1 = \frac{2g}{g-1}$.

Then, the equilibrium point of the system can be obtained when $g \neq 1$, which give expression in Eq. (40);

$$\bar{x}_1 = \frac{2g}{g - 1}, \quad \bar{x}_2 = \frac{g - 1}{2g} \tag{40}$$

The set point environment of the linearizing system can be derived as given in Eq. (41), and the characteristic polynomial of the system is given in Eq. (42):

$$\dot{x}(t) = \begin{bmatrix} -g + 2x_1x_2 & x_1^2 \\ 1 - 2x_1x_2 & -x_1^2 \end{bmatrix}_{(\bar{x}_1, \bar{x}_2)} x(t) = \begin{bmatrix} 2 - g & \frac{4g^2}{(g-1)^2} \\ -1 & \frac{-4g^2}{(g-1)^2} \end{bmatrix} x(t) \tag{41}$$

$$\Delta(g) = g^2 + \left[g - 2 + \frac{4g^2}{(g-1)^2} \right] g + \frac{4g^2}{(g-1)^2} = 0 \tag{42}$$

This expression in (42) can be resolves as given in Eq. (43).

$$\Delta(g) = g^2 + \frac{4g^2}{(g-1)^2} g + \frac{4g^2}{(g-1)} = 0 \tag{43}$$

Therefore, if the polynomial coefficient is both positive then equilibrium point is stable when $g > 1$, else is unstable when at least one eigenvalue $g < 1$.

The simulation results for the system control pathways for nonlinear system using Lyapunov function with given stability conditions $g = 2$, $g = 3$, $g = 4$, and $g = 5$ are shown in **Figure 13**.

5. Conclusions

The greenhouse control system was implemented using the Fuzzy Logic Controller design with non-linear consequent as an intelligence in the decision-making process of the system. The membership functions include two inputs (temperature and humidity) and two outputs (roof opening and water spills). The intelligent greenhouse system was designed to cater for each of the four major seasons (summer, spring, winter, and rainfall) and this was achieved by implementing different membership functions for each season. The development of an intelligent greenhouse control system was simulated and implemented in LabVIEW. These technologies, FLC and Virtual Instrumentation in LabVIEW are widely adopted to enable computing and communication to migrate out of the gray box into ordinary objects (standalone system). However, it is significant that building of an intelligent systems to model human activities or interactions is important to the agricultural technology development. The results obtained show varying performances for each season to cater for different weather conditions. Future research will be considered incorporating a heating mechanism to raise the temperature for varying conditions and hybrid intelligent techniques using optimization technique for a better system performance.

Acknowledgements

The authors acknowledge the management of the Federal University of Technology, Minna for adequate equipped Embedded and Intelligent System Laboratory which makes it suitable for conducting efficient research.

Conflict of interest

No 'conflict of interest' in this research.

Author details

Lukman Adewale Ajao[1*], Emmanuel Adewale Adedokun[2], Joseph Ebosetale Okhaifoh[3] and Habib Bello Salau[2]

1 Department of Computer Engineering, Federal University of Technology, Minna, Nigeria

2 Department of Computer Engineering, Ahmadu Bello University, Zaria, Nigeria

3 Department of Electrical and Electronics Engineering, Federal University of Petroleum Resources, Effurun, Nigeria

*Address all correspondence to: ajao.wale@futminna.edu.ng

References

[1] Ajao LA, Agajo J, Kolo JG, Maliki D, Adegboye MA. Wireless sensor network based-internet of thing for agro-climatic parameters monitoring and real-time data acquisition. *Journal of Asian Scientific Research (JASR)* 2017:7(6): 240-252. DOI:10.18488/journal.2.2017.76.240.252.

[2] Adegboye MA, Ajao LA, Folorunso TA, Mamud AY, Isah AS. Automatic fertilized irrigation control and management system. FUW Trends in Science and Technology Journal (FTSTJ), 2017:2(2):801-805.

[3] Adison R. Artificial intelligence opportunities and challenges in businesses. July 24, 2019. Retrieved: https//www.towardsdatascience.com.

[4] Ajao LA, Agajo J, Kolo JG, Inalegwu OC, Edem EA. Development of a low power consumption smart embedded wireless sensor network for the ubiquitous environmental monitoring using ZigBee module. Journal of Science Technology & Education (JOSTE), 2017:5(1):94-108.

[5] Thornton P, Dinesh D, Cramer L, Loboguerrero AM, Campbell B. Agriculture in a changing climate: Keeping our cool in the face of the hothouse, Outlook in Agriculture, 2018: 47(4); 283–290. DOI: 10.1177/0030727018815332.

[6] Beacham AM, Hand P, Barker GC, Denby KJ, Teakle GR, Walley PG, Monaghan KJ. Addressing the threat of climate change to agriculture requires improving crop resilience to short-term abiotic stress Outlook in Agriculture, 2018:3:1-7. DOI: 10.1177/00307270.

[7] Copeland BJ. Artificial intelligence, (2020). Retrieved from; https://www.britannica.com/technology/artificial-intelligence.

[8] Ajao LA, Agajo J, Mua'zu MB, Schueller JK. A scheduling-based algorithm for low energy consumption in smart agriculture precision monitoring system, Agricultural Engineering International CIGR Journal. 2020:22(3); 103-117.

[9] Treut HL, Historical overview of climate change science, in climate change 2007: The physical science basis. Contribution of Working Group 1 to the Fourth Assessment Report of the Intergovernmental Panel on Climate Change, *S. Solomon*, et al., Editors. 2007, Cambridge University Press: Cambridge, United Kingdom, and New York, NY, USA.

[10] McCarthy J. (2020) Alarming facts from the UN's new climate report, March 12, 2020. https://www.globalcitizen.org/en/content/facts-from-un-climate-report-wmo.

[11] IPCC (2007). Climate change 2007: Mitigation. (PDF) (863 pp, 24MB) exit contribution of working group III to the fourth assessment report of the intergovernmental panel on climate change [B. Metz, O.R. Davidson, P.R. Bosch, R. Dave, L.a. Meyer (eds)], Cambridge university press, Cambridge, United Kingdom and New York, NY, USA.

[12] Massey R, Ulmer A. Pork production and greenhouse gas emissions, Factsheets: Environmental Stewardship, 2015 U.S. Pork Center of Excellence, (515) 223-2641, 04/09/2010. https://porkgateway.org/resource/pork-production-and-greenhouse-gas-emissions.

[13] Singh S, Singh DR, Velmurugan A, Jaisakar I, Swarnam TP. Bio-diversity and climate change adaptation in tropical islands: Chapter 23- coping with climatic uncertainties through improved

production technologies in tropical island conditions, Elsevier, 2018: 623-666. DOI: 10.1016/B978-0-12-813064-3.00023.

[14] U.S. Environmental Protection Agency, Inventory of U.S. Greenhouse Gas Emissions and Sinks: 1990-2006. 2008.

[15] Bhagwat SD, Hulloli AI, Patil SB, Khan AA, and Kamble AS. Smart greenhouse using IoT and cloud computing. International Research Journal of Engineering and Technology (IRJET), 2018:5(3): 2330–2333.

[16] Dilbaz AS. Design and Implementation of a Smart Greenhouse, Doctoral dissertation, Middle East Technical University (2017).

[17] Mesmoudi K, Soudani A, Bournet P. Determination of the inside air temperature of a greenhouse with tomato crop under hot and arid climates Applied Science Environment Management. 2010:5(1); 117–129.

[18] Kartha S, Baer P, Athanasiou T, and Kemp-Benedict E. The greenhouse development rights framework, climate, and development, 2014:1(2); 147-165. DOI:10.3763/CDEV.2009.0010.

[19] Xu F, Chen J, Zhang L, and Zhan H. Self-tuning fuzzy logic control of greenhouse temperature using real coded genetic algorithm. In: Proceedings of the 9th International Conference on Control, Automation, Robotics and Vision, IEEE, Singapore, 2006:1–6.

[20] Pawlowski A, Guzman J, Rodríguez F, Berenguel M, Sánchez J, Dormido S. Simulation of greenhouse climate monitoring and control with wireless sensor network and event-based control. Sensors, 2009:9(1); 232–252.

[21] Ramya K, Kumar KAR. Smart controller for grid stabilization in the an optoelectric system, International Journal of Electrical Engineering Education, 2018:0(0); 1–10. DOI: 10.1177/0020720918813807.

[22] Zheng Z, Zhang K, Liu C. Greenhouse temperature monitoring system based on La2VIEW, IFIP International Federation for Information Processing, 2011:345; 23–29.

[23] Khan MI, Nazeer M, Shehzad N, Saleem A, Ahmad F. Scrutiny of entropy optimized tangent hyperbolic fluid (non-Newtonian) through perturbation and numerical methods between heated plates, Advances in Mechanical Engineering, 2020:12(12); 1–12. DOI: 10.1177/1687814020970772

[24] Ferrández-Pastor JF, Alcañiz-Lucas S, García-Chamizo JM, Platero-Horcajadas M. Smart environments design on industrial automated greenhouses. MDPI Proceedings, 2019: 31(36); 1-13. DOI:10.3390/proceedings2019031036.

[25] Li Z, Wang J, Higgs R, Zhou L, Yuan W. Design of an intelligent management system for agricultural greenhouses based on the internet of things. IEEE international conference on computational science and engineering and IEEE/IFIP international conference on embedded and ubiquitous computing. CSE and EUC, 2017:2(1); 154–160. DOI:10.1109/CSE-EUC.2017.212

[26] Imam A, Gaur D. Smart greenhouse monitoring using the internet of things. International Journal of Advanced Research in Electronics and Communication Engineering (IJARECE). 2018:7(5); 519–523.

[27] Wang X, Yu H. Research on the control system of the intelligent greenhouse of IoT based on ZigBee, Journal of Physics: Conference Series,

2019:1345(4); 1–7. DOI:10.1088/1742-6596/1345/4/042036.

[28] Ullah MW, Mortuza MG, Kabir MH, Ahmed ZU, Supta SKD, Das P, Hossain SMD. Internet of things based smart greenhouse: remote monitoring and automatic control, 2018 Joint International Conference on Energy, Ecology and Environment (ICEEE 2018) and International Conference on Electric and Intelligent Vehicles, (2018): 1-6. DOI:10.12783/dteees/iceee2018/27803.

[29] Shukla AJ, Panchal V, Patel S. Intelligent building design based on internet of things, International Journal of Emerging Trends in Electrical and Electronics, 2015:11(2); 57–61.

[30] Pavithra G. Intelligent monitoring device for agricultural greenhouse using IoT. Journal of Agricultural Science and Food Research. 2018:9(2); 2–5.

[31] Sagaya KS, and Kalpana PY. A review on context modeling techniques in context aware-computing, International Journal of Engineering Technology, 2016:8(1); 429–433.

[32] Ajao LA, Ajao FJ, Adegboye MA, Ismail AA. An embedded fuzzy logic-based application for density traffic control system. International Journal of Artificial Intelligence Research (IJAIR), 2018:2(1): 6-13.

[33] Abougarair AJ, Bakush M.M, Aramah M. Optimal estimator design based on optimal controller, Libyan International Conference on Electrical Engineering and Technologies (LICEET2018); 2018:3-7 March 2018, Tripoli, Libya.

[34] Liu, Y. Wu, F. Ban X. Dynamic output feedback control for continuous-time t-s fuzzy systems using fuzzy Lyapunov functions, IEEE Transactions on Fuzzy Systems, 2016: 1-13. DOI 10.1109/TFUZZ.2016.2598852.

[35] Mesmoudi K, Soudani A. Bournet P. Determination of the inside air temperature of a greenhouse with tomato crop under hot and arid climates Applied Science Environment Management. 2010:5(1):117–129.

[36] Hossain, MF. Solar energy integration into advanced building design for meeting energy demand and environment problem. International Journal of Energy Research, 2016:40(9): 1293-1300.

[37] Ang KH, Chong GCY, and Li Y. PID control system analysis, design, and technology, IEEE Transactions on Control Systems Technology, 2005:13 (4): 559–576. DOI:10.1109/TCST.2005.847331.

[38] Jie L, Wang W, Golnaraghi F, Kubica E. A novel fuzzy framework for nonlinear system control. Fuzzy Sets and Systems. 2010:161(21); 2746–2759. DOI:10.1016/j.fss.2010.04.009.

[39] Silva JF, Pinto SF. Linear and non-linear control switching power converters, Elsevier Power Electronics, 2018:1141-1220.

[40] Matyas, A. Nagy, Z. Lendek Z. Controller Design for Time-Delay TS Fuzzy Systems with Nonlinear Consequents, in 2020 IEEE International Conference on Fuzzy System (FUZZ-IEEE), 1–6, 2020.

[41] Faria A, Silva G, Oliverira A. Reducing the conservatism of LMI-based stabilization conditions for T-S fuzzy systems using fuzzy Lyapunov functions, International Journal of System Science, 2013:44(10); 1956-1969.

[42] Cao Y, Xu Y, Li Y, Yu, J. A lyapunov stability theory-based control strategy for three-level shunt active power filter, Energies, 2017:10, 112; doi:10.3390/en10010112.

Chapter 25

An Overview of CAN-BUS Development, Utilization, and Future Potential in Serial Network Messaging for Off-Road Mobile Equipment

Hannah M. Boland, Morgan I. Burgett, Aaron J. Etienne and Robert M. Stwalley III

Abstract

A Controller Area Network (CAN) is a serial network information technology that facilitates the passing of information between Electronic Control Units (ECUs, also known as nodes). Developed by BOSCH in 1986 to circumvent challenges in harness-connected systems and provide improved message handling in automobiles, the CAN interface allows broadcast communication between all connected ECUs within a vehicle's integrated electronic system through distributed control and decentralized measuring equipment. Since the early uses of CAN in car engine management, improvements in bitrate, bandwidth, and standardization protocols (such as ISO 11898 and SAE J1939) have led to CAN utilization in various industry applications, such as factory automation, aviation, off-highway vehicles, and telematics. Alternative wired and wireless technologies have been used to connect and network with CAN-BUS (such as Ethernet, Bluetooth, Wi-Fi, ZigBee, etc.), further expanding the diversity of applications in which the serial network is employed. In this chapter, the past, present, and prospective future developments of CAN technology, with focused attention on applications in the agricultural and off-road sectors are broadly examined. CAN technology fundamentals, standards creation, modern day uses, and potential functionalities and challenges specific to CAN in the wake of precision agriculture and smart farming are discussed in detail.

Keywords: CAN-BUS, Serial Network, Agricultural Sector, Electronic Control Units

1. Introduction and Background

A Controller Area Network (CAN) in a vehicle or machine is analogous to the nervous system of a living organism. The nervous system of the body is a neuron-based network that collects signals from sensory receptors, passes chemical messages to and from the brain, responds to stimuli, and initiates actions. Expanding the analogy, sensors in a controller circuit are the equivalent of receptors, and an electronic control unit (ECU) can be visualized as a sensory neuron system dedicated to a specific function, bridging communication between receptors and the central nervous system. CAN-BUS systems create communication pathways between the electronic control units within a vehicle, allowing the transfer and interpretation of collected data. Prior to the invention of CAN-BUS, there was no efficient means of cross-communication between ECUs. CAN-BUS is efficient by relaying the most important messages first, through a prioritization scheme of source ID-encoded messages using the binary unit system (BUS). This is an extremely robust arrangement, with a high ability to both detect signal errors and to function when hardware is cross wired. This structure is fully distributed, which allows for a single access point for all the desirable information collected. CAN-BUS is a relatively simple, low-cost system that reduces the overall harness weight and amount of wiring needed in a vehicle, improving the integrity of transmitted data in comparison to harness-connected electrical structures [1].

While CAN-BUS has been an effective communication technology in many past and present applications, future utilization of the network system continues to be a subject of research and development. In agricultural uses, this tool aids in precision agricultural applications and in the realm of data communication within larger farm systems. Vehicle autonomy is another area in which CAN-BUS may play an important role as an inter-communication system. Additionally, there is still significant untapped potential for integrating CAN-BUS messaging into both more off-road control systems and wireless technologies.

The purpose of this chapter is to familiarize the reader with the importance of CAN-BUS in commercial off-road vehicles, applications, and future potential usage. In order to fully understand the benefits of CAN-BUS, the origins of CAN-BUS and its subsequent applications will be summarized. A high-level analysis of CAN-BUS technology, standards, and communication protocols will be presented to better familiarize the reader with essential technological concepts. Current applications of CAN-BUS and a comparison with alternative electronic control systems will be provided. A final qualitative evaluation of CAN-BUS capabilities will allow for a deeper understanding of why it is the dominant technology in modern vehicles and what innovations may be needed to expand its breadth of application in the changing technological landscape of off-road equipment.

2. History

2.1 CAN-BUS development

CAN was developed in 1986 by BOSCH as a means to overcome the limitations in harness-connected control systems [2]. Their goal was greater functionality in message communication in automobiles, which could be accomplished through distributed control. A distributed control system connects multiple, specific instrumentation into a system network that facilitates the transmission of data and information, adapting to the needs of the automation control scheme used. It combines individual, decentralized measuring control equipment into a main network node, creating an interconnected network capable of controlling a larger system [3]. In developing the CAN system, the control equipment corresponded to nodes (or ECUs), which were connected to a two-wire bus, completing the network connection. The system prevented message collisions, thereby preventing the loss of crucial information, a common issue with other existing technologies at the time.

While other technologies could achieve the goal of inter-node communication, they required complex wiring systems, with each ECU individually connected to other ECUs to provide a communication pathway [1]. The point-to-point wiring of all ECUs was unnecessarily complex and caused difficulties in data and message management. In CAN-BUS implementation, all the connections are made directly on the same area network. Through utilization of microcontrollers, the system complexity decreased dramatically, allowing for a reduction in wiring, a simplified manufacturing assembly process for connecting nodes, and an overall increased system performance. Due to the improved efficiencies and system simplicity that this technology offered, CAN-BUS became a viable alternative to the complex point-to-point wiring harnesses used at the time [4].

In 1987, both Intel and Philips developed the first CAN controller chips, the Intel 82526 and the Philips 82C200, respectively [2]. The first iteration of this technology was a chip that managed messages by assigned priorities. This allowed the more important messages to be received with significantly less delay. Notably, this first system included error detection, which would automatically disconnect faulty nodes, while still allowing uninterrupted communication between working nodes [5]. The hierarchy system allowed for the most crucial information to be passed along first, making the system particularly useful in applications with high safety requirements [1].

In early CAN development, there were two hardware implementations that cover the bulk of installations: Basic CAN and Full CAN. Basic CAN utilized a single message buffer to receive and transmit messages. The standard CAN controller implemented a specified number of message buffers (usually around sixteen), wherein the programmed algorithm read the received messages and wrote messages to be transmitted [6]. In Basic CAN, the received message is passed through acceptance filtering, which then decides whether to process a message or ignore it. Software is used to control the acceptance filtering of a node in Basic CAN. Bit masks for message identifiers make it possible to ignore certain messages by ignoring specific identifiers, in order to reduce the software load requirement at the individual nodes [7].

Compared to Basic CAN, Full CAN is a bit more complex. Every transmitted or received message is accompanied by eight to sixteen memory buffers in the Full CAN scheme. Hardware, rather than software, performs acceptance filtering in this system, reducing the overall software load significantly. Individual buffers are configured to accept messages with specific identifiers, and unique buffers for individual messages allow more processing time for the messages that are received. The transmitted messages can then be better handled according to their priority levels. Data consistency is also improved through this one-on-one buffer-to-message configuration [7]. Unfortunately, Full CAN is limited in the number of frames that can be received, and it requires more computational chips at each node than Basic CAN. Early CAN controllers by Intel and Philips were constructed under the Basic CAN or Full CAN configurations, with Philips favoring the former and Intel the latter. Modern CAN controllers combine the frame handling and acceptance filtering strengths of both, so the distinction is no longer made between Basic and Full [2].

A major milestone in bringing CAN-BUS into industry was the development of the CAN-in-Automation (CiA) working group in 1992. CiA is an international organization comprised of manufacturers and users with the goal of creating developmental content based on members' interests and initiatives [2]. One year later, the International Organization for Standardization (ISO) published ISO 11898, which defined controller area network communication protocols for the automotive industry. ISO is a non-governmental organization, without corporate affiliations, comprised of individual standards organizations from 165 nations. It develops voluntary international standards and improves the world's trading potential by providing common standards across the globe [8]. The implementation of an ISO standard for CAN-BUS was an important step in bringing coherence and marketability to the serial network system.

As the bandwidth requirements of the automotive industry continued to increase, the CAN data link layer (which will be covered in later sections) needed to be updated. BOSCH began developing the CAN FD (flexible data-rate) protocol in 2011, working in conjunction with carmakers and other CAN experts. This updated protocol surmounted two of the most restrictive early CAN limitations: the data transfer rate and payload. CAN FD allows for a bit rate (transmission speed) of up to 12 Megabits per second (*Mbps*), twelve times faster than the previous maximum transmission limit. The data field message payload was expanded up to 64 bytes in length, an increase of eight times beyond the previous payload size restriction [2]. CAN FD incorporated a simple, yet powerful ideology: when only one node is transmitting data, the bit rate can increase as no nodes need to be synchronized. The nodes are then resynchronized following the data transmission and data integrity check, just prior to an acknowledgement of data acceptance [9]. By 2015, ISO 11898 had been updated to incorporate CAN FD, and it has continued to be the standard CAN system in commercial implementations [2].

2.2 Early applications of CAN technology

CAN-BUS has played a major role in industry since its debut in 1987. In the mid-1990s, companies like Infineon Technologies and Motorola began shipping large quantities of CAN-BUS controllers to European automotive manufacturers, marking the advent of CAN utilization in the automotive industry. In 1992, Mercedes-Benz was noted as the first manufacturer to implement the controller within their processes, when CAN-BUS was first incorporated in their high-end passenger cars for engine care management [2].

BMW was next to implement CAN-BUS technology in 1995. They introduced a star topology network with five electronic control units in their 7 Series cars. Then, they took the implementation even further and employed a second network for body electronics. This allowed two separate CAN-BUS networks to be associated through gateway connections. Following BMW's example, other manufacturers soon began implementing two separate systems in all their passenger cars. Today, many manufacturers have multiple CANBUS networks associated with their production vehicles [2]. An example of vehicular integration is presented in **Figure 1**.

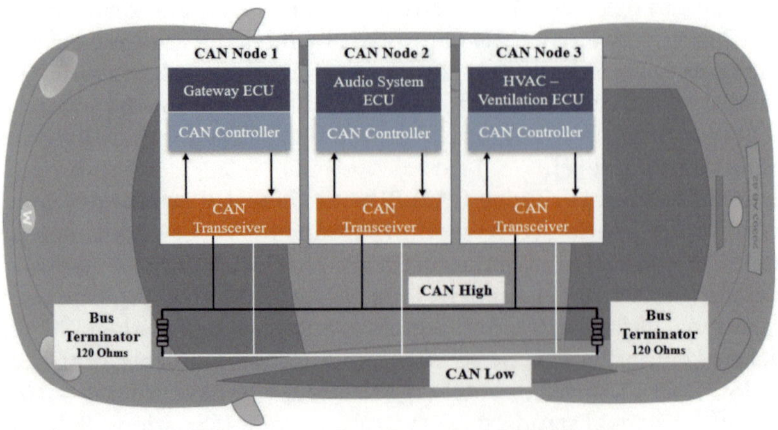

Figure 1.
Illustration showing the multiple node connections to CAN-BUS in a modern vehicle.

In 1993, a European consortium led by BOSCH prototyped a network which would later become CANopen. This project was eventually passed to CiA for further development and maintenance. In 1995, it was completely revised and became the most important standardized network in Europe within just a few years. The CANopen network protocol offers high configuration flexibility, which has allowed its installation in a multitude of applications. The networks were first used for internal machine communications, specifically in drives, but they have since been utilized in many other industries. Within the United States, CANopen has been implemented for use in forklifts, letter sorting machines, and other network processes [2].

As mentioned in the previous section, introduction of CAN-BUS into the automotive world required the standardization of protocols and testing standards to ensure CAN system conformity. ISO 11898, the first international standard for CAN, was based on the BOSCH CAN specification 2.0, and it standardized the high-speed physical layer for the system at the time [10]. As network technology continued to develop, allowing for different data transmission speeds and fault tolerances in the physical layers, new revisions to standards and interfaces for vehicle-specific applications were needed. This led to the development of SAE J1939 for heavy-duty vehicles and multiple other ISO standards (some will be covered in the CAN-BUS Standards Development section below). Due to the rapidity of CAN modification and development in the early 1990s, no error-free, complete standards or CAN specifications were available for CAN chip manufacturers. This led to the establishment of CAN conformance testing houses, where all CAN chips could be tested for compliance to the BOSCH CAN reference model using the testing plans outlined in ISO 16845 [2]. These steps were important in allowing the new technology to be widely applied in a variety of markets.

With regard to the marketing of CAN-BUS into the agricultural industry, in 2000 the German Mechanical Engineering Professional Society (VDMA) founded the Implementation Group of ISOBUS to promote the ISOBUS controller. The German Agricultural Society (Deutsche Landwirtschafts-Gesellschaft, DLG) assisted with the development of the first tests and a testing facility for ISOBUS compliance, which remains the primary test house for device compatibility. In 2009, several companies joined to form the Agricultural Industry Electronics Foundation (AEF), a non-profit organization which further promoted the use of CAN-BUS controllers, especially the implementation of ISO 11783. Since then, there have been many plug-tests organized at various locations. The first plug-test for CAN-BUS in North America was hosted by the Nebraska Tractor Test Laboratory in 2010 [11].

This review of the development of CAN-BUS and its early applications illustrates some of the current and future directions for the technology. Besides the novel use of a distributed communication network, these development efforts have truly positioned CAN-BUS as the leading serial network system in off-road vehicles. The establishment of international societies and standards has been essential in this effort. The societies are dedicated to enforcing CAN standardization across the industry and to enhancing the functionality and quality of CAN technology through research and development. These organizations will likely continue to play an important role as CAN systems are utilized in new implementations going forward.

3. Technology fundamentals

3.1 CAN utilization: messaging basics

To gain a more complete grasp on how CAN ID messaging works and how different ECUs can interpret these messages, it is helpful to understand the overall structure of CAN messages, from both a data and hardware perspective. This section covers the physical architecture of the BUS, the different components of CAN messages, CAN error-handling, a high-level breakdown of CAN layers, and provides an overview on how CAN-BUS systems support effective messaging channels.

The physical architecture or layer of a Controller Area Network includes two wires, CAN High (CAN-H) and CAN Low (CAN-L), which carry all CAN messages between ECUs and connect to BUS terminators at each end. The BUS terminators are powered and grounded, providing the necessary voltage to allow serial network operation. The most standard form of CAN wiring in modern systems is the twisted quad cabling configuration, in which a terminating bias circuit (TBC), with a power wire and ground wire, is wound together with the CAN-H and CAN-L signal wires between the two terminators [12]. Both of the signal wires have set dominant and recessive voltages that correspond to the CAN system type (high speed or low speed). The system reads the voltage difference between the two wires as a bit-value of "0" when the voltages are dominant, or a value of "1" when the voltages are recessive, creating the mechanism of sending binary messages through the system hardware [13].

A maximum of 30 ECUs can be attached to a single section of the BUS, and the overall number for ECUs connected to the network is limited to 254. The maximum number of available ECU addresses is limited to 256, because the maximum length of a data signal is 8 bytes. The 255 address is left null, and the 256 address indicates for a message to be accepted by every ECU connected to the network [12]. Since CAN-BUS is a broadcast protocol, messages are not sent to specific nodes, but rather, every ECU connected to the network receives every transmission from all other nodes on the same network. Various ECUs typically have filters on their receiving ends, so that the local computer only accepts the messages that pertain to its operational needs [14]. This open communication between all connected nodes helps to improve the manufacturing process and implementation of the system, creating vehicle-wide interconnection. Since all the nodes are linked by subsystem functions, there are no redundant connections between any two specific ECUs.

As shown in **Figure 2**, a basic CAN message has eight key parts: 1) Start of Frame (SOF); 2) CAN Identifier (CAN ID); 3) Remote Transmission Request (RTR); 4) Control; 5) Data; 6) Cyclic Redundancy Check (CRC); 7) Acknowledgement (ACK); and 8) End of Frame (EOF). It should be noted that the "CAN frame" consists of parts 2 and 5: the CAN ID and the Data [12]. The SOF is a 1 bit "dominant zero" at the beginning of a CAN message which signals that an ECU is about to send a message. This alerts other ECUs connected to the CAN to "listen" for the message transmission. The CAN ID contains information on the message priority (lower values indicate higher priority) and the source address. The identifier bit length varies by version of CAN, with CAN 2.0 being 11-bits and later versions relying on

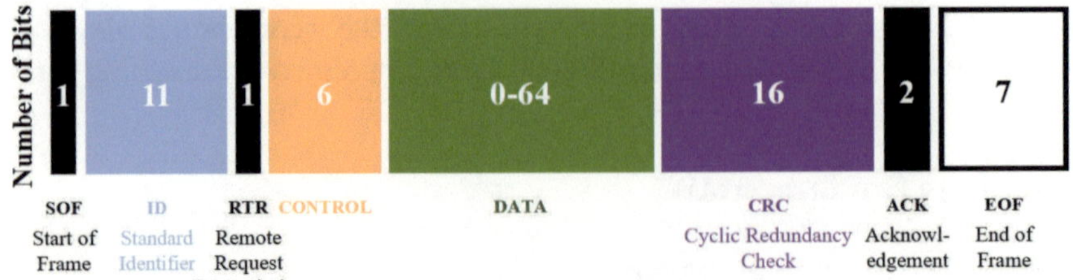

Figure 2.
CAN-BUS message structure.

extended 29-bit identifiers. The RTR is another 1-bit piece of the message indicating whether a node is sending data to or requesting data from a specific ECU. The Control portion of a CAN message is 6 bits in length, 4 of which are the data length code (DLC), which denotes the size of the data message to be transmitted (0–8 bytes) [13]. The Data segment of the CAN message makes-up the bulk of information being communicated, and it contains all the CAN signals to be extracted and decoded for use by the receiving ECUs [5].

The four message parts prior to the Data portion are all used to give the receiving ECUs adequate information on whether to receive the data being sent and what kind of data to expect. The last three parts of a CAN message are used to ensure that the data was transmitted successfully. The CRC is a 16-bit portion of the data that checks the data integrity, while the ACK is a 2-bit acknowledgement that the CRC found no issues with the data, allowing it to pass. Finally, the EOF is the 7-bit cap on a CAN message that signals the end of the transmission [13]. A breakdown of these eight parts highlights the strength of CAN messaging, in that it provides both front-end and back-end context for the data being sent. Message types used in CAN-BUS include the data frame (a data transmission message), the error frame (a message that violates CAN formatting to signal an error in data transfer), the remote frame (a message to request data), and an overload frame (a message transmitted by an overloaded node to trigger delays) [5].

System robustness and error handling are the two major benefits of the CAN-BUS system architecture. Error handling is the methodology of detecting flawed messages that come across the CAN-BUS, in which the original sender destroys a faulty message using an Error Frame, and then re-transmits the correct message. All CAN controllers connected to the BUS listen for potential transmission errors whenever a new message is sent along the BUS [15]. When an error has been identified, the node that discovered the error will transmit an Error Flag throughout the system, halting all CAN-BUS traffic. The other connected nodes will each receive the Error Flag and transmit eight recessive bits, known as an Error Delimiter signal, to clear the BUS before taking appropriate action in response to the error. The most common response to an Error Flag is to discard the erroneous message and continue to transmit and receive other messages streaming on the BUS. This allows for what is known as fault tolerance, or the ability for the system to function around an error state [15]. An example of the error handling message structure is detailed in **Figure 3**.

Each node keeps a record of detected errors through two different registers. Errors that the ECU was responsible for sending are accounted for in the Transmit Error Counter, while faults that it detected in other nodes' messages are logged in the Receive Error Counter. Several protocols have been defined which govern how recorded errors increment or decrement the counters. When a transmitter detects a fault error in a message, it increments the register for the Transmit Errors at a faster rate than the receiving nodes increment their Receive Error registers, since

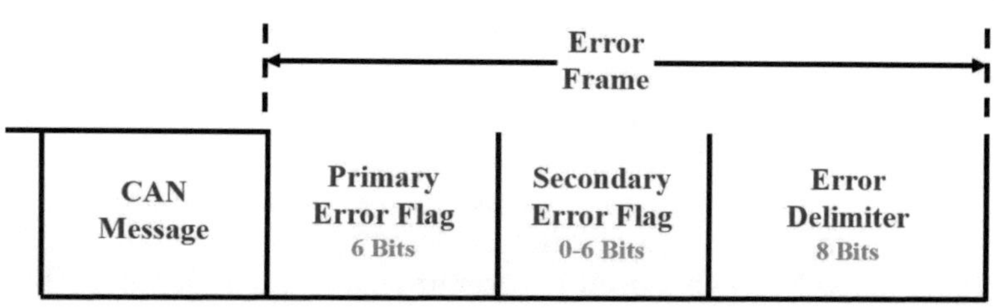

Figure 3.
A sample of an error handling message structure.

the transmitter causes system faults in most cases. When a node's Error Counter exceeds a predetermined value, the ECU enters an Error Passive state, in which its error detection activities will not be broadcast on BUS traffic for other nodes to see. When the counter rises above a second, higher preset value, it switches into a BUS-Off state, removing the ECU from participation in BUS traffic [15]. Through this process, CAN nodes can both detect faults and perform error confinement.

An Open Systems Interconnect (OSI) reference model is utilized by CAN-based network solutions. This same standard is applied across all modern communication technologies. This model is standardized in ISO/IEC 7498-1, which defines "a common basis for the coordination of standards development for the purpose of systems interconnection" [9]. The adapted CAN message model comprises three of the seven OSI layers: the first layer- the CAN physical layer, the second layer- the CAN data link layer, and the seventh layer- the CAN application layer. Typically, OSI layers 3 through 6 (network, transport, session, and presentation layers) are not explicitly implemented. It is common for the application layers in CAN to incorporate functions of network and transport layers to allow this adaption of the OSI model without sacrificing functionality [16].

Higher layer protocol functionality, which spans between the network and application layers, is an important factor in CAN network design. Network management, which includes the protocol for turning CAN nodes on and off, can be included in this functionality. Node supervision in event-driven networks is another common function in network management [17]. This supervision is required to detect nodes that are missing due to several possible fault conditions. Missing nodes could be caused from a BUS-Off state, a temporary power loss, or a permanent power loss. Application layers can search for missing nodes using one of two methods. For nodes that do not transmit messages periodically, a client/server service can be programmed so that a connected server sends a state message to the monitoring "client" after a consistent period, providing a "pulse". Any interruption to the pulse that exceeds a set time limit indicates an off-line status in that node. However, if the node does transmit messages in a periodic fashion, this detection can be done implicitly [16]. An example of this time-out utilization in error reporting is given in **Figure 4**.

One of the most significant higher-layer protocol services in CAN is breaking-up data for transmission and re-assembling it on the receiving end. While this function is typically associated with the transport layer in OSI, in CAN, this parsing of data is another role executed by the application layer. Examples of protocols that provide this service include CANopen, DeviceNet, and J1939-21 [17]. Device and network design have become simplified through the utilization of software routines that execute standardized higher-layer protocols. These protocols are typically implemented in software through protocol stacks. Standardized versions of these stacks are commonly available from a variety of manufacturers. Examples of these standardized protocol stacks include CAN Application Layer (CAL) from CiA, NMEA 2000 from the International Electrotechnical Commission (IEC), and CAN FD from CiA.

CAN-BUS, as an overarching protocol for vehicle system-to-system communication, helps the vehicle make informed decisions about component level maintenance and control by maintaining an efficient communication pathway. To facilitate effective information flow, there are often multiple levels and separate systems of CAN that control specific regions and subsystems of the vehicle. This improves information handling capacity, and it helps to simplify the system into subsets that only contain the ECUs that need to communicate with each other. There is no reason, for example, for the ECU controlling in-cab climate control to know what is happening with the left rear tire pressure sensor. These controllers are divided onto different specialized networks, enhancing system efficiency.

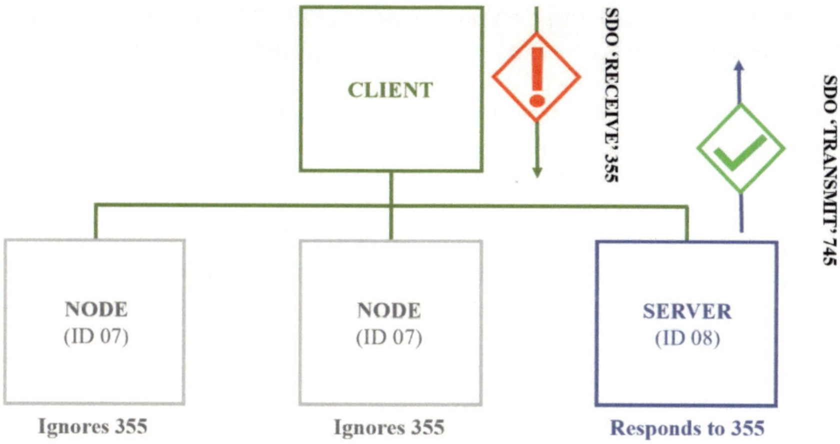

Figure 4.
Implicit message time-out reporting utilizing CANopen.

In addition to separating networks into subsystems, there are also different types of CAN-BUS systems that allow for different speeds of communication. The high-speed CAN system uses the CAN-H and CAN-L wires described above and can communicate at speeds up to 1 *Mbps*. The ECUs that require this high communication speed are safety critical systems, like the engine electronic control unit, the brake controller, and the air pollution control systems [12]. These are wired in a linear serial bus configuration terminated by resistors. The other type of CAN-BUS system commonly used is low-speed CAN, which can only reach communication speeds of up to 125 *kbps*. This is an eighth of the high-speed system rate and is appropriate for fault-tolerant or comfort systems like cab climate control or interior lights. A star serial bus configuration may be used, where multiple CAN applications are terminated at nodes [4]. By splitting-up the networks, there is a higher level of reliability for safety critical systems to get their messages broadcast across the network. This can aid in the avoidance of accidents or in notifying a driver of an in-process component failure, like the loss of engine oil pressure.

To further improve efficiency of the CAN-BUS system, every ECU on the network is also assigned an arbitration ID, or an identification number. This ID dictates which ECU is given priority in the case that there are conflicting messages or messages sent at the same time. This priority framework is a large part what makes CAN so efficient. Important messages from the engine regarding fuel input, for example, are not delayed by a message from the oil pump that oil life has decreased by one percent. In having an established priority level of messages, the system can be sure that system-critical messages are broadcast and received across all interconnected ECUs. This system of broadcasting the highest priority message has been a main contributor to the success of CAN-BUS technology and its dominance in the market.

While CANs are effective at communicating data between ECUs, they can also be utilized to record the operational metrics of a vehicle. Instead of directly measuring the data with precision instruments, approximate results can be calculated using the theoretical relationships between a specific metric and other parameters that are measured with internal sensors on the CAN. These internal sensors are commonly found in plug-and-play tools that are widely available on the market for on-board processing and diagnostics. They generally have low customizability, but they are very simple to install when compared with more specialized, auxiliary sensing equipment [18]. While estimates from these embedded controllers are inexact, very accurate measurements can be obtained via this method, by first calibrating the internal sensors with precision external sensors, as shown in Polcar, Cupera, and

Kumar's study on fuel consumption measurement [19]. This allows a reduction in both the number of sensors and the overall cost required within a vehicle's control system.

Through its methods of system interconnection and communication, CAN-BUS has revolutionized data collection and autonomy in virtually all markets, especially in the agricultural industry. By splitting-up the various subsystems to create an efficient communication pathway between the multiple electronic control systems that need to communicate, CAN-BUS has become an invaluable addition to modern agricultural equipment and continues to advance the capability for on-board real-time data collection, providing farmers with sophisticated technologies for improving their operations.

3.2 CAN-BUS standards development

Thus far, this chapter has made references to CAN standards, such as ISO 11898 and SAE J1939, but it has not given an explanation as to why there are different standards for different vehicle types. This section will discuss the purpose and need for developing such individual industry standards, as well as introduce some of the most important CAN standards in industry today, especially with respect to agricultural vehicles.

As previously mentioned, controller area networks function using a serial communication protocol, making it a useful pathway for passing digital data. However, without a standard for interpreting and forwarding the data, no useful information or actionable processes can be gleaned from it. Using the analogy of a telephone, CAN would be equivalent to the hardware and telephone lines used to connect the voices of two individuals, while the standard is the language used to make the communication meaningful [5]. Just as it is important that the individuals on opposite ends of the telephone line use the same language conventions to interpret each other's speech, the same is true with standard compatibility within a vehicle's system. Many components in a single vehicle are produced by different manufacturers, and standards allow the ECUs of these various modules to function and communicate on a common network.

The first standards were focused primarily on CAN usage in automobiles, as engine care management was the original target market for usage [2]. As off-road and heavy-duty vehicles carry-out entirely different mission profiles from passenger cars, with respect to loads, implement usage, and speed, it was not possible to apply the same "language" for priority and layer management in these vehicles. This led to the evolution of application-specific standards for the vehicle manufacturing industry. To give some more context for what these standards entail, ISO 11898, SAE J1939, and ISO 11783 will be covered briefly.

ISO 11898 was released in 1993. It was initially divided into two parts, and a third part was added later. This standard covers the data link layer, the physical layer for high-speed medium attachment (HS-PMA), and the physical layer for a fault-tolerant, low-speed, medium-dependent interface. ISO 11898-1 gives the specifications for creating an interchange of data between the modules of the CAN data link layer [10]. It also specifies the two main format options, the Classical CAN frame format and the CAN Flexible Data Rate format, the latter of which was introduced in 2012. While Classical CAN supports a maximum bit rate and payload of 1 *Mbps* and 8 bytes per frame, respectively, the Flexible Data Rate frame format extends the allowance for both bit rate and payload beyond these original limits. The general architecture of CAN is also described in this ISO standard in terms of the OSI layers mentioned previously. It contains specifications for both the logical link and medium-access control sub-layers, as well as the physical coding sub-layer [6] .

ISO 11898-2 gives the specifications for HS-PMA, which is a serial communication protocol that allows for real-time control of components in vehicle systems by multiplexing data for immediate use. The standard formalizes HS-PMAs with low-power mode and selective wake-up options [20]. ISO 11898-3 additionally covers the set-up of a data exchange between the ECUs of a vehicle utilizing CAN [21].

SAE J1939 was developed by the Society of Automotive Engineers (SAE) in 1994, and it establishes how nodes transmit data on the CAN-BUS in heavy-duty vehicles [22]. J1939 provides a common communication language across heavy equipment from different manufacturers, allowing a wide range of equipment to work with each other and enabling consistent data logging across heavy-duty and off-road vehicles. Although the first standards development papers on J1939 were drafted in 1994 (J1939–11, J1939–21, J1939–31), it was six years before the initial top-level document was published. After this, controller area networks were officially included within the language of the standard. In 2001, J1939 replaced the older standards SAE J1708 and SAE J1587. This standard, along with its accompanying documents, has since become a wider industry standard and is currently utilized for applications across multiple industries, including agricultural machinery, construction equipment, forestry machines, maritime ships, mass transportation, material handling, and military applications [1].

There are several key characteristics which define SAE J1939. Its bit rate, or the speed at which messages travel across the BUS, was originally set at 250 *kbps*. More recently, the standard was updated to support a faster bit rate of 500 *kbps*, and the identifier (ID), or unique name of each message, was extended to 29 bits. The message identifier segment, in addition to describing its data content, message priority, and indicating the source address, is also used in J1939 to specify the destination on the network [23]. The primary differentiation in message composition from other CAN systems comes from the Parameter Group Number (PGN). This 18-bit PGN is a function-specific frame sandwiched between the first 9 and last 2 bits of a traditional 11-bit CAN ID, providing more detail regarding the message content [24]. The J1939 message parameters within data bytes are identified by Suspect Parameter Numbers (SPNs). SPNs correlate to specific PGNs, with their encoded data designated by bit start position, bit length, scale, offset, and units. These PGN-specific details are used to extract desired SPN data and convert it to meaningful physical values [25]. An illustration of the J1939 message structure is shown in **Figure 5**.

Development of ISO 11783, a CAN-based agricultural bus system by Landwirtschaft Bussysteme (LBS), began in the early 1990s with the German DIN 9684 standard. The first commercially successful LBS combined the DIN 9684 virtual terminal (VT) concept with J1939 protocols and was internationally standardized as the ISO 11783 series [11]. The accompanying BUS system detailed in this standard is commonly known as ISOBUS. This standard consists of ten specific parts, including: 1) the general standard for data communication; 2) physical layer; 3) data link layer; 4) network layer; 5) network management; 6) virtual terminal; 7) implement messages applications layer; 8) power train messages; 9) tractor ECU; and 10) the task controller & management computer interface [14]. The communication protocols define messaging between the tractor and implement electronic systems through CAN. These, combined with the serial data network, regulate the methodology of data transference between actuators, control elements, display units, information storage systems, and sensors, allowing the tractor to control an implement through the virtual terminal (VT).

The VT is one of the most important features of the ISO 11783 standard, as it allows the operator to interface with the tractor and implements by both viewing real-time data and providing user inputs. The VT acts as a slave to individual ECUs,

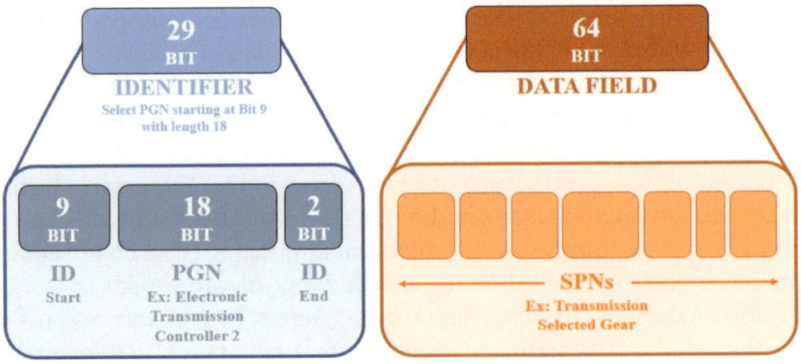

Figure 5.
SAE J1939 message structure.

each of which secure terminal connectivity to display informational data and collect operator inputs according to their individual protocol. The operator can choose which operational data to display, while each connected ECU continues to operate as if the VT were dedicated solely to its specific function [14]. This pathway makes it possible for the operator to have greater control over the functions of an implement, such as sprayer nozzle flow, combine cylinder rotational speed, or cultivator attachment height, depending on input from implement sensors. This eliminates the need for a separate control box for the implement and provides a single terminal controlling all information flow to the operator [11] . The ISOBUS is based on CAN running at 250 *kbps*. It uses twisted non-shielded quad cable and high-speed transceivers (same as ISO 11898). A 9-pin connector on the tractor is the only required point of contact between it and an implement with an ISOBUS compliant network cable.

This overview of CAN communication and standards has presented a cursory background of the technology fundamentals associated with the serial networking scheme, as well as some brief mention regarding how it is implemented. The next section will go into greater depth on how CANs have been utilized in industry, its potential connection to other network technologies, and how its usage could be expanded in the future.

4. Implementation

4.1 CAN-BUS by industry application

Although controller area network systems were originally developed for the automotive industry, they quickly became popular in other areas. CAN-utilizing industries include large over-the-road trucks, forestry, industrial factory automation, aerospace, and many others. In the aviation industry, the high-speed CAN protocol ISO 11898 is widely utilized, along with ARINC 825, a protocol created specifically for the aviation industry. The effort to create a CAN-based standard for communication in aircraft was initiated by Airbus and Boeing and was advanced by the Airlines Electronic Engineering Committee (AEEC) through their CAN Technical Working Group [26]. Several design targets were set while developing this protocol, including CAN functionality as either a main or ancillary network, an allowance for local CAN network integration into the wider aircraft network, and interoperability and interchangeability of CAN connected Line Replaceable Units (LRUs). Other design mandates were to maintain flexible configuration options; establish a simple process for adding, deleting, or modifying BUS ECUs; and simplify systems' interconnection protocols [26].

CAN-BUS systems also play an important role in both modern factory automation processes and testing facilities. Since CAN design is based on distributed control principles, it has been effectively used in manufacturing facilities to connect the essential control systems dispersed throughout a plant. Through the use of human machine interfaces (HMIs), operator inputs can be translated into instructions that a programmable logic controller (PLC) dispatches onto the BUS, allowing the remote operation of equipment ranging from sensors to actuators. This process allows the testing of new input parameters prior to execution on specific equipment and is a viable option for increasing process safety [27]. Use of CAN on assembly lines as a quality check is also becoming more common and is especially important on a line manufacturing a customizable product. Certain specifications are programmed for each checkpoint of product assembly, which are then broadcast on the CAN between machines to provide quality validation for the operators throughout the manufacturing process. CAN-BUS is also a practical option for connecting security and environmental control systems across a facility, due to both high bit-rate and inexpensive installation [27].

Returning to CAN use in the off-road vehicle market, virtually all modern agricultural machines incorporate CAN-BUS systems. Improved vehicle diagnostics, less complex design of electronic circuit controls, and advanced implement management are all benefits that CAN-BUS technology brings to the agricultural sector. CAN-BUS systems allow for high precision in machinery performance and logistics information. These metrics help to estimate operational cost and projected size in downstream operations. Specific measurement of other metrics, including fuel consumption, engine load, and average operating speed can also help supply chain managers maximize field and transport efficiency, while designing overall equipment solutions at a lower cost [28].

Displays within the cab allow the operator of the vehicle to view real-time data and information, as the vehicle is collecting it. These displays show the current location of the vehicle via GPS, the instantaneous fuel consumption rate, and other performance metrics that help the operator make intelligent decisions in order to maximize the efficiency of the vehicle. The John Deere Gen4® display shows many attributes, such as the instantaneous fuel economy and location of the vehicle within the field, but it also communicates with other vehicles in the same area to share guidance lines, coverage maps, and applied data in order to work the field efficiently [29].

The display associated with Case IH's Advanced Farming System® (AFS®) product, like the Gen4® display, is able to show the location of the vehicle within the field [30]. Using GPS and wireless data networks, it is also possible to check the performance of each vehicle from computers located away from the field. AGCO uses Fuse®, which is much like the Gen4® display and AFS®. It shows various data on how to improve the efficiency of the specific field operation, and it includes a seed and dry fertilizer monitoring system, which alerts the operator immediately, via the display, if there is a physical delivery blockage.

Aside from the role CAN-BUS plays in system-to-system communication within a vehicle, the serial network technology has also been integral in the advent of telematics. Telematics is a sector of information technology concerned with how data moves between machines over long distances. Incorporating telematics technology into a vehicle or fleet of vehicles provides the opportunity to utilize collected data outside the scope of an individual machine's operation by integrating it into a server network for wider usage and analysis. While CAN-BUS is not the sole technology responsible for telematics, it serves an important role in communicating large quantities of data that are eventually converted into valuable information for end users [31].

The general architecture of a vehicle telematics system begins with a Telematics Control Unit (TCU), a telematics cloud server, and front-end applications (Apps) through which the end user accesses captured data. The TCU is a microcontroller that manages data collection, communication, and memory through interfacing with different hardware and software modules. It provides connection ports to CAN-BUS, GPS, General Packet Radio Service (GPRS), battery, and Bluetooth modules, while maintaining a memory unit, a Central Processing Unit (CPU), and communication interfaces to Wireless Fidelity (Wi-Fi), cellular networks, and Long-Term Evolution (LTE) networks [31]. As the central component to a telematics system, the TCU accomplishes the tasks of gathering all the desired data and information from its various connections, synthesizing the information, and communicating to the cloud for use elsewhere. Focusing specifically on the CAN interface, a TCU utilizes the CAN-BUS as a pathway to collect the requested information from the ECUs, as programmed into its operating algorithm. This information acquisition could include any sensor data such as fuel consumption or vehicle speed. By converting the data from the CAN protocols, the TCU can then transfer this data to the telematics cloud server for further post-processing, after which, a user would be able to access the data.

The most common usage of telematics across all industries is within fleet management systems. This data collection process allows managers to optimize fuel usage, monitor vehicle down-time, analyze vehicle processes, and track operators driving a specific vehicle [31]. However, different companies also try to bring unique advantages to their telematics packages, which normally materialize in the form of a specialized management software. For construction and forestry equipment, Caterpillar utilizes a company-specific telematics system called ProductLink®, which has both cell and satellite transmission options, paired with their user interface VisionLink®. The focuses in these systems include the reduction of idle time and elimination of catastrophic failures through the reporting of fault codes [32]. John Deere provides customers with the option of a subscription package to the company's telematics network JDLink®, which is customizable to include mobile connections, In-Field Data Sharing®, Operations Center® (where data is synced every 30 seconds to keep it safe and secure), and other features which provide greater connective awareness of interdependent operations [33]. Case IH takes connectivity to a more automated level with their AFS® product, which has options for auto-guidance steering in tractors and combines using AFS AccuGuide® and AFS RowGuide® to aid in year-to-year repeatability. Their AFS Pro® system monitors several operational metrics and can manage ISOBUS implements [30, 34]. Utilizing CAN-BUS as a communication platform for mobile data transfer has greatly increased the capacity for utilizing data to drive decisions and functions.

In 2009, Agritechnica launched the Isomatch Tellus® VT. This allowed for the operator to observe two ISOBUS machines through one terminal, allowing for the simultaneous control of functions on different platforms. The possible connections to this terminal included a 15 pin ISOBUS, a power connector, an additional 9 pin extension connector, 4 USB interfaces, Bluetooth, Internet dongle, EIA-232 port for GPS, and others. Later, software packs such as ISO-XML were added to the VT [11]. Another example of user-focused technology is the Opus A3 CAN-BUS operator panel series from Wachendorff Elektronik, which has two CAN-BUS ports and is specifically designed for outdoor applications that include agricultural machinery [35]. As is evidenced by many of the applications in industry discussed above, different interface technology with CAN-BUS has been important in broadening its usage in a variety of fields. Further discussion of both wireless and non-wireless alternatives to and potential connection points with CAN are explored in the next section.

4.2 Alternative connectivity and networking to CAN-BUS

Different kinds of interfaces have been specifically developed to allow the conversion of CAN data into a format for Internet of Things (IoT) communication. Two specific technologies of note are CAN-Ethernet, and CAN-Bluetooth converters. A CAN-to-Ethernet converter allows the transfer of data in both directions and may be utilized in CAN-BUS monitoring, two-way remote CAN-BUS monitoring, and synchronization [36]. The firmware on such a converter contains both a communication device and a web server. The web server manages the protocol conversions, and the communication device provides the user interface. By combining two CAN-Ethernet converters, two CAN networks can be synchronized, allowing connection between CAN networks on different machines and in remote locations. This may be scaled-up further, or a custom software can be programmed to allow the converters to communicate directly to a specific IP address [36].

A CAN-to-Bluetooth gateway, unlike the ethernet connection, can transfer wireless data directly to a mobile device, using classic Bluetooth standards for Android devices and Low Energy (BLE) for Apple IOS. As with an ethernet converter, when the devices are used as a pair, a bridge for CAN data can be created for the end-user to access [37]. The ISOBlue 2.0 is an example of technology under development that utilizes Bluetooth principles. Currently being researched in the Open Ag Technology and Systems Center (OATS) at Purdue University, it is an open-source hardware product that connects agricultural machinery to the Cloud [38]. Other interfaces that allow CAN data conversion into different forms have been important tools in making telematics technology viable for off-road agricultural equipment. CAN Logger CLX000, which works between CAN and OBD2, is one such example [39].

Additional wireless technologies that have been used to interface CAN-BUS systems to IoT devices include ZigBee and Wi-Fi. These technologies also function as standalone networks for intra-vehicle and inter-vehicle communication [40]. Similar to the CAN data converters for Bluetooth and Ethernet, ZigBee and Wi-Fi converters have also been utilized to take advantage of their respective benefits in bandwidth, data transfer rate, security, and cost. More detail on each technology's specific advantages is presented in **Table 1**.

ZigBee is a globally available, wireless networking standard initially created as a home-area network for the control and monitoring of connected devices [41]. ZigBee is beneficial for sensor and vehicle network applications, due to its affordable installation and use cost, extensive battery life compared to competing devices, minimal maintenance, security and reliability, and small physical device footprint [41]. ZigBee was built on the IEEE 802.15.4 technical standard, which defines the physical layer (PHY) and medium access control (MAC) sublayer for low-data-rate wireless personal area networks (LR-WPANs) [45]. CAN-BUS-to-ZigBee conversion has demonstrated benefits in flexibility, convenience, and ease of use in system

Wireless Technology	Installation Cost	Bandwidth Capability	Data Rate	Security
ZigBee	Medium	Medium	Low	Moderately Secure
Bluetooth	Low	Low	Low	Less Secure
Wi-Fi	High	High	High	More Secure
UWB	Low	High	High	Moderately Secure

Table 1.
A comparison of wireless technologies capable of interfacing with CAN and IoT devices [41–44].

installation, adding and removing nodes, system updates, and expanded network construction [42].

Wi-Fi is a popular wireless technology for CAN-BUS interfacing and IoT communication. Wi-Fi falls under the IEEE 802.11 standard, which is part of the broader IEEE 802 technical standards for LAN and defines MAC and PHY protocols for applying wireless local area network (WLAN) computer communication [46]. This standard also specifies common radio frequency bands that Wi-Fi can communicate on. These include but are not limited to 2.4 *GHz*, 5 *GHz*, 6 *GHz*, and 60 *GHz* frequencies [46]. Wi-Fi offers a high data rate of up to 54 *Mbps* and a large bandwidth capability [43]. The most common application for Wi-Fi to CAN-BUS interaction is vehicle-to-cloud telematics services, as discussed in the previous section. On-vehicle Wi-Fi networks also allow for remote control of vehicle systems and provide capability for varying levels of autonomous control. On-vehicle Wi-Fi networks also allow for sending CAN-BUS data from vehicle-to-vehicle or across several vehicles simultaneously.

Ultra-wideband (UWB) is another wireless technology being researched for vehicle communication systems. UWB is a low-power radio protocol specifically created to improve the location accuracy of wireless technologies. UWB transmits data across a short distance and measures the time it takes for a radio signal to travel between the sending and receiving device [46]. This is similar to the time-of-flight (ToF) method used with radio detection and ranging (RADAR). A UWB transmitter sends billions of radio pulses across a wide-spectrum frequency of 7.5 *GHz*. These pulses are then translated into usable data from a UWB receiver. While UWB is not commonly used in conjunction with CANBUS, it has been studied for use in autonomous vehicle navigation and path localization [44].

The continuous development and improvement of autonomous vehicle technology necessitates an increased demand for greater bandwidth and connectivity requirements, while still providing an allowance for high system complexity. System complexity in this case could be defined as the added latency from the connected network devices. As many aspects of the interconnected vehicle networks continue to grow, management and network understanding also become more complex. Such aspects include a number of features, routing table configurations, system security, firewall protections, and others [47]. One of the most promising alternatives to vehicle CAN networks are automotive ethernet-based networks. The market for automotive ethernet is expected to increase by 22% from 2019 to 2026 [48]. High bandwidth capabilities and improved cost efficiency are two major benefits to automotive ethernet networks. Instead of a priority-based protocol, ethernet utilizes a Carrier Sense Multiple Access with Collision Detection (CSMA/CD) strategy [49]. This defines the appropriate device response when multiple control units simultaneously attempt to use a data channel and encounter a data collision. Susceptibility to radio frequency (RF) interference, the inability to provide latency at very high frequency, and synchronization issues between timing devices are potential challenges with automotive ethernet network implementation [48]. Currently, the primary consumption of Ethernet technology in vehicles is enabling personal use of the Internet. Ethernet provides rapid data transfer speed, making it ideal for data intensive applications. However, Ethernet does not adapt well to internal failure, as seen in **Table 2**. A potential associated cost with Ethernet demand increase is the expensive coated wiring needed to provide such high bandwidths.

One type of automotive network communication protocol is FlexRay. FlexRay is a network standard for automotive systems, based on a flexible high data transmission rate, high-speed bus system, like CAN FD [48]. FlexRay is designed for communication of efficiency-type applications in the vehicle. This is due to FlexRay's high complexity allowance and bandwidth. At 10 *Mbps* on two dual channels, FlexRay can provide up

Network Type	Installation Cost	Bandwidth Capability	System Complexity	Fault Tolerance
CAN	Medium	Low	High	High
FlexRay	High	Medium	High	Medium
MOST	Medium	Medium	Medium	Low
Ethernet	Low	High	Medium	Low

Table 2.
A comparison of CAN characteristics with competing technologies [48].

to 20 *Mbps* of bandwidth, making it optimal for systems such as steering and brakes. CAN shows advantages over FlexRay primarily in cost and error handling [50]. Due to FlexRay's robust complexity and bandwidth, its cost is far greater than CAN, on a value per baud rate basis. Although CAN does not generate data transfer rates as fast as FlexRay, it is better suited for the majority of smaller jobs at a far lower cost [50].

Another type of automotive network is MOST (Media Oriented System Transport). MOST provides very fast data transfer at over 24 *Mbps*. This is because the system was designed to transfer media information within luxury cars, such as GPS, radio, and video systems. MOST has comparable speeds to Ethernet and is more common in automotive applications. However, it handles much less system complexity than CAN and FlexRay, limiting its potential applications [51]. MOST is equipped with plastic optical fiber in its physical layer, which limits electromagnetic interference, thus providing faster speeds and significantly less signal jitter. CAN and MOST have comparable costs, but CAN is better suited for more versatile and sophisticated operations [48].

Overall, CAN shows the most versatility of these four main alternative systems. FlexRay is useful for safety systems, due to its high complexity allowance and multiple channel scheme, but it is a higher-cost system by a significant margin. MOST provides one of the best options for media and information transmission, with a faster data transfer rate than two of the other technologies reviewed [50]. However, MOST cannot be used for highly complex systems. Ethernet provides the fastest data transmission speeds of all the options compared, but it is limited by low complexity allowance and adaptability. CAN, while moderately priced, shows high adaptability to complex systems, while providing useful data transfer in a variety of applications [48]. An example of an interconnected system utilizing these networks in a passenger vehicle is shown in **Figure 6**.

4.3 Prospective areas for CAN technology inclusion

Currently, CAN-BUS is used in autonomous vehicle development to gather data from all electronic control sensors and consolidate it onto a single network. By gathering the data into a unified structure, the overall system controller can easily make decisions that affect multiple sub-systems at once. This data availability, combined with swift processing, is a key component in the safe operation of autonomous vehicles both on the open road and off-road. This centralized system data stream allows for advanced control of smart engine sensors, which provide more efficient management processes. The data handling capability of smart controllers is still an area in need of concentrated improvement. Present research is looking into robust solenoids and other embedded sensors to control valve timing, coolant flow rate, compression ratio, and other key processes in engine operation [52]. Integrated development of these smart controllers with CAN will be crucial to ensuring the safety of autonomous vehicle function execution and travel.

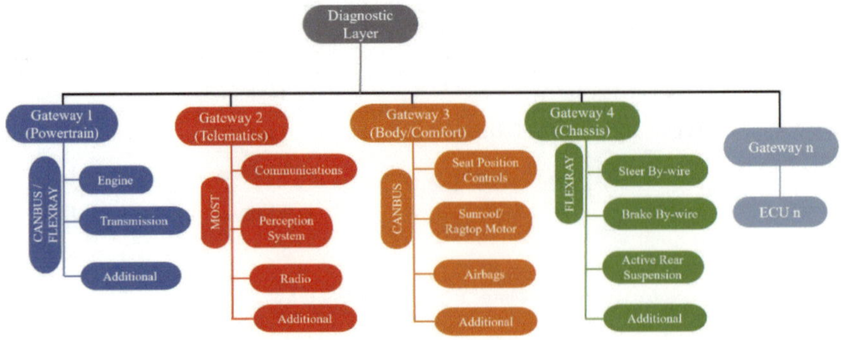

Figure 6.
An example of a FlexRay application.

While large scale agricultural mechanization has been associated with various negative environmental impacts, from soil compaction to harmful exhaust emissions, the advent of digital agriculture has played a key role in increased efficiencies and technological progress within the farming sector, reducing those detrimental elements. The utilization of CANs for improved operation is a research area where further development could have a significant impact with respect to environmental effects. For example, some of the most common technologies for limiting emissions have associated environmental costs that detract from ecological benefit. Though Exhaust Gas Recirculation (EGR) decreases NOx emissions, it simultaneously increases specific fuel consumption to lower engine efficiency. Similarly, the post-combustion treatment Selective Catalytic Reduction (SCR) results in better emissions efficiency, but consumes a urea solution that increases freshwater eutrophication risks [53].

Since fuel consumption is primarily dependent on engine speed and torque, it is possible to reliably decrease emissions with the application of alternative driving techniques optimally suited to specific drive train design and implement load [54]. However, the plausibility of deriving accurate efficiency metric assessments is limited due to present data scarcity. Current methods for Life Cycle Assessment (LCA) studies provide unreliable results because average conditions, such as soil texture, field shape, soil moisture, implement transfer difference, and engine features, have traditionally been utilized in lieu of actual conditions to estimate environmental effects [55]. CAN is advantageously positioned to help address both the data deficiency and inadequate LCA techniques, due to its data collection and communication strengths. It is possible, for example, that performance metrics could be improved through intelligent sensor solutions that can measure slippage and soil compaction at the wheels of a vehicle and attached implement [13, 54]. These sensors could communicate with sensors in the drivetrain to adjust the effective gearing ratio in real-time, reducing soil compaction and preserving the long-term viability of the soil.

An example of an instrument that, when paired with CAN-BUS communication, could be useful in achieving such operational efficiency objectives are inertial measurement units (IMUs). An inertial measurement unit functions as a sophisticated accelerometer/gyroscope combination. It boasts near zero drift between different operating conditions, and its use of magnetic fields allows it to double as an "electronic compass". The IMU allows for communication across many different CAN-BUS networks to help the tractor, or any vehicle, make decisions about how to alter the driving style for the terrain to limit "dynamic pitch and roll" through open system communication [52]. While this specific system is not currently implemented on tractors and other off-road vehicles, there is room for its introduction in the emerging field of agricultural autonomy.

Smart agriculture and digital farming practices have gained popularity in the previous decade. These techniques are precursors to a transformative implementation of information technology in the farming world. Going forward, more advanced software systems will use information collected from CAN communication devices to aid in the optimization of machinery designs and more accurate load, use-profile, and duty cycle representations of vehicles and implements [18]. Future applications for CAN-BUS technology include IoT, Edge Computing, and swarm machinery automation, as well as complex control of electrical and electric-hybrid machinery.

IoT implementation in the agricultural sector has gained enormous traction in recent years, as a result of its high potential for cross-brand interoperability, scalability, and traceability. The different types of IoT tools being applied are continuing to evolve, increasing the overall adaptability and variety of available systems to end-users [56]. IoT systems are currently being implemented on vehicles from John Deere, Case New Holland (CNH), AGCO, and others. Future IoT device use on agricultural equipment will likely be in conjunction with multiple on-board network systems. Local storage or cloud computing will be necessary to store and process the vast amount of data created by this potential technology [57]. Data processing on-board the vehicle, near the working equipment, is referred to as 'edge computing' [56, 58]. It is highly probable that agricultural vehicles will eventually be able to perform a variety of complex, agronomic tasks from a preprogrammed routing structure, through the combined utilization of both IoT and EC technologies.

In addition to on-vehicle IoT technologies, it is probable that field embedded (or in-situ) IoT sensors will also be able to communicate with larger on-farm networks [59]. Several of the previously discussed network configurations are possible whole-farm network options. These include cellular (4G, 5G, and beyond), Wi-Fi, ZigBee, and UWB. For example, real-time soil moisture can be obtained from field-based, connected sensors to create a variable-rate prescription map [60]. Utilized in conjunction with mobile soil penetrometer readings, an accurate map of soil compaction risk can be created. This could allow farmers to tailor their tillage operations to specific areas of the field, as well as control vehicle traffic.

Cutting-edge networking research is also being done with robotic and swarm machinery automation [61]. IoT technologies and improved connectivity will allow for the introduction of robotic swarm farming techniques. Swarm farming incorporates multiple, small-scale robotic platforms that perform farming operations autonomously in place of larger, manned agricultural equipment. This farming strategy, paired with a predetermined path-planning algorithm optimizing how the machines will navigate throughout the field, could allow for near-continuous field operation. Additional benefits could include a centralized command center that is controlled by a single system manager and a significant reduction in the need for skilled labor [62]. The possibility of substituting the modular vehicle design within swarm farming for traditional larger equipment will depend on cost, comparative system productivity, and accuracy. Farmers will demand a significant return on investment and the reliability that they have come to expect from their current machinery. A potential difficulty for CAN-based systems is the large bandwidth requirement for incoming and streaming data. Another potential challenge involves communication protocol differences between traditional CAN-BUS data and more memory intensive data collected from advanced machine systems, like perception engines and central processor-based codes [63]. Future developments in CAN-BUS technology should focus on addressing these weaknesses to improve adaptability to upcoming applications.

A major concern in the future of agricultural CAN use, machinery networking, and machine system automation is cybersecurity. Although increased digitization, automation, and precision services have tremendous potential to establish sustainability and profitability in farming systems, the influx of interconnected information technology simultaneously opens the market up to new areas of susceptibility, security risks, and potential targeted cyber-attacks [58]. Mission-critical systems are becoming more reliant on internet connectivity, such as controlling farming implements remotely through the ISOBUS with linked management software. Local Area Networks (LANs) have become a requirement in smart farming to enable system/device access to the data and services that control their functions [64]. This increased dependence of agricultural operations on cyber-physical systems has led to the development of new, novel threats and challenges that can be analyzed in two categories: information technology and agricultural production [58].

From an informational technology standpoint, some of the main threats are unauthorized access of resources/databases under use of falsified identity, interception of node data transfer, facility damage or downtime, malicious data attacks from malware, and compromised control systems to negatively impact decision-making [58]. Due to the nature of modern networked food systems, targeted or accidental disruption of time-sensitive agricultural processes could have a significant economic impact on a global scale. The threat of a concentrated hack on the agricultural sector has become more tangible with the analysis of cyber-security breaches in recent years, such as the 2017 infrastructure meltdown of Maersk shipping [65]. The vulnerability of Wireless Local Area Networks (WLANs) to direct cyber-attacks is already a generally recognized problem across all industries [66]. Demonstration of the damage potential in a Denial of Service (DoS) attack has been shown in the research of Sontowski et al., by disrupting in-field sensors and obstructing device network connectivity in smart farm operations [67].

Though the hacking activities of malicious actors is a highlighted concern in cyber security, there are also a number of risks associated with agricultural production that stem from physical layer vulnerabilities and limited user knowledge. The harsh environment in which agricultural equipment is used (including extreme weather conditions, dust concentration, and highly variable humidity/temperature fluctuation) can cause power failures or sensor damage [64]. Technology signal interference from other agricultural equipment, such as the high voltage pulses from Solar Insecticidal Lamps (SIL), can also lead to malfunctions and data loss [58].

However, one of the most common threats to cyber security is inadequate adoption of safety procedures by farmers who lack full awareness of device functionality. From research conducted by Nikander et al., farmers are often ill-equipped with time and resources to build LANs with appropriate network equipment, topology expansion planning, and protection software/hardware [64]. This leads to networks that are at risk of system losses due to hardware issues and human error. The adoption of countermeasures to security risks, such as authentication & access control, cryptography, key management, and intrusion detection systems, is dependent on end-users understanding the importance of cybersecurity, and better fail-safe mechanisms within hardware [58, 64]. These concerns highlight the importance of advancing security protocols in CAN-BUS systems, and it is likely that this will be a targeted focus in the future of CAN developments.

5. Conclusions

Key points from this chapter included the following:

- CAN-BUS has played a major role in industry since its debut in 1987 for its groundbreaking use of distributed network principles.

- The establishment of international societies and standards positioned CAN-BUS as the leading serial network system in all vehicles.

- CAN-BUS provides efficient and dependable communication pathways through front and back end context in messaging, error confinement, higher-layer protocols, and subsystem differentiation.

- CAN-BUS has revolutionized data collection and analysis in multiple industries, especially in the agricultural sector.

- When paired with wired or wireless technologies, CAN is an advantageous communication pathway for expanding the reach of data communication beyond point source limitations.

- Challenges for future CAN iterations include increasing bandwidth and security measures, while decreasing latency and hardware vulnerabilities.

This chapter has reviewed CAN-BUS technology including its invention, early applications, fundamentals, and standards development. Early applications of CAN-BUS came from European car manufacturers, which incorporated electronic control units for engine care management. The development of standards to allow consistent communication methods within CAN-BUS systems, such as ISO 11898, SAE J1939 and ISO 11783, were important for allowing serial networks to be applied within multiple vehicle types and industries. Modern day uses, alternative connectivity and networks, and potential future applications have also been examined. Controller area networks are responsible for the transmission, logging, and analysis of engine and machine system data currently used by vehicle manufacturers. Understanding CAN-BUS communication protocols provides insight into the advantages, uses, and future evolutions of distributed control networks.

CAN-BUS technology fundamentals, such as physical and data message structures, components, error handling, and message channel support are useful in understanding the strengths and limitations of CAN systems. Through the use of high and low speed CAN-BUS configurations, arbitration codes, and broadcast style communication, CAN-BUS can efficiently and reliably transfer messages across a vehicle's control system to ensure accurate, real-time data communication. As electronic connectivity has increased the sophistication of off-road vehicle operation management, new applications using CAN with external networks have been an important area of communications advancement within the agricultural sector. The development of converters between CAN data and other wireless data types has been important in keeping CAN-BUS integrated and relevant in the vehicle fleet telematics expansion. More research into wireless CAN may be an important direction for serial network technology going forward.

Specific CAN-BUS applications in ongoing autonomous vehicle development research include component data consolidation, embedded sensors, IoT devices, and machine-to-machine communication strategies. Future technologies that might benefit CAN-BUS technology by their incorporation include local-to-cloud data transmission, autonomous swarm vehicle management, and increased cyber security protocols. Although controller area networks face limitations within both bandwidth and latency, they still function as effective inputs to more advanced vehicle systems and more sophisticated remote networks. The potential of CAN-BUS technologies has clearly not been fully exhausted, and they will continue to play an important role in the advancement of agricultural machinery and farming practices.

Acknowledgements

We would like to acknowledge our fellow classmates from Dr. Stwalley's Fall 2020 Off-Highway Vehicle Design class at Purdue University's School of Agricultural and Biological Engineering for their contributions to the structure and content of this technical chapter.

Conflict of interest

The authors declare no conflict of interest.

Author details

Hannah M. Boland, Morgan I. Burgett, Aaron J. Etienne and Robert M. Stwalley III*
Agricultural and Biological Engineering, Purdue University, West Lafayette, Indiana, USA

*Address all correspondence to: rms3@purdue.edu

References

[1] Copperhill Technologies, "A Brief Introduction to Controller Area Network," 2020a. [Internet]. Available: https://copperhilltech.com/a-brief-introduction-to-controller-area-network/, [Last Accessed: 2021 May 15].

[2] CiA, "History of CAN technology," 2018. [Internet]. Available: https://www.can-cia.org/can-knowledge/can/can-history/, [Last Accessed: 2021 May 15].

[3] X. Gao, D. Huang, Y. Chen, W. Jin and Y. Luo, "The design of a distributed control system based on CAN bus," IEEE International Conference on Mechatronics and Automation, pp. 1118-1122, 2013, doi: 10.1109/ICMA.2013.6618071.

[4] W. Voss, A Comprehensible Guide to Controller Area Network, Greenfield: Copperhill Technologies Corporation, 2011.

[5] CSS Electronics, "CAN Bus explained – a simple intro," 2020a. [Internet]. Available: https://www.csselectronics.com/screen/page/simple-intro-to-can-bus/language/en, [Last Accessed: 2021 May 15].

[6] Kvaser, "The Kvaser Dictionary for Help in Understanding CAN Bus Systems," 2020. [Internet]. Available: https://www.kvaser.com/about-can/can-dictionary/, [Last Accessed: 2021 May 15].

[7] S. Maradana, "CAN Basics," 2012. [Internet]. Available: https://automotivetechis.wordpress.com/2012/06/01/can-basics-faq/, [Last Accessed: 2021 May 15].

[8] ISO, "About Us," 2020. [Internet]. Available: https://www.iso.org/about-us.html, [Last Accessed: 2021 May 15].

[9] CiA, "CAN-based higher-layer protocols (HLP)," 2017. [Internet]. Available: https://www.can-cia.org/can-knowledge/hlp/higher-layer-protocols/, [Last Accessed: 2021 May 15].

[10] ISO, "ISO 11898-1:2015," 2015. [Internet]. Available: https://www.iso.org/standard/63648.html, [Last Accessed: 2021 May 15].

[11] CiA, "Isobus – the CAN-based network system," CAN Newsletter, pp. 44-48, 2010.

[12] C. Goering, M. Stone, D. Smith and P. Turnquist, "Off-road vehicle engineering principles," American Society of Agricultural Engineers, 2006, ISBN: 978-1892769268.

[13] F. S. Al-Aani, "CAN bus technology for agricultural machine management," Iowa State University Capstones, Theses, and Dissertations, 2019.

[14] M. L. Stone, K. D. McKee, C. W. Formwalt and R. K. Benneweis, "ISO 11783: An Electronic Communications Protocol for Agricultural Equipment," An Electronic Communications Protocol for Agricultural Equipment, pp. 1-17, February 1999.

[15] Kvaser, "CAN Bus Error Handling," 2017. [Internet]. Available: https://www.kvaser.com/about-can/the-can-protocol/can-error-handling/, [Last Accessed: 2021 May 15].

[16] CSS Electronics , "CANopen Explained - A Simple Intro," 2020. [Internet]. Available: https://www.csselectronics.com/screen/page/canopen-tutorial-simple-intro/language/en, [Last Accessed: 2021 May 15].

[17] Kvaser, "Higher Layer Protocols," [Internet]. Available: https://www.

kvaser.com/about-can/higher-layer-protocols/, [Last Accessed: 2021 May 15].

[18] R. Rohrer, S. Pitla and J. Luck, "Tractor CAN bus interface tools and application development for real-time data analysis," Computers and Electronics in Agriculture, p. 163, 2019, doi: 10.1016/j.compag.2019.06.002.

[19] A. Polcar, J. Čupera and V. Kumbár, "Calibration and its Use in Measuring Fuel Consumption with the CAN-Bus Network," Acta Universitatis Agriculturaevet Silviculturae Mendelianae Brunensis, vol. 64, no. 2, pp. 503-507, 2016, doi: 10.11118/actaun201664020503.

[20] ISO, "ISO 11898-2:2016," 2016. [Internet]. Available: https://www.iso.org/standard/67244.html, [Last Accessed: 2021 May 15].

[21] ISO, "ISO 11898-3:2006," 2006. [Internet]. Available: https://www.iso.org/standard/36055.html, [Last Accessed: 2021 May 15].

[22] CSS Electronics, "J1939 explained - a simple intro," 2020b. [Internet]. Available: https://www.csselectronics.com/screen/page/simple-intro-j1939-explained, [Last Accessed: 2021 May 15].

[23] W. Voss, "Guide To SAE J1939 - J1939 Message Format," 2018. [Internet]. Available: https://copperhilltech.com/blog/guide-to-sae-j1939-j1939-message-format/, [Last Accessed: 2021 May 15].

[24] Copperhill Technologies, "A Brief Introduction to the SAE J1939 Protocol," 2020b. [Internet]. Available: https://copperhilltech.com/a-brief-introduction-to-the-sae-j1939-protocol/, [Last Accessed: 2021 May 15].

[25] W. Voss, "SAE J1939 Bandwidth, Busload and Message Frame Frequency," 2019. [Internet]. Available: https://copperhilltech.com/blog/sae-j1939-bandwidth-busload-and-message-frame-frequency/, [Last Accessed: 2021 May 15].

[26] A. Brehmer, "CAN Based Protocols in Avionics," 2014. [Internet]. Available: https://ieeexplore.ieee.org/stamp/stamp.jsp?arnumber=6979561, [Last Accessed: 2021 May 15].

[27] D. Fanton, "Why All the Fuss About CAN Bus?," 2020. [Internet]. Available: https://www.onlogic.com/company/io-hub/fuss-can-bus/, [Last Accessed: 2021 May 15].

[28] M. J. Darr, "CAN Bus Technology Enables Advanced Machinery Management," Agricultural and Biosystems Engineering Publications, p. 329, 2012.

[29] John Deere, "Gen 4 Command Center Display," 2020b. [Internet]. Available: http://www.deere.com/en/technology-products/precision-ag-technology/guidance/gen-4-premium-activation/, [Last Accessed: 2021 May 15].

[30] CASE IH, "AFS," 2020a. [Internet]. Available: https://www.caseih.com/apac/en-int/products/advanced-farming-system, [Last Accessed: 2021 May 15].

[31] Embitel, "Technology Behind Telematics Explained: How does a Vehicle Telematics Solution Work?," 2018. [Internet]. Available: embitel.com/blog/embedded-blog/tech-behind-telematics-explained-how-does-a-vehicle-telematics-solution-work, [Last Accessed: 2021 May 15].

[32] Caterpillar, "Telematics - Your link to equipment cost savings," 2020. [Internet]. Available: https://www.cat.com/en_US/articles/customer-stories/forestry/telematics.html, [Last Accessed: 2021 May 15].

[33] John Deere, "Operations Center," 2020a. [Internet]. Available: https://www.deere.com/en/technology-products/precision-ag-technology/data-management/operations-center, [Last Accessed: 2021 May 15].

[34] CASE IH, "AFS Connect," 2020b. [Internet]. Available: https://www.caseih.com/apac/en-int/products/advanced-farming-system/afs-connect, [Last Accessed: 2021 May 15].

[35] Topcon, "Technical Data Sheet OPUS A3 ECO Full," 2019. [Internet]. Available: http://www.motronica.com/wp-content/uploads/2019/08/TDS_A3e_FULL.pdf, [Last Accessed: 2021 May 15].

[36] Axiomatic Technologies Corporation, "Technical Datasheet #TDAX140900," 2020. [Internet]. Available: https://www.axiomatic.com/TDAX140900.pdf, [Last Accessed: 2021 May 15].

[37] Axiomatic Technologies Corporation, "Technical Datasheet #TDAX141100," 2018. [Internet]. Available: https://www.axiomatic.com/TDAX141100.pdf, [Last Accessed: 2021 May 15].

[38] B. Wallheimer, "Purdue partnering on 5G research to improve ag automation," 2019. [Internet]. Available: https://ag.purdue.edu/stories/purdue-partnering-on-5g-research-to-improve-ag-automation/, [Last Accessed: 2021 May 15].

[39] CSS Electronics, "OBD2 explained - a simple intro," 2020c. [Internet]. Available: https://www.csselectronics.com/screen/page/simple-intro-obd2-explained, [Last Accessed: 2021 May 15].

[40] S. Dorle, D. Deshpande, A. Keskar and M. Chakole, "Vehicle Classification and Communication Using Zigbee Protocol," *2010 3rd International Conference on Emerging Trends in* Engineering and Technology, pp. 106-109, 2010, doi: 10.1109/CETET.2010.170.

[41] Silicon Labs, Inc., "ZigBee-based Home Area Networks Enable Smarter Energy Management," pp. 1-4.

[42] Y. Li, C. Yu and H. Li, "The design of ZigBee communication convertor based on CAN," in 2010 International Conference on Computer Application and System Modeling (ICCASM 2010), 2010.

[43] N. Chhabra, "Comparative Analysis of Different Wireless Technologies," International Journal of Scientific Research in Network Security and Communication, vol. 1, no. 5, pp. 13-17, 2013, 2321-3256.

[44] J.-S. Lee, Y.-W. Su and C.-C. Shen, "A Comparative Study of Wireless Protocols: Bluetooth, UWB, ZigBee, and Wi-Fi," in IECON 2007 - 33rd Annual Conference of the IEEE Industrial Electronics Society, 2007 doi: 10.1109/IECON.2007.4460126.

[45] Institute of Electrical and Electronics Engineers, "IEEE 802.15.4-2020 - IEEE Standard for Low-Rate Wireless Networks," 2020.

[46] Institute of Electrical and Electronics Engineers, "IEEE 802.11-2016 - IEEE Standard for Information technology--Telecommunications and information exchange between systems Local and metropolitan area networks--Specific requirements - Part 11," 2016.

[47] M. Behringer and M. Kuehne, "Network Complexity and How to Deal with it," 2011. [Internet]. Available: https://labs.ripe.net/Members/mbehring/network-complexity-and-how-to-deal-with-it, [Last Accessed: 2020 December 20].

[48] Navixy, "CAN bus and alternatives," 2020. [Internet]. Available: https://www.navixy.com/docs/academy/can-bus/can-and-alternatives/, [Last Accessed: 2021 May 15].

[49] A. Quine, "Carrier Sense Multiple Access Collision Detect (CSMA/CD) Explained," IT Professional's Resource Center (ITPRC), 2008. [Internet]. Available: https://www.itprc.com/carrier-sense-multiple-access-collision-detect-csmacd-explained/. [Last Accessed: 2021 May 8].

[50] National Instruments Corp. , "FlexRay Automotive Communication Bus Overview," National Instruments Corp. , 2020. [Internet]. Available: https://www.ni.com/en-us/innovations/white-papers/06/flexray-automotive-communication-bus-overview.html. [Last Accessed: 2021 May 8].

[51] S. Tuohy, M. Glavin , C. Hughes, E. Jones, M. Trivedi and L. Kilmartin, "Intra-Vehicle Networks: A Review," IEEE Transactions of Intelligent Transportation Systems, pp. 1-12, 2014, doi: 10.1109/TITS.2014.2320605.

[52] M. Horton, "What can a CANbus IMU do to make an autonomous vehicle safer?," 2019. [Internet]. Available: https://www.autonomousvehicleinternational.com/opinion/what-can-a-canbus-imu-do-to-make-an-autonomous-vehicle-safer.html, [Last Accessed: 2021 May 15].

[53] J. Bacenetti, D. Lovarelli, D. Facchinetti and D. Pessina, "An environmental comparison of techniques to reduce pollutants emissions related to agricultural tractors," Biosystems Engineering, vol. 171, pp. 30-40, 2018, doi: 10.1016/j.biosystemseng.2018.04.014.

[54] M. Lindgren and P.-A. Hansson, "PM-Power and Machinery: Effects of Engine Control Strategies and Transmission Characteristics on the Exhaust Gas Emissions from an Agricultural Tractor," Biosystems Engineering, vol. 83, no. 1, pp. 55-65, 2002, doi: 10.1006/bioe.2002.0099.

[55] D. Lovarelli and J. Bacenetti, "Bridging the gap between reliable data collectiona and environmental impact for mechanised field operations," Biosystems Engineering, vol. 160, pp. 109-123, 2017, doi: 10.1016/j.biosystemseng.2017.06.002.

[56] A. Kamilaris, F. Gao, F. Prenafeta-Boldu and M. Ali, "Agri-IoT: A semantic framework for Internet of Things-enabled smart farming applications," *2016 IEEE 3rd World Forum on Internet of Things*, pp. 442-447, 2016, doi: 10.1109/WF-IoT.2016.7845467.

[57] R. Morabito, V. Cozzolino, A. Y. Ding, N. Beijar and J. Ott, "Consolidate IoT Edge Computing with Lightweight Virtualization," IEEE Network, pp. 102-111, 2018, doi: 10.1109/MNET.2018.1700175.

[58] X. Yang, L. Shu, J. Chen, M. A. Ferrag, J. Wu, E. Nurellari and K. Huang, "A Survey on Smart Agriculture: Development Modes, Technologies, and Security and Privacy Challenges," IEEE/CAA Journal of Automatica Sinica, vol. 8, no. 2, pp. 273-302, 2021, doi: 10.1109/JAS.2020.1003536.

[59] J. Liang, X. Liu and K. Liao, "Soil Moisture Retrieval Using UWB Echoes via Fuzzy Logic and Machine Learning," IEEE Internet of Things Journal, vol. 5, no. 5, pp. 3344-3352, 2018, doi: 10.1109/JIOT.2017.2760338.

[60] A. P. Atmaja, A. E. Hakim, A. P. A. Wibowo and L. A. Pratama, "Communication Systems of Smart Agriculture," *Journal of Robotics and Control*, vol. 2, no. 4, pp. 297-301, 2021, doi: 10.18196/jrc.2495.

[61] D. Albiero, A. P. Garcia, C. K. Umezu and R. L. de Paulo, "Swarm

Robots in Agriculture," *CoRR*, vol. abs/2103.06732, 2021, doi: 10.48011/asba.v2i1.1144.

[62] A. Sharda, "Is swarm farming the future of farming?," 2020. [Internet]. Available: https://www.precisionagreviews.com/post/is-swarm-farming-the-future-of-farming, [Last Accessed: 2021 May 15].

[63] J. Čupera and P. Sedlák, "The use of CAN-Bus messages of an agricultural tractor for monitoring its operation," *Research in Agricultural Engineering 57(4)*, pp. 117-127, 2011, doi: 10.17221/20/2011-RAE.

[64] J. Nikander, O. Manninen and M. Laajalahti, "Requirements for cybersecurity in agricultural communication networks," Computers and Electronics in Agriculture, vol. 179, pp. 1-10, 2020, doi: 10.1016/j.compag.2020.105776.

[65] Jahn Research Group, "Cyber Risk and Security Implications in Smart Agriculture and Food Systems," Madison, 2019.

[66] D. Kraus, E. Leitgeb, T. Plank and M. Löschnigg, "Replacement of the Controller Area Network (CAN) protocol for future automotive bus system solutions by substitution via optical networks," *18th International Conference on Transparent Optical Networks (ICTON)*, pp. 1-8, 2016, doi: 10.1109/ICTON.2016.7550335.

[67] S. Sontowski, M. Gupta, S. S. L. Chukkapalli, M. Abdelsalam, S. Mittal, A. Joshi and R. Sandhu, "Cyber Attacks on Smart Farming Infrastructure," *UMBC*, Baltimore, 2020, doi: 10.1109/CIC50333.2020.00025.